Lecture Notes in Computer Science 8747

Commenced Publication in 1973
Founding and Former Series Editors:
Gerhard Goos, Juris Hartmanis, and Jan van Leeuwen

Advanced Research in Computing and Software Science

Subline of Lectures Notes in Computer Science

T0213892

More information about this series at http://www.springer.com/series/558

Dieter Kratsch · Ioan Todinca (Eds.)

Graph-Theoretic Concepts in Computer Science

40th International Workshop, WG 2014
Nouan-le-Fuzelier, France, June 25–27, 2014
Revised Selected Papers

 Springer

Editors
Dieter Kratsch
LITA
Université de Lorraine
Metz Cedex 01
France

Ioan Todinca
Université d'Orléans
Orléans
France

ISSN 0302-9743   ISSN 1611-3349   (electronic)
ISBN 978-3-319-12339-4   ISBN 978-3-319-12340-0   (eBook)
DOI 10.1007/978-3-319-12340-0

Library of Congress Control Number: 2014953267

LNCS Sublibrary: SL1 – Theoretical Computer Science and General Issues

Springer Cham Heidelberg New York Dordrecht London

Printed on acid-free paper

Springer is part of Springer Science+Business Media (www.springer.com)

Preface

The 40th International Workshop on Graph-Theoretic Concepts in Computer Science (WG 2014) took place in Le Domaine de Chalés, near Orléans in France, during June 25–27, 2014.

The WG conference series has a long tradition. Since 1975, it has taken place twenty-two times in Germany, four times in The Netherlands, three times in France, twice in Austria and the Czech Republic, as well as once in Italy, Slovakia, Switzerland, Norway, Greece, Israel, and the UK.

The WG conferences aim to connect theory and practice by demonstrating how graph-theoretic concepts can be applied to various areas of computer science and by extracting new graph problems from applications. Their goal is to present new research results and to identify and explore directions of future research.

WG 2014 received 80 submissions. Each submission was carefully reviewed by at least three members of the Program Committee. The Program Committee accepted 32 papers for presentation at WG 2014. The WG 2014 Student Paper Award was attributed to Felix Joos for his paper on "A Characterization of Mixed Unit Interval Graphs." Furthermore, three submissions were selected by the Program Committee for a possible publication in a special section of Algorithmica. The program also included two inspiring invited talks: Reinhard Diestel (Universität Hamburg, Germany) presented "A unified duality theorem for width parameters in graphs and matroids" and Pierre Fraigniaud (LIAFA Paris, France) gave a talk on "Local distributed computing."

We would like to thank the authors of the papers submitted for possible presentation at WG 2014, the speakers of the thirty-two talks and the speakers of the two invited talks, the members of the Program Committee, and the external reviewers. Special thanks to the Local Organizing Committee from the LIFO of the Université d'Orléans; without their performance WG 2014 could not have been such a success.

We are grateful to our sponsors, the Région Centre, the Université d'Orléans, the LIFO (Laboratory of Fundamental Informatics, Orléans), and the CNRS for their financial support.

August 2014   Dieter Kratsch
                                                                           Ioan Todinca

Organization

Program Committee

Hans Bodlaender	Universiteit Utrecht, The Netherlands
Jean Cardinal	Université Libre de Bruxelles, Belgium
Josep Diaz	Universitat Politècnica de Catalunya, Spain
David Eppstein	University of California, Irvine, USA
Fedor Fomin	Universitetet i Bergen, Norway
Cyril Gavoille	Université de Bordeaux, France
Frederic Havet	CNRS/Université Nice-Sophia-Antopolis, France
Iyad Kanj	DePaul University, USA
Michael Kaufmann	Universität Tübingen, Germany
Ekki Köhler	Brandenburgische Technische Universität, Cottbus, Germany
Daniel Král'	University of Warwick, UK
Dieter Kratsch	Université de Lorraine, France
A. Marchetti-Spaccamela	Università di Roma "La Sapienza", Italy
Claire Mathieu	CNRS/École Normale Supérieure, France
George Mertzios	Durham University, UK
Marcin Pilipczuk	Universitetet i Bergen, Norway
Lorna Stewart	University of Alberta, Canada
Jayme Szwarcfiter	Universidade Federal do Rio de Janeiro, Brazil
Ioan Todinca	Université d'Orléans, France

Additional Reviewers

Pierre Aboulker	Andreas Brandstädt	Beatrice Donati
Michal Adamaszek	Hajo Broersma	Paul Dorbec
Nadia Ady	Yi Cao	Feodor Dragan
Carme Alvarez	Márcia Cappelle	Paal Groenaas Drange
Joergen Bang-Jensen	L. Sunil Chandran	Markus Sortland Dregi
Michael Barrus	Steve Chaplick	Katherine Edwards
Jesse Beisegel	Vincent Cohen-Addad	Jessica Enright
Rémy Belmonte	Derek Corneil	Elaine Eschen
Aditya Bhaskara	Dan Cranston	Louis Esperet
Marijke Bodlaender	Pierluigi Crescenzi	Luerbio Faria
Marthe Bonamy	Marek Cygan	Fabrizio Frati
Nicolas Bonichon	Simone Dantas	Philippe Gambette
Edouard Bonnet	Fabien De Montgolfier	Robert Ganian
Flavia Bonomo	Carolin Denkert	Serge Gaspers

Archontia Giannopoulou
Mordecai J. Golin
Daniel Gonalves
Corinna Gottschalk
Alexander Grigoriev
Danny Hermelin
Petr Hlineny
Bart M.P. Jansen
Matthew Johnson
Frank Kammer
Phil Klein
Dusan Knop
Mikko Koivisto
Christian Komusiewicz
Guy Kortsarz
Danny Krizanc
O-Joung Kwon
Michael Lampis
Van Bang Le
Romain Letourneur
Bernard Lidicky
Mathieu Liedloff
Vincent Limouzy
Raphael Machado
Ana Karolinna Maia
Rogers Mathew
Paul Medvedev

Wagner Meira Jr.
Ivan Mihajlin
Martin Milanic
Dieter Mitsche
Mickael Montassier
Lalla Mouatadid
Marcin Mucha Hang Zhou
Haiko Müller
Jesper Nederlof
Jan Obdrzalek
Pascal Ochem
Yoshio Okamoto
Vangelis Paschos
Christophe Paul
Daniel Paulusma
Guillem Perarnau
Stephane Perennes
Anthony Perez
Geevarghese Philip
Michal Pilipczuk
Paulo Pinto
Sheung-Hung Poon
Christoforos Raptopoulos
Dieter Rautenbach
Alexander Reich
Giovanni Rinaldi
Ignaz Rutter

Pawel Rzazewski
Piotr Sankowski
Ignasi Sau
Marcus Schaefer
Ingo Schiermeyer
Mordechai Shalom
Manuel Sorge
R. Sritharan
Martin Strehler
Ola Svensson
Nicolas Trotignon
Torsten Ueckerdt
Takeaki Uno
Pim van 't Hof
Erik Jan van Leeuwen
Jean-Marie Vanherpe
Kevin Verbeek
Jan Volec
Ton Volgenant
Jens Vygen
Egon Wanke
Mathias Weller
David R. Wood
Neal Young
Jakub Łacki

Contents

X   Contents

Unifying Duality Theorems for Width Parameters in Graphs and Matroids (Extended Abstract)

Reinhard Diestel[1](✉) and Sang-il Oum[2]

[1] Mathematisches Seminar, Universität Hamburg, Hamburg, Germany
R.Diestel@math.uni-hamburg.de
[2] Department of Mathematical Sciences, KAIST, Daejeon, South Korea

Abstract. We prove a general duality theorem for width parameters in combinatorial structures such as graphs and matroids. It implies the classical such theorems for path-width, tree-width, branch-width and rank-width, and gives rise to new width parameters with associated duality theorems. The dense substructures witnessing large width are presented in a unified way akin to tangles, as orientations of separation systems satisfying certain consistency axioms.

1 Introduction

There are a number of theorems in the structure theory of sparse graphs that assert a duality between certain 'dense objects' and an overall tree structure. For example, a graph has small tree-width if and only if it contains no large-order bramble. The aim of this paper is to prove one such theorem in a general setting, a theorem that will imply all the classical duality theorems as special cases, but with a unified and simpler proof. Our theory will give rise to new width parameters as well, with dual 'dense objects', and conversely provide dual tree-like structures for notions of dense objects that have been considered before but for which no duality theorems were known.

Amini, Mazoit, Nisse, and Thomassé [1] have also established a theory of dualities of width parameters, which pursues (and achieves) a similar aim. Our theory differs from theirs in two respects: we allow more general separations of a given ground set than just partitions, including ordinary separations of graphs; and our 'dense objects' are modelled after tangles, while theirs are modelled on brambles. Hence while our main results can both be used to deduce those classical duality theorems for width parameters, they differ in substance. And so do their corollaries for the various width parameters, even if they imply the same

This is an extended abstract of arXiv:1406.3797, which contains all the proofs omitted here. See also arXiv:1406.3798 for further work in this direction.

Sang-il Oum: Supported by Basic Science Research Program through the National Research Foundation of Korea (NRF) funded by the Ministry of Science, ICT & Future Planning (2011-0011653).

D. Kratsch and I. Todinca (Eds.): WG 2014, LNCS 8747, pp. 1–14, 2014.
DOI: 10.1007/978-3-319-12340-0_1

classical results. Moreover, while the main results of [1] can easily be deduced from ours, the converse seems less clear. And finally, our theory gives rise to duality theorems for new width parameters that can only be expressed in our setup.

All we need in our set-up is that we have a notion of 'separation' for the combinatorial structure to be considered, by which we mean an ordered pair (A, B) of subsets of some ground set V such that $A \cup B = V$.[1] For example, V might be the vertex set of a graph or the ground set of a matroid, and 'separations' would be defined as is usual for graphs and matroids. In order to apply our theorem we may need in addition that there is a submodular function defined on these separations, such as their order, but our main result can be stated without such an assumption.

Our unified treatment of 'dense objects' is gleaned from the notion of tangles in graph minor theory [10], or of ultrafilters in set theory. The idea is as follows. Consider any set S of separations of a given graph or matroid. In order to deserve its name with respect to S, we expect of a 'dense object' that for every separation in S it lies on one side but not the other. For example, if S is the set of all separations (A, B) of a graph G such that $|A \cap B| < k$, then every K_n minor of G with $n \geq k$ will have a branch set in $A \smallsetminus B$ or in $B \smallsetminus A$, but not both. Our dense object $\mathcal{D}$ therefore *orients* every separation in S by choosing exactly one of the two ordered pairs $(A, B), (B, A)$ in such cases,[2] and our paradigm is that this orientation of S is the only information about $\mathcal{D}$ that we ever use. We formalize this by *defining* 'dense objects' as certain orientations of S.

To deserve their name, 'dense objects' cannot be arbitrary orientations of S but have to satisfy some consistency rules. For example, if in a graph G we have two separations $(A, B), (C, D)$ and their inverses in S, and $A \subseteq C$ and $B \supseteq D$, then $\mathcal{D}$ should not orient $\{A, B\}$ towards A by selecting (B, A) and $\{C, D\}$ towards D by selecting (C, D). While this rule will be common to all the 'dense objects' we shall consider, there may be further rules depending on the type of object, so that we can tell them apart. These additional rules will stipulate that the orientation of S given by a dense object $\mathcal{D}$ must not contain certain subsets of S, such as the set $\{(B, A), (C, D)\}$ in the above example. Thus, each type of dense object will be specified by a collection $\mathcal{F}$ of 'forbidden' subsets of S.

The tree-like structure that is dual to a dense object $\mathcal{D}$, i.e., which will exist in a graph or matroid if and only if it contains no instance of $\mathcal{D}$, will be defined by this same collection $\mathcal{F}$ of separation sets forbidden in $\mathcal{D}$. It will typically come as a subset of S that is nested, and which thus cuts up the underlying set in a tree-like way, and the 'stars of separations' by which this tree branches will be *required* to lie in $\mathcal{F}$. Tangles, for example, are defined in this way: with $\mathcal{F}$ the set of all

---

[1] In fact, we need even less. It would be enough to consider instead of 'separations' any poset with an involution that commutes with its ordering, just as the ordering of separations introduced below satisfies $(A, B) \leq (C, D) \Leftrightarrow (B, A) \geq (D, C)$. It is only for the sake of readability that we are writing this paper in terms of separations, as readers are likely to have graphs or matroids in mind.

[2] Our notational convention will be that we think of (A, B) as pointing towards B.

triples $(A_1, B_1), (A_2, B_2), (A_3, B_3)$ of separations whose 'small' sides A_1, A_2, A_3 cover the entire graph or matroid, and branch decompositions, their dual objects, as nested sets of separations branching at precisely such triples.

The following familiar dualities between dense objects and tree structures can be captured in this way, and their duality theorems will follow from our theorem. For graphs, we can capture path-decompositions and blockages [2], tree-decompositions and brambles [11], branch-decompositions and tangles of graphs [10]. For matroids, our framework captures branch-decompositions and tangles [4,10], as well as matroid tree-decompositions [5] and their dual objects proposed by Amini, Mazoit, Nisse, and Thomassé [1]. Our framework also captures branch-decompositions and tangles of symmetric submodular functions [4,10], which includes branch-width of graphs and matroids, carving-width of graphs [12], and rank-width of graphs [8].

Since blockages and brambles are not defined in terms of orientations of sets of separations, the duality theorems we obtain when we specify S and $\mathcal{F}$ to capture path- or tree-width (of graphs or matroids) will differ from their known duality theorems. But they will be easily interderivable with these. Since S and $\mathcal{F}$ can be chosen in many other ways too, our results also imply dualities for new width parameters.

Our unifying duality theorem comes in three flavours: as *weak*, *strong*, and *general* duality. In this extended abstract we only present the Strong Duality Theorem, along with applications indicating how to derive duality theorems for all the classical width parameters.

2   Terminology and Basic Facts

A *separation* of a set V is a pair (A, B) of subsets such that $A \cup B = V$. Its *inverse* is the separation (B, A). A set S of separations is *symmetric* if $(B, A) \in S$ whenever $(A, B) \in S$, and *antisymmetric* if $(B, A) \notin S$ whenever $(A, B) \in S$. A symmetric set of separations of a set V is a *separation system* on V.

The separation (A, B) is *proper* if $A, B \neq V$, and *improper* otherwise. The separations of V are partially ordered by

$$(A, B) \leq (C, D) :\Leftrightarrow A \subseteq C \text{ and } B \supseteq D.$$

Note that this is equivalent to $(D, C) \leq (B, A)$, and that (A, B) is proper if and only if (A, B) and (B, A) are incomparable with respect to $\leq$.

Informally, we think of (A, B) as *pointing towards* B and *away from* A. Similarly, if $(A, B) \leq (C, D)$, then (A, B) *points towards* (C, D) and (D, C), while (C, D) *points away from* (A, B) and (B, A).

A set S of separations of V is *nested* if each of them is comparable with every other or its inverse. Thus, two nested separations are either comparable, or point towards each other, or point away from each other. Two separations that are not nested are said to *cross*.

A set of separations is a *star* if they point towards each other (Fig. 1). Thus, S is a *star* if $(A, B) \leq (D, C)$ for distinct $(A, B), (C, D) \in S$. In particular, stars

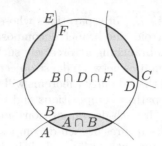

Fig. 1. The separations $(A, B), (C, D), (E, F)$ form a 3-star

are nested. They need not be antisymmetric, but if not they contain an inverse pair $(A, B), (B, A)$, then any other separation they contain must be improper.

Let $\mathcal{F} \subseteq 2^S$ be a collection of sets of separations in S, and $S^- \subseteq S$. An S-*tree over* $\mathcal{F}$ and *rooted in* S^- is a pair (T, α) of a tree T with at least one edge and a function $\alpha \colon \vec{E}(T) \to S$ from the set

$$\vec{E}(T) := \{(s, t) : \{s, t\} \in E(T)\}$$

of all *orientations* of edges of T satisfying the following:

(i) For each edge xy of T, if $\alpha(x, y) = (A, B)$ then $\alpha(y, x) = (B, A)$.
(ii) For each internal node t of T, the set $\{\alpha(s, t) : st \in E(T)\}$ is in $\mathcal{F}$.
(iii) For each leaf s of T with neighbour t, say, $\alpha(s, t) \in S^-$.

We say that the separation $\alpha(s, t)$ in (iii) is *associated with*, or simply *at*, the leaf s. The separations at leaves are the *leaf separations* of (T, α).

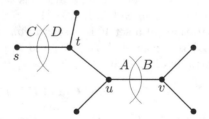

Fig. 2. An S-tree with $(C, D) = \alpha(s, t) \leq \alpha(u, v) = (A, B)$

An important example are the S-trees *over stars*: the S-trees over some $\mathcal{F}$ all whose elements are stars of separations. In such an S-tree (T, α) the map α preserves the *natural partial ordering* on $\vec{E}(T)$ defined by letting $(s, t) \leq (u, v)$ if the unique $\{s, t\}$–$\{u, v\}$ path in T starts at t and ends at u. Indeed, the images under α of the oriented stars

$$S_t = \{(s, t) : t \text{ an internal node of } T\},$$

which by (ii) are sets in $\mathcal{F}$, are then stars of separations. This means precisely that α preserves the partial ordering on the sets S_t as induced by $\vec{E}(T)$, which in turn easily implies that α preserves the ordering on all of $\vec{E}(T)$ (Fig. 2).

Let S be a separation system. An *orientation* of S is a subset $O \subseteq S$ that contains, for every $(A, B) \in S$, exactly one of (A, B) and (B, A). A *partial orientation* of S is an orientation of some symmetric subset of S.

A set P of separations is *consistent* if it contains no two separations pointing away from each other: if $(C, D) \le (A, B) \in P$ implies $(D, C) \notin P$.[3] Note that this does not imply $(C, D) \in P$: it may also happen that P contains neither (C, D) nor (D, C). Note that consistent sets of separations are antisymmetric.

If $P \subseteq S$ is consistent, it is clearly a partial orientation of S. Conversely, if P is an orientation of all of S, it is consistent if and only if it is *closed down* in the partial ordering of S, i.e., if and only if $(C, D) \in P$ whenever $(C, D) \le (A, B) \in P$ and $(C, D) \in S$.

Whenever $P \subseteq O \subseteq S$ we say that P *extends to* O, and O *extends* P.

Proposition 1. *Every consistent partial orientation of a separation system S extends to a consistent orientation of S.*

3 The (Strong) Duality Theorem

Our paradigm is to capture the notion of a 'dense object' $\mathcal{D}$ in a structure on a set V by orientations of suitable separation systems S on V. Here, 'suitable' means that for every separation in S the object $\mathcal{D}$ should 'lie on' one of its sides but not the other, and S should ideally contain all separations of V for which this is the case.

If $\mathcal{D}$ was a concrete subset X of V, for example, such as a set spanning a large complete subgraph in a graph, there would then be a unique orientation O of S that describes $\mathcal{D}$: the set $\{(A, B) \in S : X \subseteq B\}$. What makes the orientations paradigm so attractive, however, is that it is more general than this. For example, a large grid H in a graph G defines a high-order tangle $\mathcal{T}$ – for every small-order separation of G, most of H will lie on one side but not the other – yet the intersection of the 'large sides' B of all the oriented separations $(A, B) \in \mathcal{T}$ will be empty. What the existence of a large grid H in G does imply, however, is that G has no three low-order separations (A_i, B_i) $(i = 1, 2, 3)$ such that $H \subseteq G[A_1] \cup G[A_2] \cup G[A_3]$. So Robertson and Seymour [10] chose this latter property as the defining axiom for a *tangle*.

In this spirit, we seek to define our 'dense objects' as orientations of separation systems S that do not contain certain subsets of S. We say that a partial

---

[3] It is a good idea to work with this formal definition of consistency, since the more intuitive notion of 'pointing away from each other' can be counterintuitive. For example, we shall need that no consistent set of separations of V contains a separation of the form (V, A); this follows readily from the formal definition, as $(A, V) \le (V, A)$, but is less obvious from the informal.

orientation P of a separation system S *avoids* $\mathcal{F} \subseteq 2^S$ if P has no subset in $\mathcal{F}$, i.e., if $2^P \cap \mathcal{F} = \emptyset$.

Before we can state our duality theorem, we have to introduce the somewhat technical notion of 'shifting' an S-tree of a graph G across a separation (X, Y) of G to give, essentially, two S-trees of $G[X]$ and of $G[Y]$. The idea behind this is as follows.

We prove the duality theorem by inverse induction on the size of S^-. Given a separation $\{X, Y\}$ such that $\{(X, Y), (Y, X)\} \subseteq S \setminus S^-$, the induction hypothesis will give us an S-tree (T_Y, α_Y) of G rooted in $S_X^- = S^- \cup \{(X, Y)\}$ (which, if (X, Y) is indeed associated with a leaf and hence certifies that X is small, can be viewed as an S-tree of $G[Y]$), and another S-tree (T_X, α_X) of G rooted in $S_Y^- = S^- \cup \{(Y, X)\}$ (which can be viewed as an S-tree of $G[X]$). We shall then seek to combine these two trees to an S-tree (T, α) of G, rooted in the given S^-.

In order to make the separations associated with the two trees compatible (i.e., nested with each other), we have to regard the separations (of G) in the image of α_X as 'essentially separations of $G[X]$', which we shall do by adding all of Y to one of their sides. Similarly, we add X to one side of every separation in the image of α_Y. The next few paragraphs describe how exactly to do this.

Let $(X, Y) \leq (U, W)$ be elements of a set S of separations of a set V. Assume that $U \neq V$, and that (W, U) is associated with a leaf w of an S-tree (T, α) over some set $\mathcal{F} \subseteq 2^S$ of stars. Our aim is to 'shift' (T, α) to a new S-tree (T, α') based on the same tree T, by shifting the separations in the image of α over to X.

Given a separation $(A, B) \leq (U, W)$, let us define (Fig. 3, left)

$$f\!\downarrow_{(X,Y)}^{(U,W)} (A, B) := (A \cap X, B \cup Y) \quad \text{and} \quad f\!\downarrow_{(X,Y)}^{(U,W)} (B, A) := (B \cup Y, A \cap X).$$

This defines a *shifting map* $f\!\downarrow_{(X,Y)}^{(U,W)}$ on the set $S_{(U,W)}$ of separations $(A, B) \leq (U, W)$ and their inverses. Since (W, U) is a leaf separation of (T, α) and $\mathcal{F}$ consists of stars, the image of α lies in $S_{(U,W)}$ (Fig. 3, right). Hence the concatenation

$$\alpha' := f\!\downarrow_{(X,Y)}^{(U,W)} \circ \alpha$$

is well defined. However it is not clear for now whether α' takes all its images in S.

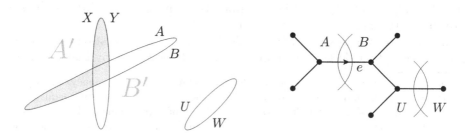

Fig. 3. Shifting $\alpha(\vec{e}) = (A, B)$ to $\alpha'(\vec{e}) = (A', B')$

What is immediate, however, is that $f\!\downarrow_{(X,Y)}^{(U,W)}$ maps stars to stars:

Lemma 1. *The map $f\!\downarrow_{(X,Y)}^{(U,W)}$ preserves the ordering $\leq$ of separations.*

Lemma 1 not only implies that $f\!\downarrow_{(X,Y)}^{(U,W)}$ maps stars to stars,[4] it also implies that all leaf separations of (T,α), other than (W,U), get smaller in the transition to α'.[5] Indeed, if $\alpha(s,t) = (A,B)$ with $s \neq w$ a leaf of T, then $(A,B) \leq (U,W)$ and hence

$$f\!\downarrow_{(X,Y)}^{(U,W)} (A,B) = (A \cap X, B \cup Y) \leq (A,B).$$

It remains to ensure that α' takes its image in S if α does. The following condition on S will ensure the existence of a separation (X,Y) for which this is the case. Let us say that $(X,Y) \in S$ is *linked* to $(U,W) \in S$ if $(X,Y) \leq (U,W)$ and

$$(A \cap X, B \cup Y) \in S$$

for all $(A,B) \in S$ with $(A,B) \leq (U,W)$. Let us call S *separable* if for every pair $(W',U') \leq (U,W)$ of separations in S there exists $(X,Y) \in S$ such that (X,Y) is linked to (U,W) and (Y,X) is linked to (U',W').

Finally, we need a condition on $\mathcal{F}$ to ensure that the shifts of stars that occur as images under α of oriented stars at nodes of T are not only again stars but are also again in $\mathcal{F}$ (see Footnote 4). Let us say that a separation $(X,Y) \in S$ is *$\mathcal{F}$-linked* to $(U,W) \in S$ with $U \neq V$ if (X,Y) is linked to (U,W) and the image under $f\!\downarrow_{(X,Y)}^{(U,W)}$ of any star $S' \subseteq S_{(U,W)}$ in $\mathcal{F}$ that contains a separation (A,B) with $(B,A) \leq (U,W)$ is again in $\mathcal{F}$. We say that S is *$\mathcal{F}$-separable* if for every pair $(W',U') \leq (U,W)$ of separations in S, with $U,U' \neq V$, there exists $(X,Y) \in S$ such that (X,Y) is $\mathcal{F}$-linked to (U,W) and (Y,X) is $\mathcal{F}$-linked to (U',W'). And a set $\mathcal{F}$ of stars in S is *closed under shifting* if whenever $(X,Y) \in S$ is linked to $(U,W) \in S$ with $U \neq V$ it is even $\mathcal{F}$-linked to (U,W).

The following observation is immediate from the definitions:

Lemma 2. *If S is separable and $\mathcal{F}$ is closed under shifting, then S is $\mathcal{F}$-separable.*

In Sect. 4 we shall see that for all sets $\mathcal{F}$ describing classical 'dense objects', such as tangles and brambles (as well as many others), the usual separation systems S are $\mathcal{F}$-separable. In many cases, $\mathcal{F}$ will even be closed under shifting, in which case we will simply prove this stronger property.

Theorem 1 (Strong Duality Theorem). *Let S be a separation system of a set V, and let $\mathcal{F} \subseteq 2^S$ be a set of stars. Let S^- be a down-closed subset of S containing all its separations of the form (A,V). If S is $\mathcal{F}$-separable, then exactly one of the following holds:*

(i) *There exists an S-tree over $\mathcal{F}$ rooted in S^-.*
(ii) *There exists a consistent $\mathcal{F}$-avoiding orientation of S extending S^-.*

---

[4] This will help us show that (T,α') is over $\mathcal{F}$ if (T,α) is.
[5] This will help us show that (T,α') is rooted in S^- if (T,α) is.

Thus, in order to derive from Theorem 1 a specific duality theorem for some given width parameter, it remains to do two things: to specify the $\mathcal{F}$ that describes this parameter, and then to show that the set S of separations we are considering is $\mathcal{F}$-separable.

We shall do this in the next section for some standard examples.

4   Applications of Strong Duality

In this section we show that the separation systems usually considered for graphs and matroids are all separable, and that the collections $\mathcal{F}$ needed to capture 'dense objects' such as tangles, brambles and blockages are closed under shifting. This will make our strong duality theorem imply the classical duality theorems for graphs and matroids. We also obtain some interesting new such theorems.

Let us call a separation system a *universe* if for any two of its separations (A, B) and (C, D) it also contains $(A \cap C, B \cup D)$. For instance, the set of all partitions of the ground set of a matroid is a universe, and so is the set of all vertex separations of a graph (which does not normally include all its vertex partitions).

We shall call a real function $(A, B) \mapsto |A, B|$ on a universe $\mathcal{U}$ an *order function* if it is symmetric and submodular, that is, if $|A, B| = |B, A|$ and

$$|A \cap C, B \cup D| + |A \cup C, B \cap D| \leq |A, B| + |C, D|$$

for all $(A, B), (C, D) \in \mathcal{U}$. We then call $|A, B|$ the *order* of the separation (A, B). Given a universe $\mathcal{U}$ with an order function, our focus will often be on the subsystem

$$S_k = \{(A, B) \in \mathcal{U} : |A, B| < k\}$$

for some positive integer k.

Lemma 3. *Every such S_k is separable.*

For the remainder of this section, whenever we consider a graph $G = (V, E)$ we let $\mathcal{U}$ be its universe of *vertex separations*, the set of pairs (A, B) of vertex sets A, B such that $A \cup B = V$ and G has no edge between $A \setminus B$ and $B \setminus A$. We then take $|A, B| := |A \cap B|$ as our order function for $\mathcal{U}$, and put

$$S_k^- := \big\{(A, B) \in S_k : |A| < k\big\}.$$

This is obviously closed down in S_k, and S_k is separable by Lemma 3.

We remark that any consistent orientation O of S_k must extend the subset of S_k^- consisting of its separations of the form (A, V). This is because otherwise O would contain (V, A), with $(A, V) \leq (V, A) \in O$ violating consistency.

4.1   Branch-Width and Tangles

Let $G = (V, E)$ be a finite graph. A *tangle of order* k in G is (easily seen to be equivalent to) an $\mathcal{F}$-avoiding orientation of S_k extending S_k^- for

$$\mathcal{F} := \big\{ \{(A_1, B_1), (A_2, B_2), (A_3, B_3)\} \subseteq S_k : G[A_1] \cup G[A_2] \cup G[A_3] = G \big\}.$$

(The three separations $(A_1, B_1), (A_2, B_2), (A_3, B_3)$ need not be distinct.) Notice that any orientation of S_k that avoids this $\mathcal{F}$ is consistent, since for any pair of separations $(C, D) \leq (A, B)$ we have $G[D] \cup G[A] \supseteq G[B] \cup G[A] = G$ and hence $\{(D, C), (A, B)\} \in \mathcal{F}$.

Since our duality theorems, so far, only work with $\mathcal{F}$ consisting of stars of separations, let us consider the set $\mathcal{F}^*$ of those sets in $\mathcal{F}$ that are stars. Using submodularity one can easily show that an $\mathcal{F}^*$-avoiding orientation of S_k in fact avoids all of $\mathcal{F}$ – but only if we assume consistency:

Lemma 4. *Every consistent $\mathcal{F}^*$-avoiding orientation O of S_k avoids $\mathcal{F}$.*

It is easy to check that $\mathcal{F}^*$ is closed under shifting, and so we have our first application:

Theorem 2. *The following are equivalent for finite graphs $G \neq \emptyset$ and $k > 0$:*

(i) G *has a tangle of order* k.
(ii) S_k *has an $\mathcal{F}$-avoiding orientation extending S_k^-.*
(iii) S_k *has a consistent $\mathcal{F}^*$-avoiding orientation extending S_k^-.*
(iv) G *has no S_k-tree over $\mathcal{F}^*$ rooted in S_k^-.*
(v) G *has branch-width at least* k, *or* $k \leq 2$ *and* G *is a disjoint union of stars and isolated vertices and has at least one edge.*

4.2   Tree-Width and Path-Width

We now apply our strong duality theorem to obtain a duality theorem for tree-width in graphs. Its dual 'dense objects' will be orientations of S_k, like tangles, and thus different from brambles (or 'screens'), the dual objects in the classical tree-width duality theorem of Seymour and Thomas [11].

This latter theorem, which ours easily implies, says that a finite graph either has a tree-decomposition of width less than $k-1$ or a bramble of order at least k, but not both. The original proof of this theorem is as mysterious as the result is beautiful. The shortest known proof is given in [3] (where we refer the reader also for definitions), but it is hardly less mysterious. A more natural, if slightly longer, proof was given recently by Mazoit [7]. The proof by our strong duality theorem, as outlined below, is perhaps the most basic proof one can have.[6]

Consider a finite graph $G = (V, E)$, with sets of vertex separations $S_k^- \subseteq S_k$ for some integer $k > 0$ as defined at the start of Sect. 4. Let

$$\mathcal{F}_k := \big\{ S \subseteq S_k \mid S = \{(A_i, B_i) : i = 0, \ldots, n\} \text{ is a star with } \big| \bigcap_{i=0}^{n} B_i \big| < k \big\}.$$

---

[6] For example, we do not need Menger's theorem, as all the other proofs do.

We have seen that S_k is separable (Lemma 3). To apply Theorem 1 we thus only need the following lemma – whose very easy proof contains the only bit of magic now left in the tree-width duality theorem:

Lemma 5. $\mathcal{F}_k$ *is closed under shifting.*

It is easy to check that G has an S_k-tree (T, α) over $\mathcal{F}_k$ rooted in S_k^- if and only if it has a tree-decomposition $(T, \mathcal{V})$ of width less than $k-1$. The translation between orientations of S_k and brambles in a graph G is more interesting. Before the notion of a bramble was introduced in [11] (under the name of 'screen'), Robertson and Seymour had looked for an object dual to small tree-width that was more akin to our orientations of S_k: maps β assigning to every set X of fewer than k vertices one component of $G - X$. The question was how to make these choices consistent, so that they would define the desired 'dense object' dual to small tree-width. The obvious consistency requirement, that $\beta(Y) \subseteq \beta(X)$ whenever $X \subseteq Y$, is easily seen to be too weak, while asking that $\beta(X) \cap \beta(Y) \neq \emptyset$ for all X, Y turned out to be too strong. In [11], Seymour and Thomas then found a requirement that worked: that any two such sets, $\beta(X)$ and $\beta(Y)$, should *touch*: that either they share a vertex or G has an edge between them. Such maps β are now called *havens*, and it is easy to show that G admits a haven of *order* k (one defined on all sets X of less than k vertices) if and only if G has a bramble of order at least k.

The notion of 'touching' was perhaps elusive because it appeals directly to the structure of G, its edges: it is not be phrased purely in terms of set containment. It turns out, however, that it can be phrased in such terms after all, as the consistency of orientations of S_k:

Lemma 6. G *has a bramble of order at least* k *if and only if* S_k *has a consistent* $\mathcal{F}_k$*-avoiding orientation extending* S_k^-*.*

The next application of our duality theorem includes the tree-width duality theorem of Seymour and Thomas [11], and extends it by the new width parameter of S_k-trees over $\mathcal{F}_k$:

Theorem 3. *The following are equivalent for all finite graphs* G *and* $k > 0$:

(i) G *has a bramble of order at least* k.
(ii) S_k *has a consistent* $\mathcal{F}_k$*-avoiding orientation extending* S_k^-.
(iii) G *has no* S_k*-tree over* $\mathcal{F}_k$ *rooted in* S_k^-.
(iv) G *has tree-width at least* $k - 1$.

We can also bound the adhesion of a tree-decomposition independently from its width. For integers $k < w$, setting

$$\mathcal{F}_w = \left\{ S \subseteq S_k \mid S = \{(A_i, B_i) : i = 0, \ldots, n\} \text{ is a star with } \left| \bigcap_{i=0}^{n} B_i \right| < w \right\}$$

yields the following new duality theorem:

Theorem 4. *The following are equivalent for all graphs* G *and* $w \geq k > 0$:

(i) S_k has a consistent $\mathcal{F}_w$-avoiding orientation extending S_k^-.
(ii) G has no S_k-tree over $\mathcal{F}_w$ rooted in S_k^-.
(iii) G has no tree-decomposition of width $< w - 1$ and adhesion $< k$.

Similarly, setting

$$\mathcal{F}_k^{(2)} := \left\{ S \subseteq S_k \mid S = \{(A_1, B_1), (A_2, B_2)\} \text{ is a star with } |B_1 \cap B_2| < k \right\}$$

yields a duality theorem for path-width:

Theorem 5. *The following are equivalent for all finite graphs G and $k > 0$:*

(i) S_k has a consistent $\mathcal{F}_k^{(2)}$-avoiding orientation extending S_k^-.
(ii) G has no S_k-tree over $\mathcal{F}_k^{(2)}$ rooted in S_k^-.
(iii) G has path-width at least $k - 1$.

4.3   Carving Width, Rank Width, and Matroid Tangles

The concepts of branch-width and tangles were introduced by Robertson and Seymour [10] not only for graphs but more generally for hypergraphs. As the order of a separation (A, B) they already considered, instead of $|A \cap B|$, also arbitrary symmetric submodular order functions $|A, B|$ and proved the relevant lemmas more generally for these. Geelen, Gerards, Robertson, and Whittle [4] applied this explicitly to a submodular connectivity function.

Our aim in this section is to derive from Theorem 1 a duality theorem for branch-width and tangles in arbitrary separation universes with an order function, as introduced at the start of Sect. 4. This will imply the above branch-width duality theorems for hypergraphs and matroids, as well as their cousins for carving width [12] and rank-width of graphs [8].

Let $\mathcal{U}$ be any universe of separations of some set E of at least two elements, with an order function $(A, B) \mapsto |A, B|$. Let $k > 0$ be an integer, and consider

$$S_k = \left\{(A, B) \in \mathcal{U} : |A, B| < k \right\} \text{ and } S_k^- = \left\{(A, B) \in S_k : |A| \le 1 \right\}.$$

Let us call an orientation of S_k a *tangle of order* k if it extends S_k^- and avoids

$$\mathcal{F} = \left\{\{(A_1, B_1), (A_2, B_2), (A_3, B_3)\} \subseteq S_k : A_1 \cup A_2 \cup A_3 = E \right\},$$

where $(A_1, B_1), (A_2, B_2), (A_3, B_3)$ need not be distinct; in particular, tangles are consistent. This extends the existing notions of tangles for hypergraphs and matroids, with their edge set or ground set as E, partitions as separations, and the appropriate order functions.

Let $\mathcal{F}^* \subseteq \mathcal{F}$ be the set of stars in $\mathcal{F}$. It is easy to prove that $\mathcal{F}^*$ is closed under shifting, and we have the following analogue of Lemma 4:

Lemma 7. *Every consistent $\mathcal{F}^*$-avoiding orientation of S_k avoids $\mathcal{F}$.*

Let us say that $\mathcal{U}$ has *branch-width* $< k$ if there exists an S_k-tree over $\mathcal{F}^*$ that is rooted in S_k^-. As before, this definition agrees with the usual ones when $\mathcal{U}$ is a hypergraph or matroid. By Lemmas 3 and 7, Theorem 1 now specializes as follows:

Theorem 6. *Given a separation universe $\mathcal{U}$ with an order function, and $k > 0$, the following assertions are equivalent:*

(i) *$\mathcal{U}$ has a tangle of order k.*
(ii) *S_k has a consistent $\mathcal{F}^*$-avoiding orientation extending S_k^-.*
(iii) *$\mathcal{U}$ does not have branch-width $< k$.*

4.4   Matroid Tree-Width

Hliněný and Whittle [5, 6] generalized the notion of tree-width from graphs to matroids.[7] Our aim in this section is to specialize our strong duality theorem to a duality theorem for tree-width in matroids.

Let $M = (E, I)$ be a matroid with rank function r. Its *connectivity function* is defined as

$$\lambda(X) := r(X) + r(E \smallsetminus X) - r(M).$$

We consider the universe $\mathcal{U}$ of all bipartitions (X, Y) of E. Since

$$|X, Y| := \lambda(X) = \lambda(Y)$$

is submodular and symmetric, it is an order function on $\mathcal{U}$.

A *tree-decomposition* of M is a pair (T, τ), where T is a tree and $\tau \colon E \to V(T)$ is any map. Let t be a node of T, and let $T_1, \ldots, T_d$ be the components of $T - t$. Then the *width* of t is the number

$$\sum_{i=1}^{d} r(E \smallsetminus F_i) - (d-1)\, r(M),$$

where $F_i = f^{-1}(V(T_i))$. (If t is the only node of T, we let its width be $r(M)$.) The *width* of (T, τ) is the maximum width of the nodes of T. The *tree-width* of M is the minimum width over all tree-decompositions of M.

Matroid tree-width generalizes the tree-width of graphs in the expected way:

Theorem 7 (Hliněný and Whittle [5, 6]). *The tree-width of a finite graph containing at least one edge equals the tree-width of its cycle matroid.*

In order to specialize Theorem 1 to a duality theorem for tree-width in matroids, we consider

$$S_k = \{\, (A, B) \in \mathcal{U} : |A, B| < k \,\} \quad \text{and} \quad S_k^- = \{\, (A, B) \in \mathcal{U} : r(A) < k \,\} \subseteq S_k.$$

---

[7] In our matroid terminology we follow Oxley [9].

Since λ is symmetric and submodular, S_k is separable by Lemma 3. Let

$$\mathcal{F}_k := \Big\{ S \subseteq \mathcal{U} \mid S = \{(A_i, B_i) : i = 0, \ldots, n\} \text{ is a star with } n \geq 1$$

$$\text{and } \sum_{i=0}^{n} r(B_i) - n\, r(M) < k \Big\}.$$

Even without requiring this in the definition, one can show that every $S \in \mathcal{F}_k$ is a subset of S_k, and that S_k is $\mathcal{F}_k$-separable.

Theorem 1 now yields the following duality theorem for matroid tree-width.

Theorem 8. *Let $M = (E, I)$ be a matroid with the rank function r, and let k be an integer. Then the following statements are equivalent:*

(i) *M has tree-width at least k.*
(ii) *M has no S_k-tree over $\mathcal{F}_k$ rooted in S_k^-.*
(iii) *S_k has a consistent $\mathcal{F}_k$-avoiding orientation extending S_k^-.*

5   Algorithms

We have not considered the algorithmic task of finding, given fixed types of S, S^- and $\mathcal{F}$, for any input graph G the right one of the two alternatives from our duality theorem: an S-tree over $\mathcal{F}$ rooted in S^-, or a consistent $\mathcal{F}$-avoiding orientation of S extending S^-. As far as we are aware, such algorithmic results do not even exist for the classical duality theorems, such as the one for tree-width and brambles. Our setup makes it possible to treat this in greater generality, which seems like a fitting challenge for this conference's academic environment.

References

1. Amini, O., Mazoit, F., Nisse, N., Thomassé, S.: Submodular partition functions. Discrete Appl. Math. **309**(20), 6000–6008 (2009)
2. Bienstock, D., Robertson, N., Seymour, P., Thomas, R.: Quickly excluding a forest. J. Combin. Theor. Ser. B **52**(2), 274–283 (1991)
3. Diestel, R.: Graph Theory, 4th edn. Springer, Heidelberg (2010)
4. Geelen, J., Gerards, B., Robertson, N., Whittle, G.: Obstructions to branch-decomposition of matroids. J. Combin. Theor. Ser. B **96**, 560–570 (2006)
5. Hliněný, P., Whittle, G.: Matroid tree-width. European J. Combin. **27**(7), 1117–1128 (2006)
6. Hliněný, P., Whittle, G.: Addendum to matroid tree-width. European J. Combin. **30**, 1036–1044 (2009)
7. Mazoit, F.: A simple proof of the tree-width duality theorem. arXiv:1309.2266 (2013)
8. Oum, S., Seymour, P.: Approximating clique-width and branch-width. J. Combin. Theor. Ser. B **96**(4), 514–528 (2006)
9. Oxley, J.: Matroid Theory. Oxford University Press, Oxford (1992)

10. Robertson, N., Seymour, P.: Graph minors. X. Obstructions to tree-decomposition. J. Combin. Theor. Ser. B **52**, 153–190 (1991)
11. Seymour, P., Thomas, R.: Graph searching and a min-max theorem for tree-width. J. Combin. Theor. Ser. B **58**(1), 22–33 (1993)
12. Seymour, P., Thomas, R.: Call routing and the ratcatcher. Combinatorica **14**(2), 217–241 (1994)

Distributedly Testing Cycle-Freeness

Heger Arfaoui[1], Pierre Fraigniaud[1(✉)], David Ilcinkas[2], and Fabien Mathieu[3]

[1] CNRS, University Paris Diderot, Paris, France
`pierref@liafa.univ-paris-diderot.fr`
[2] CNRS, University of Bordeaux, Talence, France
[3] Alcatel-Lucent Bell Labs, Marcoussis, France

Abstract. We tackle *local distributed testing* of graph properties. This framework is well suited to contexts in which data dispersed among the nodes of a network can be collected by some central authority (like in, e.g., sensor networks). In local distributed testing, each node can provide the central authority with just a few information about what it perceives from its neighboring environment, and, based on the collected information, the central authority is aiming at deciding whether or not the network satisfies some property. We analyze in depth the prominent example of checking *cycle-freeness*, and establish tight bounds on the amount of information to be transferred by each node to the central authority for deciding cycle-freeness. In particular, we show that distributedly testing cycle-freeness requires at least $\lceil \log d \rceil - 1$ bits of information per node in graphs with maximum degree d, even for connected graphs. Our proof is based on a novel version of the seminal result by Naor and Stockmeyer (1995) enabling to reduce the study of certain kinds of algorithms to order-invariant algorithms, and on an appropriate use of the known fact that every free group can be linearly ordered.

1 Introduction

1.1 Context and Objective

We are interested in *monitoring* structural properties of networks. Our setting is the one of a large-scale distributed system in which nodes are linked together so as to form a network G. Our objective is to discuss the ability of the nodes to decide whether or not the network satisfies certain structural properties, which may in turn govern the ability of the network to perform certain tasks efficiently. Examples of such structural properties are, e.g., large expansion, which governs the ability to disseminate information quickly, or cycle-freeness, which prevents communication packets to enter into infinite loops. For the purpose of deciding structural properties of the network, each of its nodes can perform some

Pierre Fraigniaud: The first and second authors receive support from the ANR project DISPLEXITY, and from the INRIA project GANG.

David Ilcinkas: Partially supported by the ANR project DISPLEXITY. This study was carried out in the frame of the program "investment for the future" of Idex Bordeaux-CPU.

© Springer International Publishing Switzerland 2014
D. Kratsch and I. Todinca (Eds.): WG 2014, LNCS 8747, pp. 15–28, 2014.
DOI: 10.1007/978-3-319-12340-0_2

local computation, and eventually produce an individual output based on structural information gathered in its vicinity, and reflecting the local structure of the network around the node. This information is then transmitted to a central authority, which is in charge of taking the final decision about G, as a combination of all individual outputs produced by the nodes. The crucial point here is that the communication channel available between the central authority and each of the nodes is supposed to be narrow, and hence the amount of information that can be transmitted from each node to the central authority is limited. Yet, we want the central authority to be able to decide whether or not G satisfies some given structural property. A typical example of such a setting is a sensor network, in which the sensed data are gathered at a distant base station, either directly, or via intermediate routers and/or other sensors.

The setting presented above shares characteristics with both *property testing* [14] and *distributed decision* [12,13]. Indeed, at a conceptual level, property testing on graphs can be viewed as: (1) querying a small number of nodes, typically $o(n)$, or even $O(1)$ nodes, in n-node networks; (2) extracting information from each query, typically $O(\log n)$ bits of information (e.g., the identity of a neighbor of the queried node); and (3) deciding whether the queried graph satisfies some given property $\mathcal{P}$, on the basis of the collection of information obtained from the queried nodes. It is the role of the tester algorithm to choose which nodes to query, and to eventually take the decision about the tested graph. The lack of information resulting from querying just a small subset of nodes is balanced by relaxing the decision requirement, which is subject to probabilistic errors, and does not impose to reject illegal instances that are "close" to legal instances.

Similarly, distributed decision in graphs [12,13] can be viewed as (1) querying *all* nodes; (2) having every node providing a *single bit* of information (true or false) based on local information gathered in its vicinity; and (3) deciding whether the queried graph satisfies some given property $\mathcal{P}$, on the basis of the collection of boolean information obtained from the queried nodes. In distributed decision, the instance is accepted if and only if the *logical conjunction* of the boolean information computed at each node is true. That is, if the input graph satisfies $\mathcal{P}$, then every node must individually accept. Otherwise, at least one node must individually reject. Distributed decision assumes no gap between, on the one hand, the instances to be accepted, and, on the other hand, the ones to be rejected. Furthermore, the decision is usually deterministic and error-free.

The formal model used in this paper for monitoring structural properties of networks relaxes property testing in the sense that, as for distributed decision, all nodes are queried. It also relaxes distributed decision in the sense that, as for property testing, the decision is made by an algorithm taking as input structured information provided by the nodes (and not only boolean information). Therefore, as far as the computational constraints are concerned, our model is very liberal, by taking the best of property testing, and of distributed decision. On the other hand, the model is very conservative regarding the final output, by allowing no errors, and by requiring perfect dichotomy between the legal

Table 1. Distributed testing

	#queried nodes	Amount of information	Decision mechanism	Gap	Error
Property testing [14]	$o(n)$	$O(\log n)$	Algorithm	ϵ-far	Yes
Distributed decision [12, 13]	n	1	Logical conjunction	None	No
Distributed testing	n	$O(\log n)$	Algorithm	None	No

instances and the illegal instances. We call this model *distributed testing*. Its main characteristics are summarized in Table 1.

One illustrative example of the differences between, on the one hand, distributed testing, and, on the other hand, property testing and distributed decision, is *cycle-freeness* (see Table 2). For 2-sided error, it is known [15] that cycle-freeness in graphs with maximum degree d can be property tested with $O(\frac{1}{\epsilon^3} + \frac{d}{\epsilon^2})$ queries returning $\Theta(\log n)$ bits per queried node, where $\epsilon \in (0, 1)$ is the gap parameter between the legal and illegal instances. For 1-sided error, $\Omega(\sqrt{n})$ queries are required [15], and this is sufficient [8]. If one does not allow errors, it is folklore that, even in connected graphs, cycle-freeness cannot be distributedly decided[1]. It is however known [18] that cycle-freeness in connected graphs can be *verified* distributedly with the help of $O(\log n)$-bit additional information (i.e., *certificates*) per node. As for distributed decision, each node just outputs a boolean, and the global decision is the logical conjunction of these booleans. Moreover, [9, 19] proved that $\Omega(\log n)$-bit certificates are necessary for verifying trees. In [3], the size of the certificates is nevertheless decreased to constant, by allowing nodes to output just 2 bits instead of 1. (We call this latter setting distributed *certification*).

Of course, there is a very simple algorithm for distributedly testing cycle-freeness in connected graphs: each node v output its degree $\deg(v)$, and the central authority accepts if and only if $\sum_v \deg(v) = 2(n - 1)$. The number of bits returned by each queried node is $\lceil \log d \rceil$ in graphs with maximum degree d. One question is: can we do better, i.e., with less bits of information transmitted from each node? Note that the answer is yes for *subdivided* graphs, where a subdivided graph [1] is a graph in which no two vertices of degree different from 2 are adjacent. Indeed, in such graphs with $n \geq 3$, the following algorithm works, with only four different kinds of outputs (i.e., 2-bit outputs): a node with degree $\neq 2$ outputs 0, and a node with degree 2 outputs 2, 3 or 4 depending on

---

[1] To see why, assume there exists a local algorithm $\mathcal{A}$ deciding cycle-freeness locally. Run $\mathcal{A}$ on the path with consecutive identities from 1 to n (nodes with identities 1 and n being the two extremities). In this configuration, the $n/2$ middle nodes output "true". Then, run $\mathcal{A}$ on the same path with identities $n/2, \ldots, n, 1, \ldots, n/2 - 1$. Again, the $n/2$ middle nodes output "true". Therefore, on the cycle with consecutive identities from 1 to n, all nodes output "true", yielding $\mathcal{A}$ to accept the cycle, a contradiction.

Table 2. Monitoring cycle-freeness in n-node max-degree-d (connected) graphs

	#queried nodes	Amount of information	Success probability	Comments
Property testing [15]	$O(\frac{1}{\epsilon^3} + \frac{d}{\epsilon^2})$	$\lceil \log n \rceil$	2-sided	ϵ-far
Property testing [8,15]	$\Theta(\sqrt{n})$	$\lceil \log n \rceil$	1-sided	ϵ-far
Distributed decision [folklore]	n	1	Impossible	–
Distributed verification [9,18,19]	n	1	Deterministic	$\Theta(\log n)$-bit certificates
Distributed certification [3]	n	2	Deterministic	$O(1)$-bit certificates
Distributed testing [this paper]	n	$\log d \pm \Theta(1)$	Deterministic	–

whether it is adjacent to 0, 1 or 2 nodes with degree $\neq 2$, respectively. The central authority then accepts if and only if the sum of the outputs equals $2(n-1)$.

In this paper, we question the existence of a distributed tester for cycle-freeness in arbitrary graphs, returning less than $\lceil \log d \rceil$ bits from each of the n queried nodes.

1.2 Our Results

We prove that every distributed tester for cycle-freeness in graphs with maximum degree d requires that at least one node outputs at least $\lceil \log d \rceil - 1$ bits. Hence, the distributed tester in which every node simply outputs its degree is essentially optimal. This tight result completes the whole picture regarding checking cycle-freeness (see Table 2). That is, if one can stand errors and slacks then property-testing enables to query just a few nodes. On the other hand, if one insists on deterministically systematically rejecting graphs with cycles, and accepting graphs without cycles, then distributed testing seems to be the right option. Indeed, it consumes moderate bandwidth resources to gather the outputs of the nodes, and needs not to provide certificates (as opposed to distributed verification, and distributed certification).

Establishing that every distributed tester for cycle-freeness must output $\lceil \log d \rceil - 1$ bits at some node requires to combine several techniques. First, we show that one can reduce our concern to *order-invariant* testers, that is, roughly, to algorithms whose output at a node does not depend on the actual *value* of the identities of the nodes in its vicinity, but solely on the *relative order* of these values. The celebrated result by Naor and Stockmeyer [20] enabling to reduce the study of certain kinds of algorithms to order-invariant algorithms cannot be applied in our context because our instances are not necessarily in the class LCL of so-called *locally checkable languages*. Nevertheless, we were able to provide a novel reduction, that does not require LCL membership, by using the *infinite version* of Ramsey Theorem.

Our second main technique is the construction, for every order-invariant distributed tester supposed to decide cycle-freeness with too few information provided by each node, of two explicit instances, one with a cycle, and one without, that cannot be distinguished by the tester. This construction is difficult because, as mentioned before, cycle-freeness can be distributedly tested with just a 2-bit output per node in subdivided graphs. Nevertheless, by an appropriate use of the known fact that every free group can be linearly ordered, we were able to construct legal and illegal instances that cannot be distinguished locally by the assumed order-invariant distributed tester.

1.3   Related Work

Local computing is a wide domain of studies in distributed network computing, and the reader is referred to the textbook [21] for an excellent introduction to local computing, providing pointers to the most relevant techniques for solving prominent problems (e.g., MIS, coloring, etc.) locally. The question of what can be computed in a constant number of communication rounds was actually introduced in the seminal work by Naor and Stockmeyer in [20]. In particular, [20] introduced the class of *locally checkable languages* (LCL), and studied the question of how to deterministically or randomly construct instances of LCL languages in a constant number of rounds.

With the objective of providing distributed network computing with a complexity theory based on decision problems, following the guidelines of classical (sequential) complexity theory, [12,13] introduced several decision classes for local computing, and studied the relationships between these classes (which are depending on the number of allowed rounds, on the potential access to oracles, on the potential use of non-determinism and/or of randomization, etc.). Paper [12] generated several following up contributions, including, e.g., studies on the impact of randomization [11], studies on the impact of node identifiers [10], studies on verification tasks where certificates include node IDs [17], etc. See also [16] for other forms of local checking, and for their impact on distributed graph-optimization problems.

Local distributed testing was introduced in [3] (although [3] does not use this terminology). Beside introducing complexity classes related to local distributed testing, and studying the relationships between these classes, [3] focused attention to verification, and on the size of the certificates involved in the verification. Our paper is also very much related to [4,5], which use models that resemble local distributed testing, but where nodes are restricted to perform just one round of communication before outputting a value on $O(\log n)$ bits. In the restricted setting of [4,5], even checking the presence of a 4-cycle in the network may not be feasible. Instead, in local distributed testing, the number of communication rounds is just restricted to be constant, but one aims at producing smaller output values, e.g., on $O(1)$ bits.

2   Local Distributed Testing

Let us consider a network modeled as a simple *connected* graph G. We are interested in the task consisting, for the nodes of G, to collectively decide whether G satisfies some given property $\mathcal{P}$, like, say, being planar, being a tree, etc. For this purpose, nodes can exchange information, so that every node eventually produces an output. The computational model considered in this paper is the classical $\mathcal{LOCAL}$ model [21], which is a standard distributed computing model capturing the essence of locality. In this model, nodes have pairwise distinct identities (the identity of node v is denoted by $\mathrm{id}(v) \in \mathbb{N}$). They are woken up simultaneously, and computation proceeds in fault-free synchronous *rounds* during which every node exchanges messages of unlimited size with its neighbors in the underlying network G, and performs arbitrary individual computations on its data. The running time of an algorithm is defined as the maximum number of rounds it takes to terminate at all nodes, over all possible networks, and all possible identity assignments for the nodes in these networks. Similarly to [20], we consider algorithms whose running time is independent of the size of the network, and independent of the size of the identities. That is, they run in constant time.

Let $\mathsf{out}_{\mathcal{A}}(G, \mathrm{id}, v)$ denotes the output of node $v \in V(G)$ running Algorithm $\mathcal{A}$ in G with identity assignment id. We denote by $\mathsf{out}_{\mathcal{A}}(G, \mathrm{id})$ the global output, that is,

$$\mathsf{out}_{\mathcal{A}}(G, \mathrm{id}) = \{\mathsf{out}_{\mathcal{A}}(G, \mathrm{id}, v), v \in V(G)\}$$

is the multiset of all individual outputs (the same individual output may appear more than once in $\mathsf{out}_{\mathcal{A}}(G, \mathrm{id})$). By "collectively decide" a graph property $\mathcal{P}$, we mean the following. To each output corresponds a global state of the system. We question the ability to define two classes of global states, one called *accept*, and one called *reject*, so that the following holds. If G satisfies $\mathcal{P}$, then the nodes must compute outputs that yields the system to be in an accept state, while if G does not satisfy $\mathcal{P}$, then the nodes must compute outputs that yields the system to be in a reject state. More specifically, assume that each node of an n-node network can output one of the different values in a set S. Let $M_{n,S} = \left(\binom{S}{n}\right)$ be the set of all multisets of cardinality n, with elements taken from S. We say that a graph property $\mathcal{P}$ can be distributedly tested with output set S if there exists a local distributed algorithm $\mathcal{D}$, and a decomposition of $M_{n,S}$ for every $n \geq 1$, into two computable sets Y_n (the "yes"-set, or accept set) and $M_{n,S} \backslash Y_n$ (the reject set) such that, for every n-node graph G, and for every identity assignment id to the nodes in G, the following holds:

$$G \text{ satisfies } \mathcal{P} \iff \mathsf{out}_{\mathcal{D}}(G, \mathrm{id}) \in Y_n.$$

In other words, a distributed tester for $\mathcal{P}$ consists in a local distributed algorithm $\mathcal{D}$ producing an output at every node, coupled with a sequential algorithm $\mathcal{S}$ which takes as input the collection of all outputs produced by the nodes, and accepts or rejects, under the constraint that it must accept if and only if G satisfies $\mathcal{P}$.

An example (borrowed from [13]) of a graph property that can be distributedly tested is: clique-width at most 2. Indeed, a graph has clique-width at most 2 if and only if it is a cograph – see [7]. Now, cographs have also been characterized as the family of P_4-free graphs (i.e., the graphs which do not contain the path on 4 nodes as an induced subgraph) – see [6]. Hence, to decide clique-width at most 2, each node can simply communicate at bounded distance to check whether it contains an induced P_4 in its vicinity, and output 0 if it is the case, and 1 otherwise. The accept set is simply defined as $Y_n = \{\{x_1, \ldots, x_n\} \in M_{n,\{0,1\}} : \prod_{i=1}^{n} x_i = 1\}$. Distributed testing restricted to this latter class of accept sets actually reduces to distributed decision [13]. (See [2] for the difficulty of property testing cographs).

The size of S needs not be constant, and may actually vary with n. This is for instance the case of the aforementioned task of testing cycle-freeness, for which every node simply returns its degree. In this case, $Y_n = \{\{x_1, \ldots, x_n\} \in M_{n,[n]} : \sum_{i=1}^{n} x_i = 2(n-1)\}$, where $[n] = \{1, \ldots, n\}$.

Note that every computable graph property $\mathcal{P}$ can be distributedly tested in this model. Indeed, each node can just output its identity, and the set of all the identities of its neighbors. In other words, each node v outputs its identity and its adjacency list L_v in the current graph G. In this case,

$$Y_n = \{\{L_1, \ldots, L_n\} : \text{graph } (L_1, \ldots, L_n) \text{ satisfies } \mathcal{P}\}$$

would enable to distinguish graphs that satisfy $\mathcal{P}$ from those that do not. However, such a trivial solution involves individual outputs of size $\Omega(n \log n)$ bits in dense graphs. Our objective is to study the ability of distributedly testing graph properties with individual outputs having size as small as possible, ideally constant, independent of the network size, and of the range of identities. We define the *output size* of a distributed tester in a graph family $\mathcal{G}$ as the maximum, taken over all instances (G, id) where $G \in \mathcal{G}$, of the maximum number of bits outputted by a node in this instance.

3 Order-Invariance Revisited

Our first result in the paper is a key ingredient for the proof of our main result. This ingredient may have its interest on its own, and it is worth dedicating an entire section to it. We show that, w.l.o.g., one can consider only *order-invariant* distributed testers. Recall that an order-invariant distributed algorithm is a distributed algorithm for which the output at any given node does not depend on the actual values of the identities of the nodes in its vicinity, but only on the relative order of these identities. More precisely, let $B_G(v, t)$ be the ball of radius t around node v in graph G, that is, $B_G(v, t)$ is the subgraph of G induced by all nodes at distance at most t from v, excluding the edges between the nodes at distance exactly t from v. An algorithm $\mathcal{A}$ is order-invariant if the following holds: for any graph G, for any node v, and for any two identity assignments id and id$'$ of the nodes in G, if the ordering of the nodes in $B_G(v, t)$ induced by id, and the one induced by id' are identical, then the output of $\mathcal{A}$ at node v is the same in both (G, id) and (G, id').

A *distributed language* $\mathcal{L}$ is defined by a collection of labeled graphs, or *configurations* (G, ℓ) where G is a connected graph, and $\ell : V(G) \to \{0, 1\}^*$ is a function that labels each node v with the label $\ell(v)$. The *construction* task defined by a language $\mathcal{L}$ consists, for every node v of every graph G, to compute $\mathsf{out}(v)$ such that the global output satisfies $(G, \mathsf{out}) \in \mathcal{L}$. In their seminal paper, Naor and Stockmeyer [20] consider the subclass LCL of *locally checkable languages*. Languages in LCL are distributed languages that are defined on graph families with constant maximum degree, and with constant label size at every node (i.e., $|\ell(v)| = O(1)$ for every node v). A language $\mathcal{L}$ is locally checkable, or, alternatively, is in LD according to the terminology of [13], if there exists a distributed algorithm performing in a constant number of rounds such that, for any $(G, \ell) \in \mathcal{L}$, all nodes accept, and, for any $(G, \ell) \notin \mathcal{L}$, at least one node rejects. (See [13, 20] for more details). Theorem 3.3 in [20] establishes that, for every language $\mathcal{L} \in$ LCL, if there exists a construction algorithm for $\mathcal{L}$ performing in $t = O(1)$ rounds, then there exists a t-round order-invariant construction algorithm for $\mathcal{L}$.

In the context of this paper, the language corresponding to a distributed tester $(\mathcal{D}, \mathcal{S})$ is determined by the accept set, that is by the set of multisets of outputs that $\mathcal{S}$ accepts. In general, the language corresponding to a distributed tester $(\mathcal{D}, \mathcal{S})$ for a property $\mathcal{P}$ is

$$\mathcal{L} = \{(G, \ell) : \mathcal{S} \text{ accepts } \ell \iff G \text{ satisfies } \mathcal{P}\}$$

where ℓ is the collection of values $\ell(v)$, $v \in V(G)$, and $\ell(v)$ is the value owned by node v. In particular, the language corresponding to the distributed tester $(\mathcal{D}, \mathcal{S})$ for cycle-freeness where $\mathcal{D}$ outputs $\deg(v)$ at each node v is

$$\mathcal{L}_{\text{cycle-free}} = \{(G, \ell) : \sum_{v \in VG} \ell(v) = 2(n-1) \iff G \text{ is an n-node tree}\}.$$

Observe that such languages are not necessarily locally checkable. For instance, $\mathcal{L}_{\text{cycle-free}} \notin$ LCL (even if restricted on graphs with maximum degree d, for some constant d). Hence, Theorem 3.3 in [20] does not apply to our setting. The result below extends this latter theorem to non locally checkable languages. We define the *domain* of a language $\mathcal{L}$ as the set of all values taken by the labels $\ell(v)$ in $\mathcal{L}$, for all graphs G, and all nodes v of G. Note that, in a construction task defined by a distributed language $\mathcal{L}$, nodes may be a priori provided with inputs. In this context, $x(v)$ denotes the input to node v, and every node v has to compute an output $y(v)$ such that $(G, (x, y)) \in \mathcal{L}$.

Theorem 1. *For every non-negative integers k, t, d, and every language $\mathcal{L}$ defined on connected graphs with maximum degree d, and k-valued domain, if there exists a t-round construction algorithm $\mathcal{A}$ for $\mathcal{L}$, then there is a t-round order-invariant construction algorithm $\mathcal{A}'$ for $\mathcal{L}$.*

Proof. For any set X, and any positive integer r, let us denote by $X^{(r)}$ the set of all subsets of X with size exactly r. Let X be a countably infinite set, let r

and s be two positive integers, and let $c : X^{(r)} \to [s]$ be a "coloring" of each set in $X^{(r)}$ by an integer in $[s] = \{1, \ldots, s\}$. Recall that (the infinite version of) Ramsey's Theorem states that there exists an infinite set $Y \subseteq X$ such that the image by c of $Y^{(r)}$ is a singleton (that is, all sets in $Y^{(r)}$ are colored the same by c). We make use of this theorem as follows.

Let us consider the collection $\mathcal{B}$ of all graphs isomorphic to some ball $B_G(v, t)$ of radius t, centered at some node v in some graph G with maximum degree d. If the language $\mathcal{L}$ is described by labels encoding input-output relations, then $\mathcal{B}$ is the collection of all labeled graphs isomorphic to some labeled ball $B_G(v, t)$. Since the domain of the labels has k values, there are at most k different inputs, and thus there is a finite number β of pairwise non-isomorphic balls in $\mathcal{B}$.

We enumerate these (labeled) balls from 1 to β, and let n_i be the number of vertices in the ith ball, for $i = 1, \ldots, \beta$. For every i, the vertices of the ith ball can be ordered in $n_i!$ different manners, corresponding to the $n_i!$ permutations in Σ_{n_i}. We consider the $N = \sum_{i=1}^{\beta} n_i!$ ordered balls $B_{i,\sigma}$, for $i = 1, \ldots, \beta$, and $\sigma \in \Sigma_{n_i}$, and we enumerate these ordered balls as $\mathbf{B}_1, \ldots, \mathbf{B}_N$ in an arbitrary order. Using these balls, we define an infinite set $\mathcal{I}$ of identities as follows.

Let $X_0 = \mathbb{N}$, and assume that we have already secured the existence of a sequence of infinite sets $X_0 \supseteq X_1 \supseteq \cdots \supseteq X_j$, $0 \le j < N$, such that, for every i, $1 \le i \le j$, the output of $\mathcal{A}$ at the center of $\mathbf{B}_i$ is the same for all possible identity assignments to the nodes in $\mathbf{B}_i$ with values in X_i, and respecting the ordering of the nodes in $\mathbf{B}_i$. We define the coloring $c : X_j^{(r)} \to [k]$ where r is the number of nodes in $\mathbf{B}_{j+1}$, as follows: for each r-element set $I \in X_j^{(r)}$, assign r pairwise distinct identities to the nodes of $\mathbf{B}_{j+1}$ using the r values in I, and respecting the order of the nodes in $\mathbf{B}_{j+1}$. Then, define $c(I)$ as the output of Algorithm $\mathcal{A}$ at the center of $\mathbf{B}_{j+1}$ under this identity assignment to the nodes of $\mathbf{B}_{j+1}$. By Ramsey's Theorem, there exists an infinite set $Y_j \subseteq X_j$ such that all r-element sets $I \in Y_j^{(r)}$ are given the same color. We set $X_{j+1} = Y_j$. We proceed that way until we exhaust all balls $\mathbf{B}_i$, $i = 1, \ldots, N$, and we set $\mathcal{I} = X_N$.

By construction, the set $\mathcal{I}$ satisfies that, for every ball $B_{i,\sigma}$, for $i = 1, \ldots, \beta$, and $\sigma \in \Sigma_{n_i}$, the output of $\mathcal{A}$ at the center of $B_{i,\sigma}$ is the same for all identity assignments to the nodes of $B_{i,\sigma}$ with identities taken from $\mathcal{I}$ and assigned to the nodes in the order σ.

We now define the order-invariant algorithm $\mathcal{A}'$ as follows. Every node v inspects its radius-t ball $B_G(v, t)$ around it in the actual graph G. In particular, it collects the identities of the nodes in that ball. Let σ be the ordering of the nodes in $B_G(v, t)$ induced by their identities. Node v simulates $\mathcal{A}$ by reassigning identities to the nodes of $B_G(v, t)$ using the $r = |B_G(v, t)|$ smallest values in $\mathcal{I}$, in the order specified by σ, and outputs what would have outputted $\mathcal{A}$ if nodes were given these identities.

$\mathcal{A}'$ is well defined, as nodes can be provided with the $\nu = \sum_{i=0}^{t} d^i$ smallest integers in the set $\mathcal{I}$. (I.e., nodes do not need to know the entire set $\mathcal{I}$, but only a finite number of values in $\mathcal{I}$). Also, by construction, $\mathcal{A}'$ is order-invariant. To establish that $\mathcal{A}'$ is correct, let us consider some n-node input graph G, with nodes provided with pairwise distinct identities in $\mathcal{I}$, and let $\mathsf{out} = \{\mathsf{out}(v), v \in V(G)\}$

be the output of $\mathcal{A}$ in this context. This output is precisely the multi-set outputted by $\mathcal{A}'$ in G. Indeed, every node v relabels its radius-t ball with identities in $\mathcal{I}$, respecting the order induced by the original identities in $\mathcal{I}$, and $\mathcal{I}$ is precisely defined so that the output of v will be the same in both cases. In other words, the output of $\mathcal{A}'$ is precisely the output of $\mathcal{A}$ if nodes were assigned identities restricted to be in $\mathcal{I}$. Hence, since $\mathcal{A}$ is correct, it follows that $\mathcal{A}'$ is correct as well.   □

4 Distributedly Testing Cycle-Freeness

In this section, we prove our main result:

Theorem 2. *For any positive even integer d, every distributed tester for cycle-freeness in connected graphs with maximum degree at most d has output size at least $\lceil \log d \rceil - 1$ bits.*

The rest of the section is entirely dedicated to prove this result. The proof is by contradiction. Let d be a positive even integer. We assume the existence of a distributed tester $(\mathcal{D}, \mathcal{S})$ for cycle-freeness in connected graphs with maximum degree d, where $\mathcal{D}$ runs in t rounds, for some constant $t \geq 0$, and outputs at most $\lceil \log d \rceil - 2$ bits at each node. We first start by shrinking the set of candidate algorithms $\mathcal{D}$. Indeed, as a direct consequence of Theorem 1, we get the following:

Corollary 1. *If there exists a t-round distributed tester $(\mathcal{D}, \mathcal{S})$ for cycle-freeness with k-valued outputs in connected graphs with maximum degree d, then there is distributed tester $(\mathcal{D}', \mathcal{S})$ satisfying the same, but where $\mathcal{D}'$ is order-invariant.*

Based on this latter result, we now show that every distributed tester $(\mathcal{D}, \mathcal{S})$ for cycle-freeness in connected graphs with maximum degree at most d, where $\mathcal{D}$ is order-invariant, has output size at least $\lceil \log d \rceil - 1$ bits.

The intuition is as follows. We will focus our attention on so-called type-i nodes, with i being an even integer between 2 and d. Intuitively, such nodes are defined as nodes of degree i that only "see" a tree of nodes of degree i in their neighborhood up to distance t, and with a particular ordering of their identities. We will construct two (connected) graphs, with their corresponding identity assignments, such that only one of this two graphs is a tree, and the multi-set of the local views gathered by the nodes in the two graphs will only differ by their numbers of type-i and type-j nodes, for some $i \neq j$. Any distributed tester for cycle-freeness has to distinguish the two graphs and has thus to give different output values to type-i and type-j nodes. This will prove that any distributed tester for cycle-freeness must have at least $d/2$ different output values, thus proving our main theorem.

We now define formally two families of trees, which will be used as building blocks in our constructions. (See Fig. 1). Let i, $1 \leq i \leq d$, be an even integer. We define the trees T_i and T_i' as follows. For T_i, we start from one single node, called the *downtown* node (this node is considered as a leaf). For T_i', we start from $i + 1$ nodes organized as a star (i.e., with one center and i leaves), and also

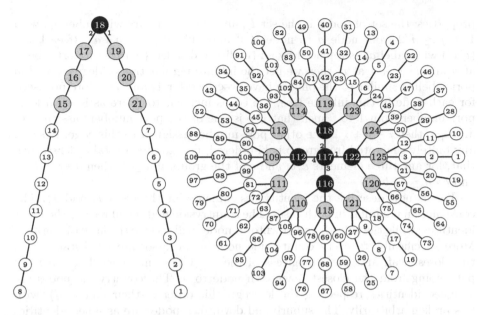

Fig. 1. The trees T_2, for $t = 3$, and T_4', for $t = 1$. The downtown, suburb, and countryside nodes are depicted as black, grey, and white nodes respectively. Port numbers are only indicated for the central downtown node. The other port numbers can be inferred using the same cyclic ordering of port numbers around the nodes.

called *downtown* nodes. Then, we replace each of the leaves of these two "seeds" by a $(i-1)$-ary tree of height t. As a consequence, all the internal nodes of the resulting two trees are of degree i. Moreover, the downtown node closest to every leaf is at distance exactly $t+1$. The internal nodes of the resulting trees that are not downtown nodes are called *suburb* nodes. For the ease of description of our constructions, and for simplifying our arguments, we assign numbers to the edges incident to the internal nodes of these two trees. More specifically, for each internal node v, a distinct label between 1 and $\deg(v)$ is assigned to every edge e incident to v. This label is called the *port number* of the edge e at node v. The port numbers are assigned in such a way that, for each edge whose extremities are two nodes with the same degree i, if $p \in [1, i]$ is one of the port numbers assigned to the edge, then $i - p + 1$ is the other port number assigned to this edge. Then, we apply another transformation, which consists in replacing every edge of these two trees that is incident to a leaf by a path of length $2t + 1$. The trees T_i and T_i' are the trees resulting from this second transformation. In these two trees, all the nodes that are neither downtown nodes, nor suburb nodes, are called *countryside* nodes.

Before describing the identity assignments for these trees, let us make the following observations. An infinite regular tree of degree i can be viewed as the Cayley graph of the free group of rank $i/2$. More precisely, let $\{a_1, a_2, \ldots, a_{i/2}\}$ be the set of the $i/2$ generators of the free group $(F, \star)$ of rank $i/2$. The Cayley graph associated to this group is a directed arc-labeled graph with the following properties: the set of nodes is the set F, and there is an arc with label a_p, with $1 \leq p \leq i/2$ from node g to node g' if and only if $g' = g \star a_p$. If each arc (g, g') with label a_p is replaced by an (undirected) edge $\{u, v\}$ with port label p at u and $i - p + 1$ at v, then we get the infinite regular tree of degree i with a port-labeling similar to the one we have described for T_i and T_i'. More precisely, for each node in this infinite tree, the edges incident to it are assigned a local port number from 1 to i such that if p is one of the port numbers assigned to an edge, then $i - p + 1$ is the other port number assigned to this edge. Besides, any finitely generated free group is bi-ordered, i.e., admits a total order $\preceq$ such that, for any three elements a, b, and c of the group, if $a \preceq b$, then $a \star c \preceq b \star c$ and $c \star a \preceq c \star b$.

Let us now describe the identity assignments for the trees T_i and T_i'. The construction is similar in both cases. The countryside nodes will receive the lower identities, while the suburb and downtown nodes will receive the larger identities. More specifically, to every countryside node u, we associate its distance j to the closest leaf, and the sequence s of the $t + 1$ port numbers describing the path going from the closest downtown node to u. The countryside nodes are assigned identities respecting the lexicographic order of their pair (s, j), with ties broken arbitrarily. The suburb and downtown nodes are assigned identities that are compatible with the total order $\preceq$ of the corresponding free group. Again, see Fig. 1 for examples of these identity assignments.

A node having the same local view up to distance t as the local view up to distance t of a downtown node of T_i, except for actual identity values but respecting the order of these identities, is called a type-i node.

For the purpose of contradiction, assume that there exists an order-invariant distributed tester for cycle-freeness using less than $d/2$ output values. Therefore there must exist two even integers i, j with $i < j \leq d$, such that the distributed tester outputs the same value for type-i and type-j nodes. Hence, let G_1' be the graph formed by the disjoint union of one copy of T_i' and $j - 1$ copies of T_j, with disjoint ranges of identity values assigned to the nodes in these copies. Connect $j - 1$ disjoint pairs of leaves by an edge to make the graph connected. We denote by G_1 the resulting graph. Note that G_1 is a tree. Similarly, let G_2' be the graph formed by the disjoint union of $i - 1$ copies of T_i and one copy of T_j', with disjoint ranges of identities assigned to the nodes in these copies. Connect $i - 1$ disjoint pairs of leaves by as many edges to make the graph connected. Further, connect $j - i$ other pairs of leaves to create cycles. We denote by G_2 the resulting graph. Note that G_2 is connected but is not a tree.

The multiset of local views up to distance t, ignoring the identity values but taking into account their relative order, is exactly the same in both graphs G_1 and G_2, with the only exception that G_1 has two more type-i nodes than G_2, and two less type-j nodes than G_2. The distributed tester $(\mathcal{D}, \mathcal{S})$ will thus take the same decision for both graphs, which contradicts the fact that it is a correct distributed tester for cycle-freeness. This contradiction completes the proof of the theorem.   □

References

1. Alon, N.: Subdivided graphs have linear ramsey numbers. J. Graph Theor. **18**(4), 343–347 (1994)
2. Alon, N., Shapira, A.: A characterization of easily testable induced subgraphs. Comb. Probab. Comput. **15**(6), 791–805 (2006)
3. Arfaoui, H., Fraigniaud, P., Pelc, A.: Local decision and verification with bounded-size outputs. In: Higashino, T., Katayama, Y., Masuzawa, T., Potop-Butucaru, M., Yamashita, M. (eds.) SSS 2013. LNCS, vol. 8255, pp. 133–147. Springer, Heidelberg (2013)
4. Becker, F., Kosowski, A., Nisse, N., Rapaport, I., Suchan, K.: Allowing each node to communicate only once in a distributed system: shared whiteboard models. In: Proceediings 24th ACM Sympsium on Parallelism in Algorithms and Architectures (SPAA), pp. 11–17 (2012)
5. Becker, F., Matamala, M., Nisse, N., Rapaport, I., Suchan, K., Todinca, I.: Adding a referee to an interconnection network: what can(not) be computed in one round. In: Proceedings 25th IEEE International Sympusim on Parallel and Distributed Processing (IPDPS), pp. 508–514 (2011)
6. Corneil, D., Lerchs, H., Burlingham, L.: Complement reducible graphs. Discrete Appl. Math. **3**(3), 163–174 (1981)
7. Courcelle, B., Olariu, S.: Upper bounds to the clique width of graphs. Discrete Appl. Math. **101**(13), 77–144 (2000)
8. Czumaj, A., Goldreich, O., Ron, D., Seshadhri, C., Shapira, A., Sohler, C.: Finding cycles and trees in sublinear time. Electron. Colloq. Comput. Complex. (ECCC) **19**, 35 (2012)
9. Dolev, S., Gouda, M.G., Schneider, M.: Memory requirements for silent stabilization. Acta Inf. **36**(6), 447–462 (1999)
10. Fraigniaud, P., Göös, M., Korman, A., Suomela, J.: What can be decided locally without identifiers? In: Proceedings of the 32nd ACM Symposium on Principles of Distributed Computing (PODC), pp. 157–165 (2013)
11. Fraigniaud, P., Korman, A., Parter, M., Peleg, D.: Randomized distributed decision. In: Aguilera, M.K. (ed.) DISC 2012. LNCS, vol. 7611, pp. 371–385. Springer, Heidelberg (2012)
12. Fraigniaud, P., Korman, A., Peleg, D.: Local distributed decision. In: Proceedings of the 52nd Annual IEEE Symposium on Foundations of Computer Science (FOCS), pp. 708–717 (2011)
13. Fraigniaud, P., Korman, A., Peleg, D.: Towards a complexity theory for local distributed computing. J. ACM **60**(5), 35 (2013)
14. Goldreich, O. (ed.): Property Testing. LNCS, vol. 6390. Springer, Heidelberg (2010)
15. Goldreich, O., Ron, D.: Property testing in bounded degree graphs. Algorithmica **32**(2), 302–343 (2002)
16. Göös, M., Hirvonen, J., Suomela, J.: Lower bounds for local approximation. J. ACM **60**(5), 39 (2013)
17. Göös, M., Suomela, J.: Locally checkable proofs. In: Proceedings of the 30th ACM Sympusim on Principles of Distributed Computing (PODC), pp. 159–168 (2011)
18. Itkis, G., Levin, L.A.: Fast and lean self-stabilizing asynchronous protocols. In: 35th IEEE Sympsium on Foundations of Computer Science (FOCS), pp. 226–239 (1994)
19. Korman, A., Kutten, S., Peleg, D.: Proof labeling schemes. Distrib. Comput. **22**(4), 215–233 (2010)

20. Naor, M., Stockmeyer, L.: What can be computed locally? SIAM J. Comput. **24**(6), 1259–1277 (1995)
21. Peleg, D.: Distributed Computing: A Locality-Sensitive Approach. SIAM, Philadelphia (2000)

DMVP: Foremost Waypoint Coverage of Time-Varying Graphs

Eric Aaron[1], Danny Krizanc[2]($^{\boxtimes}$), and Elliot Meyerson[2]

[1] Computer Science Department, Vassar College, Poughkeepsie, NY, USA
eaaron@cs.vassar.edu
[2] Department of Mathematics and Computer Science,
Wesleyan University, Middletown, CT, USA
{dkrizanc,ekmeyerson}@wesleyan.edu

Abstract. We consider the Dynamic Map Visitation Problem (DMVP), in which a team of agents must visit a collection of critical locations as quickly as possible, in an environment that may change rapidly and unpredictably during the agents' navigation. We apply recent formulations of time-varying graphs (TVGs) to DMVP, shedding new light on the computational hierarchy $\mathcal{R} \supset \mathcal{B} \supset \mathcal{P}$ of TVG classes by analyzing them in the context of graph navigation. We provide hardness results for all three classes, and for several restricted topologies, we show a separation between the classes by showing severe inapproximability in $\mathcal{R}$, limited approximability in $\mathcal{B}$, and tractability in $\mathcal{P}$. We also give topologies in which DMVP in $\mathcal{R}$ is fixed parameter tractable, which may serve as a first step toward fully characterizing the features that make DMVP difficult.

1 Introduction

In navigation-oriented application domains such as autonomous mobile robots, wireless sensor networks, security, surveillance, mechanical inspection, and more, graph representations are commonly employed for formulating and analyzing the central navigation or area inspection problems. Many approaches to coverage problems [11–13] are based on static graph representations, as are visitation problems [1] or related combinatorial optimization problems such as the k-Chinese Postman Problem [3,7] and k-Traveling Repairman Problem [14,15]. But static graph structures do not represent the dynamic environments that can occur in applications of autonomous robots or non-player characters in video games and virtual worlds. In this paper, we present the *Dynamic Map Visitation Problem* (DMVP), applying recent formulations of *highly dynamic graphs* (or *time-varying graphs* (TVGs)) [9,23] to an essential graph navigation problem: In DMVP, a team of agents must inspect a collection of critical locations on a map (represented as a graph) as quickly as possible, but the agents' environment may change during navigation.

The application of TVG models is essential to DMVP. In applications such as planetary exploration [26], search and rescue in hazardous environments

© Springer International Publishing Switzerland 2014
D. Kratsch and I. Todinca (Eds.): WG 2014, LNCS 8747, pp. 29–41, 2014.
DOI: 10.1007/978-3-319-12340-0_3

(e.g., natural disasters, areas of armed conflict), or even ad-hoc network inspection, many aspects of the structure of graph waypoints and edges governing navigation can change during agent navigation, and TVG models can capture variation in graph structure in ways that static graphs cannot. Our paper presents new results about DMVP complexity and demonstrates distinctions among classes of TVGs; details of our main results are summarized in Sect. 1.2.

When incorporating dynamics into a problem such as DMVP, there are many options for how to constrain/model the dynamics of the graph. Dynamics can be deterministic (e.g., [8,19,20,24,27]) or stochastic (e.g., [5,10]). In this paper, to provide a foundation for future work, and exemplify the aspects of topologies and dynamics that make our problem easy or hard, we focus on the deterministic case. The deterministic approach is also particularly relevant for situations in which some prediction of changes is feasible. Quite a bit of this previous work has required that the graph be connected *at all times* [10,20,22]. Indeed, for complete map visitation to be possible, every critical location must be eventually reachable. However, in application environments such as those outlined above, at any given time the waypoint graph may be disconnected. Our model must be general enough to allow for this phenomenon.

We adopt three classes of TVGs, each of which places constraints on edge dynamics. In $\mathcal{R}$, edges must reappear eventually; in $\mathcal{B}$, edges must appear within some time bound; in $\mathcal{P}$ edge appearances are periodic. These classes have proven to be critical to the TVG taxonomy [9]. They have been studied with respect to problems such as broadcast [8] and exploration [16,19], with results relating to feasibility of computation and bounds on broadcast and exploration time. $\mathcal{R}$, $\mathcal{B}$, and $\mathcal{P}$ place intuitive constraints on the nature of dynamic navigation domains. Even the assumption of periodicity of edges has applications to navigation of transportation networks [16,19], as well as environments periodically patrolled by other agents, who can prohibit or guarantee safe traversal of an edge.

In this paper, we shed further light on the computational hierarchy of $\mathcal{R}$, $\mathcal{B}$, and $\mathcal{P}$ [8], by analyzing them in the context of DMVP, a natural but difficult problem in global navigation. We provide hardness results for all three classes. For several restricted topologies, we demonstrate separation between the classes by showing severe inapproximability in $\mathcal{R}$, limited approximability in $\mathcal{B}$, and tractability in $\mathcal{P}$. We also give topologies in which DMVP in $\mathcal{R}$ is tractable and fixed parameter tractable, which may serve as a first step towards fully characterizing the topological features that make DMVP difficult. Because our goal in this paper is to cleanly differentiate the classes of dynamics we are exploring, rather than explore the interactions between multiple agents, our results here focus on the case of a single agent.

1.1 Definitions and TVG Concepts

As a foundation for our work, we adopt the definitions below from Santoro et al. [9].

Definition 1. *A TVG (time-varying graph, dynamic graph, or dynamic network) is a five-tuple $\mathcal{G} = (V, E, \mathcal{T}, \rho, \zeta)$, where $\mathcal{T} \subseteq \mathbb{T}$ is the* lifetime *of the system,* presence function $\rho(e, t) = 1 \iff$ *edge $e \in E$ is available at time $t \in \mathcal{T}$, and* latency function $\zeta(e, t)$ *gives the time it takes to cross e if starting at time t. The graph $G = (V, E)$ is called the* underlying graph *of $\mathcal{G}$, with $|V| = n$.*

In the most general case, $\mathbb{T}$ can be $\mathbb{R}$, and edges can be directed. However, in our work we consider the discrete case in which $\mathbb{T} = \mathbb{N}$, edges are undirected, and all edges have uniform travel cost $\zeta(e, t) = 1$ at all times. If agent a is at u, and edge (u, v) is available at time τ, then a can take (u, v) during this time step, visiting v at time $\tau + 1$. As a traverses $\mathcal{G}$ we say a both *visits* and *covers* the vertices in its traversal, and we will henceforth use these terms interchangeably. A *temporal subgraph* of a TVG $\mathcal{G}$ results from restricting the lifetime $\mathcal{T}$ of $\mathcal{G}$ to some $\mathcal{T}' \subseteq \mathcal{T}$.

Definition 2. $\mathcal{J} = \{(e_1, t_1), ..., (e_k, t_k)\}$ *is a* journey $\iff \{e_1, ..., e_k\}$ *is a walk in G (called the* underlying walk *of $\mathcal{J}$), $\rho(e_i, t_i) = 1$ and $t_{i+1} \geq t_i + \zeta(e_i, t_i)$ for all $i < k$. The* topological length *of $\mathcal{J}$ is k, the number of edges traversed. The* temporal length *is the duration of the journey: (arrival date)−(departure date).*

Given a date t, a journey from u to v departing on or after t whose arrival date t' is soonest is called *foremost*; whose topological length is minimal is called *shortest*; and whose temporal length is minimal is called *fastest*.

In [9], a hierarchy of thirteen classes of TVG's is presented. In related work on exploration [16] and broadcast [8], focus is primarily on the chain $\mathcal{R} \supset \mathcal{B} \supset \mathcal{P}$ defined below. We adopt these classes into our domain, which we believe enforce natural constraints in our application environments.

Definition 3 *(Recurrent edges).* $\mathcal{R}$ *is the class of all TVG's $\mathcal{G}$ such that G is connected, and $\forall e \in E, \forall t \in \mathcal{T}, \exists t' > t$ s.t. $\rho(e, t') = 1$.*

Definition 4 *(Time-bounded recurrent edges).* $\mathcal{B}$ *is the class of all TVG's $\mathcal{G}$ such that G is connected, and $\forall e \in E, \forall t \in \mathcal{T}, \exists t' \in [t, t + \Delta)$ s.t. $\rho(e, t') = 1$, for some integer Δ.*

Definition 5 *(Periodic edges).* $\mathcal{P}$ *is the class of all TVG's $\mathcal{G}$ such that G is connected, and $\forall e \in E, \forall t \in \mathcal{T}, \forall k \in \mathbb{N}, \rho(e, t) = \rho(e, t + kp)$ for some integer p. p is called the* period *of $\mathcal{G}$.*

As much as possible, we also take standard notation and terms from the graph theory literature. We rely on several underlying graph topologies. A *star* is a tree in which at most one vertex has degree greater than one. The leaves of a star are called *points*. A *spider* is a tree in which at most one vertex has degree greater than two. In other words, a spider consists of a set of vertex-disjoint paths, called *arms*, each of which has exactly one endpoint connected to the common central vertex c. A *comb* is a max-degree 3 tree, in which there exists a simple path containing every vertex of degree 3. Such a path is called a *backbone* of the comb. Paths edge-disjoint to the backbone are called *arms*. A leaf distance

1 from the backbone is called a *tooth*. An *r-almost-tree* is a connected graph with $|V| + r - 1$ edges, that is, r edges can be removed to produce a tree.

Problem. *Given a TVG $\mathcal{G}$ and a set of starting locations S for k agents in G, the TVG foremost coverage or dynamic map visitation problem (DMVP) is the task of finding journeys starting at time 0 for each of these k agents such that every node in V is in some journey, and the maximum temporal length among all k journeys is minimized. The decision variant asks whether these journeys can be found such that no journey ends after time t.*

We think of the input $\mathcal{G}$ as a temporal subgraph of some TVG $\mathcal{G}_\infty$ with lifetime $\mathbb{N}$ and the same edge constraints as $\mathcal{G}$. Thus, the limited information provided in $\mathcal{G}$ is used to compute complete solutions for agents covering $\mathcal{G}_\infty$. When unspecified, assume that DMVP refers to DMVP for a single agent.

1.2   Main Results

Our results are summarized in Table 1. We show that DMVP in $\mathcal{R}$ is NP-hard to approximate within any factor, when the underlying graph G is restricted to a star or tree of max degree 3. We show that in $\mathcal{B}$ this problem is NP-hard to approximate within any factor less than Δ, when G is restricted to a spider or tree of max degree 3. We show that in $\mathcal{P}$, DMVP is NP-complete when $p = 1$, and that there is a nontrivial class of graphs for which $p = 2$ is NP-hard, but $p = 1$ is not.

We show that in $\mathcal{R}$, DMVP is solvable in $O(T)$ when G is a path, $O(Tn)$ when G is a cycle, and $O(Tn^3 + n^2 2^n)$ for general graphs, where T is the duration of $\mathcal{G}$, as defined in Sect. 2. Furthermore, in $\mathcal{R}$, DMVP is fixed parameter tractable when G is an m-leaf $O(1)$-almost tree, and poly-time solvable when $m = O(\lg n)$. In $\mathcal{B}$, we demonstrate a tight Δ-approximation for trees, and a 2Δ-approximation for general graphs. We demonstrate a class of problems which are NP-hard in $\mathcal{B}$, but solvable by an online algorithm in $\mathcal{P}$. We show that DMVP in $\mathcal{P}$ is solvable in polynomial time when G is a spider, for fixed p, and we show that when $p = 2$, DMVP is solvable in linear time for general trees.

Table 1. DMVP separations and results by TVG class and graph class

DMVP separations			
TVG class	spiders	max-degree 3 trees	general trees
$\mathcal{R}$	no approx.	no approx.	no approx.
$\mathcal{B}$	tight Δ-approx.	tight Δ-approx.	tight Δ-approx.
$\mathcal{P}$	in P, for fixed p	$O(n)$ exact, for $p = 2$	$O(n)$ exact, for $p = 2$
$\exists$ graph class s.t. DMVP NP-hard in $\mathcal{P}$ with $p = 2$, easy with $p = 1$.			
Complexity of exact algorithms in $\mathcal{R}$			
path	cycle	general graphs	m-leaf c-almost trees	$O(\lg n)$-leaf c-almost trees
$O(T)$	$O(Tn)$	$O(Tn^3 + n^2 2^n)$	in FPT	in P

The remainder of this paper is organized as follows: preliminaries (2), lower bounds (3), upper bounds (4), open problems and discussion (5). Details of all missing proofs can be found in the full version of this paper [2].

2   Preliminaries

For the minimization problem DMVP($\mathcal{G}, S$) and the corresponding decision problem DMVP($\mathcal{G}, S, t$), input is viewed as a sequence of graphs G_i each represented as an adjacency matrix, with an associated integer duration t_i, i.e. $\mathcal{G} = (G_1, t_1), (G_2, t_2), ..., (G_m, t_m)$, where G_1 appears initially at time zero. Let $T = \sum_{i=1}^{m} t_i$. Note that since each t_i can be encoded in $O(\lg t_i)$ space, it is possible for T to be exponential in the size of $\mathcal{G}$. The following observation is required to show that the number of time steps of $\mathcal{G}$ that need to be considered for DMVP is in fact polynomial in the size of $\mathcal{G}$.

Observation 1. *When computing DMVP over $\mathcal{G}$, it is not necessary to consider each static temporal subgraph (G_i, t_i) for more than $2n - 3$ time steps.*

The idea is that on a static graph anything that can be accomplished in more than $2n-3$ steps can be accomplished in $2n-3$ steps or fewer. By Observation 1, for any $t_i > 2n - 3$, when computing DMVP, all time steps after the first $2n - 3$ can be ignored (skipped). DMVP over $\mathcal{G}$ can be computed by computing DMVP over $\mathcal{G}' = (G_1, \min(t_1, 2n - 3)), ..., (G_m, \min(t_m, 2n - 3))$, and adding back the cumulative time skipped before completion. $\mathcal{G}'$ can clearly be derived from $\mathcal{G}$ in $O(m)$ time. The total duration of $\mathcal{G}'$ is $T' = \sum_{i=1}^{m} \min(t_i, 2n - 3) < 2nm - 3m$, which is polynomial in $|\mathcal{G}|$. Let $\epsilon(\tau)$ be the time skipped through time τ. $\epsilon(\tau)$ can be simply calculated for all $\tau \le T'$ in $O(T')$ time. A similar $O(T')$ preprocessing step can be run to associate each time $\tau \in T'$ with the corresponding available static graph G_i, enabling $O(1)$ edge presence lookups $\rho(e, \tau)$.

Since all of the algorithms we present run in $\Omega(T')$ time, we can run these preprocessing steps for every instance of DMVP and not affect the asymptotic running time. Therefore, for the sake of simplicity, for the rest of our results we assume that this preprocessing has taken place, i.e., we think of $\mathcal{G}$ as $\mathcal{G}'$ and T as T', thereby avoiding the exponential nature of T. Note also that for the case of $\mathcal{P}$, the constraint of periodicity implies that it is only necessary to look at p consecutive time steps of the input.

3   Lower Bounds

As motivation for many of the results in this paper, it is important to note that MVP for a single agent is solvable in linear time on trees [1]. To characterize the difficulty of DMVP in $\mathcal{R}$, we first show inapproximability over stars. A similar theorem was independently discovered in [25].

Theorem 1. *DMVP for a single agent in $\mathcal{R}$ is NP-hard to approximate within any factor, even when the underlying graph is a star.*

This inapproximability also holds over the restriction of underlying graphs to trees of max-degree 3, in particular, combs.

Theorem 2. *DMVP for a single agent in $\mathcal{R}$ is NP-hard to approximate within any factor, even when the underlying graph is a comb.*

We have a similar set of lower bounds for the case of $\mathcal{B}$, but with *some* ability to approximate. We later show (Theorem 11) that these approximation bounds are indeed tight for all trees.

Theorem 3. *DMVP for a single agent in $\mathcal{B}$ is NP-hard to approximate within any factor less than Δ, even when the underlying graph is a spider, $\forall \Delta > 1$.*

Theorem 4. *DMVP for a single agent in $\mathcal{B}$ is NP-hard to approximate within any factor less than Δ, even when the underlying graph is a comb, $\forall \Delta > 1$.*

As is shown in Sect. 4, there is a much greater potential for tractability of DMVP in $\mathcal{P}$ than in $\mathcal{B}$ or $\mathcal{R}$. However, the next result follows immediately via reduction from hamiltonian path by simply restricting t to $n - 1$.

Theorem 5. *DMVP for a single agent in $\mathcal{P}$ is NP-complete, when $p = 1$.*

DMVP in $\mathcal{P}$ for $p = 1$ is then also NP-complete for all classes of graphs for which hamiltonian path is NP-complete, in particular, planar graphs of maximum degree 3, bridgeless undirected planar 3-regular bipartite graphs, and 3-connected 3-regular bipartite graphs [4]. To show that $\mathcal{P}$ is an interesting dynamics class for DMVP in its own right, it is important to show that DMVP yields different hardness results over $\mathcal{P}$ than over static graphs. Thus, we construct a class of graphs for the following result:

Theorem 6. *There is an infinite class of graphs C such that DMVP for a single agent in $\mathcal{P}$ over graphs in C is NP-complete when $p = 2$, but trivial when $p = 1$.*

4   Upper Bounds

In this section, we map out a class of graphs over which DMVP in $\mathcal{R}$ is solvable in polynomial time. We first start with a very useful lemma. Note that a related observation (about turning around on a ring) was made in [20].

Lemma 1 (Turning around lemma). *There is always an optimal solution J that never turns around at a degree 2 vertex of the edge-induced subgraph of J in G.*

See Fig. 1. The idea is that if an agent turns around at such a vertex, that vertex must also be reached at some other time in J. We apply Lemma 1 to get the following solvability results for restricted classes of underlying graphs.

Theorem 7. *DMVP for a single agent in $\mathcal{R}$ on a path is solvable in $O(T)$ time.*

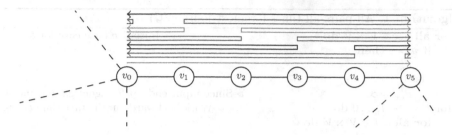

Fig. 1. The 7 ways, satisfying Lemma 1, of covering the vertices of a length 5 path with degree 2 intermediate nodes.

Theorem 8. *DMVP for a single agent in $\mathcal{R}$ on a cycle is solvable in $O(Tn)$ time.*

Now we show that despite the severe inapproximability of DMVP over $\mathcal{R}$, we can always compute an optimal solution in exponential time.

Theorem 9. *DMVP for a single agent in $\mathcal{R}$ is solvable in $O(Tn^3 + n^2 2^n)$ time.*

Proof. The proposed algorithm first computes all-pairs-all-times-foremost-journey for input TVG $\mathcal{G}$, using a straightforward dynamic programming algorithm, then uses this information to run another dynamic programming algorithm, conceived along the lines of a standard method for TSP [6].

Let d_{uv}^t be the length of the foremost journey from u to v, starting at time t. Algorithm 1 computes d_{uv}^t for all vertex pairs (u,v), and times $t \in \mathcal{T}$ for a given TVG $\mathcal{G}$.

At all times t, for all vertices $u \in V$, d_{uu}^t is clearly 0. At time T, the time limit has been reached, so an agent cannot move to another vertex in any guaranteed time, and thus we set $d_{uv}^T = \infty$ for all $u \neq v$. For all $T - 1 \geq t \geq 0$, in the worst case an agent can wait at u for one step, and take the foremost journey to v starting at time $t + 1$. If there is a better journey than this, it must consist of not waiting, rather taking one of the edges available at time t from u to some vertex k. Subsequently taking the foremost journey from k to v starting at time $t+1$ results in an optimal journey through k. Algorithm 1 clearly runs in $O(Tn^3)$ time, and uses $O(Tn^2)$ space.

Algorithm 2 uses the d_{uv}^t values computed by Algorithm 1 to compute the cost of a minimal solution to DMVP for a single agent in $\mathcal{R}$. Let $V' \subseteq V$ and $c(V', v)$ be the minimal time it takes to visit all vertices in V' starting at vertex s at time 0 and ending at vertex $v \in V'$.

After initializing the minimal costs for visiting subsets up to size 2, the algorithm repeatedly uses the minimal costs for size i subsets to calculate $c(V', v)$ for each size $i + 1$ subset V' and $v \neq s \in V'$. Once computed up to size n, the algorithm returns the minimal cost among journeys that cover all vertices. This is an optimal solution to DMVP as it is the minimum cost of taking foremost journeys between vertices that results in a complete cover. There are 2^n subsets

Algorithm 1. All-pairs-all-times-foremost-journey($\mathcal{G}$)

---

for all $u, v \in V \times V$ **do**   ▷ Initialize base case for $t = T$.
  if $u = v$ **then**
    $d_{uv}^T = 0$
  else
    $d_{uv}^T = \infty$ ▷ Since input ends at T, agent cannot move.
for $t = T - 1, ..., 0$ **do**   ▷ Work backwards until start time $t = 0$.
  for all $u, v \in V \times V$ **do**
    if $u = v$ **then**
      $d_{uv}^t = 0$
    else
      $d_{uv}^t = d_{uv}^{t+1} + 1$ ▷ In worst case, just wait at u.
      for all $k \in V$ **do**
        if $\rho((u, k), t) = 1$ **then**   ▷ Check for better route.
          $d_{uv}^t = \min(d_{uv}^t, d_{kv}^{t+1} + 1)$

Algorithm 2. $DMVP(\mathcal{G}, \{s\})$

$c(\{s\}, s) = 0$ ▷ Initialize subset of size 1.
for all $v \neq s \in V$ **do**   ▷ Initialize subsets of size 2.
  $c(\{s, v\}, v) = d_{sv}^0$
for i = 3,...,n **do**   ▷ Build up to subsets of size n.
  for all $S \subseteq V s.t. |S| = i$ **do**
    for all $v \neq s \in V$ **do**
      $c(V', v) = \min_{u \neq s \in V' \setminus \{v\}} (c(V' \setminus \{v\}, u) + d_{uv}^{c(V' \setminus \{v\}, u)})$
return $\min_{v \neq s \in V} (c(V, v))$

---

of V, and so $n2^n$ subset-vertex pairs of the form (V', v). For each of these, the algorithm computes the minimum of $O(n)$ values. So, Algorithm 2 has running time $O(n^2 2^n)$. Since it saves one cost for each subset-vertex pair, Algorithm 2 also uses $O(n2^n)$ space. Sequentially running Algorithm 1 followed by Algorithm 2, we have a complete algorithm for DMVP for a single agent in $\mathcal{R}$, with combined running time $O(Tn^3 + n^2 2^n)$.   □

We use Theorem 9 to generalize Theorems 7 and 8 with the following:

Theorem 10. *DMVP in $\mathcal{R}$ is fixed parameter tractable, when G is an m-leaf c-almost-tree, for fixed parameter m, and c constant.*

Proof. First, consider the restricted case where G is an m-leaf tree. Since every leaf must be visited, and visiting all leaves implies coverage of the entire tree, there is a minimal solution that can be thought of as an ordering of the set of leaves of G, and the foremost journeys between them. In this case, there is only one *way* to visit any node, namely, on the way to a leaf. Using this observation and Algorithm 2 from the proof of Theorem 9, we see that we only need to consider all orderings of leaves, instead of all orderings of vertices, yielding a run time of $O(Tn^3 + m^2 2^m)$, which is indeed fixed parameter tractable for parameter m.

Suppose the underlying graph G of $\mathcal{G}$ is an m-leaf c-almost-tree. Consider all edges e such that removing e from G results in a $(c-1)$-almost-tree. Each of these edges lies on some path P such that removing any edge of P will similarly result in a $(c-1)$-almost-tree, and every intermediate vertex on the path has degree 2. Suppose P is the path $v_0...v_l$. Since G is an m-leaf c-almost-tree, there are $O(m)$ paths of this type. The edge-induced subgraph G' of the underlying walk of an optimal covering of $\mathcal{G}$ can be any $(c-c')$-almost-tree $\subseteq G$, for $0 \leq c' \leq c$. For each c', a solution involves selecting c' paths, each of $O(n)$ length, from which to remove an edge. So, there are $O(m^{c'}n^{c'})$ possible choices of $(c-c')$-almost-trees, and thus $O(\sum_{c'=0}^{c}(m^{c'}n^{c'})) = O(m^cn^c)$ choices for G'. Every G' has no more than $m+2c$ leaves. Since every edge of G' is covered, by Lemma 1, there are at most 2 ways to cover each of the remaining $O(m)$ paths $v_0...v_l$ of intermediate vertex degree 2, namely: entering at v_0 and exiting at v_l, or entering at v_l and exiting at v_0. Augment the set of leaves to be ordered in a solution with the selected ways of covering these paths, that is, select one of the consecutive subsequences v_0v_l or v_lv_0 to be in the ordering. With this augmentation, we still have a set of $O(m)$ elements to be ordered, the optimal ordering of which can be computed via Theorem 9 in $O(Tn^3 + m^22^m)$ time. The minimum over all ways of covering G' can then be computed in $O(2^m)O(Tn^3 + m^22^m) = O(Tn^32^m + m^22^{2m})$. The overall minimum cost for covering $\mathcal{G}$ can then be computed by taking the minimum cost over all $O(m^cn^c)$ edge-induced subgraphs in $O(m^{c'}n^c)O(Tn^32^m + m^22^{2m}) = O(Tn^{3+c}f(m))$ time.   □

The following result follows immediately for the case when $m = O(\lg n)$.

Corollary 1. *DMVP in $\mathcal{R}$ is solvable in polynomial time, if $\mathcal{G}$ is an $O(\lg n)$-leaf c-almost-tree, for c constant.*

We conjecture (see Sect. 5) that the maximal class of graphs over which DMVP in $\mathcal{R}$ is poly-time solvable is the class of all graphs with polynomially many spanning trees, all of which have $O(\lg n)$ leaves.

Since DMVP in $\mathcal{B}$ is bounded by $2\Delta n$, the running time of the algorithm in Theorem 9 on TVGs over $\mathcal{B}$ reduces to $O(\Delta n^4 + n^22^n)$. We also see that we are able to greatly improve on approximation from $\mathcal{R}$ to $\mathcal{B}$:

Theorem 11. *DMVP in $\mathcal{B}$ over a tree can be Δ-approximated in $O(n)$ time. This approximation is tight.*

Theorem 12. *DMVP in $\mathcal{B}$ can be 2Δ-approximated by any online spanning tree traversal of G.*

These approximation upper bounds derive from static solutions [1], but waiting at most $\Delta-1$ steps for each edge to appear. Theorems 3 and 4 show the tightness of Theorem 11. Here, $\mathcal{B}$ is starkly differentiated from $\mathcal{R}$ in that we have at least some ability to approximate in $\mathcal{B}$. See Sect. 5 for a further discussion of the relationship between these two classes.

Similar to the case for $\mathcal{B}$, DMVP in $\mathcal{P}$ is bounded by $2pn$, so the running time of the algorithm in Theorem 9 reduces to $O(pn^4 + n^22^n)$. To exemplify the differences between $\mathcal{P}$ and $\mathcal{B}$, and motivate interest in the tractability of DMVP over $\mathcal{P}$, we first give the following simple example:

Theorem 13. *For any p, there is a class of problems over combs, for which DMVP in $\mathcal{B}$ is NP-hard, but in $\mathcal{P}$ is solvable by the online algorithm: take arms when you get to them.*

The quality of $\mathcal{P}$ we take advantage of above is that if the fastest journey between two nodes takes d steps, the foremost journey can take no longer than $d + (p-1)$, while in $\mathcal{B}$ it can be as bad as $d\Delta$. We again harness this effect in the following result, a stronger theorem in the context of our inapproximability results for $\mathcal{R}$ and $\mathcal{B}$ (Theorems 1 and 3):

Theorem 14. *DMVP in $\mathcal{P}$ over a spider is solvable in polynomial time, for fixed p.*

Proof Sketch: Each arm can be classified into one of $O(p^3)$ equivalence classes, based on the return time and cost above fastest of taking that arm for all time $t \equiv i \mod p$. A solution is an ordering of arms by when they are traversed. Suppose S is an optimal solution. Every length p subsequence of S must contain a shortest subsequence (called a *pattern*) that begins and ends at an equivalent time $t, t' \equiv j \mod p$. Patterns can be moved to any location beginning at some $t'' \equiv j \mod p$ without changing the cost of the solution. We can then cluster patterns by start time, without changing the cost of the solution, so that pairs of consecutive clusters are separated by no more than $p-1$ arms. There are only $O((p-1)! p^5 n^{(p^3)^{p+1}}) = O(n^{(p^3)^{p+1}})$ solutions of this form, one of which must be minimal. □

This polynomial runtime can be significantly improved for the case of $p = 2$.

Theorem 15. *DMVP in $\mathcal{P}$ over a tree is solvable in $O(n)$ time, when $p = 2$.*

Proof Sketch: Consider the problem for an agent starting at root o at time 0. We can show by induction that there is always an optimal solution that never enters any of the subtrees of o's children more than once. We then show each subtree is of one of three types: (11) fastest coverage is always available, (10) fastest coverage is available at even times, and (01) fastest coverage is available at odd times, with both 01 and 10 subtrees taking odd time to cover fastest, and 11's taking even time. Alternating between 10's and 01's, and then taking the remaining subtrees in any order, before ending at a *furthest* leaf, results in an optimal solution, as we maximize how many subtrees are traversed optimally. Using dynamic programming, we can compute bottom-up the type and cost of the maximal subtree rooted at each node v, in $O(\deg(v))$ time for each. □

5   Open Problems and Discussion

This paper presents significant advances towards isolating the maximal class of graphs over which DMVP in $\mathcal{R}$ is solvable in polynomial time. We conjecture that this maximal class is the class of all graphs with polynomially many spanning trees, all of which have $O(\lg n)$ leaves. Furthermore, we conjecture that this class is equivalent for $\mathcal{R}$ and $\mathcal{B}$. But we are very interested in expanding this class with respect to $\mathcal{P}$, motivated by our solvability results for $\mathcal{P}$ over subclasses of trees. We have shown that for the case of $p = 2$, DMVP for a single agent over general trees can be computed in linear time. This result relies on the fact that we know how to optimally piece together patterns with period 2. New methods for finding optimal pattern sequences could greatly reduce computation for cases of $p > 2$. We are hopeful that DMVP in $\mathcal{P}$ will be shown to be poly-time solvable over arbitrary trees or at least bounded degree trees, for greater values p both fixed and not fixed.

Considering $\mathcal{B}$ and $\mathcal{R}$, $\mathcal{B}$ is clearly differentiated from $\mathcal{R}$ in that we have at least some ability to approximate in $\mathcal{B}$. There remains, however, an important open question: Is there any class C of underlying graphs such that DMVP is NP-hard over C in $\mathcal{R}$, but not in $\mathcal{B}$? We are particularly interested in whether or not DMVP in $\mathcal{B}$ is NP-hard when the underlying graph is a star and Δ is fixed, in particular, when $\Delta = 2$. Note: The proof of Theorem 1 implies it is hard when Δ is some relatively small function of the input. We conjecture that even for $\Delta = 2$ this problem is NP-hard, but the highly-restricted nature of the input makes an answer to this problem more elusive than some of the others we have results for. Towards an answer to this question, we give the following observation:

Observation 2. *DMVP in $\mathcal{R}$ over a spider with arms of uniform length l, e.g., a star (when $l = 1$), can be decided in polynomial time, when t disallows waiting, i.e., $t = 2n - l - d$, where d is topological distance from s to c.*

Proof Idea: We can reduce this problem to the problem of finding a perfect bipartite matching between arms and blocks of time, for which there are many known efficient polynomial time algorithms, e.g., [18].

Overall, our results show some instances where DMVP is tractable as well as showing that DMVP faces difficult computational challenges for some natural classes of underlying topologies and dynamics. These challenges motivate research into online, multi-agent solutions to the problem, since in many cases having a complete global view of the present and future does not appear to be very helpful; moreover, in agent-oriented applications ranging from software agents to mobile robots, the information available to teams of agents can be bounded both temporally and geographically, and such online, multi-agent approaches could be well suited to agent dynamics without diminishing tractability.

References

1. Aaron, E., Kranakis, E., Krizanc, D.: On the complexity of the multi-robot, multi-depot map visitation problem. In: IEEE MASS, pp. 795–800 (2011)
2. Aaron, E., Krizanc, D., Meyerson, E.: DMVP: foremost waypoint coverage of time-varying graphs (2014). http://arxiv.org/abs/1407.7279

3. Ahr, D., Reinhelt, G.: A tabu search algorithm for the min-max k-Chinese postman problem. Comput. Oper. Res. **33**(12), 3403–3422 (2006)
4. Akiyama, T., Nishizeki, T., Saito, N.: NP-completeness of the Hamiltonian cycle problem for bipartite graphs. J. Inf. Process. **3**(2), 73–76 (1980)
5. Baumann, H., Crescenzi, P., Fraigniaud, P.: Parsimonious flooding in dynamic graphs. Distrib. Comput. **24**(1), 31–44 (2011)
6. Bellman, R.: Dynamic programming treatment of the travelling salesman problem. JACM **9**(1), 61–63 (1962)
7. Blum, A., Chalasani, P., Coppersmith, D., Pulleyblank, B., Raghavan, P., Sudan, M.: The minimum latency problem. In: Proceedings of 26th STOC, pp. 163–171 (1994)
8. Casteigts, A., Flocchini, P., Mans, B., Santoro, N.: Deterministic computations in time-varying graphs: broadcasting under unstructured mobility. In: Calude, C.S., Sassone, V. (eds.) TCS 2010. IFIP AICT, vol. 323, pp. 111–124. Springer, Heidelberg (2010)
9. Casteigts, A., Flocchini, P., Quattrociocchi, W., Santoro, N.: Time-varying graphs and dynamic networks. IJPED **27**(5), 387–408 (2012)
10. Avin, C., Koucký, M., Lotker, Z.: How to explore a fast-changing world (cover time of a simple random walk on evolving graphs). In: Aceto, L., Damgård, I., Goldberg, L.A., Halldórsson, M.M., Ingólfsdóttir, A., Walukiewicz, I. (eds.) ICALP 2008, Part I. LNCS, vol. 5125, pp. 121–132. Springer, Heidelberg (2008)
11. Choset, H.: Coverage for robotics: a survey of recent results. Ann. Math. AI **31**, 113–126 (2001)
12. Correll, N., Rutishauser, S., Martinoli, A.: Comparing coordination schemes for miniature robotic swarms. In: Springer Tracts in Advanced Robotics, vol. 39, pp. 471–480 (2008)
13. Easton, K., Burdick, J.: A coverage algorithm for multi-robot boundary inspection. In: Proceedings of ICRA, pp. 727–734 (2005)
14. Edmonds, J., Johnson, E.: Matching, Euler tours and the Chinese postman problem. Math. Program. **5**, 88–124 (1973)
15. Fakcharoenphol, J., Harrelson, C., Rao, S.: The k-traveling repairman problem. In: Proceedings of 39th STOC (2007)
16. Flocchini, P., Mans, B., Santoro, N.: On the exploration of time-varying networks. Theor. Comput. Sci. **469**, 53–68 (2013)
17. Garey, M., Johnson, D.: Computers and Intractability: A Guide to the Theory of NP-Completeness. W.H. Freeman, New York (1979)
18. Hopcroft, J., Karp, R.: An $n^{5/2}$ algorithm for maximum matchings in bipartite graphs. SIAM J. Comput. **2**(4), 225–231 (1973)
19. Ilcinkas, D., Wade, A.M.: On the power of waiting when exploring public transportation systems. In: Fernàndez Anta, A., Lipari, G., Roy, M. (eds.) OPODIS 2011. LNCS, vol. 7109, pp. 451–464. Springer, Heidelberg (2011)
20. Ilcinkas, D., Wade, A.M.: Exploration of the T-interval-connected dynamic graphs: the case of the ring. In: Moscibroda, T., Rescigno, A.A. (eds.) SIROCCO 2013. LNCS, vol. 8179, pp. 13–23. Springer, Heidelberg (2013)
21. Karp, R.: Reducibility among combinatorial problems. In: Miller, R.E., Thatcher, J.W. (eds.) Complexity of Computer Computations, pp. 85–103. Plenum, New York (1972)
22. Kuhn, F., Lynch, N., Oshman, R.: Distributed computation in dynamic networks. In: ACM Symposium on Theory of Computing (2010)
23. Kuhn, F., Oshman, R.: Dynamic networks: models and algorithms. ACM SIGACT News **42**(1), 82–96 (2011)

24. Mans, B., Mathieson, L.: On the treewidth of dynamic graphs. In: Du, D.-Z., Zhang, G. (eds.) COCOON 2013. LNCS, vol. 7936, pp. 349–360. Springer, Heidelberg (2013)
25. Michail, O., Spirakis, P.G.: Traveling salesman problems in temporal graphs. In: Csuhaj-Varjú, E., Dietzfelbinger, M., Ésik, Z. (eds.) MFCS 2014, Part II. LNCS, vol. 8635, pp. 553–564. Springer, Heidelberg (2014)
26. Wagner, A., Lindenbaum, M., Bruckstein, A.: Distributed covering by ant-robots using evaporating traces. IEEE Trans. Robot. Autom. 15(5), 918–933 (1999)
27. Xuan, B., Ferreira, A., Jarry, A.: Computing shortest, fastest, and foremost journeys in dynamic networks. IJ Found. Comput. Sci. 14(02), 267–285 (2003)

Linear Rank-Width of Distance-Hereditary Graphs

Isolde Adler[1], Mamadou Moustapha Kanté[2], and O-joung Kwon[3]([⊠])

[1] Institut für Informatik, Goethe-Universität, Frankfurt, Germany
iadler@informatik.uni-frankfurt.de
[2] Clermont-Université, Université Blaise Pascal, LIMOS, CNRS,
Clermont-Ferrand, France
mamadou.kante@isima.fr
[3] Department of Mathematical Sciences, KAIST,
291 Daehak-ro, Yuseong-gu, Daejeon 305-701, South Korea
ojoung@kaist.ac.kr

Abstract. We present a characterization of the linear rank-width of distance-hereditary graphs. Using the characterization, we show that the linear rank-width of every n-vertex distance-hereditary graph can be computed in time $\mathcal{O}(n^2 \cdot \log(n))$, and a linear layout witnessing the linear rank-width can be computed with the same time complexity. For our characterization, we combine modifications of canonical split decompositions with an idea of [Megiddo, Hakimi, Garey, Johnson, Papadimitriou: The complexity of searching a graph. JACM 1988], used for computing the path-width of trees. We also provide a set of distance-hereditary graphs which contains the set of distance-hereditary vertex-minor obstructions for linear rank-width. The set given in [Jeong, Kwon, Oum: Excluded vertex-minors for graphs of linear rank-width at most k. STACS 2013: 221–232] is a subset of our obstruction set.

1 Introduction

Rank-width [18] is a graph parameter introduced by Oum and Seymour with the goal of efficient approximation of the *clique-width* [5] of a graph. *Linear rank-width* can be seen as the linearized variant of rank-width, similar to path-width, which in turn can be seen as the linearized variant of tree-width. While path-width is a well-studied notion, much less is known about linear rank-width. Computing linear rank-width is NP-complete in general (this follows from [10]). Therefore it is natural to ask which graph classes allow for an efficient computation. Until now, the only (non-trivial) known such result is for forests [2].

Isolde Adler: Supported by the German Research Council, Project GalA, AD 411/1-1.
Mamadou Moustapha Kanté: Supported by the French Agency for Research under the DORSO project.
O-joung Kwon: Supported by Basic Science Research Program through the National Research Foundation of Korea (NRF) funded by the Ministry of Education, Science and Technology (2011-0011653).

© Springer International Publishing Switzerland 2014
D. Kratsch and I. Todinca (Eds.): WG 2014, LNCS 8747, pp. 42–55, 2014.
DOI: 10.1007/978-3-319-12340-0_4

A graph G is *distance-hereditary*, if for any two vertices u and v of G, the distance between u and v in any connected, induced subgraph of G that contains both u and v, is the same as the distance between u and v in G. Distance-hereditary graphs are exactly the graphs of rank-width ≤ 1 [17]. They include co-graphs (i.e. graphs of clique-width 2), complete (bipartite) graphs and forests.

We show that the linear rank-width of n-vertex distance-hereditary graphs can be computed in time $\mathcal{O}(n^2 \cdot \log(n))$ (Theorem 5). Moreover, we show that a layout of the graph witnessing the linear rank-width can be computed with the same time complexity (Corollary 2). Given that computing the path-width of distance-hereditary graphs is NP-complete [15], this is indeed surprising. We give a new characterization of linear rank-width of distance-hereditary graphs (Theorem 4), which we use for our algorithm. We also provide, for each k, a set Ψ_k of distance-hereditary graphs such that any distance-hereditary graph of linear rank-width at least $k+1$ contains a vertex-minor isomorphic to a graph in Ψ_k. The set Ψ_k generalizes the set of obstructions given in [14] and we conjecture a subset of it to be the set of distance-hereditary vertex-minor obstructions for linear rank-width k.

Our characterization makes use of the special structure of canonical split decompositions [6] of distance-hereditary graphs. Roughly, these decompositions decompose the distance-hereditary graph in a tree-like fashion into cliques and stars, and our characterization is recursive along the subtrees of the decomposition. While a similar idea has been exploited in [2,9,16], here we encounter a new problem: The decomposition may have vertices that are not present in the original graph. It is not at all obvious how to deal with these vertices in the recursive step. We handle this by introducing *limbs* of canonical split decompositions, that correspond to certain vertex-minors of the original graphs, and have the desired properties to allow our characterization. We think that the notion of limbs may be useful in other contexts, too, and hopefully, it can be extended to other graph classes and allow for further new efficient algorithms.

The paper is structured as follows. Section 2 introduces the basic notions, in particular linear rank-width, vertex-minors and split decompositions. In Sect. 3, we define limbs and show some important properties. We use them in Sect. 4 for our characterization of linear rank-width of distance-hereditary graphs. Finally, Sect. 5 presents the algorithm for computing the linear rank-width of distance-hereditary graphs and we discuss vertex-minor obstructions in Sect. 6.

2   Preliminaries

For a set A, we denote the power set of A by 2^A. We let $A \backslash B := \{x \in A \mid x \notin B\}$ denote the *difference* of two sets A and B. For a subset X of a ground set A, let $\overline{X} := A \backslash X$.

In this paper, graphs are finite, simple and undirected, unless stated otherwise. Our graph terminology is standard, see for instance [8]. Let G be a graph. We denote the vertex set of G by $V(G)$ and the edge set by $E(G)$. An edge between x and y is written xy (equivalently yx). If X is a subset of the vertex set of G, we denote the subgraph of G induced by X by $G[X]$, and we let

$G\backslash X := G[V(G)\backslash X]$. For a vertex $x \in V(G)$ we let $N_G(x) := \{y \in V(G) \mid x \neq y,\ xy \in E(G)\}$ denote the set of *neighbors* of x (in G). The *degree* of x (in G) is $\deg_G(x) := |N_G(x)|$. A partition of $V(G)$ into two sets X and Y is called a *cut* in G. We denote it by (X, Y).

A *tree* is a connected, acyclic graph. A *leaf* of a tree is a vertex of degree one. A *path* is a tree where every vertex has degree at most two. The *length* of a path is the number of its edges. A *rooted tree* is a tree with a distinguished vertex r, called the *root*. A *complete* graph is the graph with all possible edges. A graph G is called *distance-hereditary* (or DH for short) if for every two vertices x and y of G the distance of x and y in G equals the distance of x and y in any connected induced subgraph containing both x and y [3]. A *star* is a tree with a distinguished vertex, called its *center*, adjacent to all other vertices.

2.1   Linear Rank-Width and Vertex-Minors

Linear rank-width. For sets R and C an (R, C)-*matrix* is a matrix where the rows are indexed by elements in R and columns indexed by elements in C. (Since we are only interested in the rank of matrices, it suffices to consider matrices up to permutations of rows and columns.) For an (R, C)-matrix M, if $X \subseteq R$ and $Y \subseteq C$, we let $M[X, Y]$ be the submatrix of M where the rows and the columns are indexed by X and Y respectively.

Let A_G be the adjacency $(V(G), V(G))$-matrix of G over the binary field. For a graph G, let $x_1, \ldots, x_n$ be a linear layout of $V(G)$. Every index $i \in \{1, \ldots, n\}$ induces a cut $(X_i, \overline{X_i})$, where $X_i := \{x_1, \ldots, x_i\}$ (and hence $\overline{X_i} = \{x_{i+1}, \ldots, x_n\}$). The *cutrank* of the ordering $x_1, \ldots, x_n$ is defined as

$$\mathrm{cutrk}_G(x_1, \ldots, x_n) := \max\{\mathrm{rank}(A_G[X_i, \overline{X_i}]) \mid i \in \{1, \ldots, n\}\}.$$

The *linear rank-width* of G is defined as

$$\mathrm{lrw}(G) := \min\{\mathrm{cutrk}_G(x_1, \ldots, x_n) \mid x_1, \ldots, x_n \text{ is a linear layout of } V(G)\}.$$

Disjoint unions of caterpillars have linear rank-width ≤ 1. Ganian [11] gives an alternative characterization of the graphs of linear rank-width ≤ 1 as *thread graphs*. It is proved in [2] that linear rank-width and path-width coincide on trees. It is easy to see that the linear rank-width of a graph is the maximum over the linear rank-widths of its connected components.

Vertex-minors. For a graph G and a vertex x of G, the *local complementation at x* of G consists in replacing the subgraph induced on the neighbors of x by its complement. The resulting graph is denoted by $G * x$. If H can be obtained from G by a sequence of local complementations, then G and H are called *locally equivalent*. A graph H is called a *vertex-minor* of a graph G if H is a graph obtained from G by applying a sequence of local complementations and deletions of vertices.

For an edge xy of G, let $W_1 := N_G(x) \cap N_G(y)$, $W_2 = (N_G(x)\backslash N_G(y))\backslash\{y\}$, and $W_3 = (N_G(y)\backslash N_G(x))\backslash\{x\}$. *Pivoting on xy of G, denoted by $G \wedge xy$, is* the operation which consists in complementing the adjacencies between distinct sets W_i and W_j, and swapping the vertices x and y. It is known that $G \wedge xy = G * x * y * x = G * y * x * y$ [17].

Lemma 1 [17]. *Let G be a graph and let x be a vertex of G. Then for every subset X of $V(G)$, we have $\mathrm{cutrk}_G(X) = \mathrm{cutrk}_{G*x}(X)$. Therefore, every vertex-minor H of G satisfies $\mathrm{lrw}(H) \leq \mathrm{lrw}(G)$.*

2.2   Split Decompositions and Local Complementations

Split decompositions. We will follow the definitions in [4]. Let G be a connected graph. A *split* in G is a cut (X, Y) in G such that $|X|, |Y| \geq 2$ and $\mathrm{rank}(A_G[X, Y]) = 1$. In other words, (X, Y) is a split in G if $|X|, |Y| \geq 2$ and there exist non-empty sets $X' \subseteq X$ and $Y' \subseteq Y$ such that $\{xy \in E(G) \mid x \in X, y \in Y\} = \{xy \mid x \in X', y \in Y'\}$. Notice that not all connected graphs have a split, and those that do not have a split are called *prime* graphs.

A *marked graph* D is a connected graph D with a distinguished set of edges $M(D)$, called *marked edges*, that form a matching, and such that every edge in $M(D)$ is a *bridge*, i.e., its deletion increases the number of components. The ends of the marked edges are called *marked vertices*, and the components of $D \backslash M(D)$ are called *bags* of D. If (X, Y) is a split in G, we construct a marked graph D with vertex set $V(G) \cup \{x', y'\}$ for two distinct new vertices $x', y' \notin V(G)$ and edge set $E(G[X]) \cup E(G[Y]) \cup \{x'y'\} \cup E'$ where we define $x'y'$ as marked and

$$E' := \{x'x \mid x \in X \text{ and there exists } y \in Y \text{ such that } \mathrm{xy} \in \mathrm{E}(\mathrm{G})\} \cup$$
$$\{y'y \mid y \in Y \text{ and there exists } x \in X \text{ such that } xy \in E(G)\}.$$

The marked graph D is called a *simple decomposition of G*. A *decomposition* of a connected graph G is a marked graph D defined inductively to be either G or a marked graph defined from a decomposition D' of G by replacing a component H of $D' \backslash M(D')$ by a simple decomposition of H. We call the transformation of D' into D a *refinement of D'*. Notice that in a decomposition of a connected graph G, the two ends of a marked edge do not have a common neighbor. For a marked edge xy in a decomposition D, the *recomposition of D along xy* is the decomposition $D' := (D \wedge xy) \backslash \{x, y\}$. For a decomposition D, we let $\hat{D}$ denote the connected graph obtained from D by recomposing all marked edges. Note that if D is a decomposition of G, then $\hat{D} = G$. Since marked edges of a decomposition D are bridges and form a matching, if we contract all the unmarked edges in D, we obtain a tree called the *decomposition tree of G associated with D* and denoted by T_D. Obviously, the vertices of T_D are in bijection with the bags of D, and we will also call them bags.

A decomposition D of G is called a *canonical split decomposition* if each bag of D is either prime, or a star or a complete graph, and D is not the refinement of a decomposition with the same property. Shortly, we call it a *canonical decomposition*. The following is due to Cunningham and Edmonds [6], and Dahlhaus [7].

Theorem 1 [6,7]. *Every connected graph G has a unique canonical decomposition, up to isomorphism, that can be computed in time $\mathcal{O}(|V(G)| + |E(G)|)$.*

For a given connected graph G, by Theorem 1, we can talk about only one canonical decomposition of G because all canonical decompositions of G are isomorphic.

Let D be a decomposition of G with bags that are either primes, or complete graphs or stars (it is not necessarily a canonical decomposition). The *type of a bag* of D is either P, or K or S depending on whether it is a prime, or a complete graph or a star. The *type of a marked edge* uv is AB where A and B are the types of the bags containing u and v respectively. If $A = S$ or $B = S$, we can replace S by S_p or S_c depending on whether the end of the marked edge is a leaf or the center of the star.

Theorem 2 [4]. *Let D be a decomposition of a graph with bags of types P or K or S. Then D is a canonical decomposition if and only if it has no marked edge of type KK or S_pS_c.*

We will use the following characterization of distance-hereditary graphs.

Theorem 3 [4]. *A connected graph is a distance-hereditary graph if and only if each bag of its canonical decomposition is of type K or S.*

Local complementations in decompositions. We now relate the decompositions of a graph and the ones of its locally equivalent graphs. Let D be a decomposition. A vertex v of D *represents* an unmarked vertex x (or is a *representative* of x) if $v = x$ or there is a path from v to x in D starting with a marked edge such that marked edges and unmarked edges appear alternately in the path. Two unmarked vertices x and y are *linked* in D if there is a path from x to y in D such that unmarked edges and marked edges appear alternately in the path.

Lemma 2. *Let D be a decomposition of a graph. Let v' and w' be two marked vertices in a same bag of D, and let v and w be two unmarked vertices of D represented by v' and w', respectively. Then v and w are linked in D if and only if $vw \in E(\widehat{D})$ if and only if $v'w' \in E(D)$.*

A *local complementation* at an unmarked vertex v in a decomposition D, denoted by $D * v$, is the operation which consists in replacing each bag B containing a representative w of v with $B * w$. Observe that $D * v$ is a decomposition of $\widehat{D} * v$, and that $M(D) = M(D*v)$. Two decompositions D and D' are *locally equivalent* if D can be obtained from D' by applying a sequence of local complementations.

Lemma 3 [4]. *Let D be the canonical decomposition of a graph and let v be an unmarked vertex of D. Then $D * v$ is the canonical decomposition of $\widehat{D} * v$.*

Let v and w be linked unmarked vertices in a decomposition D, and let B_v and B_w be the bags containing v and w, respectively. Note that if B is a bag of type S in the path from B_v to B_w in T_D, then the center of B is a representative of either

v or w. *Pivoting on* vw *of* D, denoted by $D \wedge vw$, is the decomposition obtained as follows: for each bag B on the path from B_v to B_w in T_D, if $v', w' \in V(B)$ represent v and w in D, respectively, then we replace B with $B \wedge v'w'$. (Note that by Lemma 2, we have $v'w' \in E(B)$, hence $B \wedge v'w'$ is well-defined).

Lemma 4. *Let D be a decomposition of a distance-hereditary graph, and let $xy \in E(\widehat{D})$. Then $D \wedge xy = D * x * y * x$.*

The proof of Lemma 4, as well as all omitted proofs, can be found in the appendix. As a corollary of Lemmas 3 and 4, we get the following.

Corollary 1. *Let D be the canonical decomposition of a distance-hereditary graph and $xy \in E(\widehat{D})$. Then $D \wedge xy$ is the canonical decomposition of $\widehat{D} \wedge xy$.*

3 Limbs in Canonical Decompositions

In this section we define the notion of *limb* that is the key ingredient in our characterization. Intuitively, a limb in the canonical decomposition of a distance-hereditary graph G is a subtree of the decomposition with the property that the linear rank-width of the graph obtained from the subtree by recomposing all marked edges is invariant under taking local complementations.

Let D be the canonical decomposition of a distance-hereditary graph. We recall from Theorem 2 that each bag of D is of type K or S, and marked edges of types KK or S_pS_c do not occur. Given a bag B of D, an unmarked vertex y of D represented by some marked vertex $w \in V(B)$, let T be the component of $D\backslash V(B)$ containing y and let $v \in V(T)$ be the neighbor of w in D. We define the *limb* $\mathcal{L} := \mathcal{L}[D, B, y]$ as follows:

1. if B is of type K, then $\mathcal{L} := T * v\backslash v$,
2. if B is of type S and w is a leaf, then $\mathcal{L} := T\backslash v$,
3. if B is of type S and w is the center, then $\mathcal{L} := T \wedge vy\backslash v$.

Note that in T, v becomes an unmarked vertex, so a limb is well-defined. While T is a canonical decomposition, $\mathcal{L}$ may not be a canonical decomposition at all, because deleting v may create a bag of size 2. Suppose a bag B' of size 2 appears in $\mathcal{L}$. If B' has one neighbor bag B_1 and a marked vertex $v_1 \in B_1$ is adjacent to a marked vertex of B' and r is the unmarked vertex of B' in $\mathcal{L}$, then we can transform the limb into a canonical decomposition by removing the bag B' and replacing v_1 with r. If B' has two neighbor bags B_1 and B_2 and two marked vertices $v_1 \in B_1$ and $v_2 \in B_2$ are adjacent to the marked vertices of B', then we can first transform the limb into a decomposition by removing B' and adding a marked edge v_1v_2. However, the new marked edge v_1v_2 still could be of type KK or S_pS_c. Then by recomposing along v_1v_2, we finally transform the limb into a canonical decomposition.

Let $\widetilde{\mathcal{L}} = \widetilde{\mathcal{L}}[D, B, y]$ be the canonical decomposition obtained from $\mathcal{L}[D, B, y]$, and let $\widehat{\mathcal{L}} = \widehat{\mathcal{L}}[D, B, y]$ be the graph obtained from $\mathcal{L}[D, B, y]$ by recomposing all marked edges. See Fig. 1 for an example. If the original canonical decomposition D is clear from the context, we remove D in the notation $\mathcal{L}[D, B, y]$.

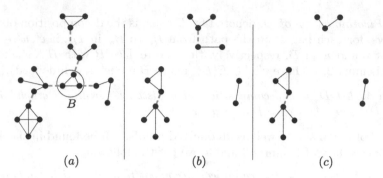

Fig. 1. In (a), we have a canonical decomposition D of a distance-hereditary graph and a bag B of D. The dashed edges are marked edges in D. In (b), we have limbs $\mathcal{L}$ associated with the components of $D\backslash V(B)$. The canonical decompositions $\widetilde{\mathcal{L}}$ associated with limbs $\mathcal{L}$ are shown in (c).

Lemma 5. *Let B be a bag of D. If two unmarked vertices x and y are represented by a marked vertex $w \in V(B)$, then $\widehat{\mathcal{L}}[B, x]$ is locally equivalent to $\widehat{\mathcal{L}}[B, y]$.*

For a bag B in D and a component T of $D\backslash V(B)$, we define $f(D, B, T)$ as the linear rank-width of $\widehat{\mathcal{L}}[D, B, y]$ for some unmarked vertex $y \in V(T)$. In fact, by Lemma 5, $f(D, B, T)$ does not depend on the choice of y. As in the notation $\mathcal{L}[D, B, x]$, if the canonical decomposition D is clear from the context, we remove D in the notation $f(D, B, T)$.

Proposition 1. *Let B be a bag of D. Let $x \in V(\widehat{D})$ and let y be an unmarked vertex represented in D by $v \in V(B)$. If y' is represented by v in $D * x$, then $\widehat{\mathcal{L}}[D, B, y]$ is locally equivalent to $\widehat{\mathcal{L}}[D * x, B, y']$. Therefore, $f(D, B, T) = f(D * x, B, T_x)$ where T_x is the component of $(D * x)\backslash V(B)$ containing y.*

Proposition 2. *Let B_1 and B_2 be two bags of D. Let T_1 be a component of $D\backslash V(B_1)$ such that T_1 does not contain the bag B_2, and let T_2 be the component of $D\backslash V(B_2)$ such that T_2 contains the bag B_1. Then $f(B_1, T_1) \leq f(B_2, T_2)$.*

4   Characterizing the Linear Rank-Width of DH Graphs

In this section, we prove the main theorem of the paper, which characterizes distance-hereditary graphs of linear rank-width k.

Theorem 4 (Main Theorem). *Let k be a positive integer and let D be the canonical decomposition of a distance-hereditary graph. Then $\mathrm{lrw}(\widehat{D}) \leq k$ if and only if for each bag B of D, D has at most two components T of $D\backslash V(B)$ such that $f(B, T) = k$, and for all the other components T' of $D\backslash V(B)$, $f(B, T') \leq k - 1$.*

To prove the converse direction, we use the following technical lemmas. Let k be a positive integer and let D be the canonical decomposition of a distance-hereditary graph.

Proposition 3. *Let B be a bag of D with two unmarked vertices x, y. If for every component T of $D \backslash V(B)$, $f(B, T) \leq k - 1$, then the graph $\widehat{D}$ has a linear layout of width at most k such that the first vertex and the last vertex of it are x and y, respectively.*

Lemma 6. *Suppose for each bag B of D, there are at most two components T of $D \backslash V(B)$ satisfying $f(B, T) = k$ and for all the other components T' of $D \backslash V(B)$, $f(B, T') \leq k - 1$. Then T_D has a path P such that for each bag B in P and a component T of $D \backslash V(B)$ not containing a bag of P, $f(B, T) \leq k - 1$.*

We are now ready to prove Theorem 4.

Proof (of Theorem 4). For the forward direction, it is sufficient to show that if B is a bag of D such that $D \backslash V(B)$ has at least three components T_1, T_2, T_3 such that $f(B, T_i) = k$, then $\mathrm{lrw}(\widehat{D}) \geq k + 1$. The proof idea is the same as the one used in [9]. We start from a linear layout assumed to have width k and we prove using Lemmas 1, 3 and Proposition 1 and tools from linear algebra that there exists $i \in \{1, 2, 3\}$ such that $f(B, T_i) \leq k - 1$, contradicting that $f(B, T_i) = k$. The details are omitted due to space constraints.

Now we prove the converse direction. Let $P := B_0 - B_1 - \cdots - B_n - B_{n+1}$ be the path in T_D such that for each bag B in P and a component T of $D \backslash V(B)$ not containing a bag of P, $f(B, T) \leq k - 1$ (such a path exists by Lemma 6). If B_0 does not have an unmarked vertex, then we add one unmarked vertex to B_0 and we call it a_0. Similarly for B_{n+1}, but the added unmarked vertex is called b_{n+1}.

Now for each $0 \leq i \leq n$, let b_i be the marked vertex of B_i adjacent to B_{i+1} and let a_{i+1} be the marked vertex of B_{i+1} adjacent to b_i. And for each $0 \leq i \leq n + 1$, let D_i be the subdecomposition of D induced on the bag B_i and the components of $D \backslash V(B_i)$ which do not contain a vertex of P. Notice that the vertices a_i and b_i are unmarked vertices in D_i. Since every component T of $D_i \backslash V(B_i)$ is such that $f(D_i, B_i, T) \leq k - 1$, by Proposition 3, $\widehat{D_i}$ has a linear layout L'_i of width k such that the first vertex of it is a_i and the last vertex of it is b_i. For each $1 \leq i \leq n$, let L_i be the linear layout obtained from L'_i by removing a_i and b_i. Let L_1 and L_{n+1} be obtained from L'_1 and L'_{n+1} by removing b_0 and a_{n+1}, respectively, and also the vertices a_0 and b_{n+1}, respectively, if they were added. Then we can easily check that $L := L_0 \oplus L_1 \oplus \cdots \oplus L_{n+1}$ is a linear layout of $\widehat{D}$ having width at most k. Therefore $\mathrm{lrw}(\widehat{D}) \leq k$.   □

5   Computing the Linear Rank-Width of DH Graphs

In this section, we describe an algorithm to compute the linear rank-width of distance-hereditary graphs. Since the linear rank-width of a graph is the maximum linear rank-width over all its connected components, we will focus on connected distance-hereditary graphs.

Theorem 5. *The linear rank-width of any connected graph with n vertices can be computed in time $\mathcal{O}(n^2 \cdot \log n)$.*

We say that a canonical decomposition D is *rooted* if we distinguish either a bag of D or a marked edge of D, and call it the *root of D*. In a rooted canonical decomposition with the root bag, the parent of a bag is defined analogously as in rooted trees, and when the root is a marked edge, every bag has a parent according to the convention below: if the marked edge between two bags B_1 and B_2 is the root, then we call B_2 the *artificial parent* of B_1, and similarly B_1 is also called the *artificial parent* of B_2. We remark that the (artificial) parent will be used to define certain limbs. For two bags B and B' in D, B is called a *descendant* of B' if B' is on the unique path from B to the root in T_D. Two bags in D are called *comparable* if one bag is a descendant of the other bag. Otherwise, they are called *incomparable*. If two canonical decompositions D_1 and D_2 are locally equivalent and B is the root bag of D_1, then we say $D_2[V(B)]$ is also the root of D_2. Similarly, if a marked edge e is the root of D_1, then we say e is also the root of D_2.

To easy the understanding and to avoid the choice of y in the definition of limbs, we will deal with a set of pairwise locally equivalent canonical decompositions. For a canonical decomposition D of a distance-hereditary graph, we define Γ_D as the set of all canonical decompositions locally equivalent to D. We remark that for $D_1, D_2 \in \Gamma_D$ and $B \subseteq V(D)$, B induces a bag in D_1 if and only if B induces a bag in D_2. We also have $M(D_1) = M(D_2)$.

For a bag B of a canonical decomposition D and a marked edge e adjacent to B in D, let $\mathcal{G}(\Gamma_D, B, e)$ be the set of all canonical decompositions $\widetilde{\mathcal{L}}[D', D'[V(B)], y]$ where $D' \in \Gamma_D$, T is the component of $D' \backslash V(B)$ incident with e, and $y \in V(T)$ is an unmarked vertex represented by a vertex of $D'[V(B)]$ in D'.

Proposition 4. $\mathcal{G}(\Gamma_D, B, e) = \Gamma_{D'}$ *for some canonical decomposition D'.*

Let D be the rooted canonical decomposition of a distance-hereditary graph G with the root R. We introduce two ways to take a set of limbs from the decompositions in Γ_D. Let B be a non-root bag of D and let B' be the (possibly artificial) parent of B and let e be the marked edge connecting B and B' in D.

1. Let $\Gamma_1(\Gamma_D, B) := \mathcal{G}(\Gamma_D, B', e)$ and $\mathcal{F}_1(\Gamma_D, B) := \mathrm{lrw}(\widehat{D'})$ for $D' \in \Gamma_1(\Gamma_D, B)$.
2. Let $\Gamma_2(\Gamma_D, B) := \mathcal{G}(\Gamma_D, B, e)$ and $\mathcal{F}_2(\Gamma_D, B) := \mathrm{lrw}(\widehat{D'})$ for $D' \in \Gamma_2(\Gamma_D, B)$.

By Proposition 4, $\Gamma_i(\Gamma_D, B) = \Gamma_{D'}$ for some canonical decomposition D' and so we can apply this function recursively, for instance, $\Gamma_2(\Gamma_1(\Gamma_D, B_1), B_2)$.

In the algorithm, we will compute decompositions in $\Gamma_1(\Gamma_D, B)$ or $\Gamma_2(\Gamma_D, B)$. As explained in Sect. 3, we need sometimes to merge two bags to be able to turn a limb into a canonical decomposition. Whenever a merging operation on two bags B_1 and B_2 appears, if B_2 is a descendant of B_1, then we regard the merged bag as B_1, and if they are incomparable, then we regard it as a new one.

For $D_i \in \Gamma_i(\Gamma_D, B)$, we define the root R' of D_i as follows. If the root R of D exists in D_i, then let $R' := R$. Assume the root R does not exist in D_i. In this case, some bag, which was either the root R itself or incident with the root edge R, is removed, and two children of it are merged or linked by a marked edge. If two children of the removed bag are merged, then let R' be the merged bag, and if otherwise, let R' be the marked edge between them. We have the following.

Lemma 7. *Let B be a non-root bag of D and let $D_i \in \Gamma_i(\Gamma_D, B)$. If B' is a non-root bag of D_i, then B' is a non-root bag of D (for $i = 1, 2$).*

Our algorithm uses methods of the algorithm for vertex separation of trees [9]. Our algorithm works bottom-up on D, and computes $\mathcal{F}_1(\Gamma_D, B)$ for all bags B in D using dynamic programming. Let B be a bag of D, and let $B_1, B_2, \ldots, B_m$ be the children of B in D. Let $k := \max_{1 \leq i \leq m} \mathcal{F}_1(\Gamma_D, B_i)$. We can easily observe that $k \leq \mathcal{F}_1(\Gamma_D, B) \leq k + 1$. We discuss now how to determine $\mathcal{F}_1(\Gamma_D, B)$. A bag B of D is called *k-critical* if $\mathcal{F}_1(\Gamma_D, B) = k$ and B has two children B_1 and B_2 such that $\mathcal{F}_1(\Gamma_D, B_1) = \mathcal{F}_1(\Gamma_D, B_2) = k$. We first observe the following which can be derived from Theorem 4 and Proposition 2.

Proposition 5. *Let $k = \max\{\mathcal{F}_1(\Gamma_D, B) \mid B$ is a non-root bag of $D\}$. Assume that D has neither a bag B having at least three children B' such that $\mathcal{F}_1(\Gamma_D, B') = k$ nor two incomparable bags B_1 and B_2 with a k-critical bag B_1 and $\mathcal{F}_1(\Gamma_D, B_2) = k$. Let B be a k-critical bag of D. Then B is the unique k-critical bag of D. Moreover, $\mathrm{lrw}(G) = k + 1$ if and only if $\mathcal{F}_2(\Gamma_D, B) = k$.*

By Proposition 5, the computation of $\mathcal{F}_1(\Gamma_D, B)$ is reduced to the computation of $\mathcal{F}_2(\Gamma_1(\Gamma_D, B), B_c)$ if $D' \in \Gamma_1(\Gamma_D, B)$ has the unique k-critical bag B_c. In order to compute $\mathcal{F}_2(\Gamma_1(D, B), B_c)$, we can recursively call the algorithm. However, we will prove that these recursive calls are not needed if we compute more than the linear rank-width, and it is the key for the $\mathcal{O}(n^2 \cdot \log(n))$ time algorithm (Table 1).

Table 1. Examples of $PD(B, j)$ and $LD(B, j)$.

j	$PD(B, j)$	$LD(B, j)$	Status
10	8	9	$D' \in D(B, 10)$ has no 10-critical bags.
9	8	9	$D' \in D(B, 9)$ has no 9-critical bags.
8	8	9	$D' \in D(B, 8)$ has the unique 8-critical bag B_c and the maximum $\mathcal{F}_1$ value over all bags B' except the root in $\Gamma_1(D', B_c, v)$ is 7.
7	7	8	$D' \in D(B, 7)$ has a bag having three children B' such that $\mathcal{F}_1(D', B') = 7$. Thus, $LD(B, 7) = 8$.
6	-	-	Once we have $LD(B, \ell) = \ell + 1$, it is unnecessary to compute $D(B, j)$ where $j < \ell$.

For each bag B of D and $0 \leq j \leq \lfloor \log|V(G)| \rfloor$, we recursively define a set $D(B, j)$ of canonical decompositions. The integer j will be at most the linear rank-width. The choice of $j \leq \lfloor \log|V(G)| \rfloor$ comes from the following fact.

Lemma 8. *For a distance-hereditary graph G, $\mathrm{lrw}(G) \leq \log|V(G)|$.*

Let $D(B, \lfloor\log|V(G)|\rfloor) := \Gamma_1(\Gamma_D, B)$. For each bag B, j and $D' \in D(B,j)$, let $PD(B,j)$ be the maximum $\mathcal{F}_1(\Gamma_{D'}, B')$ over all non-root bags B' in D', and let $LD(B,j) := \mathrm{lrw}(\widehat{D'})$.

1. Let $D(B, \lfloor\log|V(G)|\rfloor) := \Gamma_1(\Gamma_D, B)$.
2. For all $1 \leq j \leq \lfloor\log|V(G)|\rfloor$, if $PD(B,j) \neq j$, let $D(B, j-1) := D(B,j)$. If $PD(B,j) = j$, then for $D' \in D(B,j)$,
   (a) if (D' has a bag with 3 children B_1 such that $LD(B_1, j) = j$) or (D' has two incomparable bags B_1 and B_2 with a j-critical bag B_1 and $LD(B_2, j) = j$) or (D' has no j-critical bags), then let $D(B, j-1) := D(B,j)$,
   (b) if D' has the unique j-critical bag B_c, then let $D(B, j-1) := \Gamma_2(D(B,j), B_c)$.

The essential cases are when $PD(B,j) = j$, and in these cases, we want to determine whether $LD(B,j) = j$ or $j+1$. We prove the following.

Proposition 6. *Let B be a non-root bag of D. Let i be an integer such that $0 \leq i \leq \lfloor\log|V(G)|\rfloor$ and $PD(B,i) \leq i$. Let $D' \in D(B,i)$ and let B' be a non-root bag of D'. Then B' is also a non-root bag of D and $PD(B', i) \leq i$. Moreover, $\Gamma_1(D(B,i), B') = D(B', i)$. Therefore, $\mathcal{F}_1(D(B,i), B') = LD(B', i)$.*

Now we describe the algorithm explicitly. For convenience, we modify the given decomposition as follows. For the canonical decomposition D' of a distance-hereditary graph G, we modify D' into a canonical decomposition D by adding a bag R adjacent to a bag R' in D so that $f(D, R, D') = \mathrm{lrw}(G)$. So, if we regard R as the root bag of D, then $\mathcal{F}_1(\Gamma_D, R') = \mathrm{lrw}(G) = LD(R', \lfloor\log|V(G)|\rfloor)$. The basic strategy is to compute $LD(B,i)$ for all non-root bags B of D and integers i such that $PD(B,i) \leq i$. If B is a non-root leaf bag of D, then clearly $\mathcal{F}_1(\Gamma_D, B) = 1$, so let $LD(B,i) = 1$ for all $0 \leq i \leq \lfloor\log|V(G)|\rfloor$. For convenience, let $t = \lfloor\log|V(G)|\rfloor$.

1. Compute the canonical decomposition D' of G, and obtain a canonical decomposition D from D' by adding a root bag R adjacent to a bag R' in D so that $\mathrm{lrw}(G) = LD(R', t)$.
2. For all non-root leaf bags B in D, set $LD(B,j) := 1$ for all $0 \leq j \leq t$.
3. While (D has a non-root bag B such that $LD(B,t)$ is not computed).
   (a) Choose a non-root bag B in D such that for every child B' of B, $LD(B', t)$ is computed.
   (b) Compute a decomposition D_t in $\Gamma_1(\Gamma_D, B) = D(B,t)$.
   (c) Compute $k := PD(B,t)$ and set $D_k := D_t$ and $i := k$.
   (d) Let S be a stack.
   (e) While (true) do.
      i. If either (D_i has a bag with at least 3 children B_1 such that $LD(B_1, i) = i$) or (D_i has two incomparable bags B_1 and B_2 with B_1 an i-critical bag and $LD(B_2, i) = i$) or (D_i has no i-critical bags), then stop this loop.
      ii. Find the unique i-critical bag in D_i.

    iii. Compute $D_{i-1} \in D(B, i-1)$ and push(S, i).

    iv. Set $j := i - 1$ and $i := PD(B, i-1)$ and $D_i := D_j$.

  (f) If either (D_i has a bag with at least 3 children B_1 such that $LD(B_1, i) = i$) or (D_i has two incomparable bags B_1 and B_2 with B_1 an i-critical bag and $LD(B_2, i) = i$), then set $LD(B, i) := i + 1$, else, $LD(B, i) := i$.

  (g) While ($S \neq \emptyset$) do.

    i. Set $j := \text{pull}(S)$.

    ii. If $LD(B, i) = j$, then $LD(B, j) := j + 1$, else $LD(B, j) := j$.

    iii. For $\ell = i + 1$ to $j - 1$, set $LD(B, \ell) := LD(B, i)$.

    iv. Set $i := j$.

  (h) Set $LD(B, j) := LD(B, k)$ for all $k < j \leq t$.

4. Return $LD(R', t)$.

Proof (of Theorem 5). By Propositions 5 and 6 the steps of the algorithm outlined above computes the linear rank-width of every connected distance-hereditary graph G. Let us now analyze its running time. Let n and m be the number of vertices and edges of G. Its canonical decomposition D' can be computed in time $\mathcal{O}(n + m)$ by Theorem 1, and one can of course add a new bag to obtain a new canonical decomposition D and root it in constant time. The number of bags in D is bounded by $\mathcal{O}(n)$ (see [12, Lemma 2.2]). For each bag B, $LD(B, j)$ for all $0 \leq j \leq t$ can be computed in time $\mathcal{O}(n \cdot \log(n))$. In fact, Steps 3(a-c) can be done in time $\mathcal{O}(n)$. The loop in 3(e) runs $\log(n)$ times since $k \leq \log(n)$, and all the steps in 3(e) can be implemented in time $\mathcal{O}(n)$. Since Steps 3(f-h) can be done in time $\mathcal{O}(n)$, we conclude that this algorithm runs in time $\mathcal{O}(n^2 \cdot \log n)$. □

Corollary 2. *For every connected distance-hereditary graph G, we can compute in time $\mathcal{O}(n^2 \cdot \log(n))$ a layout of the vertices of G witnessing* $\mathrm{lrw}(G)$.

6   Obstructions

A graph H is a *vertex-minor obstruction* for (linear) rank-width k if it has (linear) rank-width $k+1$ and every proper vertex-minor of H has (linear) rank-width at most k. The set of pairwise locally non-equivalent vertex-minor obstructions for (linear) rank-width k is not known, but for rank-width k a bound on their size is known [17], which is not the case for linear rank-width k. For $k = 1$, Adler, Farley, and Proskurowski [1] characterized the distance-hereditary vertex-minor obstructions for linear rank-width at most 1 by two pairwise locally non-equivalent graphs. For general k, Jeong, Kwon, and Oum recently provided a $2^{\Omega(3^k)}$ lower bound on the number of pairwise locally non-equivalent distance-hereditary vertex-minor obstructions for linear rank-width at most k [14]. Using our characterization, we generalize the construction in [14] and conjecture a subset of the given set to be the set of distance-hereditary vertex-minor obstructions.

    We will use the notion of *one-vertex extensions* introduced in [13]. We call a graph G' an *one-vertex extension* of a distance-hereditary graph G if G' is a graph obtained from G by adding a new vertex v with some edges and G' is again distance-hereditary. For convenience, if D and D' are canonical decompositions of G and G', respectively, then D' is also called an *one-vertex extension* of D. For example, any one-vertex extension of K_2 is isomorphic to either K_3 or $K_{1,2}$. For a set $\mathcal{D}$ of canonical decompositions, we define

$$\mathcal{D}^+ = \mathcal{D} \cup \{D' | D' \text{ is an one vertex extension of } D \in \mathcal{D}\}.$$

For a set $\mathcal{D}$ of canonical decompositions, we define a new set $\Delta(\mathcal{D})$ of canonical decompositions D as follows:

– Choose three decompositions D_1, D_2, D_3 in $\mathcal{D}$ and take one-vertex extensions D_i' of D_i with new vertices w_i for each i. We introduce a new bag B of type K or S having three vertices v_1, v_2, v_3 and
  1. if v_i is in a complete bag, then $D_i'' = D_i' * w_i$,
  2. if v_i is the center of a star bag, then $D_i'' = D_i' \wedge w_i z_i$ for some z_i linked to w_i in D',
  3. if v_i is a leaf of a star bag, then $D_i'' = D_i'$.
  Let D be the canonical decomposition obtained by the disjoint union of D_1'', D_2'', D_3'' and B by adding the marked edges $v_1 w_1, v_2 w_2, v_3 w_3$.

For each non-negative integer k, we construct the sets Ψ_k and Φ_k of canonical decompositions as follows.

1. $\Psi_0 = \Phi_0 := \{K_2\}$ (K_2 is the canonical decomposition of the graph K_2).
2. For $k \geq 0$, let $\Psi_{k+1} := \Delta(\Psi_k^+)$.
3. For $k \geq 0$, let $\Phi_{k+1} := \Delta(\Phi_k)$.

We prove the following.

Theorem 6. *Let $k \geq 0$ and let G be a distance-hereditary graph such that* $\mathrm{lrw}(G) \geq k + 1$. *Then there exists a canonical decomposition D in Ψ_k such that G contains a vertex-minor isomorphic to $\hat{D}$.*

In order to prove that Ψ_k is the set of canonical decompositions of distance-hereditary vertex-minor obstructions for linear rank-width at most k, we need to prove that for every $D \in \Psi_k$, $\hat{D}$ has linear rank-width $k + 1$ and every of its proper vertex-minors has linear rank-width $\leq k$. However, we were not able to prove it, and we showed this property for Φ_k instead of Ψ_k.

Proposition 7. *Let $k \geq 0$ and let $D \in \Phi_k$. Then $\mathrm{lrw}(\hat{D}) = k + 1$ and every proper vertex-minor of $\hat{D}$ has linear rank-width at most k.*

One can observe that the obstructions constructed in [1,14] are contained in Φ_k for all $k \geq 1$.

We leave open the question to identify a set $\Phi_k \subset \Theta_k \subset \Psi_k$ that forms the set of canonical decompositions of distance-hereditary vertex-minor obstructions for linear rank-width k.

References

1. Adler, I., Farley, A.M., Proskurowski, A.: Obstructions for linear rank-width at most 1. Discrete Appl. Math. **168**, 3–13 (2014)
2. Adler, I., Kanté, M.M.: Linear rank-width and linear clique-width of trees. In: Brandstädt, A., Jansen, K., Reischuk, R. (eds.) WG 2013. LNCS, vol. 8165, pp. 12–25. Springer, Heidelberg (2013)
3. Bandelt, H.-J., Mulder, H.M.: Distance-hereditary graphs. J. Comb. Theory, Ser. B **41**(2), 182–208 (1986)
4. Bouchet, A.: Transforming trees by successive local complementations. J. Graph Theory **12**(2), 195–207 (1988)
5. Courcelle, B., Olariu, S.: Upper bounds to the clique width of graphs. Discrete Appl. Math. **101**(1–3), 77–114 (2000)
6. Cunnigham, W.H., Edmonds, J.: A combinatorial decomposition theory. Can. J. Math. **32**, 734–765 (1980)
7. Dahlhaus, E.: Parallel algorithms for hierarchical clustering, and applications to split decomposition and parity graph recognition. J. Graph Algorithms **36**(2), 205–240 (2000)
8. Diestel, R.: Graph Theory. Graduate texts in mathematics, vol. 173, 3rd edn. Springer, Heidelberg (2005)
9. Ellis, J.A., Sudborough, I.H., Turner, J.S.: The vertex separation and search number of a graph. Inf. Comput. **113**(1), 50–79 (1994)
10. Fellows, M.R., Rosamond, F.A., Rotics, U., Szeider, S.: Clique-width is np-complete. SIAM J. Discrete Math. **23**(2), 909–939 (2009)
11. Ganian, R.: Thread graphs, linear rank-width and their algorithmic applications. In: Iliopoulos, C.S., Smyth, W.F. (eds.) IWOCA 2010. LNCS, vol. 6460, pp. 38–42. Springer, Heidelberg (2011)
12. Gavoille, C., Paul, C.: Distance labeling scheme and split decomposition. Discrete Math. **273**(1–3), 115–130 (2003)
13. Gioan, E., Paul, C.: Split decomposition and graph-labelled trees: characterizations and fully dynamic algorithms for totally decomposable graphs. Discrete Appl. Math. **160**(6), 708–733 (2012)
14. Jeong, J., Kwon, O.-J., Oum, S.-I.: Excluded vertex-minors for graphs of linear rank-width at most k. In: Portier, N., Wilke, T. (eds.) STACS. LIPIcs, vol. 20, pp. 221–232. Schloss Dagstuhl - Leibniz-Zentrum fuer Informatik (2013)
15. Kloks, T., Bodlaender, H.L., Müller, H., Kratsch, D.: Computing treewidth and minimum fill-in: All you need are the minimal separators. In: Lengauer, T. (ed.) ESA 1993. LNCS, vol. 726, pp. 260–271. Springer, Heidelberg (1993)
16. Megiddo, N., Louis Hakimi, S., Garey, M.R., Johnson, D.S., Papadimitriou, C.H.: The complexity of searching a graph. J. ACM **35**(1), 18–44 (1988)
17. Oum, S.: Rank-width and vertex-minors. J. Comb. Theory, Ser. B **95**(1), 79–100 (2005)
18. Oum, S., Seymour, P.D.: Approximating clique-width and branch-width. J. Comb. Theory, Ser. B **96**(4), 514–528 (2006)

Vertex Contact Graphs of Paths on a Grid

Nieke Aerts[✉] and Stefan Felsner

Institut Für Mathematik, Technische Universität Berlin, Berlin, Germany
{aerts,felsner}@math.tu-berlin.de

Abstract. We study Vertex Contact representations of Paths on a Grid (VCPG). In such a representation the vertices of G are represented by a family of interiorly disjoint grid-paths. Adjacencies are represented by contacts between an *endpoint* of one grid-path and an *interior point* of another grid-path. Defining $u \to v$ if the path of u ends on path of v we obtain an orientation on G from a VCPG. To get hand on the bends of the grid path the orientation is not enough. We therefore consider pairs (α, ψ): a 2-orientation α and a flow ψ in the angle graph. The 2-orientation describes the contacts of the ends of a grid-path and the flow describes the behavior of a grid-path between its two ends. We give a necessary and sufficient condition for such a pair (α, ψ) to be realizable as a VCPG.

Using realizable pairs we show that every planar $(2,2)$-tight graph admits a VCPG with at most 2 bends per path and that this is tight. Using the same we show that simple planar $(2,1)$-sparse graphs have a 4-bend representation and simple planar $(2,0)$-sparse graphs have 6-bend representation. We do not believe that the latter two are tight, we conjecture that simple planar $(2,0)$-sparse graphs have a 3-bend representation.

1 Introduction

Outline of results. In this paper we consider the question whether a planar graph G admits a VCPG, i.e. a Vertex Contact representation of Paths on a Grid. In such a representation the vertices are represented by a family of interiorly disjoint grid-paths. An *endpoint* of one grid-path coincides with an *interior point* of another grid-path if and only if the two represented vertices are adjacent.

A VCPG induces a unique orientation of the edges of G: Orienting the edge uv as $u \to v$ if the grid-path of u ends on grid-path of v we obtain an orientation of G. As each grid-path has two ends, in the induced orientation each vertex has outdegree at most two. We denote such an orientation simply *2-orientation*.

On the other hand, for a planar graph, every 2-orientation induces a VCPG (Sect. 1.1). However, a 2-orientation of G defines the representation of the edges in a VCPG but not how the grid-paths behave (e.g. how many bends a grid-path has). To get a description of the behavior of the grid-paths between its endpoints, we introduce a flow network in the angle graph (Sect. 2.1).

To obtain a full combinatorial description of a VCPG we consider a pair (α, ψ): a 2-orientation α in the graph and a flow ψ in the angle graph. Our main

© Springer International Publishing Switzerland 2014
D. Kratsch and I. Todinca (Eds.): WG 2014, LNCS 8747, pp. 56–68, 2014.
DOI: 10.1007/978-3-319-12340-0_5

contribution is a necessary and sufficient condition for such a pair (α, ψ) to be realizable as a VCPG. We will then use such realizable pairs to give bounds on the number of bends needed for certain graph classes.

When the number of bends of each path is at most k we denote the representation by B_k-VCPG and when every path has precisely k bends we speak about strict B_k-VCPG.

Related results. Contact and intersection representations of graphs and particularly of planar graphs have been studied for decades. The by now best known result in the area may be the Koebe-Andreev-Thurston circle packing theorem. A more recent highlight in the area is a result of Chalopin and Gonçalves: every planar graph is an intersection graph of segments in the plane. This boosted the study of intersection and contact graphs of restricted classes of curves. In [9] Kobourov, Ueckerdt and Verbeek show that all planar Laman graphs admit an L-contact representation, i.e. a strict B_1-VCPG. A graph $G = (V, E)$ is Laman if $|E| = 2|V| - 3$ and every subset of k vertices induces at most $2k - 3$ edges. It is immediate that every subgraph of a planar Laman graph also admits a strict B_1-VCPG. There are graphs that are not Laman that admit a strict B_1-VCPG, e.g. K_4. In [9] the question was posed which conditions are necessary and sufficient for a graph to have such a representation.

Vertex intersection graphs of paths on a grid (VPG-graphs) have been investigated by Asinowski et al. [2]. They showed that all planar graphs are B_3-VPG, i.e., each vertex is represented by a path with at most three bends and the edges are intersections of two grid-paths. They conjectured that this bound was tight. Chaplick and Ueckerdt disproved this by showing that every planar graph is B_2-VPG [3].

In orthogonal graph drawing there have been many results on minimizing bend numbers, i.e. vertices are points in the plane and edges are grid-paths between these points and the number of bends is minimized. Note that in this setting vertices have at most degree four, or as a workaround, the vertices can be represented as boxes. An early result of Tamassia gives an algorithm to obtain an orthogonal drawing with minimal bend number which preserves the embedding [11]. Optimizing the bend number locally (for each path) has gotten much attention too, Schäffter gives an algorithm to draw 4-regular graphs in a grid with at most two bends per edge (which is tight when not restricted to planar graphs) [10]. For orthogonal drawings without degree restriction, Fößmeier, Kant and Kaufmann have shown that every plane graph has an orthogonal drawing preserving the embedding with at most one bend per edge [7].

Outline of the paper. The remainder of this Section we will give the definitions and show some necessary conditions based on 2-orientations. In Sect. 2 we will introduce the flow network. We then give the necessary and sufficient condition for a pair, a 2-orientation and a flow, to be realizable as a VCPG. In Sects. 3 and 4 we show how to use realizable pairs to give bounds on the number of bends in a VCPG.

1.1   Preliminaries: On $(2, l)$-Sparse Graphs

In an orientation that is induced by a VCPG, all edges of a graph $G = (V, E)$ are oriented and each vertex has outdegree at most two. Therefore the number of edges of G is at most twice the number of vertices: $|E| \leq 2|V|$. Moreover, this bound must hold for all induced subgraphs as well.

Definition 1 (Sparse and Tight Graphs). *Let $G = (V, E)$ a graph and $k, l \geq 0$ integers. If $\forall W \subseteq V : |E_W| \leq k|W| - l$, where E_W is the set of edges induced by W, then G is called (k, l)-sparse. If also $|E| = k|V| - l$ holds, the graph is called (k, l)-tight.*

Graphs that admit a VCPG must be planar and $(2, 0)$-sparse. In this paper we focus on $(2, 0)$-tight graphs, $(2, 1)$-tight graphs and $(2, 2)$-tight graphs which are simple and planar. Note that for every $(2, l)$-sparse graph H there exists a $(2, l)$-tight graph G such that H is a subgraph of G.

First we show that every planar $(2, l)$-tight graph, $l \geq 0$, admits a VCPG.

Lemma 1. *Every planar $(2, l)$-tight graph has a 2-orientation.*

Proof. Let $G = (V, E)$ a planar $(2, l)$-tight graph. Suppose there is a subset W of the vertices of G that has less than $2|W|$ incident edges. Then $G[V - W]$ must induce at least $2|V| - l - (2|W| - 1) = 2|V - W| - l + 1$ edges, which contradicts $(2, l)$-tightness. Hence every subset W of the vertices of G has at least $2|W|$ incident edges. Now we construct a bipartite graph B. The first vertex class, V_1, consists of two copies of the vertices of G. When $l \neq 0$, l copies of one vertex are removed from V_1. The second class, V_2, contains all the edges of G. The edge set of B is defined by the incidences in G. By $(2, l)$-tightness of G the graph B satisfies Hall's marriage condition and hence it has a perfect matching. A perfect matching defines a 2-orientation of G.   □

When $l \neq 0$ there must be at least one vertex that has outdegree less than 2. An end of a grid-path that does not end on another grid-path is denoted by *free end*. Let v a vertex that has outdegree $p < 2$ in the 2-orientation that comes from a VCPG R. Then the grid-path that represents v in R has $2 - p$ free ends. Free ends require special attention. Therefore we consider a 2-orientation such that the vertices with outdegree less than 2 are on the boundary of the outer face.

When a planar graph has a 2-orientation it easily follows that it has a VCPG.

Lemma 2. *Let $l \geq 0$. Every planar $(2, l)$-tight graph admits a VCPG.*

Proof. Consider an embedding of a planar $(2, l)$-tight graph G and a 2-orientation α of G. Subdivide each loop twice. Every pair of vertices is connected by at most two edges (since the graph is $(2, l)$-tight and $l \geq 2$) and if so one of the multiple edges is subdivided. The result is a simple plane graph, which has a straight line drawing by Fàry's theorem. Replace each straight line edge in such a drawing by an axis-aligned grid-path leaving the start and endpoint intact and such that two grid-paths starting in the same point only coincide in this point. The subdivided edges are merged without changing the grid-paths. A vertex is identified with its outgoing edge(s). The last step is to perturb the last straight part of a grid-path p_v that ends on a grid-path p_w in such a way that this point is not used by any grid-path other than p_v and p_w. This procedure gives a VCPG of G that realizes the chosen embedding.   □

An obvious question is: how many bends are needed in a VCPG of a certain graph. As mentioned before, Kobourov, Ueckerdt and Verbeek have shown that all planar $(2,3)$-tight graphs (Laman graphs) admit a strict B_1-VCPG [9]. They also gave the following bound on the number of edges of a graph that admits a (strict) B_1-VCPG.

Proposition 1. *If $G = (V, E)$ admits a B_1-VCPG then*

$$\forall W \subseteq V : |E_W| \leq 2|W| - 2 . \tag{1}$$

Proof: cf. appendix.

Therefore the candidate graphs that admit a B_1-VCPG are planar $(2,2)$-sparse graphs. In this paper we show that planar $(2,2)$-tight graphs admit a B_2-VCPG and that this is tight. Thus condition (1) is necessary but not sufficient. The proof is based on realizable pairs.

2   Realizable Pairs

A VCPG is not completely described by a plane graph G and a 2-orientation. Therefore we introduce a flow network. We will use a flow in such a network to obtain a full description of a VCPG. We denote the 2-orientation by α and the flow by ψ. In this section we identify a property of a pair (α, ψ) that comes from a VCPG of a plane graph G, hence this property is necessary. On the other hand, not every pair (α, ψ) on G induces a VCPG of G (an example is shown in Fig. 1(c)). We call a pair (α, ψ) *realizable* when it does. We will prove that the necessary property is also sufficient, hence realizable pairs are in bijection to VCPGs. Our proof method is algorithmic, it shows how one can construct a VCPG (the geometric setting) from a realizable pair (the combinatorial setting).

2.1   The Flow Network

From here on, we consider the graph to be simple and 2-connected. Note that any $(2, l)$-tight graph can easily be extended to a 2-connected $(2, l)$-tight graph by adding an appropriate number of degree two vertices. The angle graph $A(G)$ of a plane 2-connected graph G is a plane bipartite graph that arises from G by setting the union of the vertices and faces of G as the vertices of $A(G)$ and the edges of $A(G)$ are the pairs vf, $v \in V(G), f \in F(G)$, such that v is a vertex on f in G. The angle graph is a plane maximal bipartite graph.

Intuitively, a unit of flow in the angle graph from f_1 to f_2 through v is a bend of p_v (the grid-path that represents v) such that the convex angle of this bend

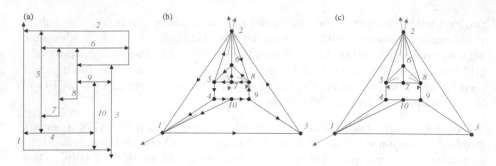

Fig. 1. Given a graph G, the figure shows (a) a VCPG of G, the arrows correspond to the orientation of the edges; (b) the agreeing realizable pair, the orientation is given on the edges, the flow is given by the red arrows between faces through vertices; (c) a flow ψ such that there is no 2-orientation α for which (α, ψ) is realizable, as the orientation must orient $5 \rightarrow 7$ and $8 \rightarrow 7$ hence 7 can only have outdegree one (Color figure online).

lies in f_1 and the concave angle lies in f_2 (see e.g. Fig. 1(a) and (b)). More precise, a flow ψ is a weighted directed graph, with as underlying graph the angle graph $A(G)$. The face-vertices of $A(G)$ can be a source or a sink, depending on the degree. The vertex-vertices of $A(G)$ are neither sources nor sinks. The capacity of the edges is unbounded. The number of bends prescribed by the flow ψ for a vertex v is denoted with $\psi(v)$, which is the sum of incoming flow, which in turn is equal to the sum of outgoing flow. We define $c(f)$ to be the excess of an interior face-vertex f of $A(G)$. The excess prescribes the amount of outgoing flow minus the amount of incoming flow of this face.

Following the boundary of an interior region of a VCPG and adding the changes in direction one should obtain 2π. Each edge is represented as a proper contact and therefore changes the direction with $\pi/2$. A convex angle changes the direction with $\pi/2$ as well and a concave angle changes the direction with $-\pi/2$. By the following equation: $2\pi = \pi/2 \cdot |f| + \pi/2 \cdot c(f)$ where $|f|$ is the number of edges on the boundary of f, the excess of each interior face is:

$$c(f) = 4 - |f|. \tag{2}$$

For the outer face f_∞ of a $(2, l)$-tight graph we set the excess $c(f_\infty) = (2l - 4) - |f_\infty|$. With $|f_\infty|$ we denote the number of vertices on the outer face.

Let ψ be a flow in $A(G)$ that satisfies the excess of each face, then the value of the flow ψ is

$$w(\psi) = \sum_{v \in V(G)} \psi(v). \tag{3}$$

The sum of the excess over all faces cancels out and there is no capacity restraint on the edges, therefore there exists a flow that satisfies the excess of every face. A vertex cannot absorb any flow, as having a convex corner means having a concave corner on the other side. Therefore the minimum value of a flow that satisfies the facial excesses is a lower bound on the number of bends needed for a VCPG[1]. As shown by Fig. 1(c) not every flow that satisfies the facial excesses is related to a VCPG.

Necessary and Sufficient Condition. Given a simple, plane, 2-connected $(2, l)$-tight graph, a 2-orientation α and a flow ψ that satisfies the facial excesses. We will give a necessary and sufficient condition on the pair (α, ψ). When this condition is satisfied, there exists a VCPG that maintains the embedding such that:

(a) The grid-path of u ends on the grid-path of v if and only if the edge uv is oriented from u to v in α, and,
(b) The grid-path of v has precisely $\psi(v)$ bends.

We denote a pair that satisfies the condition *realizable*. Let $A[N_{A(G)}[v]]$ denote the angle graph induced by the closed neighborhood of a vertex v, i.e. induced by v and all its neighbors in $A(G)$.

Let n_1, n_2 be the neighbors of v along the outgoing edges of v in α. If v has outdegree 0, then n_1, n_2 are its neighbors on the outer face. If v has outdegree 1, then n_1 is its neighbor along the outgoing edge. The vertex n_2 is the neighbor of v on the outer face, chosen such that the units of flow are equally distributed on the clockwise and counterclockwise side of the path n_1, v, n_2. Informally, a unit of flow through a vertex v represents a bend of the grid-path of v. Following the grid-path from n_1 to n_2, looking left and right, the bends are met at the same time. This implies that the flow through a vertex must be laminar, i.e., non-crossing.

Definition 2 (Realizability Condition). *The pair (α, ψ) satisfies the realizability condition at vertex v if and only if, given $A[N_{A(G)}[v]]$ and the flow in this subgraph, (see Fig. 3)*

(a) There is a decomposition of the flow into non-crossing paths, and,
(b) Every path of such a decomposition crosses the path n_1, v, n_2.

When the pair (α, ψ) satisfies the realizability condition at each vertex we say that the pair is realizable.

Theorem 1. *The realizable pairs are in bijection with VCPGs.*

The remainder of this section is dedicated to the proof of Theorem 1. First we will show that a VCPG induces a realizable pair (α, ψ) and then the converse, i.e. we will construct a VCPG from a realizable pair.

Lemma 3. *A pair (α, ψ) that comes from a VCPG is realizable.*

---

[1] Note that there might be different bounds for different embeddings of a graph.

Proof. First note that a VCPG of G describes an embedding of G. If there is a grid-path with one free end, then before proceeding we reduce all unnecessary bends, i.e. if a grid-path has bends between its last neighbor and its free end, these bends are removed. A 2-orientation can be constructed from a VCPG by orienting an edge $u \to v$ if and only if the grid-path of u ends on the grid-path of v. Consider the grid-path that represents a vertex v. If this path has no bends, the realizability condition is satisfied at this vertex. Suppose the path has k bends. Draw an arrow from the face containing a convex corner to the face in which the associated concave corner lies. Now the set of arrows represents the flow $\psi(v)$. This flow is non-crossing through v and every unit of flow is cut by the grid-path of v. When these arrows are introduced for all bends of all grid-paths, the flow given by these arrows satisfies the excess of each face. Contract the strictly interior steps of the grid-path to a vertex and every unit of the non-crossing flow through v is now cut by the outermost two segments of the grid-path, which correspond to the outgoing edges of v, or to the outgoing edge and the location of the last incoming edge before the free end of the grid-path. Hence the realizability condition is satisfied at each vertex, therefore the pair obtained from the VCPG is realizable.   □

To prove the converse we will show how to construct a VCPG given an realizable pair. Note that an embedding follows from the map $A(G)$ (in which the flow ψ is defined). Consider a realizable pair (α, ψ) (see Fig. 2(a)). The proof consists of four steps, which we first outline here:

Step 1: First we expand the vertices that have k units of flow going through them, to a path of length k. We obtain a bipartite graph.

Step 2: We introduce help edges and vertices in the bipartite graph to construct a quadrangulation (see Fig. 2(b)).

Step 3: We then find a segment contact representation of the quadrangulation. It has been shown that the 2-orientations of maximal bipartite planar graphs are in bijection with separating decompositions of this graph (e.g. [4]). In turn a separating decomposition induces a segment contact representation (cf. [5,6,8]). Hence we can construct a segment contact representation where the representation of the edges is in bijection with the given 2-orientation. An example is shown in Fig. 2(c).

Step 4: Last we will show that the extra edges that have been introduced to make a quadrangulation of the bipartite graph can be deleted in order to obtain a VCPG of G (see Fig. 2(d)).

Step 1. Given a realizable pair (α, ψ) for G. We expand all vertices with non-zero flow. The plane graph we obtain is denoted by $\tilde{G}$. For every vertex v for which $\psi(v) \neq 0$, expanding v denotes the following steps (see Fig. 3):

1. Expand v to a circle, we will denote this the *bag* of v.
2. Between the two outgoing edges of v, or the special incoming and the outgoing edge of v if v has outdegree one, inside the circle, add a path with $\psi(v) + 1$ vertices.

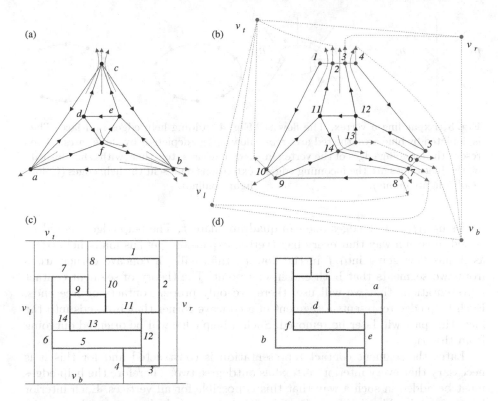

Fig. 2. From a realizable pair to a VCPG: (a) a plane $(2,0)$-tight graph with a realizable pair (ψ in red); (b) expanding the vertices according to the flow (in blue) and extending the bipartite graph to a quadrangulation (in green); (c) a segment contact representation of the quadrangulation with the segments that belong to the original graph highlighted; (d) a VCPG (Color figure online).

3. Connect the edges that end on the circle to the path vertex in such a way that the flow between two faces only crosses an edge of the path.

Step 2. After all the expansions have been done, we obtain a graph where all faces have even length. Each face gets $|4 - |f|| + 2k$ extra vertices due to the expanding step, where k is the amount of flow proceeding through the face. The resulting faces in $\tilde{G}$ have size $|f| + |4 - |f|| + 2k$, for $|f| = 3$ this gives $4 + 2k$ and for $|f| > 3$ this gives $2|f| - 4 + 2k$, both are even. So in $\tilde{G}$ all faces have even length and therefore $\tilde{G}$ is a bipartite graph. Now we add help edges to extend $\tilde{G}$ to a quadrangulation. We denote the quadrangulation G_Q. We will also orient the new edges to obtain a 2-orientation of G_Q. In order to explain how the help edges are added, we need the following lemma.

Lemma 4. *Every interior face $\tilde{f}$ of $\tilde{G}$ has $(|\tilde{f}| - 4)/2$ units of incoming flow.*

Proof: cf. appendix.

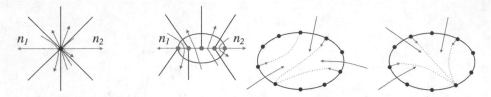

Fig. 3. Expanding a vertex. The flow ψ is depicted by blue arcs, the red arcs represent the outgoing edges of the vertex in α, the black lines are the incoming edges (Color figure online).

Fig. 4. Adding help edges in a face. The flow ψ is depicted by blue arcs. The red half-arcs together with the dashed extensions represent the help edges (Color figure online).

Using $(|\tilde{f}| - 4)/2$ edges one can quadrangulate $\tilde{f}$. The help edges should be added in such a way that every bag (vertex expansion) gets as many help edges as it has flow going into $\tilde{f}$ in the flow ψ. Informally, a concave corner arises from two segments that both end in one point. The theory of segment contact representations that we will use, there are only proper contacts or free ends. Each help edge represents a segment of a concave corner that proceeds into the face, this part will later be removed. Such a help edge will be oriented outgoing from the bag.

Later the segment contact representation is constructed and for this it is necessary that every interior vertex has outdegree two. Therefore the help-edges must be added in such a way that this is possible for all vertices. Each interior bag vertex should gain precisely two outgoing arcs (which are not edges in the original graph) and the vertices on the boundary of the bag gain precisely one outgoing arc that is not in the original graph. The help edges are added along flow from a vertex into a face, this will give the correct amount of new edges for every bag.

Lemma 5. *Each inner face $\tilde{f}$ of $\tilde{G}$ can be quadrangulated in such a way that each bag through which k units of flow enter $\tilde{f}$ gets k new outgoing arcs. The outer face of $\tilde{G}$ can be quadrangulated using four help-vertices (v_t, v_r, v_b, v_l), in such a way that each bag through which k units of flow enter the outer face gets k new outgoing arcs.*

Proof: cf. appendix.

An example of quadrangulating an interior face is depicted in Fig. 4. The flow ψ is given by the blue arrows. First half-arcs are added, the red solid arcs. Then these half-arcs are subsequently connected in such a way that they close one 4-face (red dashed lines). The quadrangulation of the outer face is based on the same idea.

To obtain a quadrangulation G_Q with a 2-orientation, we still need to orient the edges that are strictly inside the bags and the four boundary edges. The orientation of all other edges comes from α. Each bag b_v contains $|b_v| - 1 = \psi(v)$ edges which are not yet oriented, all others are oriented and such that b_v has outdegree $|b_v| + 1$.

Lemma 6. *Each bag b_v in G_Q has precisely outdegree $|b_v| + 1$ and each vertex $(\in b_v)$ has outdegree at most two.*

Proof: cf. appendix.

Orient $(v_l, v_t), (v_r, v_t), (v_l, v_b)$ and (v_r, v_b) towards v_t, respectively v_b, the two poles of the 2-orientation. The remaining not oriented edges can be oriented greedily.

Lemma 7. *The path edges can be oriented (greedily) such that the resulting orientation is a 2-orientation of G_Q.*

Proof: cf. appendix.

Step 3. From G and the realizable pair (α, ψ) we constructed the quadrangulation G_Q with a 2-orientation $\hat{\alpha}$ (v_t and v_b are the only two vertices with outdegree zero instead of outdegree two). We construct a segment contact representation, i.e. the vertices of the two color classes become horizontal respectively vertical segments and the edges are proper contacts between the segments satisfying $\hat{\alpha}$ (cf. [5,6,8]).

Step 4. Last to show is that this segment representation of G_Q is equivalent to a VCPG of G where the path of a vertex v is given by its outgoing arcs in α and $\psi(v)$ denotes the number of bends of the path of v. For this we need the following lemma, which shows that the sets of segments ending on different sides of a segment s can be moved independently.

Lemma 8. *Given a horizontal segment in a segment contact representation, then its (vertical) bottom neighbors can be shifted independently from its (vertical) top neighbors. The same holds for a vertical segment and its left resp. right incoming neighbors.*

Proof: cf. appendix.

Theorem 2. *The segment representation of G_Q obtained from an realizable pair (α, ψ) induces a VCPG of G.*

Sketch of proof. The complete proof is moved to the appendix. Consider a segment representation of G_Q.

First identify the grid-path of each vertex v of the original graph by highlighting the parts of the segments that belong to v. Then by shifting we make sure that all edges are contacts on highlighted segments. Deleting the not highlighted segments gives a VCPG of G.

With the four steps we have obtained a VCPG from an realizable pair. This completes the proof of Theorem 1. In the remainder of this paper we will use realizable pairs to give bounds on the number of bends for certain graph classes.

3   Not All $(2, 2)$-Tight Graphs Admit a B_1-VCPG

In a B_1-VCPG each vertex is represented by a grid-path with at most one bend. Strict B_1-VCPG is a proper subclass of B_1-VCPG (this is shown in the full

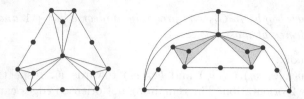

Fig. 5. Two embeddings of a planar $(2, 2)$-tight graph that does not admit a B_1-VCPG.

version of this paper [1]). Suppose a graph admits a B_1-VCPG, then there must be a realizable pair (α, ψ) such that $0 \leq \psi(v) \leq 1$ for all vertices v. Consider the plane graph on the right of Fig. 5. The two grey-colored K_4 subgraphs both need at least three units of flow leaving the K_4. So there must be 6 units of flow going out, but there are only 5 different vertices bounding the two K_4 subgraphs. We conclude that there is no flow that satisfies all excesses such that there is at most one unit of flow going through each vertex.

4   Locally Minimizing Bends

For simple $(2, 2)$-tight graphs we show that for every 2-orientation α (with sink(s) on the outer face), there exists a flow ψ such that $\psi(v) \leq 2$ for all vertices v and the pair (α, ψ) is realizable. Sinks are the vertices with outdegree less than two.

Theorem 3. *Every planar (2, 2)-tight graph admits a B_2-VCPG.*

Proof. Let G a planar $(2,2)$-tight graph and $\mathcal{E}$ a planar embedding of G. Hence we have a dual graph G^*. The excess of the face-vertices is given by $c(f) = 4 - |f|$ and for the outer face it is $c(f_\infty) = -|f_\infty|$. Consider a flow in the dual graph where each edge has capacity one and at each face the excess is satisfied, then given any orientation of the edge, this flow can be translated into a flow in the angle graph through the vertex for which this edge is outgoing.

We will show that for every subset of face-vertices H there are at least $|\sum_{f \in H} c(f)|$ edges leaving H in the dual graph. Let b the number of edges leaving H, i.e. the number of boundary edges in the primal w.r.t. H and e_H the number of edges induced by H. Note that we only count interior faces (when we use Euler's formula).

$$|\sum_{f \in H} c(f)| = |\sum_{f \in H} 4 - |f|| = |4|H| - 2e_H + b| \leq b \qquad (4)$$

Hence we can satisfy all excesses in the dual graph by using every edge at most once. Let ϕ be such an edge-disjoint flow in the dual graph. Consider any orientation α of G such that every vertex has outdegree 2 except for one sink, and such that the sink is on the outer face. We construct the flow ψ in the angle graph as follows: If there is a flow in ϕ from f_1 to f_2 crossing edge uv, then if $u \to v$ we add $f_1 \to u$ and $u \to f_2$ to ψ. If $v \to u$ we add $f_1 \to v$ and $v \to f_2$ to ψ.

Since ϕ is edge-disjoint and each vertex has at most two outgoing edges we have $\psi(v) \leq 2$ for all vertices v. At each vertex the flow cuts off the outgoing edge, hence the realizability condition is satisfied at each vertex.

We conclude that the pair (α, ψ) is realizable, which concludes the proof. □

Similarly we obtain the following results (cf. appendix for the proofs).

Theorem 4. *Every simple planar (2, 1)-tight graph admits a B_4-VCPG.*

Theorem 5. *Every simple planar (2, 0)-tight graph admits a B_6-VCPG. If the graph has no separating triangle it admits a B_4-VCPG.*

5   Conclusion

We have shown that there exist planar $(2, 2)$-tight graphs that do not admit a B_1-VCPG. In such a graph there are a number of tight subsets V_i of the vertices, i.e. the induced graphs have $2|V_i| - 2$ edges, which have precisely one vertex in common. Do all planar $(2, 2)$-tight graphs that have no vertex in the intersection of different tight sets admit a B_1-VCPG?

We have obtained bounds for simple $(2, 1)$-tight and $(2, 0)$-tight planar graphs, however we believe that these bounds are not tight. Three bends are necessary, as shown by the octahedron minus one edge and the octahedron respectively.

Conjecture 1. Simple planar $(2, 0)$-tight graphs admit a B_3-VCPG.

The bounds that we have shown do not depend on a chosen 2-orientation (i.e. the bounds hold for every 2-orientation). There are easy examples of a $(2, 0)$-tight graph with a particular 2-orientation such that there is a vertex represented by a 4-bend path for every flow. Hence it would be interesting to find a sufficient condition on a flow such that, when satisfied, there exists a 2-orientation such that the pair is realizable. Is there a way to construct an realizable pair simultaneously? Is there a way to find that minimal flow that belongs to an realizable pair?

References

1. Aerts, N., Felsner, S.: Vertex Contact graphs of Paths on a Grid. http://page.math.tu-berlin.de/~aerts/pubs/vcpg.pdf
2. Asinowski, A., Cohen, E., Golumbic, M.C., Limouzy, V., Lipshteyn, M., Stern, M.: Vertex intersection graphs of paths on a grid. J. Graph Algorithms Appl. **16**, 129–150 (2012)
3. Chaplick, S., Ueckerdt, T.: Planar graphs as VPG-graphs. J. Graph Algorithms Appl. **17**, 475–494 (2013)
4. de Fraysseix, H., de Mendez, P.O.: On topological aspects of orientations. Discrete Math. **229**, 57–72 (2001)
5. de Fraysseix, H., de Mendez, P.O., Pach, J.: A left-first search algorithm for planar graphs. Discrete Comput. Geom. **13**, 459–468 (1995)

6. Felsner, S.: Rectangle and square representations of planar graphs. In: Pach, J. (ed.) Thirty Essays on Geometric Graph Theory, pp. 213–248. Springer, New York (2013)
7. Fößmeier, U., Kant, G., Kaufmann, M.: 2-visibility drawings of planar graphs. In: North, Stephen C. (ed.) GD 1996. LNCS, vol. 1190, pp. 155–168. Springer, Heidelberg (1997)
8. Hartman, I.B.-A., Newman, I., Ziv, R.: On grid intersection graphs. Discrete Math. **87**, 41–52 (1991)
9. Kobourov, S.G., Ueckerdt, T., Verbeek, K.: Combinatorial and geometric properties of planar laman graphs. In: Khanna, S. (ed.) SODA, pp. 1668–1678. SIAM (2013)
10. Schäffter, M.W.: Drawing graphs on rectangular grids. Discrete Appl. Math. **63**, 75–89 (1995)
11. Tamassia, R.: On embedding a graph in the grid with the minimum number of bends. SIAM J. Comput. **16**, 421–444 (1987)

Deciding the Bell Number
for Hereditary Graph Properties
(Extended Abstract)

Aistis Atminas[✉], Andrew Collins, Jan Foniok, and Vadim V. Lozin

DIMAP and Mathematics Institute, University of Warwick, Coventry CV4 7AL, UK
A.Atminas@warwick.ac.uk

Abstract. The paper [J. Balogh, B. Bollobás, D. Weinreich, A jump to the Bell number for hereditary graph properties, *J. Combin. Theory Ser. B* 95 (2005) 29–48] identifies a jump in the speed of hereditary graph properties to the Bell number B_n and provides a partial characterisation of the family of minimal classes whose speed is at least B_n. In the present paper, we give a complete characterisation of this family. Since this family is infinite, the decidability of the problem of determining if the speed of a hereditary property is above or below the Bell number is questionable. We answer this question positively for properties defined by finitely many forbidden induced subgraphs. In other words, we show that there exists an algorithm which, given a finite set $\mathcal{F}$ of graphs, decides whether the speed of the class of graphs containing no induced subgraphs from the set $\mathcal{F}$ is above or below the Bell number.

Keywords: Hereditary class of graphs · Speed of hereditary properties · Bell number · Decidability

1 Introduction

A *graph property* (or a *class of graphs*[1]) is a set of graphs closed under isomorphism. Given a property $\mathcal{X}$, we write $\mathcal{X}_n$ for the number of graphs in $\mathcal{X}$ with vertex set $\{1, 2, \ldots, n\}$ (that is, we are counting *labelled* graphs). Following [5], we call $\mathcal{X}_n$ the *speed* of the property $\mathcal{X}$.

A property is *hereditary* if it is closed under taking induced subgraphs. It is well-known (and can be easily seen) that a graph property $\mathcal{X}$ is hereditary if and only if $\mathcal{X}$ can be described in terms of forbidden induced subgraphs. More formally, for a set $\mathcal{F}$ of graphs we write Free($\mathcal{F}$) for the class of graphs containing no induced subgraph isomorphic to any graph in the set $\mathcal{F}$. A property $\mathcal{X}$ is

---

This research was supported by DIMAP: the Centre for Discrete Mathematics and its Applications at the University of Warwick, and by EPSRC, grant EP/I01795X/1.
[1] Throughout the paper we use the two terms – graph property and class of graphs – interchangeably.

D. Kratsch and I. Todinca (Eds.): WG 2014, LNCS 8747, pp. 69–80, 2014.
DOI: 10.1007/978-3-319-12340-0_6

hereditary if and only if $\mathcal{X} = \text{Free}(\mathcal{F})$ for some set $\mathcal{F}$. We call $\mathcal{F}$ a set of *forbidden induced subgraphs* for $\mathcal{X}$ and say that graphs in $\mathcal{X}$ are $\mathcal{F}$-*free*.

The speeds of hereditary properties and their asymptotic structure have been extensively studied, originally in the special case of a single forbidden subgraph [9–11,13–15], and more recently in general [1,4–7,16]. These studies showed, in particular, that there is a certain correlation between the speed of a property $\mathcal{X}$ and the structure of graphs in $\mathcal{X}$, and that the rates of the speed growth constitute discrete layers. The first four lower layers have been distinguished in [16]: these are constant, polynomial, exponential, and factorial layers. In other words, the authors of [16] showed that some classes of functions do not appear as the speed of any hereditary property, and that there are discrete jumps, for example, from polynomial to exponential speeds.

Independently, similar results were obtained by Alekseev in [2]. Moreover, Alekseev provided the first four layers with the description of all minimal classes, that is, he identified in each layer the family of all classes every proper hereditary subclass of which belongs to a lower layer (see also [5] for some more involved results). In each of the first four lower layers the set of minimal classes is finite and each of them is defined by finitely many forbidden induced subgraphs. This provides an efficient way of determining whether a property $\mathcal{X}$ belongs to one of the first three layers.

One more jump in the speed of hereditary properties was identified in [7] and it separates – within the factorial layer – the properties with speeds strictly below the Bell number B_n from those whose speed is at least B_n. The importance of this jump is due to the fact that all the properties below the Bell number are well-structured. In particular, all of them have bounded clique-width [3] and all of them are well-quasi-ordered by the induced subgraph relation [12]. From the results in [5,12] it follows that every hereditary property below the Bell number can be characterised by finitely many forbidden induced subgraphs and hence the membership problem for each of them can be decided in polynomial time.

Even so, very little is known about the boundary separating the two families, that is, very little is known about the *minimal* classes on or above the Bell number. Paper [7] distinguishes two cases in the study of this question: the case where a certain parameter associated with each class of graphs is finite and the case where this parameter is infinite. In the present paper, we call this parameter *distinguishing number*. For the case where the distinguishing number is infinite, [7] provides a complete description of minimal classes, of which there are precisely 13. For the case where the distinguishing number is finite, [7] mentions only one minimal class above the Bell number (linear forests) and leaves the question of characterising other minimal classes open.

In the present paper, we give a complete answer to the above open question: we provide a structural characterisation of all minimal classes above the Bell number with a finite distinguishing number. This family of minimal classes is infinite, which makes the problem of deciding whether a hereditary class is above or below the Bell number questionable. Nevertheless, for properties defined by *finitely many* forbidden induced subgraphs, our characterisation allows us to prove decidability of this problem: we show that there exists an algorithm which, given a finite set $\mathcal{F}$ of graphs, decides whether the class Free($\mathcal{F}$) is above or below the Bell number.

2   Preliminaries and Preparatory Results

2.1   Basic Notation and Terminology

All graphs we consider are undirected without multiple edges. The graphs in our hereditary classes have no loops; however, we allow loops in some auxiliary graphs, called "density graphs" and denoted usually by H, that are used to represent the global structure of our hereditary classes.

If G is a graph, $V(G)$ stands for its vertex set, $E(G)$ for its edge set and $|G|$ for the number of vertices (the *order*) of G. The edge joining two vertices u and v is uv (we do not use any brackets); uv is the same edge as vu.

If $W \subseteq V(G)$, then $G[W]$ is the subgraph of G induced by W. For W_1, W_2 disjoint subsets of $V(G)$ we define $G[W_1, W_2]$ to be the bipartite subgraph of G with vertex set $W_1 \cup W_2$ and edge set $\{uv : u \in W_1,\ v \in W_2,\ uv \in E(G)\}$. The *bipartite complement* of $G[W_1, W_2]$ is the bipartite graph in which two vertices $u \in W_1$, $v \in W_2$ are adjacent if and only if they are not adjacent in $G[W_1, W_2]$.

The *neighbourhood* $N(u)$ of a vertex u in G is the set of all vertices adjacent to u, and the *degree* of u is the number of its neighbours. Note that if (and only if) there is a loop at u then $u \in N(u)$.

As usual, P_n, C_n and K_n denote the path, the cycle and the complete graph with n vertices, respectively. Furthermore, $K_{1,n}$ is a star (i.e., a tree with $n+1$ vertices one of which has degree n), and $G_1 + G_2$ is the disjoint union of two graphs. In particular, mK_n is the disjoint union of m copies of K_n.

A *forest* is a graph without cycles, i.e., a graph every connected component of which is a tree. A *star forest* is a forest every connected component of which is a star, and a *linear forest* is a forest every connected component of which is a path.

A *quasi-order* is a binary relation which is reflexive and transitive. A *well-quasi-order* is a quasi-order which contains neither infinite strictly decreasing sequences nor infinite antichains (sets of pairwise incomparable elements). That is, in a well-quasi-order any infinite sequence of elements contains an infinite increasing subsequence.

Recall that the Bell number B_n, defined as the number of ways to partition a set of n labelled elements, satisfies the asymptotic formula $\ln B_n / n = \ln n - \ln \ln n + \Theta(1)$.

Balogh, Bollobás and Weinreich [7] showed that if the speed of a hereditary graph property is at least $n^{(1-o(1))n}$, then it is actually at least B_n; hence we call any such property a *property above the Bell number*. Note that this includes hereditary properties whose speed is exactly equal to the Bell numbers (such as the class of disjoint unions of cliques).

2.2 (ℓ, d)-graphs and Sparsification

Given a graph G and two vertex subsets $U, W \subset V(G)$, define $\Delta(U, W) = \max\{|N(u) \cap W|, |N(w) \cap U| : u \in U, w \in W\}$. With $\overline{N}(u) = V(G) \backslash (N(u) \cup \{u\})$, let $\overline{\Delta}(U, W) = \max\{|\overline{N}(u) \cap W|, |\overline{N}(w) \cap U| : w \in W, u \in U\}$. Note that $\Delta(U, U)$ is simply the maximum degree in $G[U]$.

Definition 2.1. *Let G be a graph. A partition $\pi = \{V_1, V_2, \ldots, V_{\ell'}\}$ of $V(G)$ is an (ℓ, d)-partition if $\ell' \leq \ell$ and for each pair of not necessarily distinct integers $i, j \in \{1, 2, \ldots, \ell'\}$ either $\Delta(V_i, V_j) \leq d$ or $\overline{\Delta}(V_i, V_j) \leq d$. We call the sets V_i bags. A graph G is an (ℓ, d)-graph if it admits an (ℓ, d)-partition.*

It should be clear that, given an (ℓ, d)-partition $\{V_1, V_2, \ldots, V_{\ell'}\}$ of $V(G)$, for each $x \in V(G)$ and $i \in \{1, 2, \ldots, \ell'\}$ either $|N(x) \cap V_i| \leq d$ or $|\overline{N}(x) \cap V_i| \leq d$. In the former case we say that x is d-*sparse* with respect to V_i and in the latter case we say x is d-*dense* with respect to V_i. Similarly, if $\Delta(V_i, V_j) \leq d$, we say V_i is d-*sparse* with respect to V_j, and if $\overline{\Delta}(V_i, V_j) \leq d$, we say V_i is d-*dense* with respect to V_j. We will also say that the pair (V_i, V_j) is d-*sparse* or d-*dense*, respectively. Note that if the bags are large enough (i.e., $\min\{|V_i|\} > 2d + 1$), the terms d-dense and d-sparse are mutually exclusive.

Definition 2.2. *A strong (ℓ, d)-partition is an (ℓ, d)-partition each bag of which contains at least $5 \times 2^\ell d$ vertices; a strong (ℓ, d)-graph is a graph which admits a strong (ℓ, d)-partition.*

Given any strong (ℓ, d)-partition $\pi = \{V_1, V_2, \ldots, V_{\ell'}\}$ we define an equivalence relation $\sim$ on the bags by putting $V_i \sim V_j$ if and only if for each k, either V_k is d-dense with respect to both V_i and V_j, or V_k is d-sparse with respect to both V_i and V_j. Let us call a partition π *prime* if all its $\sim$-equivalence classes are of size 1. If the partition π is not prime, let $p(\pi)$ be the partition consisting of unions of bags in the $\sim$-equivalence classes for π.

    In the full version of this paper we prove that the partition $p(\pi)$ of a strong (ℓ, d)-graph is an $(\ell, \ell d)$-partition whose dense (sparse) pairs correspond to the dense (sparse) pairs of π, and that it does not depend on the choice of a strong (ℓ, d)-partition π:

Theorem 2.3. *Let G be a strong (ℓ, d)-graph with strong (ℓ, d)-partitions π and π'. Then $p(\pi) = p(\pi')$.* $\square$

With any strong (ℓ, d)-partition $\pi = \{V_1, V_2, \ldots, V_{\ell'}\}$ of a graph G we can associate a *density graph* (with loops allowed) $H = H(G, \pi)$: the vertex set of H is $\{1, 2, \ldots, \ell'\}$ and there is an edge joining i and j if and only if (V_i, V_j) is a d-dense pair (so there is a loop at i if and only if V_i is d-dense).

    For a graph G, a vertex partition $\pi = \{V_1, V_2, \ldots, V_{\ell'}\}$ of G and a graph with loops allowed H with vertex set $\{1, 2, \ldots, \ell'\}$, we define (as in [5]) the H, π-*transform* $\psi(G, \pi, H)$ to be the graph obtained from G by replacing $G[V_i, V_j]$ with its bipartite complement for every pair (V_i, V_j) for which ij is an edge of H, and replacing $G[V_i]$ with its complement for every V_i for which there is a loop at the vertex i in H.

Moreover, if π is a strong (ℓ, d)-partition we define $\phi(G, \pi) = \psi(G, \pi, H(G, \pi))$; recall that both $p(\pi)$ and $p(\pi')$ are (not necessarily strong) $(\ell, \ell d)$-partitions of G. Note that π is a strong (ℓ, d)-partition for $\phi(G, \pi)$ and each pair (V_i, V_j) is d-sparse in $\phi(G, \pi)$. We now show that the result of this "sparsification" does not depend on the initial strong (ℓ, d)-partition.

Proposition 2.4. *Let G be a strong (ℓ, d)-graph. Then for any two strong (ℓ, d)-partitions π and π', the graph $\phi(G, \pi)$ is identical to $\phi(G, \pi')$.*

Proof. Suppose that $\pi = \{U_1, U_2, \ldots, U_{\hat{\ell}}\}$ and $\pi' = \{V_1, V_2, \ldots, V_{\hat{\ell}'}\}$. By Theorem 2.3, $p(\pi) = p(\pi') = \{W_1, W_2, \ldots, W_{\hat{\ell}''}\}$. Consider two vertices x, y of G. Let i, j, i', j', i'', j'' be the indices such that $x \in U_i$, $x \in V_{i'}$, $x \in W_{i''}$, $y \in U_j$, $y \in V_{j'}$, $y \in W_{j''}$. As the partitions have at least $5 \times 2^\ell d$ vertices in each bag, ℓd-dense and ℓd-sparse are mutually exclusive properties. Hence the pair (U_i, U_j) is d-sparse if and only if $(W_{i''}, W_{j''})$ is ℓd-sparse if and only if $(V_{i'}, V_{j'})$ is d-sparse; and analogously for dense pairs. Therefore xy is an edge of $\phi(G, \pi)$ if and only if it is an edge of $\phi(G, \pi')$. $\qquad\square$

Proposition 2.4 motivates the following definition, originating from [5].

Definition 2.5. *For a strong (ℓ, d)-graph G, its sparsification is $\phi(G) = \phi(G, \pi)$ for any strong (ℓ, d)-partition π of G.*

2.3 Distinguishing Number $k_\mathcal{X}$

Given a graph G and a set $X = \{v_1, \ldots, v_t\} \subseteq V(G)$, we say that the disjoint subsets $U_1, \ldots, U_m$ of $V(G)$ are *distinguished* by X if for each i, all vertices of U_i have the same neighbourhood in X, and for each $i \neq j$, vertices $x \in U_i$ and $y \in U_j$ have different neighbourhoods in X. We also say that X *distinguishes* the sets $U_1, U_2, \ldots, U_m$.

Definition 2.6. *Given a hereditary property $\mathcal{X}$, we define the* distinguishing *number $k_\mathcal{X}$ as follows:*

(a) *If for all $k, m \in \mathbb{N}$ we can find a graph $G \in \mathcal{X}$ that admits some $X \subset V(G)$ distinguishing at least m sets, each of size at least k, then put $k_\mathcal{X} = \infty$.*

(b) *Otherwise, there must exist a pair (k, m) such that any vertex subset of any graph $G \in \mathcal{X}$ distinguishes at most m sets of size at least k. We define $k_\mathcal{X}$ to be the minimum value of k in all such pairs.*

In [5] Balogh, Bollobás and Weinreich show that the speed of any hereditary property $\mathcal{X}$ with $k_\mathcal{X} = \infty$ is above the Bell number. To study the classes with $k_\mathcal{X} < \infty$ in the next sections we will use the following results from their paper:

Lemma 2.7 ([5], **Lemma 27**). *If $\mathcal{X}$ is a hereditary property with finite distinguishing number $k_\mathcal{X}$, then there exist absolute constants $\ell_\mathcal{X}$, $d_\mathcal{X}$ and $c_\mathcal{X}$ such that for all $G \in \mathcal{X}$, the graph G contains an induced subgraph G' such that G' is a strong $(\ell_\mathcal{X}, d_\mathcal{X})$-graph and $|V(G) \backslash V(G')| < c_\mathcal{X}$.* $\qquad\square$

Theorem 2.8 ([5], **Theorem 28**). *Let $\mathcal{X}$ be a hereditary property with $k_\mathcal{X} < \infty$. Then $\mathcal{X}_n \geq n^{(1+o(1))n}$ if and only if for every m there exists a strong $(\ell_\mathcal{X}, d_\mathcal{X})$-graph G in $\mathcal{X}$ such that its sparsification $\phi(G)$ has a component of order at least m.*   □

3   Structure of Minimal Classes Above Bell

In this section, we describe minimal classes with speed above the Bell number. In [7], Balogh, Bollobás and Weinreich characterised all minimal classes with infinite distinguishing number. In Sect. 3.1 we report this result and show additionally that each of these classes can be characterised by finitely many forbidden induced subgraphs. Then in Sect. 3.2 we move on to the case of finite distinguishing number, which had been left open in [7].

3.1   Infinite Distinguishing Number

Theorem 3.1 (Balogh–Bollobás–Weinreich [7]). *Let $\mathcal{X}$ be a hereditary graph property with $k_\mathcal{X} = \infty$. Then $\mathcal{X}$ contains at least one of the following (minimal) classes:*

(a) *the class $\mathcal{K}_1$ of all graphs each of whose connected components is a clique;*

(b) *the class $\mathcal{K}_2$ of all star forests;*

(c) *the class $\mathcal{K}_3$ of all graphs whose vertex set can be split into an independent set I and a clique Q so that every vertex in Q has at most one neighbour in I;*

(d) *the class $\mathcal{K}_4$ of all graphs whose vertex set can be split into an independent set I and a clique Q so that every vertex in I has at most one neighbour in Q;*

(e) *the class $\mathcal{K}_5$ of all graphs whose vertex set can be split into two cliques Q_1, Q_2 so that every vertex in Q_2 has at most one neighbour in Q_1;*

(f) *the class $\mathcal{K}_6$ of all graphs whose vertex set can be split into two independent sets I_1, I_2 so that the neighbourhoods of the vertices in I_1 are linearly ordered by inclusion (that is, the class of all* chain graphs*);*

(g) *the class $\mathcal{K}_7$ of all graphs whose vertex set can be split into an independent set I and a clique Q so that the neighbourhoods of the vertices in I are linearly ordered by inclusion (that is, the class of all* threshold graphs*);*

(h) *the class $\overline{\mathcal{K}_i}$ of all graphs whose complement belongs to $\mathcal{K}_i$ as above, for some $i \in \{1, 2, \ldots, 6\}$ (note that the complementary class of $\mathcal{K}_7$ is $\mathcal{K}_7$ itself).*

Before showing the characterisation of the classes $\mathcal{K}_1$–$\mathcal{K}_6$ in terms of forbidden subgraphs, we introduce some of the less commonly appearing graphs: the *claw $K_{1,3}$, the 3-fan F_3, the diamond K_4^-*, and the *H-graph H_6* (Fig. 1).

Theorem 3.2. *Each of the classes of Theorem 3.1 is defined by finitely many forbidden induced subgraphs, namely*

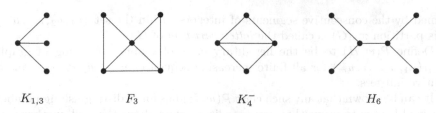

$K_{1,3}$ F_3 K_4^- H_6

Fig. 1. Some small graphs

(a) $\mathcal{K}_1 = \mathrm{Free}(P_3)$,
(b) $\mathcal{K}_2 = \mathrm{Free}(K_3, P_4, C_4)$,
(c) $\mathcal{K}_3 = \mathrm{Free}(\mathcal{F})$ for $\mathcal{F} = \{2K_2, C_4, C_5, K_{1,3}, F_3\}$,
(d) $\mathcal{K}_4 = \mathrm{Free}(\mathcal{F})$ for $\mathcal{F} = \{2K_2, C_4, C_5, K_4^-\}$,
(e) $\mathcal{K}_5 = \mathrm{Free}(\mathcal{F})$ for $\mathcal{F} = \{3K_1, C_5, \overline{P_4 + K_1}, \overline{2K_2 + K_1}, \overline{C_4 + K_2}, \overline{C_4 + 2K_1}, H_6\}$,
(f) $\mathcal{K}_6 = \mathrm{Free}(2K_2, K_3, C_5)$ [17],
(g) $\mathcal{K}_7 = \mathrm{Free}(2K_2, P_4, C_4)$ [8],
(h) $\overline{\mathrm{Free}(\mathcal{F})} = \mathrm{Free}(\overline{\mathcal{F}})$.

3.2 Finite Distinguishing Number

In this section we provide a characterisation of the minimal classes for the case of finite distinguishing number $k_{\mathcal{X}}$. It turns out that these minimal classes consist of $(\ell_{\mathcal{X}}, d_{\mathcal{X}})$-graphs, that is, the vertex set of each graph is partitioned into at most $\ell_{\mathcal{X}}$ bags and dense pairs are defined by a density graph H (see Lemma 2.7). The condition of Theorem 2.8 is enforced by long paths (indeed, an infinite path in the infinite universal graph). Thus actually $d_{\mathcal{X}} \leq 2$ for the minimal classes $\mathcal{X}$.

Let A be a finite alphabet. A *word* is a mapping $w : S \to A$, where $S = \{1, 2, \ldots, n\}$ for some $n \in \mathbb{N}$ or $S = \mathbb{N}$; $|S|$ is the *length* of w, denoted by $|w|$. We write w_i for $w(i)$, and we often use the notation $w = w_1 w_2 w_3 \ldots w_n$ or $w = w_1 w_2 w_3 \ldots$ For $n \leq m$ and $w = w_1 w_2 \ldots w_n$, $w' = w_1' w_2' \ldots w_m'$ (or $w' = w_1' w_2' \ldots$), we say that w is a *factor* of w' if there exists a non-negative integer s such that $w_i = w_{i+s}'$ for $1 \leq i \leq n$; w is an *initial segment* of w' if we can take $s = 0$.

Let H be an undirected graph with loops allowed and with vertex set $V(H) = A$, and let w be a (finite or infinite) word over the alphabet A. For any increasing sequence $u_1 < u_2 < \cdots < u_m$ of positive integers such that $u_m \leq |w|$, define $G_{w,H}(u_1, u_2, \ldots, u_m)$ to be the graph with vertex set $\{u_1, u_2, \ldots, u_m\}$ and an edge between u_i and u_j if and only if

– either $|u_i - u_j| = 1$ and $w_{u_i} w_{u_j} \notin E(H)$,
– or $|u_i - u_j| > 1$ and $w_{u_i} w_{u_j} \in E(H)$.

Let $G = G_{w,H}(u_1, u_2, \ldots, u_m)$ and define $V_a = \{u_i \in V(G) : w_{u_i} = a\}$ for any $a \in A$. Then $\pi = \pi_w(G) = \{V_a : a \in A\}$ is an $(|A|, 2)$-partition, and so G is an $(|A|, 2)$-graph. Moreover, $\psi(G, \pi, H)$ is a linear forest whose paths are formed by the consecutive segments of integers within the set $\{u_1, u_2, \ldots, u_m\}$. This partition $\pi_w(G)$ is called the *letter partition* of G.

Define $\mathcal{P}(w, H)$ to be the hereditary class of graphs consisting of graphs $G_{w,H}(u_1, u_2, \ldots, u_m)$ for all finite increasing sequences $u_1 < u_2 < \cdots < u_m$ of positive integers.

It can be shown that any such class $\mathcal{P}(w, H)$ has finite distinguishing number. Our goal here is to prove that any hereditary class above the Bell number with finite distinguishing number contains the class $\mathcal{P}(w, H)$ for some word w and graph H. Moreover, we describe sufficient conditions on the graph H and the word w so that $\mathcal{P}(w, H)$ is a minimal class above the Bell number.

Definition 3.3. *If* $u_1, u_2, \ldots, u_m$ *is a sequence of consecutive integers (i.e., $u_{k+1} = u_k + 1$ for each k), we call the graph $G_{w,H}(u_1, u_2, \ldots, u_m)$ an $|H|$-factor. Notice that each $|H|$-factor is an $(|H|, 2)$-graph; if its letter partition is a strong $(|H|, 2)$-partition, we call it a strong $|H|$-factor.*

Note that if $G = G_{w,H}(u_1, u_2, \ldots, u_m)$ is a strong ℓ-factor, then its sparsification $\phi(G) = \psi(G, \pi_w(G), H)$ is an induced path of length $m - 1$.

Proposition 3.4. *If w is an infinite word over a finite alphabet A and H is a graph on A, with loops allowed, then the class $\mathcal{P}(w, H)$ is above the Bell number.*

Proof. We may assume that every letter of A appears in w infinitely many times: otherwise we can remove a sufficiently long starting segment of w to obtain a word w' satisfying this condition, replace H with its induced subgraph H' on the alphabet A' of w', and obtain a subclass $\mathcal{P}(w', H')$ of $\mathcal{P}(w, H)$. Then for sufficiently large k, the $|A|$-factor $G_k = G_{w,H}(1, \ldots, k)$ is a strong $|A|$-factor; thus $\phi(G_k)$ is an induced path of length $k - 1$. Hence by Theorem 2.8, the class $\mathcal{P}(w, H)$ is above the Bell number.   □

Definition 3.5. *A word w is called* almost periodic *if for any factor f of w there is a constant k_f such that any factor of w of size at least k_f contains f as a factor.*

The next theorem asserts that any class with finite distinguishing number, if it is above Bell, contains one of the classes $\mathcal{P}(w, H)$. Consequently any minimal class will be of the form $\mathcal{P}(w, H)$.

Theorem 3.6. *Suppose $\mathcal{X}$ is a hereditary class above the Bell number with $k_\mathcal{X}$ finite. Then $\mathcal{X} \supseteq \mathcal{P}(w, H)$ for an infinite almost periodic word w and a graph H of order at most $\ell_\mathcal{X}$ with loops allowed.*

Sketch of proof. From Theorem 2.8 it follows that for each m there is a graph $G_m \in \mathcal{X}$ which admits a strong $(\ell_\mathcal{X}, d_\mathcal{X})$-partition $\{V_1, V_2, \ldots, V_{\ell_m}\}$ with $\ell_m \leq \ell_\mathcal{X}$ such that the sparsification $\phi(G_m)$ has a connected component C_m of order at least $(\ell_\mathcal{X} d_\mathcal{X})^m$. Fix an arbitrary vertex v of C_m. As C_m is an induced subgraph of $\phi(G_m)$, the maximum degree in C_m is bounded by $d = \ell_\mathcal{X} d_\mathcal{X}$. Therefore C_m contains an induced path $v = v_1, v_2, \ldots, v_m = v'$ of length $m - 1$. Then the induced subgraph $G_m[v_1, v_2, \ldots, v_m]$ is an $\ell_\mathcal{X}$-factor of order m contained in $\mathcal{X}$.

The existence of arbitrarily large $\ell_{\mathcal{X}}$-factors in $\mathcal{X}$ implies that $\mathcal{X}$ contains arbitrarily large *strong* $\ell_{\mathcal{X}}$-factors: It can be shown that by removing the small bags (repeatedly, if necessary) we cannot decrease the size of the $\ell_{\mathcal{X}}$-factor too much.

Having established that each class $\mathcal{X}$ with speed above the Bell number with finite distinguishing number $k_{\mathcal{X}}$ contains an infinite set $\mathcal{S}$ of strong $\ell_{\mathcal{X}}$-factors of increasing order, we can assume that each of the strong $\ell_{\mathcal{X}}$-factors is of the form $G_{w,H}(1, \ldots, m)$ for some prime graph H and that its letter partition is prime. For each H on $\{1, 2, \ldots, \ell\}$ with $1 \leq \ell \leq \ell_{\mathcal{X}}$ let $\mathcal{S}_H = \{G_{w,H}(1, \ldots, m) \in \mathcal{S}\}$ be the set of all $\ell_{\mathcal{X}}$-factors in $\mathcal{S}$ whose adjacencies are defined using the density graph H. Then for some (at least one) fixed graph H_0 the set $\mathcal{S}_{H_0}$ is infinite. Hence also $L = \{w : G_{w,H_0}(1, \ldots, m) \in \mathcal{X}\}$ is an infinite language. As $\mathcal{X}$ is a hereditary class, the language L is closed under taking word factors (it is a *factorial language*).

It is not hard to prove that any infinite factorial language contains a minimal infinite factorial language. So let $L' \subseteq L$ be a minimal infinite factorial language. It follows from minimality that L' is well quasi-ordered by the factor relation, because removing one word from any infinite antichain and taking all factors of the remaining words would generate an infinite factorial language strictly contained in L'. Thus there exists an infinite chain $w^{(1)}, w^{(2)}, \ldots$ of words in L' such that for any $i < j$, the word $w^{(i)}$ is a factor of $w^{(j)}$. More precisely, for each i there is a non-negative integer s_i such that $w_k^{(i)} = w_{k+s_i}^{(i+1)}$. Let $g(i, k) = k + \sum_{j=1}^{i-1} s_j$. Now we can define an infinite word w by putting $w_k = w_{g(i,k)}^{(i)}$ for the least value of i for which the right-hand side is defined. (Without loss of generality we get that w is indeed an infinite word; otherwise we would need to take the reversals of all the words $w^{(i)}$.)

Observe that any factor of w is in the language L'; if w is not almost periodic, then there exists a factor f of w such that there are arbitrarily long factors f' of w not containing f. These factors f' generate an infinite factorial language $L'' \subseteq L'$ which does not contain $f \in L'$, contradicting the minimality of L'.

Because any factor of w is in L, any $G_{w,H}(u_1, \ldots, u_m)$ is an induced subgraph of some $\ell_{\mathcal{X}}$-factor in $\mathcal{X}$. Therefore $\mathcal{P}(w, H) \subseteq \mathcal{X}$.   □

As a matter of fact, we can also show that if H is a graph with loops allowed and w is an almost periodic infinite word, then $\mathcal{P}(w, H)$ is a minimal property above the Bell number. This implies the following characterisation.

Theorem 3.7. *Let $\mathcal{X}$ be a class of graphs with $k_{\mathcal{X}} < \infty$. Then $\mathcal{X}$ is a minimal hereditary class above the Bell number if and only if there exists a finite graph H with loops allowed and an infinite almost periodic word w over $V(H)$ such that $\mathcal{X} = \mathcal{P}(w, H)$.*

4 Decidability of the Bell Number

Our main goal is to provide an algorithm that decides for an input consisting of a finite number of graphs $F_1, \ldots, F_n$ whether the speed of $\mathcal{X} = \text{Free}(F_1, \ldots, F_n)$ is above the Bell number. That is, we are interested in the following problem.

Problem 4.1. INPUT: *A finite set of graphs* $\mathcal{F} = \{F_1, F_2, \ldots, F_n\}$
OUTPUT: *Yes, if the speed of* $\mathcal{X} = \text{Free}(\mathcal{F})$ *is above the Bell number; no otherwise.*

Our algorithm, following the characterisation of minimal classes above the Bell number, distinguishes two cases depending on whether the distinguishing number $k_{\mathcal{X}}$ is finite or infinite. First we show how to discriminate between these two cases.

Problem 4.2. INPUT: *A finite set of graphs* $\mathcal{F} = \{F_1, F_2, \ldots, F_n\}$
OUTPUT: *Yes, if* $k_{\mathcal{X}} = \infty$ *for* $\mathcal{X} = \text{Free}(\mathcal{F})$; *no otherwise.*

Theorem 4.3. *There is a polynomial-time algorithm that solves Problem 4.2.*

Proof. By Theorem 3.1, $k_{\mathcal{X}} = \infty$ if and only if $\mathcal{X}$ contains one of the thirteen minimal classes listed there. By Theorem 3.2, each of the minimal classes is defined by finitely many forbidden induced subgraphs; thus membership can be tested in polynomial time. Then the answer to Problem 4.2 is no if and only if each of the minimal classes given by Theorem 3.1 contains at least one of the graphs in $\mathcal{F}$, which can also be tested in polynomial time. $\square$

By Theorem 3.7, the minimal hereditary classes with finite distinguishing number with speed above the Bell number can be described as $\mathcal{P}(w, H)$ with an almost periodic infinite word w. Here we give a more precise characterisation restricted to classes defined by finitely many forbidden induced subgraphs.

Definition 4.4. *Let* $w = w_1 w_2 \ldots$ *be an infinite word over a finite alphabet* A. *If there exists some* p *such that* $w_i = w_{i+p}$ *for all* $i \in \mathbb{N}$, *we call the word* w periodic *and the number* p *its* period. *If, moreover, for some period* p *the letters* $w_1, w_2, \ldots, w_p$ *are all distinct, we call the word* w cyclic.
  If w *is a finite word, then* $w' = (w)^{\infty}$ *is the periodic word obtained by concatenating infinitely many copies of the word* w; *thus* $w'_i = w_k$ *for* $k = i \bmod |w'|$.
  A class $\mathcal{X}$ *of graphs is called a* periodic class *(*cyclic class, respectively*) if there exists a graph* H *with loops allowed and a periodic (cyclic, respectively) word* w *such that* $\mathcal{X} = \mathcal{P}(w, H)$.

Definition 4.5. *Let* $A = \{1, 2, \ldots, \ell\}$ *be a finite alphabet,* H *a graph on* A *with loops allowed, and* M *a positive integer. Define a graph* $S_{H,M}$ *with vertex set* $V(S_{H,M}) = A \times \{1, 2, \ldots, M\}$ *and an edge between* (a, j) *and* (b, k) *if and only if one of the following holds:*

- $ab \in E(H)$ *and either* $|a - b| \neq 1$ *or* $j \neq k$;
- $ab \notin E(H)$ *and* $|a - b| = 1$ *and* $j = k$.

The graph $S_{H,M}$ *is called an* (ℓ, M)-strip.

Notice that a strip can be viewed as the graph obtained from the union of M disjoint paths $(1, j)-(2, j)-\cdots-(\ell, j)$ for $j \in \{1, 2, \ldots, M\}$ by swapping edges with non-edges between vertices (a, j) and (b, k) if $ab \in E(H)$.

Theorem 4.6. *Let* $\mathcal{X} = \text{Free}(F_1, F_2, \ldots, F_n)$ *with the distinguishing number* $k_{\mathcal{X}}$ *finite. Then the following conditions are equivalent:*

(a) The speed of $\mathcal{X}$ *is above the Bell number.*

(b) $\mathcal{X}$ *contains a periodic class.*

(c) For every $p \in \mathbb{N}$, $\mathcal{X}$ *contains a cyclic class with period at least* p.

(d) There exists a cyclic word w *and a graph* H *on the alphabet of* w *such that* $\mathcal{X}$ *contains the* ℓ-*factor* $G_{w,H}(1, 2, \ldots, 2\ell m)$ *with* $\ell = |V(H)|$ *and* $m = \max\{|F_i| : i \in \{1, 2, \ldots, n\}\}$.

(e) For any positive integers ℓ, m, *the class* $\mathcal{X}$ *contains an* (ℓ, m)-*strip.*

We omit the proof of Theorem 4.6 here (it can be found in the full version of this paper). Finally, we are ready to tackle the decidability of Problem 4.1.

Algorithm 4.7. INPUT: A finite set of graphs $\mathcal{F} = \{F_1, F_2, \ldots, F_n\}$
OUTPUT: Yes, if the speed of $\mathcal{X} = \text{Free}(\mathcal{F})$ is above the Bell number; no otherwise.

(1) Using Theorem 4.3, decide whether $k_{\mathcal{X}} = \infty$. If it is, output *yes* and stop.

(2) Set $m := \max\{|F_1|, |F_2|, \ldots, |F_n|\}$ and $\ell := 1$.

(3) Loop:

(3a) For each graph (with loops allowed) H on $\{1, 2, \ldots, \ell\}$ construct the (ℓ, ℓ)-strip $S_{H,\ell}$. Check if some F_i is an induced subgraph of $S_{H,\ell}$. If for each H the strip $S_{H,\ell}$ contains some F_i, output *no* and stop.

(3b) For each graph (with loops allowed) H on $\{1, 2, \ldots, \ell\}$ and for each word w consisting of ℓ distinct letters from $\{1, 2, \ldots, \ell\}$ check if the ℓ-factor $G_{w^\infty, H}(1, 2, \ldots, 2\ell m)$ contains some F_i as an induced subgraph. If one of these ℓ-factors contains no F_i, output *yes* and stop.

(3c) Set $\ell := \ell + 1$ and repeat.

It remains to prove the correctness of this algorithm.

Theorem 4.8. *Algorithm 4.7 correctly solves Problem 4.1.*

Proof. We show that if the algorithm stops, it gives the correct answer, and furthermore that it will stop on any input without entering an infinite loop. First, if it stops in step (1), the answer is correct by [7], since any class with infinite distinguishing number has speed above the Bell number.

Assume that the algorithm stops in step (3a) and outputs *no*. This is because every (ℓ, ℓ)-strip contains some forbidden subgraph F_i, hence no (ℓ, ℓ)-strip belongs to $\mathcal{X}$. By Theorem 4.6(e), the speed of $\mathcal{X}$ is below the Bell number.

Next suppose that the algorithm stops in step (3b) and answers *yes*. Then $\mathcal{X}$ contains the ℓ-factor $G_{w^\infty, H}(1, 2, \ldots, 2\ell m)$, where w^∞ is a cyclic word. Hence by Theorem 4.6(d) the speed of $\mathcal{X}$ is above the Bell number.

Finally, if $k_{\mathcal{X}} = \infty$ the algorithm stops in step (1). If $k_{\mathcal{X}} < \infty$ and the speed of $\mathcal{X}$ is above the Bell number, then by Theorem 4.6(d) the algorithm will stop in step (3b). If, on the other hand, the speed of $\mathcal{X}$ is below the Bell number, then by

Theorem 4.6(e) there exist positive integers ℓ, M such that $\mathcal{X}$ contains no (ℓ, M)-strip. Let $N = \max\{\ell, M\}$. Obviously, $\mathcal{X}$ contains no (N, N)-strip, because any (N, N)-strip contains some (many) (ℓ, M)-strips as induced subgraphs and $\mathcal{X}$ is hereditary. Therefore the algorithm will stop in step (3a) after finitely many steps.   □

Our result leaves many open questions. For instance, what is the computational complexity of Problem 4.1?

References

1. Alekseev, V.E.: Range of values of entropy of hereditary classes of graphs. Diskret. Mat. **4**(2), 148–157 (1992). (in Russian); translation in Discrete Math. Appl. **3**(2), 191–199 (1993)
2. Alekseev, V.E.: On lower layers of a lattice of hereditary classes of graphs. Diskretn. Anal. Issled. Oper. Ser. 1 **4**(1), 3–12 (1997). (in Russian)
3. Allen, P., Lozin, V., Rao, M.: Clique-width and the speed of hereditary properties. Electron. J. Combin. **16**(1), Research Paper 35, 11 p. (2009)
4. Alon, N., Balogh, J., Bollobás, B., Morris, R.: The structure of almost all graphs in a hereditary property. J. Combin. Theory Ser. B **101**(2), 85–110 (2011)
5. Balogh, J., Bollobás, B., Weinreich, D.: The speed of hereditary properties of graphs. J. Combin. Theory Ser. B **79**(2), 131–156 (2000)
6. Balogh, J., Bollobás, B., Weinreich, D.: The penultimate rate of growth for graph properties. Eur. J. Combin. **22**(3), 277–289 (2001)
7. Balogh, J., Bollobás, B., Weinreich, D.: A jump to the Bell number for hereditary graph properties. J. Combin. Theory Ser. B **95**(1), 29–48 (2005)
8. Chvátal, V., Hammer, P.L.: Set-packing and threshold graphs. Res. Report CORR 73–21, Comp. Sci. Dept. Univ. of Waterloo (1973)
9. Erdős, P., Frankl, P., Rödl, V.: The asymptotic number of graphs not containing a fixed subgraph and a problem for hypergraphs having no exponent. Graphs Combin. **2**(2), 113–121 (1986)
10. Erdős, P., Kleitman, D.J., Rothschild, B.L.: Asymptotic enumeration of K_n-free graphs. Colloquio Internazionale sulle Teorie Combinatorie (Rome, 1973), Tomo II, number 17 in Atti dei Convegni Lincei, pp. 19–27. Accad. Naz. Lincei, Rome (1976)
11. Kolaitis, P.G., Prömel, H.J., Rothschild, B.L.: K_{l+1}-free graphs: asymptotic structure and a 0–1 law. Trans. Amer. Math. Soc. **303**(2), 637–671 (1987)
12. Korpelainen, N., Lozin, V.: Two forbidden induced subgraphs and well-quasi-ordering. Discrete Math. **311**(16), 1813–1822 (2011)
13. Prömel, H.J., Steger, A.: Excluding induced subgraphs: quadrilaterals. Random Struct. Algorithms **2**(1), 55–71 (1991)
14. Prömel, H.J., Steger, A.: Excluding induced subgraphs. III. A general asymptotic. Random Struct. Algorithms **3**(1), 19–31 (1992)
15. Prömel, H.J., Steger, A.: Excluding induced subgraphs. II. Extremal graphs. Discrete Appl. Math. **44**(1–3), 283–294 (1993)
16. Scheinerman, E.R., Zito, J.S.: On the size of hereditary classes of graphs. J. Combin. Theory Ser. B **61**(1), 16–39 (1994)
17. Yannakakis, M.: The complexity of the partial order dimension problem. SIAM J. Algebraic Discrete Methods **3**(3), 351–358 (1982)

Boxicity and Separation Dimension

Manu Basavaraju[1], L. Sunil Chandran[2], Martin Charles Golumbic[3]([✉]),
Rogers Mathew[3], and Deepak Rajendraprasad[3]

[1] University of Bergen, Postboks 7800, 5020 Bergen, Norway
manu.basavaraju@ii.uib.no
[2] Department of Computer Science and Automation, Indian Institute of Science,
Bangalore 560012, India
sunil@csa.iisc.ernet.in
[3] Department of Computer Science, Caesarea Rothschild Institute,
University of Haifa, 31905 Haifa, Israel
golumbic@cs.haifa.ac.il, {rogersmathew,deepakmail}@gmail.com

Abstract. A family $\mathcal{F}$ of permutations of the vertices of a hypergraph
H is called *pairwise suitable* for H if, for every pair of disjoint edges in
H, there exists a permutation in $\mathcal{F}$ in which all the vertices in one edge
precede those in the other. The cardinality of a smallest such family
of permutations for H is called the *separation dimension* of H and is
denoted by $\pi(H)$. Equivalently, $\pi(H)$ is the smallest natural number k
so that the vertices of H can be embedded in $\mathbb{R}^k$ such that any two
disjoint edges of H can be separated by a hyperplane normal to one of
the axes. We show that the separation dimension of a hypergraph H
is equal to the *boxicity* of the line graph of H. This connection helps
us in borrowing results and techniques from the extensive literature on
boxicity to study the concept of separation dimension.

Keywords: Separation dimension · Boxicity · Scrambling permutation ·
Line graph · Acyclic chromatic number

1 Introduction

Let $\sigma : U \to [n]$ be a permutation of elements of an n-set U. For two disjoint
subsets A, B of U, we say $A \prec_\sigma B$ when every element of A precedes every
element of B in σ, i.e., $\sigma(a) < \sigma(b), \forall (a, b) \in A \times B$. Otherwise, we say $A \not\prec_\sigma B$.
We say that σ *separates* A and B if either $A \prec_\sigma B$ or $B \prec_\sigma A$. We use $a \prec_\sigma b$ to
denote $\{a\} \prec_\sigma \{b\}$. For two subsets A, B of U, we say $A \preceq_\sigma B$ when $A \setminus B \prec_\sigma$
$A \cap B \prec_\sigma B \setminus A$.

In this paper, we introduce and study a notion called *pairwise suitable family*
of permutations for a hypergraph H and the *separation dimension* of H.

Manu Basavaraju: Supported by the European Research Council under the European
Union's Seventh Framework Programme (FP/2007–2013)/ERC Grant Agreement n.
267959.

Rogers Mathew and Deepak Rajendraprasad: Supported by VATAT Postdoctoral
Fellowship, Council of Higher Education, Israel.

D. Kratsch and I. Todinca (Eds.): WG 2014, LNCS 8747, pp. 81–92, 2014.
DOI: 10.1007/978-3-319-12340-0_7

Definition 1. *A family $\mathcal{F}$ of permutations of $V(H)$ is* pairwise suitable *for a hypergraph H if, for every two disjoint edges $e, f \in E(H)$, there exists a permutation $\sigma \in \mathcal{F}$ which separates e and f. The cardinality of a smallest family of permutations that is pairwise suitable for H is called the* separation dimension *of H and is denoted by $\pi(H)$.*

A family $\mathcal{F} = \{\sigma_1, \ldots, \sigma_k\}$ of permutations of a set V can be seen as an embedding of V into $\mathbb{R}^k$ with the i-th coordinate of $v \in V$ being the rank of v in the σ_i. Similarly, given any embedding of V in $\mathbb{R}^k$, we can construct k permutations by projecting the points onto each of the k axes and then reading them along the axis, breaking the ties arbitrarily. From this, it is easy to see that $\pi(H)$ is the smallest natural number k so that the vertices of H can be embedded into $\mathbb{R}^k$ such that any two disjoint edges of H can be separated by a hyperplane normal to one of the axes. This motivates us to call such an embedding a *separating embedding* of H and $\pi(H)$ the *separation dimension* of H.

The notion of separation dimension introduced here seems so natural but, to the best of our knowledge, has not been studied in this generality before. The authors of [15] provide suggested applications motivating the study of permutation covering and separation problems on event sequencing of tasks. Apart from that, a major motivation for us to study this notion of separation is its interesting connection with a certain well studied geometric representation of graphs. In fact, we show that $\pi(H)$ is same as the *boxicity* of the intersection graph of the edge set of H, i.e., the line graph of H.

An axis-parallel k-dimensional box or a k-box is a Cartesian product $R_1 \times \cdots \times R_k$, where each R_i is a closed interval on the real line. For example, a line segment lying parallel to the X axis is a 1-box, a rectangle with its sides parallel to the X and Y axes is a 2-box, a rectangular cuboid with its sides parallel to the X, Y, and Z axes is a 3-box and so on. A *box representation* of a graph G is a geometric representation of G using axis-parallel boxes as follows.

Definition 2. *The k-box representation of a graph G is a function f that maps each vertex in G to a k-box in $\mathbb{R}^k$ such that, for all vertices u, v in G, the pair $\{u, v\}$ is an edge if and only if $f(u)$ intersects $f(v)$. The* boxicity *of a graph G, denoted by* boxicity(G), *is the minimum positive integer k such that G has a k-box representation.*

The concept of boxicity was introduced by F.S. Roberts in 1969 [20]. He showed that every graph on n vertices has an $\lfloor n/2 \rfloor$-box representation. The n-vertex graph whose complement is a perfect matching is an example of a graph whose boxicity is equal to $n/2$. Upper bounds for boxicity in terms of other graph parameters like maximum degree, treewidth, minimum vertex cover, degeneracy etc. are available in literature. Studies on box representations of special graph classes too are available in abundance. Scheinerman showed that every outerplanar graph has a 2-box representation [21] while Thomassen showed that every planar graph has a 3-box representation [23]. Results on boxicity of series-parallel graphs [8], Halin graphs [12], chordal graphs, AT-free graphs, permutation graphs [14], circular arc graphs [7], chordal bipartite graphs [11] etc. can be found in literature. Here we are interested in boxicity of the line graph of hypergraphs. The *line graph* of a hypergraph H, denoted by $L(H)$, is the graph with vertex set $V(L(H)) = E(H)$ and edge set $E(L(H)) = \{\{e, f\} : e, f \in E(H), e \cap f \neq \emptyset\}$.

For the line graph of a graph G with maximum degree Δ, it was shown by Chandran, Mathew and Sivadasan that its boxicity is in $O(\Delta \log \log \Delta)$ [13]. It was in their attempt to improve this result that the authors stumbled upon pairwise suitable family of permutations and its relation with the boxicity of the line graph of G. In an arXiv preprint version of this paper available at [6], we improve the upper bound for boxicity of the line graph of G to $2^{9 \log^* \Delta} \Delta$, where $\log^* \Delta$ denotes the iterated logarithm of Δ to the base 2, i.e. the number of times the logarithm function (to the base 2) has to be applied so that the result is less than or equal to 1. In a recent joint work with Noga Alon, we have shown that there exist graphs of maximum degree Δ whose line graphs have boxicity in $\Omega(\Delta)$. Bounds for separation dimension of a graph based on its treewidth, degeneracy etc. are also established in the arXiv version.

1.1 Outline of the Paper

The remainder of this paper[1] is organised as follows. A brief note on some standard terms and notations used throughout this paper is given in Sect. 1.2. Section 2 demonstrates the equivalence of separation dimension of a hypergraph H and boxicity of the line graph of H. In Sect. 3.1, we characterize graphs of separation dimension 1. Using a probabilistic argument, in Sect. 3.2, we prove a tight (up to constants) upper bound for separation dimension of a graph based on its size. Section 3.3 relates separation dimension with acyclic chromatic number. In Sect. 3.4, using Schnyder's celebrated result on planar drawing, we show that the separation dimension of a planar graph is at most 3. This bound is the best possible as we know of series-parallel graphs (that are subclasses of planar graphs) of separation dimension 3. In Sect. 3.5, we prove the theorem that yields a non-trivial lower bound to the separation dimension of a graph. This theorem and its corollaries are used in establishing the tightness of the upper bounds proved. Moreover, the theorem is used to prove a lower bound for the separation dimension of a random graph in Sect. 3.6.

Once again, in Sect. 4.1, we use a probabilistic argument to show an upper bound on the separation dimension of a rank-r hypergraph based on its size. This is followed by an upper bound based on maximum degree in Sect. 4.2. We get this upper bound as a consequence of a non-trivial result in the area of boxicity. In Sect. 4.3, we prove a lower bound on the separation dimension of a complete r-uniform hypergraph by extending the lower bounding technique used in the context of graphs. Finally, in Sect. 5, we conclude with a discussion of a few open problems that we find interesting.

---

[1] The full version of this paper, which includes all the proofs, is available at http://arxiv.org/abs/1404.4486.

1.2   Notational Note

A *hypergraph* H is a pair (V, E) where V, called the *vertex set*, is any set and E, called the *edge set*, is a collection of subsets of V. The vertex set and edge set of a hypergraph H are denoted respectively by $V(H)$ and $E(H)$. The *rank* of a hypergraph H is $\max_{e \in E(H)} |e|$ and H is called k-*uniform* if $|e| = k, \forall e \in E(H)$. The *degree* of a vertex v in H is the number of edges of H which contain v. The *maximum degree* of H, denoted as $\Delta(H)$ is the maximum degree over all vertices of H. All the hypergraphs considered in this paper are finite.

A *graph* is a 2-uniform hypergraph. For a graph G and any $S \subseteq V(G)$, the subgraph of G induced by the vertex set S is denoted by $G[S]$. For any $v \in V(G)$, we use $N_G(v)$ to denote the neighbourhood of v in G, i.e., $N_G(v) = \{u \in V(G) : \{v, u\} \in E(G)\}$.

A *closed interval* on the real line, denoted as $[i, j]$ where $i, j \in \mathbb{R}$ and $i \leq j$, is the set $\{x \in \mathbb{R} : i \leq x \leq j\}$. Given an interval $X = [i, j]$, define $l(X) = i$ and $r(X) = j$. We say that the closed interval X has *left end-point* $l(X)$ and *right end-point* $r(X)$. For any two intervals $[i_1, j_1], [i_2, j_2]$ on the real line, we say that $[i_1, j_1] < [i_2, j_2]$ if $j_1 < i_2$.

For any finite positive integer n, we shall use $[n]$ to denote the set $\{1, \ldots, n\}$. A permutation of a finite set V is a bijection from V to $[|V|]$. The logarithm of any positive real number x to the base 2 and e are respectively denoted by $\log(x)$ and $\ln(x)$.

2   Pairwise Suitable Family of Permutations and a Box Representation

In this section we show that a family of permutations of cardinality k is pairwise suitable for a hypergraph H (Definition 1) if and only if the line graph of H has a k-box representation (Definition 2). Before we proceed to prove it, let us state an equivalent but more combinatorial definition for boxicity.

Lemma 3 (Roberts [20]). *For every graph G, boxicity$(G) \leq k$ if and only if there exist k interval graphs $I_1, \ldots, I_k$, with $V(I_1) = \cdots = V(I_k) = V(G)$ such that $G = I_1 \cap \cdots \cap I_k$.*

From the above lemma, we get an equivalent definition of boxicity.

Definition 4. *The* boxicity *of a graph G is the minimum positive integer k for which there exist k interval graphs $I_1, \ldots, I_k$ such that $G = I_1 \cap \cdots \cap I_k$.*

Note that if $G = I_1 \cap \cdots \cap I_k$, then each I_i is a supergraph of G. Moreover, for every pair of vertices $u, v \in V(G)$ with $\{u, v\} \notin E(G)$, there exists some $i \in [k]$ such that $\{u, v\} \notin E(I_i)$. Now we are ready to prove the main theorem of this section.

Theorem 5. *For a hypergraph H, $\pi(H) = $ boxicity$(L(H))$.*

Proof. First we show that $\pi(H) \leq \text{boxicity}(L(H))$. Let $\text{boxicity}(L(H)) = b$. Then, by Lemma 3, there exists a collection of b interval graphs, say $\mathcal{I} = \{I_1, \ldots, I_b\}$, whose intersection is $L(H)$. For each $i \in [b]$, let f_i be an interval representation of I_i. For each $u \in V(H)$, let $E_H(u) = \{e \in E(H) : u \in e\}$ be the set of edges of H containing u. Consider an $i \in [b]$ and a vertex $u \in V(H)$. The closed interval $C_i(u) = \bigcap_{e \in E_H(u)} f_i(e)$ is called the *clique region* of u in f_i. Since any two edges in $E_H(u)$ are adjacent in $L(H)$, the corresponding intervals have non-empty intersection in f_i. By the Helly property of intervals, $C_i(u)$ is non-empty. We define a permutation σ_i of $V(H)$ from f_i such that $\forall u, v \in V(H)$, $C_i(u) < C_i(v) \implies u \prec_{\sigma_i} v$. It suffices to prove that $\{\sigma_1, \ldots, \sigma_b\}$ is a family of permutations that is pairwise suitable for H.

Consider two disjoint edges e, e' in H. Hence $\{e, e'\} \notin E(L(H))$ and since $L(H) = \bigcap_{i=1}^{b} I_i$, there exists an interval graph, say $I_i \in \mathcal{I}$, such that $\{e, e'\} \notin E(I_i)$, i.e., $f_i(e) \cap f_i(e') = \emptyset$. Without loss of generality, assume $f_i(e) < f_i(e')$. For any $v \in e$ and any $v' \in e'$, since $C_i(v) \subseteq f_i(e)$ and $C_i(v') \subseteq f(e')$, we have $C_i(v) < C_i(v')$, i.e. $v \prec_{\sigma_i} v'$. Hence $e \prec_{\sigma_i} e'$. Thus the family $\{\sigma_1, \ldots, \sigma_b\}$ of permutations is pairwise suitable for H.

Next we show that $\text{boxicity}(L(H)) \leq \pi(H)$. Let $\pi(H) = p$ and let $\mathcal{F} = \{\sigma_1, \ldots, \sigma_p\}$ be a pairwise suitable family of permutations for H. From each permutation σ_i, we shall construct an interval graph I_i such that $L(H) = \bigcap_{i=1}^{p} I_i$. Then by Lemma 3, $\text{boxicity}(L(H)) \leq \pi(H)$.

For a given $i \in [p]$, to each edge $e \in E(H)$, we associate the closed interval

$$f_i(e) = \left[\min_{v \in e} \sigma_i(v) \, , \, \max_{v \in e} \sigma_i(v) \right],$$

and let I_i be the intersection graph of the intervals $f_i(e), e \in E(H)$. Let $e, e' \in V(L(H))$. If e and e' are adjacent in $L(H)$, let $v \in e \cap e'$. Then $\sigma_i(v) \in f_i(e) \cap f_i(e')$, $\forall i \in [p]$. Hence e and e' are adjacent in I_i for every $i \in [p]$. If e and e' are not adjacent in $L(H)$, then there is a permutation $\sigma_i \in \mathcal{F}$ such that either $e \prec_{\sigma_i} e'$ or $e' \prec_{\sigma_i} e$. Hence by construction $f_i(e) \cap f_i(e') = \emptyset$ and so e and e' are not adjacent in I_i. This completes the proof. $\qquad\square$

3   Separation Dimension of Graphs

3.1   Characterizing Graphs of Separation Dimension 1

"When is $\pi(G) = 0$?" Clearly, if $\pi(G) = 0$, then G may have at most one non-trivial connected component and every pair of edges must share an endpoint. The following is a simple exercise answering the question:

Proposition 6. *For a graph G, $\pi(G) = 0$ if and only if G has at most one connected component of size greater than one and this component is either a clique of size at most 3 or a star.*

A *caterpillar* is a tree consisting of a chordless path $[v_1, v_2, \ldots, v_k]$ called the *spine*, plus an unlimited number of pendant vertices. A *caterpillar with single humps* is formed from a caterpillar by adding at most one new vertex x_i adjacent to v_i and v_{i+1} for every $i = 1, \ldots, k-1$. Without loss of generality, we may assume that the first and last vertex of the spine have no pendent vertices (i.e., the spine is longest possible.) The *diamond*, denoted here by D, is the graph with 4 vertices and 5 edges; the 3-net N_3 consists of a triangle with a pendant vertex attached to each of its vertices; the graph T_2 is the tree with 6 edges $\{cx, cy, cz, xx', yy', zz'\}$; and the graph C_k $(k \geq 4)$ denotes the cycle of size k.

Theorem 7. *Let G be a graph. The following conditions are equivalent:*

(i) $\pi(G) \leq 1$,
(ii) G *is a disjoint union of caterpillars with single humps,*
(iii) G *has no partial subgraph C_k $(k \geq 4)$, N_3 or T_2,*
(iv) G *is a chordal graph with no induced subgraph D, K_4, T_2, N_3, G_1, G_2 or*
  G_3, *where $G_1 = T_2 \cup \{cx'\}$, $G_2 = G_1 \cup \{cy'\}$ and $G_3 = G_2 \cup \{cz'\}$,*
(v) *The line graph $L(G)$ is an interval graph.*

The proof of Theorem 7 suggests a linear time algorithm for recognizing whether a graph G has separation dimension 1 and constructing its representation as a caterpillar with single humps: (1) Using either Lexicographic Breadth First Search or Maximum Cardinality Search, obtain an ordering of the vertices $a_1, a_2, \ldots, a_n$ (but do not bother to test whether it is a perfect elimination ordering[2]; (2) Starting with a_n and proceeding in reverse order, follow the rules in the proof of $(iii) \Rightarrow (ii)$ to construct the spine, pendant vertices and the humps. If either (1) or (2) fails, then $\pi(G) > 1$.

3.2   Separation Dimension and the Size of a Graph

For graphs, sometimes we work with a notion of suitability that is stronger than the pairwise suitability of Definition 1. This will come in handy in proving certain results later in this article.

Definition 8. *For a graph G, a family $\mathcal{F}$ of permutations of G is 3-mixing if, for every two adjacent edges $\{a, b\}, \{a, c\} \in E(G)$, there exists a permutation $\sigma \in \mathcal{F}$ such that either $b \prec_\sigma a \prec_\sigma c$ or $c \prec_\sigma a \prec_\sigma b$.*

Notice that a family of permutations $\mathcal{F}$ of $V(G)$ is pairwise suitable and 3-mixing for G if, for every two edges $e, f \in E(G)$, there exists a permutation $\sigma \in \mathcal{F}$ such that either $e \preceq_\sigma f$ or $f \preceq_\sigma e$. Let $\pi^*(G)$ denote the cardinality of a smallest family of permutations that is pairwise suitable and 3-mixing for G. From their definitions, $\pi(G) \leq \pi^*(G)$.

---

[2] If G is chordal, any LexBFS or MCS ordering will be a perfect elimination ordering, but testing whether each v_i has exactly one forward neighbor or two connected forward neighbors will be enough.

Observation 9. $\pi(G)$ *and* $\pi^*(G)$ *are monotone increasing properties.*

The following theorem is the special case of Theorem 25 when the rank-r hypergraph under consideration is a graph. Theorem 25 yields a bound of $\pi(G) \leq 9.596 \log n$.

Theorem 10. *For a graph G on n vertices, $\pi(G) \leq \pi^*(G) \leq 6.84 \log n$.*

Proof. From the definitions of $\pi(G)$ and $\pi^*(G)$ and Observation 9, we have $\pi(G) \leq \pi^*(G) \leq \pi^*(K_n)$, where K_n denotes the complete graph on n vertices. Here we prove that $\pi^*(K_n) \leq 6.84 \log n$.

Choose r permutations, $\sigma_1, \ldots, \sigma_r$, independently and uniformly at random from the $n!$ distinct permutations of $[n]$. Let e, f be two distinct edges of K_n. The probability that $e \preceq_{\sigma_i} f$ is $1/6$, for each $i \in [r]$. (4 out of 4! outcomes are favourable when e and f are non-adjacent and 1 out of 3! outcomes is favourable otherwise.) Therefore, the probability that $e \preceq_{\sigma_i} f$ or $f \preceq_{\sigma_i} e$ is $1/3$. Let $B(e, f)$ denote the "bad" event of $e \not\preceq_{\sigma_i} f$ and $f \not\preceq_{\sigma_i} e$ for all $i \in [r]$. Then, $Pr[B(e, f)] = (2/3)^r$. Taking union bound over all distinct pairs of edges e and f, we get

$$Pr\left[\bigcup_{\forall \text{ pairs of distinct edges } e, f} B(e, f) \right] < n^4 \left(\frac{2}{3} \right)^r$$

When $r = 6.84 \log n$, the left hand side of the above inequality is a quantity less than 1. That is, there exists a family of permutations of $V(K_n)$ of cardinality at most $6.84 \log n$ which is pairwise suitable and 3 mixing for K_n. $\qquad \square$

Tightness of Theorem 10. Let K_n denote a complete graph on n vertices. Since $\omega(K_n) = n$, it follows from Corollary 21 that $\pi(K_n) \geq \log \lfloor n/2 \rfloor$. Hence the bound proved in Theorem 10 is tight up to a constant factor.

3.3 Acyclic and Star Chromatic Number

Definition 11. *The* acyclic chromatic number *of a graph G, denoted by $\chi_a(G)$, is the minimum number of colours needed to do a proper colouring of the vertices of G such that the graph induced on the vertices of every pair of colour classes is acyclic. The* star chromatic number *of a graph G, denoted by $\chi_s(G)$, is the minimum number of colours needed to do a proper colouring of the vertices of G such that the graph induced on the vertices of every pair of colour classes is a star forest.*

We know that that a star forest is a disjoint union of stars. Therefore, $\chi_s(G) \geq \chi_a(G) \geq \chi(G)$, where $\chi(G)$ denotes the chromatic number of G. In order to bound $\pi(G)$ in terms of $\chi_a(G)$ and $\chi_s(G)$, we first bound $\pi(G)$ for forests and star forests. Then the required result follows from an application of Lemma 14.

Since forests are outerplanar graphs, the following lemma follows directly from the discussion on outerplanar graphs in Sect. 3.4.

Lemma 12. *For a forest G, $\pi(G) \leq 2$.*

Lemma 13. *For a star forest G, $\pi(G) = 1$.*

Proof. Follows directly from Theorem 7.   □

Lemma 14. *Let $P_G = \{V_1, \ldots, V_r\}$ be a partitioning of the vertices of a graph G, i.e., $V(G) = V_1 \uplus \cdots \uplus V_r$. Let $\hat{\pi}(P_G) = \max_{i,j \in [r]} \pi(G[V_i \cup V_j])$. Then, $\pi(G) \leq 13.68 \log r + \hat{\pi}(P_G) r$.*

Theorem 15. *For a graph G, $\pi(G) \leq 2\chi_a(G) + 13.68 \log(\chi_a(G))$. Further, $\pi(G) \leq \chi_s(G) + 13.68 \log(\chi_s(G))$.*

Proof. The theorem follows directly from Lemmas 12, 13, and 14.   □

This, together with some existing results from literature, gives us a few easy corollaries. Alon, Mohar, and Sanders have showed that a graph embeddable in a surface of Euler genus g has an acyclic chromatic number in $O(g^{4/7})$ [5]. It is noted by Esperet and Joret in [17], using results of Nesetril, Ossona de Mendez, Kostochka, and Thomassen, that graphs with no K_t minor have an acyclic chromatic number in $O(t^2 \log t)$. Hence the following corollary.

Corollary 16. *(i) For a graph G with Euler genus g, $\pi(G) \in O(g^{4/7})$; and (ii) for a graph G with no K_t minor, $\pi(G) \in O(t^2 \log t)$.*

3.4 Planar Graphs

Since planar graphs have acyclic chromatic number at most 5 [9], it follows from Theorem 15 that, for every planar graph G, $\pi(G) \leq 42$. Using Schnyder's celebrated result on non-crossing straight line plane drawings of planar graphs we improve this bound to the best possible.

Theorem 17 (Schnyder, Theorem 1.1 in [22]). *Let λ_1, λ_2, λ_3 be three pairwise non-parallel straight lines in the plane. Then, each plane graph has a straight line embedding in which any two disjoint edges are separated by a straight line parallel to λ_1, λ_2 or λ_3.*

This immediately gives us the following tight bound for planar graphs.

Theorem 18. *Separation dimension of a planar graph is at most 3. Moreover there exist planar graphs with separation dimension 3.*

Outerplanar and Series-Parallel Graphs. We know that outerplanar graphs form a subclass of series-parallel graphs which in turn form a subclass of planar graphs. It is not difficult to see that the separation dimension of outerplanar graphs is at most 2. The idea is to take one permutation by reading the vertices from left to right along the spine in a one page embedding of the graph and the second permutation in the order in which we see the vertices when we recursively peel off the outermost edge till every vertex is enlisted. As for series-parallel graphs, we know of series-parallel graphs that require separation dimension 3.

3.5   Lower Bounds

The tightness of many of the upper bounds we showed in the previous section relies on the lower bounds we derive in this section. First, we show that if a graph contains a uniform bipartite subgraph, then it needs a large separation dimension. This immediately gives a lower bound on separation dimension for complete bipartite graphs and hence a lower bound for every graph G in terms $\omega(G)$. The same is used to obtain a lower bound on the separation dimension for random graphs of all density. Finally, it is used as a critical ingredient in proving a lower bound on the separation dimension for complete r-uniform hypergraphs.

Theorem 19. *For a graph G, let $V_1, V_2 \subsetneq V(G)$ such that $V_1 \cap V_2 = \emptyset$. If there exists an edge between every s_1-subset of V_1 and every s_2-subset of V_2, then $\pi(G) \geq \min \left\{ \log \frac{|V_1|}{s_1}, \log \frac{|V_2|}{s_2} \right\}$.*

The next two corollaries are immediate.

Corollary 20. *For a complete bipartite graph $K_{m,n}$ with $m \leq n$, $\pi(K_{m,n}) \geq \log(m)$.*

Corollary 21. *For a graph G, $\pi(G) \geq \log \left\lfloor \frac{\omega(G)}{2} \right\rfloor$, where $\omega(G)$ is the size of a largest clique in G.*

3.6   Random Graphs

Definition 22 (Erdős-Rényi model). *$\mathcal{G}(n, p)$, $n \in \mathbb{N}$ and $0 \leq p \leq 1$, is the discrete probability space of all simple undirected graphs G on n vertices with each pair of vertices of G being joined by an edge with a probability p independent of the choice for every other pair of vertices.*

Definition 23. *A property P is said to hold for $\mathcal{G}(n, p)$ asymptotically almost surely (a.a.s) if the probability that P holds for $G \in \mathcal{G}(n, p)$ tends to 1 as n tends to ∞.*

Theorem 24. *For $G \in \mathcal{G}(n, p(n))$*

$$\pi(G) \geq \log(np(n)) - \log \log(np(n)) - 2.5 \ a.a.s.$$

Note that the expected average degree of a graph in $\mathcal{G}(n, p)$ is $\mathbb{E}_p[\bar{d}] = (n-1)p$. And hence the above bound can be written as $\log \mathbb{E}_p[\bar{d}] - \log \log \mathbb{E}_p[\bar{d}] - 2.5$.

4   Separation Dimension of Hypergraphs

4.1   Separation Dimension and Size of a Hypergraph

Using a direct probabilistic argument similar to the one used in Theorem 10 we obtain the following theorem.

Theorem 25. *For any rank-r hypergraph H on n vertices*

$$\pi(H) \leq \frac{e \ln 2}{\pi \sqrt{2}} 4^r \sqrt{r} \log n.$$

Tightness of Theorem 25. Let K_n^r denote a complete r-uniform graph on n vertices. Then by Theorem 27, $\pi(K_n^r) \geq \frac{1}{2^7} \frac{4^r}{\sqrt{r-2}} \log n$ for n sufficiently larger than r. Hence the bound in Theorem 25 is tight by factor of $64r$.

4.2   Maximum Degree

Theorem 26. *For any rank-r hypergraph H of maximum degree D, $\pi(H) \leq O\left(rD \log^2(rD)\right)$.*

Proof. This is a direct consequence of the nontrivial fact that boxicity$(G) \in O\left(\Delta \log^2 \Delta\right)$ for any graph G of maximum degree Δ [1]. □

It is known that there exist graphs of maximum degree Δ whose boxicity can be as high as $c\Delta \log \Delta$ [1], where c is a small positive constant. Let G be one such graph. Consider the following hypergraph H constructed from G. Let $V(H) = E(G)$ and $E(H) = \{E_v : v \in V(G)\}$ where E_v is the set of edges incident on the vertex v in G. It is clear that $G = L(H)$. Hence $\pi(H) = $ boxicity$(G) \geq c\Delta(G) \log \Delta(G)$. Note that the rank of H is $r = \Delta(G)$ and the maximum degree of H is 2. Thus $\pi(H) \geq cr \log(r)$ and hence the dependence on r in the upper bound cannot be considerably brought down in general.

4.3   Lower Bound

Now we illustrate one method of extending the above lower bounding technique from graphs to hypergraphs. Let K_n^r denote the complete r-uniform hypergraph on n vertices. We show that the upper bound of $O\left(4^r \sqrt{r} \log n\right)$ obtained for K_n^r from Theorem 25 is tight up to a factor of r. The lower bound argument below is motivated by an argument used by Radhakrishnan to prove a lower bound on the size of a family of scrambling permutations [19]. From Corollary 21 we know that the separation dimension of K_n, the complete graph on n vertices, is in $\Omega(\log n)$. Below we show that given any separating embedding of K_n^r in $\mathbb{R}^d$, the space $\mathbb{R}^d$ contains $\binom{2r-4}{r-2}$ orthogonal subspaces such that the projection of the given embedding on to these subspaces gives a separating embedding of a K_{n-2r+4}.

Theorem 27. *Let K_n^r denote the complete r-uniform hypergraph on n vertices with $r > 2$. Then*

$$c_1 \frac{4^r}{\sqrt{r-2}} \log n \le \pi(K_n^r) \le c_2 4^r \sqrt{r} \log n,$$

for n sufficiently larger than r and where $c_1 = \frac{1}{2^7}$ and $c_2 = \frac{e \ln 2}{\pi \sqrt{2}} < \frac{1}{2}$.

5   Discussion and Open Problems

Since $\pi(G)$ is the boxicity of the line graph of G, it is interesting to see how it is related to boxicity of G itself. But unlike separation dimension, boxicity is not a monotone parameter. For example the boxicity of K_n is 1, but deleting a perfect matching from K_n, if n is even, blows up its boxicity to $n/2$. Yet we couldn't find any graph G such that boxicity$(G) > 2^{\pi(G)}$. Hence we are curious about the following question: Does there exist a function $f : \mathbb{N} \to \mathbb{N}$ such that boxicity$(G) \le f(\pi(G))$? Note that the analogous question for $\pi^*(G)$ has an affirmative answer. If there exists a vertex v of degree d in G, then any 3-mixing family of permutations of $V(G)$ should contain at least $\log d$ different permutations because any single permutation will leave $\lceil d/2 \rceil$ neighbours of v on the same side of v. Hence $\log \Delta(G) \le \pi^*(G)$. From [1], we know that boxicity$(G) \in O\left(\Delta(G) \log^2 \Delta(G)\right)$ and hence boxicity$(G) \in O\left(2^{\pi^*(G)} (\pi^*(G))^2\right)$.

Another interesting direction of enquiry is to find out the maximum number of hyperedges (edges) possible in a hypergraph (graph) H on n vertices with $\pi(H) \le k$. Such an extremal hypergraph H, with $\pi(H) \le 0$, is seen to be a maximum sized intersecting family of subsets of $[n]$. A similar question for order dimension of a graph has been studied [3,4] and has found applications in ring theory. We can also ask a three dimensional analogue of the question answered by Schnyder's theorem in two dimensions. Given a collection P of non-parallel planes in $\mathbb{R}^3$, can we embed a graph G in $\mathbb{R}^3$ so that every pair of disjoint edges is separated by a plane parallel to one in P? Then $|P|$ has to be at least $\pi(G)$ for this to be possible. This is because the permutations induced by projecting such an embedding onto the normals to the planes in P gives a pairwise suitable family of permutations of G of size $|P|$. Can $|P|$ be upper bounded by a function of $\pi(G)$?

We know that Theorem 7 yields a linear time algorithm for recognizing graphs of separation dimension at most 1. This gives rise to a very natural question. Is it possible to recognize graphs of separation dimension at most 2 in polynomial time?

References

1. Adiga, A., Bhowmick, D., Chandran, L.S.: Boxicity and poset dimension. In: Thai, M.T., Sahni, S. (eds.) COCOON 2010. LNCS, vol. 6196, pp. 3–12. Springer, Heidelberg (2010)

2. Adiga, A., Bhowmick, D., Chandran, L.S.: The hardness of approximating the box-icity, cubicity and threshold dimension of a graph. Discrete Appl. Math. **158**(16), 1719–1726 (2010)
3. Agnarsson, G.: Extremal graphs of order dimension 4. Math. Scand. **90**(1), 5–12 (2002)
4. Agnarsson, G., Felsner, S., Trotter, W.T.: The maximum number of edges in a graph of bounded dimension, with applications to ring theory. Discrete Math. **201**(1–3), 5–19 (1999)
5. Alon, N., Mohar, B., Sanders, D.P.: On acyclic colorings of graphs on surfaces. Isr. J. Math. **94**(1), 273–283 (1996)
6. Basavaraju, M., Chandran, L.S., Mathew, R., Rajendraprasad, D.: Pairwise suitable family of permutations and boxicity. arXiv preprint arXiv:1212.6756 (2012)
7. Bhowmick, D., Chandran, L.S.: Boxicity of circular arc graphs (2008) (preprint). http://arxiv.org/abs/0810.5524
8. Bohra, A., Chandran, L.S., Raju, J.K.: Boxicity of series parallel graphs. Discrete Math. **306**(18), 2219–2221 (2006)
9. Borodin, O.V.: On acyclic colorings of planar graphs. Discrete Math. **25**(3), 211–236 (1979)
10. Chalermsook, P., Laekhanukit, B., Nanongkai, D.: Graph products revisited: Tight approximation hardness of induced matching, poset dimension and more. In: SODA, vol. 13, pp. 1557–1576 (2013)
11. Chandran, L.S., Francis, M.C., Mathew, R.: Chordal bipartite graphs with high boxicity. Graphs Comb. **27**(3), 353–362 (2011)
12. Chandran, L.S., Francis, M.C., Suresh, S.: Boxicity of Halin graphs. Discrete Math. (2008). doi:10.1016/j.disc.2008.09.037
13. Chandran, L.S., Mathew, R., Sivadasan, N.: Boxicity of line graphs. Discrete Math. **311**(21), 2359–2367 (2011)
14. Chandran, L.S., Sivadasan, N.: Boxicity and treewidth. J. Comb. Theory, Ser. B **97**(5), 733–744 (2007)
15. Chee, Y.M., Colbourn, C.J., Horsley, D., Zhou, J.: Sequence covering arrays. SIAM J. Discrete Math. **27**(4), 1844–1861 (2013)
16. Cozzens, M.B.: Higher and multidimensional analogues of interval graphs. Ph.D. thesis, Rutgers University, New Brunswick, NJ (1981)
17. Esperet, L., Joret, G.: Boxicity of graphs on surfaces. Graphs Comb., 1–11 (2011)
18. Kratochvil, J.: A special planar satisfiability problem and a consequence of its NP-completeness. Discrete Appl. Math. **52**, 233–252 (1994)
19. Radhakrishnan, J.: A note on scrambling permutations. Random Struct. Algorithms **22**(4), 435–439 (2003)
20. Roberts, F.S.: On the boxicity and cubicity of a graph. In: Tutte, W.T. (ed.) Recent Progresses in Combinatorics, pp. 301–310. Academic Press, New York (1969)
21. Scheinerman, E.R.: Intersection classes and multiple intersection parameters. Ph.D. thesis, Princeton University (1984)
22. Schnyder, W.: Embedding planar graphs on the grid. In: Proceedings of the First Annual ACM-SIAM Symposium on Discrete Algorithms, pp. 138–148. Society for Industrial and Applied Mathematics (1990)
23. Thomassen, C.: Interval representations of planar graphs. J. Comb. Theory, Ser. B **40**, 9–20 (1986)

Maximal Induced Matchings
in Triangle-Free Graphs

Manu Basavaraju, Pinar Heggernes, Pim van 't Hof[⊠], Reza Saei,
and Yngve Villanger

Department of Informatics, University of Bergen, Bergen, Norway
{manu.basavaraju,pinar.heggernes,pim.vanthof,
reza.saeidinvar,yngve.villanger}@ii.uib.no

Abstract. An induced matching in a graph is a set of edges whose
endpoints induce a 1-regular subgraph. It is known that every n-vertex
graph has at most $10^{n/5} \approx 1.5849^n$ maximal induced matchings, and this
bound is best possible. We prove that every n-vertex triangle-free graph
has at most $3^{n/3} \approx 1.4423^n$ maximal induced matchings, and this bound
is attained by every disjoint union of copies of the complete bipartite
graph $K_{3,3}$. Our result implies that all maximal induced matchings in an
n-vertex triangle-free graph can be listed in time $O(1.4423^n)$, yielding
the fastest known algorithm for finding a maximum induced matching in
a triangle-free graph.

1 Introduction

A celebrated result due to Moon and Moser [8] states that every graph on n
vertices has at most $3^{n/3} \approx 1.4423^n$ maximal independent sets. Moon and Moser
also proved that this bound is best possible by characterizing the extremal graphs
as follows: a graph on n vertices has exactly $3^{n/3}$ maximal independent sets if
and only if it is the disjoint union of $n/3$ triangles. Given the structure of these
extremal graphs, it is natural to investigate how many maximal independent sets
a triangle-free graph can have. Hujter and Tuza [6] showed that a triangle-free
graph on n vertices has at most $2^{n/2} \approx 1.4143^n$ maximal independent sets; this
bound is attained by every 1-regular graph. Later, Byskov [1] gave an algorithmic
proof of the same result, along with more general results.

More recently, Gupta, Raman, and Saurabh [4] showed that for any fixed
non-negative integer r, there exists a constant $c < 2$ such that every graph on
n vertices has at most c^n maximal r-regular induced subgraphs. The aforementioned result by Moon and Moser implies that if $r = 0$, then $c = 3^{1/3}$ is the
best possible upper bound. Gupta et al. [4] complement this by proving tight
upper bounds for the case where $r \in \{1, 2\}$. In particular, their result for $r = 1$

This work is supported by the Research Council of Norway (project SCOPE,
197548/F20), and by the European Research Council under the European
Union's Seventh Framework Programme (FP/2007–2013)/ERC Grant Agreement
no. 267959.

© Springer International Publishing Switzerland 2014
D. Kratsch and I. Todinca (Eds.): WG 2014, LNCS 8747, pp. 93–104, 2014.
DOI: 10.1007/978-3-319-12340-0_8

shows that every n-vertex graph has at most $10^{n/5} \approx 1.5849^n$ maximal induced matchings, and this upper bound is attained by every disjoint union of complete graphs on five vertices. The structure of these extremal graphs again raises the question how much the upper bound can be improved for triangle-free graphs. We answer this question by proving the following result.

Theorem 1. *Every triangle-free graph on n vertices contains at most $3^{n/3}$ maximal induced matchings, and this bound is attained by every disjoint union of copies of $K_{3,3}$.*

We would like to mention some implications of the above theorem. There exist algorithms that list the maximal independent sets of any graph with polynomial delay [7,9], which means that the time spent between the output of two successive maximal independent sets is polynomial in the size of the graph. Together with the aforementioned upper bounds on the number of maximal independent sets, this implies that the maximal independent sets of an n-vertex graph G can be listed in time $O^*(3^{n/3})$, or in time $O^*(2^{n/2})$ in case G is triangle-free.[1]

Cameron [2] observed that the maximal induced matchings of a graph G are exactly the maximal independent sets in the square of the line graph of G. Consequently, the maximal induced matchings of any graph can be listed with polynomial delay. Combining this with the aforementioned upper bound by Gupta et al. [4] yields an algorithm for listing all maximal induced matchings of an n-vertex graph in time $O^*(10^{n/5}) = O(1.5849^n)$. Gupta et al. [4] also obtained an algorithm for finding a maximum induced matching in an n-vertex graph in time $O(1.4786^n)$, which is the current fastest algorithm for solving this problem. Theorem 1 implies that we can do better on triangle-free graphs, as the following two results show. We point out that the problem of finding a maximum induced matching remains NP-hard on subcubic planar bipartite graphs [5], a small subclass of triangle-free graphs.

Corollary 1. *For every triangle-free graph on n vertices, all its maximal induced matchings can be listed in time $O^*(3^{n/3}) = O(1.4423^n)$ with polynomial delay.*

Corollary 2. *For every triangle-free graph G on n vertices, a maximum induced matching in G can be found in time $O^*(3^{n/3}) = O(1.4423^n)$.*

2   Definitions and Notations

All graphs we consider are finite, simple and undirected. We refer the reader to the monograph by Diestel [3] for graph terminology and notation not defined below.

Let G be a graph. For a vertex $v \in V(G)$, we write $N_G(v)$ and $N_G[v]$ to denote open and closed neighborhoods of v, respectively. Let $A \subseteq V(G)$. The closed neighborhood of A is defined as $N_G[A] = \bigcup_{v \in A} N_G[v]$, and the open

---

[1] We use the O^*-notation to suppress polynomial factors, i.e., we write $O^*(f(n))$ instead of $O(f(n) \cdot n^{O(1)})$ for any function f.

neighborhood of A is $N_G(A) = N_G[A] \setminus A$. We write $G[A]$ to denote the subgraph of G induced by A, and we write $G - A$ to denote the graph $G[V(G) \setminus A]$. If $A = \{v\}$, then we simply write $G - v$ instead of $G - \{v\}$. For any non-negative integer r, we say that G is r-*regular* if the degree of every vertex in G is r. A 3-regular graph is called *cubic*. A cycle C with vertices $v_1, v_2 \ldots, v_k$ and edges $v_1 v_2, \ldots, v_{k-1} v_k, v_k v_1$ is denoted by $C = v_1 v_2 \cdots v_k$.

A *matching* in G is a subset $M \subseteq E(G)$ such that no two edges in M share an endpoint. For a matching M in G and a vertex $v \in V(G)$, we say that M *covers* v if v is an endpoint of an edge in M. A matching M is called *induced* if the subgraph induced by endpoints of the edges in M is 1-regular. An induced matching M in G is *maximal* if there exists no induced matching M' in G such that $M \subsetneq M'$. We write $\mathcal{M}_G$ to denote the set of all maximal induced matchings in G. Let X and Y be two disjoint subsets of $V(G)$. We define $\mathcal{M}_G(X, Y)$ to be the set of all maximal induced matchings of G that cover no vertex of X and every vertex of Y. Clearly, $\mathcal{M}_G = \mathcal{M}_G(\emptyset, \emptyset)$. When there is no ambiguity we omit subscripts from the notations.

3   Twins and Maximal Induced Matchings

Let G be a graph. Two vertices $u, v \in V(G)$ are *(false) twins* if $N_G(u) = N_G(v)$. In this paper, whenever we write twin, we mean false twin. For every vertex $u \in V(G)$, the *twin set* of u is defined as $T_G(u) = \{v \in V(G) \mid N_G(u) = N_G(v)\}$, i.e., $T_G(u)$ consists of the vertex u and all its twins. All the twin sets together form a partition of the vertex set of G, and we write $\tau(G)$ to denote the number of sets in this partition, i.e., $\tau(G)$ denotes the number of twin sets in G.

Definition 1. *Let G be a graph. For any two non-adjacent vertices $u, v \in V(G)$, we define $G_{u \to v}$ to be the graph obtained from G by making u into a twin of v by deleting the edge ux for every $x \in N_G(u) \setminus N_G(v)$ and adding the edge uy for every $y \in N_G(v) \setminus N_G(u)$.*

The following lemma identifies certain pairs of vertices u and v for which the operation in Definition 1 does not decrease the number of maximal induced matchings in the graph. This lemma will play a crucial role in the proof of our main result. Note that this lemma holds for general graphs G, and not only for triangle-free graphs.

Lemma 1. *Let G be a graph and let $u, v \in V(G)$. If no maximal induced matching in G covers both u and v, then $|\mathcal{M}_{G_{u \to v}}| \geq |\mathcal{M}_G|$ or $|\mathcal{M}_{G_{v \to u}}| \geq |\mathcal{M}_G|$.*

Proof. Without loss of generality, we assume that the number of matchings in $\mathcal{M}_G$ that cover u is greater than or equal to the number of matchings in $\mathcal{M}_G$ that cover v, i.e., $|\mathcal{M}_G(\emptyset, \{u\})| \geq |\mathcal{M}_G(\emptyset, \{v\})|$. Since every matching in $\mathcal{M}_G$ that covers u does not cover v due to the assumption that $\mathcal{M}_G(\emptyset, \{u, v\}) = \emptyset$, it holds that $\mathcal{M}_G(\emptyset, \{u\}) = \mathcal{M}_G(\{v\}, \{u\})$. By symmetry, we also have that $\mathcal{M}_G(\emptyset, \{v\}) = \mathcal{M}_G(\{u\}, \{v\})$. This implies that $|\mathcal{M}_G(\{v\}, \{u\})| \geq |\mathcal{M}_G(\{u\}, \{v\})|$. We now use this fact to prove that $|\mathcal{M}_{G_{v \to u}}| \geq |\mathcal{M}_G|$.

For convenience, we write $G' = G_{v \to u}$. The set M_G of all maximal induced matchings in G can be partitioned as follows:

$$M_G = M_G(\{v\}, \{u\}) \uplus M_G(\{u\}, \{v\}) \uplus M_G(\emptyset, \{u, v\}) \uplus M_G(\{u, v\}, \emptyset).$$

We can partition $M_{G'}$ in the same way:

$$M_{G'} = M_{G'}(\{v\}, \{u\}) \uplus M_{G'}(\{u\}, \{v\}) \uplus M_{G'}(\emptyset, \{u, v\}) \uplus M_{G'}(\{u, v\}, \emptyset).$$

We claim that $M_G(\{v\}, \{u\}) = M_{G'}(\{v\}, \{u\})$. Let $M \in M_G(\{v\}, \{u\})$. We claim that $M \in M_{G'}(\{v\}, \{u\})$. It is easy to verify that M is an induced matching in G', as we only change edges incident with v when transforming G into G', and M does not cover v. For contradiction, suppose M is not a maximal induced matching in G'. Then there is an edge $xy \in E(G')$ such that $M \cup \{xy\}$ is an induced matching in G'. Since u and v are twins in G' and M covers u, we find that $v \notin \{x, y\}$. This implies that $xy \in E(G)$, so $M \cup \{xy\}$ is a matching in G that does not cover v. In fact, $M \cup \{xy\}$ is an induced matching in G, since every edge in $E(G) \setminus E(G')$ is incident with v. This contradicts the maximality of M in G. Hence we have that $M_G(\{v\}, \{u\}) \subseteq M_{G'}(\{v\}, \{u\})$. To show why $M_{G'}(\{v\}, \{u\}) \subseteq M_G(\{v\}, \{u\})$, let $M' \in M_{G'}(\{v\}, \{u\})$. For similar reasons as before, M' is an induced matching in G. To show that M' is maximal in G, suppose for contradiction that there is an edge $xy \in E(G)$ such that $M' \cup \{xy\}$ is an induced matching in G. Then $v \notin \{x, y\}$, this time due to the assumption that no maximal induced matching in G covers both u and v. Now we can use similar arguments as before to conclude that $M' \cup \{x, y\}$ is an induced matching in G', yielding the desired contradiction.

By assumption, we have $M_G(\emptyset, \{u, v\}) = \emptyset$. Since u and v are twins in G' by construction, we also know that $M_{G'}(\emptyset, \{u, v\}) = \emptyset$ and $M_{G'}(\{v\}, \{u\}) = M_{G'}(\{u\}, \{v\})$. Recall that $|M_G(\{v\}, \{u\})| \geq |M_G(\{u\}, \{v\})|$, which implies that $|M_{G'}(\{u\}, \{v\})| \geq |M_G(\{u\}, \{v\})|$. Hence, in order to show that $|M_{G'}| \geq |M_G|$, it suffices to show that $|M_{G'}(\{u, v\}, \emptyset)| \geq |M_G(\{u, v\}, \emptyset)|$.

Let $M \in M_G(\{u, v\}, \emptyset)$. We claim that $M \in M_{G'}(\{u, v\}, \emptyset)$. It is easy to see that M is an induced matching in G', as the only edges that are modified are incident with v and M does not cover v. Suppose, for contradiction, that M is not a maximal induced matching in G'. Then there exists an edge $xy \in E(G')$ such that $M \cup \{xy\}$ is an induced matching in G'. If $v \notin \{x, y\}$, then $M \cup \{xy\}$ is also an induced matching in G, contradicting the maximality of M. Thus we have $v \in \{x, y\}$. Without loss of generality, suppose $x = v$. Let $M' = M \cup \{vy\}$. Now consider $M'' = M \cup \{uy\}$. Since M' is induced matching and u and v are twins in G', we infer that M'' is also an induced matching in G'. Note that the edge uy is also present in G, so M'' is an induced matching in G. This contradicts the maximality of M, implying that $M \in M_{G'}(\{u, v\}, \emptyset)$ and consequently $M_G(\{u, v\}, \emptyset) \subseteq M_{G'}(\{u, v\}, \emptyset)$. This completes the proof of Lemma 1. $\square$

For our purposes, we need to extend Definition 1 as follows.

Definition 2. *Let G be a graph. For any two non-adjacent vertices $u, v \in V(G)$, the graph $G_{T_G(u) \to v}$ is the graph obtained from G by making each vertex of $T_G(u)$ into a twin of v as follows: for every $u' \in T_G(u)$, delete the edge $u'x$ for every $x \in N_G(u) \setminus N_G(v)$ and add the edge $u'y$ for every $y \in N_G(v) \setminus N_G(u)$.*

The following lemma is an immediate corollary of Lemma 1, since we can repeatedly apply the operation in Definition 1 on all the vertices in $T_G(u)$.

Lemma 2. *Let G be a graph and let $u, v \in V(G)$. If no maximal induced matching in G covers both u and v, then $|M_{G_{T_G(u) \to v}}| \geq |M_G|$ or $|M_{G_{T_G(v) \to u}}| \geq |M_G|$.*

We also need the following two lemmas in the proof of our main result.

Lemma 3. *Let G be a triangle-free graph. For any two non-adjacent vertices $u, v \in V(G)$, the graph $G_{T_G(u) \to v}$ is triangle-free.*

Proof. Let $u, v \in V(G)$. For contradiction, suppose that $G_{T_G(u) \to v}$ contains a triangle C. Observe that every edge that was added to G in order to create $G_{T_G(u) \to v}$ is incident with a vertex in $T_G(u)$ and a vertex in $N_G(v) \setminus N_G(u)$. Hence, C contains an edge $u'x$ such that $u' \in T_G(u)$ and $x \in N_G(v) \setminus N_G(u)$. Let y be the third vertex of C. Since G is triangle-free, $N_G(v)$ forms an independent set in both G and $G_{T_G(u) \to v}$. This implies in particular that y is not adjacent to v in $G_{T_G(u) \to v}$, and since we did not delete any edge incident with v when creating $G_{T_G(u) \to v}$, it holds that y is not adjacent to v in G either. Moreover, since both u' and y do not belong to $N_G(v) \setminus N_G(u)$, the edge $u'y$ is present in G. But then, by Definition 2, the edge $u'y$ should have been deleted when G was transformed into $G_{T_G(u) \to v}$. This yields the desired contradiction. $\square$

Lemma 4. *Let G be a triangle-free graph and let $u, v \in V(G)$ be two non-adjacent vertices. If u and v are not twins, then $\tau(G_{T_G(u) \to v}) < \tau(G)$.*

Proof. Suppose u and v are not twins. Then $T_G(u)$ and $T_G(v)$ are two different twin sets in G. By Definition 2, the vertices of $T_G(u) \cup T_G(v)$ all belong to the same twin set in $G_{T_G(u) \to v}$, namely the twin set $T_{G_{T_G(u) \to v}}(u) = T_{G_{T_G(u) \to v}}(v)$. Let $x \in V(G) \setminus (T_G(u) \cup T_G(v))$. We prove that all the vertices in $T_G(x)$ belong to the same twin set in $G_{T_G(u) \to v}$, which implies that $\tau(G_{T_G(u) \to v}) < \tau(G)$.

Suppose there is a vertex $y \in T_G(x)$ such that x and y are not twins in $G_{T_G(u) \to v}$. Without loss of generality, suppose there is a vertex $z \in N_{G_{T_G(u) \to v}}(y) \setminus N_{G_{T_G(u) \to v}}(x)$. Since x and y are twins in G, we either have $xz, yz \in E(G)$ or $xz, yz \notin E(G)$. In the first case, the edge xz is deleted from G when $G_{T_G(u) \to v}$ is created, which implies that $x \in N_G(u) \setminus N_G(v)$ by Definition 2. However, since x and y are twins in G, it holds that $y \in N_G(u) \setminus N_G(v)$ as well, implying that the edge yz should not exist in $G_{T_G(u) \to v}$. This contradicts the definition of z. If $xz, yz \notin E(G)$, then we can use similar argument to conclude that xz should be an edge in $G_{T_G(u) \to v}$, again yielding a contradiction. $\square$

4 Proof of Theorem 1

This section is devoted to proving Theorem 1. We first prove that every triangle-free graph on n vertices has at most $3^{n/3}$ maximal induced matchings. At the end of the section, we show why the bound in Theorem 1 is best possible.

A triangle-free graph on n vertices that has more than $3^{n/3}$ maximal induced matchings is called a *counterexample*. For contradiction, let us assume that there exists a counterexample. Then there exists a counterexample G such that for every counterexample G', it holds that either $|V(G')| > |V(G)|$, or $|V(G')| = |V(G)|$ and $\tau(G') \geq \tau(G)$. Let $n = |V(G)|$. By definition of a counterexample, $|M_G| > 3^{n/3}$. We will prove a sequence of structural properties of G, and finally conclude that G does not exist, yielding the desired contradiction.

Lemmas that appear without proofs below are repeated with proofs in a separate appendix at the end.

Lemma 5. *G is connected and has at least three vertices.*

Lemma 6. *Let $u, v \in V(G)$. If there is no maximal induced matching in G that covers both u and v, then u and v are twins.*

Proof. Suppose there is no maximal induced matching in G that covers both u and v. In particular, this implies that u and v are not adjacent. Let $G' = G_{T_G(u) \to v}$ and $G'' = G_{T_G(v) \to u}$. By Lemma 2, we have that $|M_{G'}| \geq |M_G|$ or $|M_{G''}| \geq |M_G|$. Without loss of generality, suppose $|M_{G'}| \geq |M_G|$. The graph G' is triangle-free due to Lemma 3. This, together with the fact that $|M_{G'}| \geq |M_G| > 3^{n/3}$, implies that G' is a counterexample. But by Lemma 4, it holds that $\tau(G') < \tau(G)$, which contradicts the choice of G. □

Lemma 7. *For every edge $uv \in E(G)$ and every set $X \subseteq V(G) \setminus \{u, v\}$, it holds that $|M_G(X, \{u, v\})| \leq 3^{(n - |X \cup N[\{u,v\}]|)/3}$.*

Proof. Let $G' = G - (X \cup N_G[\{u, v\}])$. We first show that for every matching $M \in M_G(X, \{u, v\})$, it holds that $M \setminus \{uv\} \in M_{G'}$. Let $M \in M_G(X, \{u, v\})$. Since $uv \in E(G)$ and M covers both u and v, the edge uv belongs to M. Since M does not cover any vertex in X, it is clear that the set $M' = M \setminus \{uv\}$ is an induced matching in G'. We show that M' is maximal. For contradiction, suppose there exists an edge $xy \in E(G')$ such that $M' \cup \{xy\}$ is an induced matching in G'. Since neither x nor y belongs to the set $X \cup N_G[\{u, v\}]$, we have in particular that there is no edge between the sets $\{x, y\}$ and $\{u, v\}$. Hence, adding the edge xy to M yields an induced matching in G, contradicting the assumption that M is a maximal induced matching in G.

We now know that for every matching $M \in M_G(X, \{u, v\})$, it holds that $M \setminus \{uv\} \in M_{G'}$. Note that, for any two matchings $M_1, M_2 \in M_G(X, \{u, v\})$ with $M_1 \neq M_2$, the sets $M_1 \setminus \{uv\}$ and $M_2 \setminus \{uv\}$ are not equal, as both M_1 and M_2 contain the edge uv. Hence we have that $|M_G(X, \{u, v\})| \leq |M_{G'}|$. Since G' has less vertices than G and is thus not a counterexample, we have that $|M_{G'}| \leq 3^{|V(G')|/3} = 3^{(n - |X \cup N[\{u,v\}]|)/3}$. We conclude that $|M_G(X, \{u, v\})| \leq |M_{G'}| \leq 3^{(n - |X \cup N[\{u,v\}]|)/3}$. □

Lemma 8. *G has no vertex of degree less than 2.*

Lemma 9. *G has no 5-cycle containing two non-adjacent vertices of degree 2.*

Lemma 10. *G has no 4-cycle containing exactly one vertex of degree 2.*

Lemma 11. *G has no two adjacent vertices of degree 2.*

Proof. For contradiction, suppose there are two vertices u and v such that $d(u) = d(v) = 2$ and $uv \in E(G)$. Let a and b denote the other neighbors of u and v, respectively. Since G is triangle-free, we have that $a \neq b$. We first show that $ab \notin E(G)$. For contradiction, assume that $ab \in E(G)$ and both a and b have degree 2. Then G is isomorphic to C_4, implying that $|M_G| = 4 \leq 3^{4/3}$. This contradicts the fact that G is a counterexample. Hence a or b has degree more than 2. Assume without loss of generality that $d(a) \geq 3$. Then a and v are not twins, and there is no matching in M_G covering both a and v. This contradiction to Lemma 6 implies that $ab \notin E$.

We now partition M_G into three sets $M(\emptyset, \{a\})$, $M(\{a\}, \{b\})$, and $M(\{a, b\}, \emptyset)$, and find an upper bound on the size of each of these sets.

We first consider $M(\emptyset, \{a\})$. Clearly, $|M(\emptyset, \{a\})| = \sum_{p \in N(a)} |M(\emptyset, \{a, p\})|$. Let $p = u$. Since $|N[\{a, u\}]| = d(a) + 2$, from Lemma 7 we have $|M(\emptyset, \{a, u\})| \leq 3^{(n-(d(a)+2))/3}$. Now consider the case that $p \neq u$. In this case, $|N[\{a, p\}]| = d(a) + d(p)$ and $d(p) \geq 2$ due to Lemma 8, and thus Lemma 7 implies $|M(\emptyset, \{a, p\})| \leq 3^{(n-(d(a)+2))/3}$. Consequently, we obtain

$$|M(\emptyset, \{a\})| = \sum_{p \in N(a)} |M(\emptyset, \{a, p\})| \leq d(a) \cdot 3^{\frac{n-(d(a)+2)}{3}}.$$

We now find an upper bound on $|M(\{a\}, \{b\})|$. Since no matching in the set $M(\{a\}, \{b\})$ covers u, it holds that $M(\{a\}, \{b\}) = M(\{a, u\}, \{b\})$. We use the fact that $|M(\{a, u\}, \{b\})| = \sum_{q \in N(b)} |M(\{a, u\}, \{b, q\})|$. If $q = v$, then $|M(\{a, u\}, \{b, v\})| \leq 3^{(n-(d(b)+3))/3}$ due to Lemma 7 and the fact that $d(v) = 2$ and $a \notin N[\{b, v\}]$. Let now $q \neq v$. First suppose q is adjacent to a. Then $qauvb$ is a 5-cycle, and hence Lemma 9 implies that $d(q) \geq 3$. Consequently, $|N[\{b, q\}]| = d(b) + d(q) \geq d(b) + 3$, and since $u \notin N[\{b, q\}]$, we find that $|M(\{a, u\}, \{b, q\})| \leq 3^{(n-(d(b)+4))/3}$ due to Lemma 7. Now suppose that q is not adjacent to a. Then $N[\{b, q\}]$ contains neither a nor u. Hence, Lemma 7 and the fact that $|N[\{b, q\}] \geq d(b) + 2$ imply that $|M(\{a, u\}, \{b, q\})| \leq 3^{(n-(d(b)+4))/3}$. We conclude that

$$|M(\{a\}, \{b\})| \leq 3^{\frac{n-(d(b)+3)}{3}} + (d(b) - 1) \cdot 3^{\frac{n-(d(b)+4)}{3}}.$$

Finally, we consider $M(\{a, b\}, \emptyset)$. Every matching in $M(\{a, b\}, \emptyset)$ is maximal and covers neither a nor b, so it must contain edge uv. Hence, it holds that $M(\{a, b\}, \emptyset) = M(\{a, b\}, \{u, v\})$. Since $|N[\{u, v\}]| = 4$, Lemma 7 gives

$$|M(\{a, b\}, \emptyset)| = |M(\{a, b\}, \{u, v\})| \leq 3^{\frac{n-4}{3}}.$$

Combining the obtained upper bounds, we find that

$$|M_G| \leq f(d(a), d(b)) \cdot 3^{\frac{n}{3}},$$

where the function f is defined as follows:

$$f(d(a), d(b)) = d(a) \cdot 3^{-\frac{d(a)+2}{3}} + 3^{-\frac{d(b)+3}{3}} + (d(b) - 1) \cdot 3^{-\frac{d(b)+4}{3}} + 3^{-\frac{4}{3}}.$$

Recall that both a and b have degree at least 2 due to Lemma 8. We observe that $f(2, 2) < 0.965$, yielding an upper bound of $0.965 \cdot 3^{n/3}$ on $|M_G|$ in case $d(a) = d(b) = 2$. Now consider the case where $d(a) = 2$ and $d(b) \geq 3$. Then the function f is decreasing with respect to $d(b)$. Since $f(2, 3) < 0.959$, we find that $|M_G| < 0.959 \cdot 3^{n/3}$ in this case. By using similar arguments, we find that $|M_G| < 0.984 \cdot 3^{n/3}$ when $d(b) = 2$ and $d(a) \geq 3$. Finally, when both $d(a) \geq 3$ and $d(b) \geq 3$, then the function f is decreasing with respect to both variables $d(a)$ and $d(b)$ and is maximum when $d(a) = d(b) = 3$. Since $f(3, 3) < 0.978$, we find that $|M_G| < 0.978 \cdot 3^{n/3}$ whenever $d(a) \geq 3$ and $d(b) \geq 3$. Summarizing, we obtain a contradiction to the assumption that $|M_G| > 3^{n/3}$ in each case, which completes the proof of this case. □

Lemma 12. *Let $u \in V(G)$. If u has degree 2, then both its neighbors have degree 3.*

Lemma 13. *G has no vertex of degree more than 4.*

Lemma 14. *G has no 4-cycle containing a vertex of degree 2.*

Lemma 15. *G has no vertex of degree 2.*

Lemma 16. *G is cubic.*

Proof. Due to Lemmas 8, 13, and 15, every vertex in G has degree 3 or 4. Hence, in order to prove Lemma 16, it suffices to prove that G has no vertex of degree 4. For contradiction, suppose there exists a vertex u such that $d(u) = 4$. Let v be a neighbor of u. To find an upper bound on $|M_G|$, we partition M_G into two sets $M(\emptyset, \{v\})$ and $M(\{v\}, \emptyset)$ and find upper bounds on the sizes of these sets.

Observe that $|M(\emptyset, \{v\})| = \sum_{q \in N(v)} |M(\emptyset, \{v, q\})|$. If $q = u$, then $|N[\{v, q\}]| = d(v) + 4$ and hence $|M(\emptyset, \{u, v\})| \leq 3^{(n-(d(v)+4))/3}$ by Lemma 7. For any vertex $q \in N(v) \setminus \{u\}$, the fact that $|N[\{q, v\}]| \geq d(v) + 3$ together with Lemma 7 implies that $|M(\emptyset, \{q, v\})| \leq 3^{(n-(d(v)+3))/3}$. Hence we find that

$$|M(\emptyset, \{v\})| \leq 3^{\frac{n-(d(v)+4)}{3}} + (d(v) - 1) \cdot 3^{\frac{n-(d(v)+3)}{3}}.$$

Since $M(\{v\}, \emptyset) = M_{G-v}$ and $G - v$ is not a counterexample, we have that

$$|M(\{v\}, \emptyset)| \leq 3^{\frac{n-1}{3}}.$$

Hence we conclude that

$$|M_G| \leq 3^{\frac{n-(d(v)+4)}{3}} + (d(v) - 1) \cdot 3^{\frac{n-(d(v)+3)}{3}} + 3^{\frac{n-1}{3}} .$$

For every fixed value of $d(v) \in \{3,4\}$, it can easily be verified that $|M_G| \leq 3^{n/3}$, yielding the desired contradiction. □

Lemma 17. *Let $u,v \in V(G)$. If u and v are contained in a 5-cycle C, then u and v have no common neighbor in $V(G) \setminus V(C)$.*

Lemma 18. *G contains at least one 4-cycle.*

Lemma 19. *G is isomorphic to $K_{3,3}$.*

Proof. Let uv be an edge of G such that no edge in $E(G) \setminus \{uv\}$ is contained in more 4-cycles than uv is. Since u and v are adjacent and G is triangle-free, u and v have no common neighbor. Recall that G is cubic due to Lemma 16. Let $N(u) = \{a,d\}$ and $N(v) = \{b,c\}$. It is easy to see that edge uv is contained in at most four 4-cycles.

If uv is contained in exactly four 4-cycles, then G is isomorphic to $K_{3,3}$ and the lemma holds. Suppose uv is contained in at most three 4-cycles. Due to Lemma 18 and the choice of uv, edge uv belongs to at least one 4-cycle. Hence, there is at least one edge between sets $\{a,d\}$ and $\{b,c\}$. Note that $ad \notin E(G)$ and $bc \notin E(G)$, as G is triangle-free. We distinguish four cases, depending on the adjacencies between vertices in $\{a,d\}$ and $\{b,c\}$.

> *Case 1: $ab \in E(G)$ and $ac, db, dc \notin E(G)$.*
> *Case 2: $ab, ac \in E(G)$ and $db, dc \notin E(G)$.*
> *Case 3: $ab, cd \in E(G)$ and $ac, db \notin E(G)$.*
> *Case 4: $ab, ac, bd \in E(G)$ and $dc \notin E(G)$.*

Note that uv belongs to exactly one 4-cycle in Case 1, to exactly two 4-cycles in Cases 2 and 3, and to exactly three 4-cycles in Case 4.

Observe that M_G is equal to

$$M(\emptyset, \{a\}) \uplus M(\{a\}, \{b\}) \uplus M(\{a,b\}, \{c\}) \uplus M(\{a,b,c\}, \{d\}) \uplus M(\{a,b,c,d\}, \emptyset) .$$

In Claims 1–5 below, we prove upper bounds on the sizes of the five sets in the above expression. We then combine these five upper bounds in order to obtain an upper bound on $|M_G|$.

Claim 1. $|M(\emptyset, \{a\})| \leq 3 \cdot 3^{(n-6)/3}$.

Since G is cubic due to Lemma 16, the closed neighborhood of any of the three edges incident with a consists of six vertices. Hence Lemma 7 ensures that $|M(\emptyset, \{a,q\})| \leq 3^{(n-6)/3}$ for every $q \in N(a)$, implying the upper bound given in Claim 1.

Claim 2. $|M(\{a\}, \{b\})| \leq 2 \cdot 3^{(n-6)/3}$.

Note that in all four cases, ab belongs to $E(G)$. By definition, there is no matching in $M(\{a\}, \{b\})$ that contains ab. For any of the other two edges incident with b, its closed neighborhood has size 6. Hence the correctness of the claimed upper bound again follows from Lemma 7.

Claim 3. $|M(\{a, b\}, \{c\})| \le 3^{(n-6)/3} + 3^{(n-7)/3}$.

First we consider Case 1. In this case, the closed neighborhood of cv contains vertex b and it does not contain a. Therefore, $|\{a, b\} \cup N[\{c, v\}]| = 7$ and consequently, by Lemma 7 we have that $|M(\{a, b\}, \{c, v\})| \le 3^{(n-7)/3}$. Let cq be one of the other edges incident with c. Recall that uv belongs to exactly one 4-cycle in Case 1. Hence, vertex q is not adjacent to b, as otherwise bv belongs to two 4-cycles, contradicting the choice of uv. We claim that q is not adjacent to a. For contradiction, suppose q is adjacent to a. Then $qabvc$ is a 5-cycle containing a and v, so a and v cannot have a common neighbor in $V(G) \setminus \{q, a, b, v, c\}$ due to Lemma 17. The fact that both a and v are adjacent to u gives the desired contradiction. Hence, for any $q \in N(c) \setminus \{v\}$, we have that $|\{a, b\} \cup N[\{c, q\}]| = 8$, and thus Lemma 7 implies that $|M(\{a, b\}, \{c, q\})| \le 3^{(n-8)/3}$. We obtain that $|M(\{a, b\}, \{c\})| \le 3^{(n-7)/3} + 2 \cdot 3^{(n-8)/3}$, which is strictly smaller than $3^{(n-6)/3} + 3^{(n-7)/3}$.

Let us now consider Case 2. Observe that no matching in $M(\{a, b\}, \{c\})$ contains edge ac. Hence $|M(\{a, b\}, \{c\})| = |M(\{a, b\}, \{c, v\})| + |M(\{a, b\}, \{c, q\})|$, where q is the neighbor of c other than v and a. Both vertices a and b belong to $N[\{c, v\}]$ and hence $|\{a, b\} \cup N[\{c, v\}]| = 6$. Therefore, Lemma 7 guarantees that $|M(\{a, b\}, \{c, v\})| \le 3^{(n-6)/3}$. Note that $a \notin N[\{c, q\}]$. We claim that $b \notin N[\{c, q\}]$. For contradiction, suppose $b \in N[\{c, q\}]$. Then b is adjacent to q, and hence bv belongs to three 4-cycles, namely $bvua$, $bvcq$, and $bvca$. Since uv belongs to only two 4-cycles in Case 2, this contradicts the choice of uv. Hence, we have that $|\{a, b\} \cup N[\{c, q\}]| = 7$ and consequently $|M(\{a, b\}, \{c, q\})| \le 3^{(n-7)/3}$ by Lemma 7. We can now conclude that $|M(\{a, b\}, \{c\})| \le 3^{(n-6)/3} + 3^{(n-7)/3}$.

For Case 3, let q be the neighbor of c other than d and v. Since $b \in N[\{c, v\}]$ and $a \notin N[\{c, v\}]$ in Case 3, we have that $|\{a, b\} \cup N[\{c, v\}]| = 7$ and therefore $|M(\{a, b\}, \{c, v\})| \le 3^{(n-7)/3}$ due to Lemma 7. Moreover, since $N[\{c, d\}]$ contains neither a nor b, Lemma 7 implies that $|M(\{a, b\}, \{c, d\})| \le 3^{(n-8)/3}$. We now consider edge cq. Since no matching in $M(\{a, b\}, \{c, q\})$ covers u, it holds that $M(\{a, b\}, \{c, q\}) = M(\{a, b, u\}, \{c, q\})$. Recall that $N(u) = \{v, a, d\}$, so $q \notin N[u]$. For contradiction, suppose $a \in N[q]$. Then $qauvc$ is a 5-cycle containing two vertices, namely u and c, that have a common neighbor, namely d, in the set $V(G) \setminus \{q, a, u, v, c\}$. This contradicts Lemma 17, so we conclude that $a \notin N[q]$. Consequently, we have that $|\{a, b, u\} \cup N[\{c, p\}]| = 8$, so Lemma 7 implies that $|M(\{a, b\}, \{c, p\})| \le 3^{(n-8)/3}$. We conclude that $|M(\{a, b\}, \{c\})| \le 3^{(n-7)/3} + 2 \cdot 3^{(n-8)/3} < 3^{(n-6)/3} + 3^{(n-7)/3}$.

Finally, we consider Case 4. Let $N(c = \{a, v, q\}$. Since no matching in $M(\{a, b\}, \{c\})$ covers a, we have that $|M(\{a, b\}, \{c\})| = |M(\{a, b\}, \{c, v\})| + |M(\{a, b\}, \{c, q\})|$. The fact that $|N[\{c, v\}]| = 6$ together with Lemma 7 implies that $|M(\{a, b\}, \{c, v\})| \le 3^{(n-6)/3}$. Since $N(a) = \{u, b, c\}$ and $N(b) = \{v, a, c\}$ in this case, neither a nor b belongs to $N[\{c, q\}]$. Hence, $|\{a, b\} \cup N[\{c, q\}]| = 8$ and using Lemma 7, we deduce that $|M(\{a,b\},\{c,q\})| \le 3^{(n-8)/3}$. We can therefore conclude that $|M(\{a,b\},\{c\})| \le 3^{(n-6)/3} + 3^{(n-8)/3} < 3^{(n-6)/3} + 3^{(n-7)/3}$.

Claim 4. $|M(\{a,b,c\},\{d\})| \le 3^{(n-7)/3} + 3^{(n-8)/3}$.

First we consider Cases 1 and 2. In both of these cases, the closed neighborhood of du contains a, but neither b nor c belong to $N[\{d,u\}]$. Therefore, $|\{a,b,c\} \cup N[\{d,u\}]| = 8$ and consequently, by Lemma 7, we have that $|M(\{a,b,c\},\{d,u\})| \le 3^{(n-8)/3}$. Let $q \in N(d) \setminus \{u\}$. Since no matching in the set $M(\{a,b,c\},\{d,q\})$ covers vertex v, we have that $M(\{a,b,c\},\{d,q\}) = M(\{a,b,c,v\},\{d,q\})$. Edge au belongs to all the 4-cycles to which uv belongs. By the choice of uv, edge au does not belong to any other 4-cycle. In particular, the vertices $\{a,q,d,u\}$ do not induce a C_4, which implies that $a \notin N[q]$. We claim that b is also not adjacent to q. For contradiction, suppose $b \in N[q]$. Then vertices b and u are contained in a 5-cycle, namely $bpduv$, so the fact that they are adjacent to a contradicts Lemma 17. Finally, we observe that $v \notin N[q]$, since $N(v) = \{u,b,c\}$. From this, we deduce that $|\{a,b,c,v\} \cup N[\{d,q\}]| = 9$, and hence $|M(\{a,b,c\},\{d,q\})| \le 3^{(n-9)/3}$ due to Lemma 7. We conclude that $|M(\{a,b,c\},\{d\})| \le 3^{(n-8)/3} + 2 \cdot 3^{(n-9)/3}$, which is less than $3^{(n-7)/3} + 3^{(n-8)/3}$.

Now we consider Case 3. Since no matching in $M(\{a,b,c\},\{d\})$ contains edge dc, we have that $|M(\{a,b,c\},\{d\})| = |M(\{a,b,c\},\{d,u\})| + |M(\{a,b,c\},\{d,q\})|$, where q is the neighbor of d other than c and u. In Case 3, both vertices a and c are in $N[\{d,u\}]$ and $b \notin N[\{d,u\}]$. Hence $|\{a,b,c\} \cup N[\{d,u\}]| = 7$, and therefore Lemma 7 implies that $|M(\{a,b,c\},\{d,u\})| \le 3^{(n-7)/3}$. From the observation that no matching in $M(\{a,b,c\},\{d\})$ covers v, it follows that $M(\{a,b,c\},\{d,q\}) = M(\{a,b,c,v\},\{d,q\})$. We claim that neither v nor b belong to $N[\{d,q\}]$. The fact that neither v nor b belongs to $N[d]$ follows from the triangle-freeness of G and the fact that we are in Case 3. Moreover, since $N(v) = \{u,b,c\}$, we have that $v \notin N[q]$. For contradiction, suppose $b \in N[q]$. Then the vertices $\{q,b,v,u,d\}$ induce a 5-cycle. Vertices b and u lie on this 5-cycle and have a common neighbor, namely a, in $V(G) \setminus \{q,b,v,u,d\}$. This contradiction to Lemma 17 implies that $\{v,b\} \cap N[\{d,q\}] = \emptyset$. Hence $|\{a,b,c,v\} \cup N[\{d,q\}]| \ge 8$, and we can use Lemma 7 to find that $M(\{a,b,c\},\{d,q\})| \le 3^{(n-8)/3}$. Consequently, we have that $|M(\{a,b,c\},\{d\})| \le 3^{(n-7)/3} + 3^{(n-8)/3}$.

It remains to consider Case 4. Let q be the neighbor of d other than b and u. Since edge db is not contained in any of the matchings in $M(\{a,b,c\},\{d\})$, we have that $|M(\{a,b,c\},\{d\}| = |M(\{a,b,c\},\{d,u\})| + |M(\{a,b,c\},\{d,q\})|$. Since $c \notin N[\{d,u\}]$, we have that $|\{a,b,c\} \cup N[\{d,u\}]| = 7$ and hence $|M(\{a,b,c\}, \{d,u\})| \le 3^{(n-7)/3}$ due to Lemma 7. Observe that vertex v is not covered by any matching in $M(\{a,b,c\},\{d,q\})$, which implies that $M(\{a,b,c\},\{d,q\}) = M(\{a,b,c,v\},\{d,q\})$. Since $N(a) = N(v) = \{u,b,c\}$, we have that neither a nor v belongs to $N[\{d,q\}]$. We now show that c does not belong to $N[\{d,q\}]$ either. For contradiction, suppose otherwise. Since c is not adjacent to d in Case 4, vertex c must be adjacent to q. Hence the vertices $\{c,q,d,u,v\}$ induce a 5-cycle. By Lemma 17, no two vertices on this cycle have a common neighbor outside the cycle, contradicting the fact that both u and c are adjacent to a. This implies that $|\{a,b,c,v\} \cup N[\{d,q\}]| \ge 9$, so $|M(\{a,b,c\},\{d,q\})| \le 3^{(n-9)/3}$ due to Lemma 7.

We conclude that $|M(\{a,b,c\},\{d\})| \leq 3^{(n-7)/3} + 3^{(n-9)/3}$, which is clearly upper bounded by $3^{(n-7)/3} + 3^{(n-8)/3}$.

Claim 5. $|M(\{a,b,c,d\},\emptyset)| \leq 3^{(n-6)/3}$.

Every maximal induced matching in G that does not cover any vertex in $\{a,b,c,d\}$ contains edge uv. Therefore, $M(\{a,b,c,d\},\emptyset) = M(\{a,b,c,d\},\{u,v\})$. Since $|N[\{u,v\}]| = 6$ due to the fact that G is cubic by Lemma 16, it follows from Lemma 7 that $|M(\{a,b,c,d\},\{u,v\})| \leq 3^{(n-6)/3}$. This completes the proof of Claim 5.

Combining the upper bounds in Claims 1–5 yields the following:

$$|M_G| \leq 7 \cdot 3^{\frac{n-6}{3}} + 2 \cdot 3^{\frac{n-7}{3}} + 3^{\frac{n-8}{3}} < 3^{\frac{n}{3}}.$$

This contradicts the assumption that G is a counterexample.   □

Lemma 19 states that G is isomorphic to $K_{3,3}$, so in particular $n = 6$. Since every maximal induced matching in $K_{3,3}$ consists of a single edge, we have that $|M_G| = |E(K_{3,3})| = 9 = 3^{n/3}$, contradicting the assumption that G is a counterexample. This contradiction implies that every triangle-free graph on n vertices has at most $3^{n/3}$ maximal induced matchings.

It remains to show that the bound in Theorem 1 is best possible. Let G be the disjoint union of p copies of $K_{3,3}$ for some positive integer p. Every maximal induced matching in G contains exactly one edge of each connected component of G, which implies that $|M_G| = 9^p = 9^{n/6} = 3^{n/3}$. This completes the proof of Theorem 1.

References

1. Byskov, J.M.: Enumerating maximal independent sets with applications to graph colouring. Oper. Res. Lett. **32**, 547–556 (2004)
2. Cameron, K.: Induced matchings. Discr. Appl. Math. **24**, 97–102 (1989)
3. Diestel, R.: Graph Theory. Electronic Edn. Springer, Heidelberg (2005)
4. Gupta, S., Raman, V., Saurabh, S.: Maximum r-regular induced subgraph problem: fast exponential algorithms and combinatorial bounds. SIAM J. Disc. Math. **26**, 1758–1780 (2012)
5. Hocquard, H., Ochem, P., Valicov, P.: Strong edge-colouring and induced matchings. Inf. Proc. Lett. **113**, 836–843 (2013)
6. Hujter, M., Tuza, Z.: The number of maximal independent sets in triangle-free graphs. SIAM J. Disc. Math. **6**, 284–288 (1993)
7. Johnson, D.S., Papadimitriou, C.H., Yannakakis, M.: On generating all maximal independent sets. Inf. Process. Lett. **27**, 119–123 (1988)
8. Moon, J.W., Moser, L.: On cliques in graphs. Israel J. Math. **3**, 23–28 (1965)
9. Tsukiyama, S., Ide, M., Ariyoshi, H., Shirakawa, I.: A new algorithm for generating all the maximal independent sets. SIAM J. Comput. **6**, 505–517 (1977)

Independent Set Reconfiguration in Cographs

Paul Bonsma[✉]

Faculty of EEMCS, University of Twente, PO Box 217,
7500 AE Enschede, The Netherlands
p.s.bonsma@ewi.utwente.nl

Abstract. We study the following independent set reconfiguration problem: given two independent sets I and J of a graph G, both of size at least k, is it possible to transform I into J by adding and removing vertices one-by-one, while maintaining an independent set of size at least k throughout? This problem is known to be PSPACE-hard in general. For the case that G is a cograph on n vertices, we show that it can be solved in polynomial time. More generally, we show that for a graph class $\mathcal{G}$ that includes all chordal and claw-free graphs, the problem can be solved in polynomial time for graphs that can be obtained from a collection of graphs from $\mathcal{G}$ using disjoint union and complete join operations.

1 Introduction

Reconfiguration problems have been studied often in recent years. These arise in settings where the goal is to transform feasible solutions to a problem in a step-by-step manner, while maintaining a feasible solution throughout. A *reconfiguration problem* is obtained by defining *feasible solutions* (or configurations) for *instances* of the problem, and a (symmetric) *adjacency relation* between solutions. This defines a *solution graph* for every instance, which is usually exponentially large in the input size. Usually, it is assumed that *adjacency* and *being a feasible solution* can be tested in polynomial time. Typical questions that are studied are deciding the existence of a path between two given solutions *(reachability)*, finding shortest paths between solutions, deciding whether the solution graph is connected or giving sufficient conditions for this, and giving bounds on its diameter. For example, the literature contains such results on the reconfiguration of vertex colorings [1,3,7,9–11], boolean assignments that satisfy a given formula [16], independent sets [17,20,22,24], matchings [20], shortest paths [4,5,21], subsets of a (multi-)set of integers [14,19], etc. Techniques for many different reconfiguration problems are discussed in [20,24]. See the recent survey by Van den Heuvel [18] for an overview of and introduction to reconfiguration problems, and a discussion of their various applications.

One of the most well-studied problems of this kind is the reconfiguration of *independent sets* (which are sets of pairwise nonadjacent vertices). For a graph G

Supported by the European Community's Seventh Framework Programme (FP7/2007–2013), grant agreement n° 317662.

D. Kratsch and I. Todinca (Eds.): WG 2014, LNCS 8747, pp. 105–116, 2014.
DOI: 10.1007/978-3-319-12340-0_9

and integer k, the independent sets of size at least/exactly k of G form the feasible solutions. Independent sets are also called *token configurations*, where the independent set vertices are viewed as *tokens*. Three types of adjacency relations have been studied in the literature: in the *token jumping (TJ)* model [20,22], a token can be moved from any vertex to any other vertex. In the *token sliding (TS)* model, tokens can be moved along edges of the graph [17,22]. In the *token addition and removal (TAR)* model [20,22], tokens can be removed and added in arbitrary order, though at least k tokens should remain at any time (k is the *token lower bound*). Of course, in all of these cases, an independent set should be maintained.

The *reachability problem* has received the most attention in this context: given two independent sets I and J of a graph G, and possibly a token lower bound $k \leq \min\{|I|, |J|\}$, is there a path (or *reconfiguration sequence*) from I to J in the solution graph? We call this problem *TJ-Reachability, TS-Reachability* or *TAR-Reachability*, depending on the adjacency relation that is used. Kamiński et al. [22] showed that the TAR-Reachability problem generalizes the TJ-Reachability problem. For all three adjacency relations, this problem is PSPACE-hard, even in perfect graphs [22], and even in planar graphs of maximum degree 3 [17] (see also [7]). In [20] an alternative, simple PSPACE-hardness proof is given. In addition, in [22], the problem of deciding whether there exists a path of length at most l between two solutions is shown to be strongly NP-hard, for all three adjacency models.

On the positive side, these problems can be solved in polynomial time for various restricted graph classes. The result on matching reconfiguration by Ito et al. [20] implies that for line graphs, TJ-Reachability and TAR-Reachability can be solved efficiently. This result has recently been generalized to claw-free graphs, also for TS-Reachability [8]. Kamiński et al. [22] give an efficient algorithm for TS-Reachability in cographs, and show that for TJ-Reachability in even-hole-free graphs, a reconfiguration sequence of length $|I \backslash J|$ exists between every pair of independent sets I and J. TAR-Reachability has also been studied under the name *Vertex Cover Reconfiguration* in [24], where parameterized complexity results for the problem are given. (Recall that I is an independent set of G if and only if $V(G) \backslash I$ is a vertex cover of G.)

New results and techniques. In this paper, we show that TAR-Reachability can be solved in polynomial time for cographs. Using [22], it follows that the same holds for TJ-Reachability. This answers an open question from [22]. Recall that a graph is a *cograph* iff it has no induced path on four vertices. Alternatively, cographs can be defined as graphs that can be obtained from a collection of trivial (one vertex) graphs by repeatedly applying *(disjoint) union* and *(complete) join* operations. The order of these operations can be described using a rooted *cotree*. This characterization allows efficient dynamic programming (DP) algorithms for various NP-hard problems. Our algorithm is also a DP algorithm over the cotree, albeit more complex than many known DP algorithms on cographs. For both solutions I and J, certain values are computed, using first a *bottom up* DP phase, and next a *top down* DP phase over the cotree. Using these values, we can conclude whether J is reachable from I. Because of this method, we in fact obtain a stronger result: TJ- and TAR-Reachability can be decided efficiently for any graph that can be obtained using join and union operations, when starting with a collection of base graphs from a graph class $\mathcal{G}$ that satisfies the following properties: (i) For any graph in $\mathcal{G}$, the TAR-Reachability problem can be decided efficiently, and (ii) for any graph in $\mathcal{G}$ and independent set I, the size of a maximum independent set that is TAR-reachable from I can be computed efficiently, for all token lower bounds $k \leq |I|$. Results from [8,15,22,23,25] can easily be combined to show that chordal graphs and claw-free graphs satisfy these properties. In all, this yields quite a rich graph class for which this PSPACE-hard problem can be solved efficiently. Considering the fact that TAR-Reachability is PSPACE-hard for perfect graphs [22], the boundary between hard and easy graph classes for this problem starts to become clear.

This paper presents one of the first nontrivial examples of how dynamic programming over graph decompositions can be used to solve reconfiguration problems. (We remark that a DP approach has also been used to show that the PSPACE-hard Shortest Path Reconfiguration problem can be solved in polynomial time on planar graphs [4], using a problem-specific layer decomposition of the graph.) This is especially interesting since cographs form the base class for various graph width measures: cographs are exactly the graphs of cliquewidth at most two, and exactly the graphs of modular-width two [13]. We expect that our method forms a first step towards efficiently solving various reconfiguration problems for graphs of bounded modular-width, and provides useful concepts for addressing other graph classes/decompositions. However, for graphs of bound cliquewidth, similar efficient algorithms should not be expected, since it was shown very recently that many reconfiguration problems, including TAR-Reachability and TJ-Reachability, remain PSPACE-hard for graphs of bandwidth/treewidth/cliquewidth at most k, for some constant k [27].

Our DP algorithm for the TAR-Reachability problem is presented in Sects. 3–5. First, in Sect. 3, an example is given, the proof of this statement is outlined, and a detailed overview of Sects. 4 and 5 is given. In Sect. 6, examples of graph classes are given to which this algorithm applies. We start in Sect. 2 with precise definitions, and end in Sect. 7 with a discussion. Statements for which additional proof details can be found in the full version of this paper [6] are marked with a star.

2 Preliminaries

By $\alpha(G)$ we denote the maximum size of an independent set in G. In this paper, we use the *token addition and removal (TAR)* model for independent set reconfiguration. For a graph G and integer k, the vertex set of the graph $\mathrm{TAR}_k(G)$ is the set of all independent sets of size at least k in G. Two distinct independent sets I and J are adjacent in $\mathrm{TAR}_k(G)$ if there exists a vertex $v \in V(G)$ such that $I \cup \{v\} = J$ or $I = J \cup \{v\}$. Vertices from independent sets will also be called *tokens*, and we will also say that J is obtained from I by *adding one token on v* resp. *removing one token from v*.

For an integer k and two independent sets I and J of G with $|I| \geq k$ and $|J| \geq k$, we write $I \leftrightarrow_k^G J$ if $\text{TAR}_k(G)$ contains a walk from I to J. Such a walk in $\text{TAR}_k(G)$ (a sequence of independent sets) is also called a k-TAR-sequence *for* G *from* I *to* J. To avoid discussing trivial cases in our proofs, we allow that a k-TAR-sequence contains consecutive sets that are identical. Observe that $I \leftrightarrow_0^G J$ always holds, and that the relation $\leftrightarrow_k^G$ is an equivalence relation, for all G and k. The superscript G is omitted if the graph in question is clear. If G and k are clear from the context, we will also simply say that J *is reachable from* I.

A *generalized cotree* is a binary tree T with root r, together with

– a partition of the nonleaf vertices into *union nodes* and *join nodes*, and
– a graph G_u for every leaf u of T, such that for any two leaves u and v, the graphs G_u and G_v are vertex and edge disjoint.

Vertices of T are called *nodes*. With every node $u \in V(T)$ we associate a graph G_u in the following way: for leaves u, G_u is as given. Otherwise, u has two child nodes; denote these by v and w. If u is a union node, then G_u is the *disjoint union* of G_v and G_w. If u is a join node, then G_u is obtained by taking the *complete join* of G_v and G_w. This operation is defined as follows: start with the disjoint union of G_v and G_w, and add edges yz for every combination of $y \in V(G_v)$ and $z \in V(G_w)$. For a node $u \in V(T)$, we denote $V_u = V(G_u)$. A generalized cotree T is called a *cotree* if for every leaf $v \in V(T)$, the graph G_v consists of a single vertex. Such a leaf is called a *trivial leaf*. (See Fig. 1(d) for an example.)

Let T be a (generalized) cotree, with root r. For a graph G, we say that T is a *(generalized) cotree for G* if $G_r = G$. A graph G is called a *cograph* if there exists a cotree for G. Let $\mathcal{G}$ be a graph class. We say that a generalized cotree T for a graph G is a *cotree decomposition of G into $\mathcal{G}$-graphs* if for every leaf $v \in V(T)$, the graph $G_v \in \mathcal{G}$.

3   Example and Proof Outline

In Fig. 1, three independent sets A, B and C are shown for a cograph G. In order to go from A to B in $\text{TAR}_5(G)$, an independent set must be visited which has no tokens on the component G_x, and therefore at least five tokens on the other two components. The only such independent set of G is C. Using similar observations, it can be verified that the *shortest* 5-TAR-sequence from A (or B) to C is unique up to symmetries, and has length twelve (six additions and deletions). Hence the shortest 5-TAR-sequence from A to B has length 24. In general, deciding whether $A \leftrightarrow_k^G B$ requires computing the following values $\lambda_k^I(v)$, which indicate the minimum number of vertices of V_v that must be contained in any independent set reachable from I.

Definition 1. *Let T be a generalized cotree for a graph G, I be an independent set of G, and $k \leq |I|$. For $v \in V(T)$, define $\lambda_k^I(v) = \min |J \cap V_v|$ over all independent sets J of G with $I \leftrightarrow_k^G J$.*

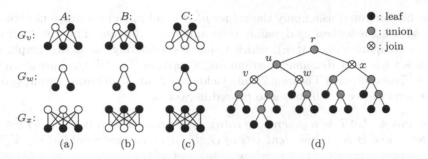

Fig. 1. (a), (b), (c): A cograph G, with independent sets A, B and C indicated by the white vertices. Any 5-TAR-sequence from A to B must visit C and use all vertices of G. (d): A cotree of G with root r, with join nodes v, w and x corresponding to the components of G.

For instance, in the example from Fig. 1, $\lambda_5^A(x) = 0 = \lambda_5^B(x)$, and this fact is essential for concluding that $A \leftrightarrow_5^G B$ in this case. The following theorem characterizes whether B is reachable from A, using the values from Definition 1.

Theorem 2. *Let T be a generalized cotree for a graph G. Let A and B be two independent sets of G of size at least k. Then $A \leftrightarrow_k^G B$ if and only if*

1. *for all nodes $u \in V(T)$, $\lambda_k^A(u) = \lambda_k^B(u)$, and*
2. *for all leaves $u \in V(T)$, $(A \cap V_u) \leftrightarrow_\ell^{G_u} (B \cap V_u)$, where $\ell = \lambda_k^A(u)$.*

The forward direction of the proof is straightforward: if $A \leftrightarrow_k^G B$, then since $\leftrightarrow_k^G$ is an equivalence relation, any independent set J is reachable from A if and only if it is reachable from B. It follows that $\lambda_k^A(v) = \lambda_k^B(v)$ for all $v \in V(T)$. The second property follows by restricting all independent sets in a k-TAR-sequence from A to B to the subgraph G_u for any leaf $u \in V(T)$. By definition, these all have size at least $\ell = \lambda_k^A(u)$, so this yields an ℓ-TAR-sequence from $A \cap V_u$ to $B \cap V_u$ for G_u. In Sect. 5, the backward direction of the proof is given.

In order to efficiently decide whether $A \leftrightarrow_k^G B$, it remains to compute the values $\lambda_k^I(v)$ for all $v \in V(T)$ and $I = A, B$. In the example from Fig. 1, it holds that $\lambda_5^A(x) = 0$. This is because on the subgraph G_u, which is the disjoint union of components G_v and G_w (see Fig. 1(d)), it is possible to reconfigure from the initial independent set A to an independent set with at least five tokens on G_u, while keeping at least two tokens on G_u throughout. This indicates that in order to compute the values $\lambda_k^I(v)$, the following values must be computed.

Definition 3. *Let T be a generalized cotree for G, and let I be an independent set of G. For $v \in V(T)$ and $\ell \in \{0, \ldots, |I \cap V_v|\}$, denote by $\mu_\ell^I(v)$ the maximum of $|J|$ over all independent sets J of G_v with $(I \cap V_v) \leftrightarrow_\ell^{G_v} J$.*

Note that $\mu_0^I(v) = \alpha(G_v)$ (regardless of the choice of I). The value $\mu_\ell^I(v)$ depends only on the situation in the subgraph G_v; not on the entire graph. This is in contrast to the values $\lambda_k^I(v)$. So the values $\mu_\ell^I(u)$ for a node u with children v and

w can be computed using only the values $\mu_{\ell'}^I(v)$ and $\mu_{\ell'}^I(w)$ for different choices of ℓ', so using a *bottom up* dynamic programming algorithm. This can then be used to compute values $\lambda_k^I(u)$, which requires considering the entire graph, so this uses a *top down* dynamic programming algorithm. The DP rules are given in Sect. 4. Together with Theorem 2, this yields our main algorithmic result, given in the next theorem, which is also proved in Sect. 5.

Theorem 4. *Let T be a generalized cotree for a graph G on n vertices, let $k \in \mathbb{N}$ and let A and B be independent sets of G. If for every nontrivial leaf $v \in V(T)$ and relevant integer ℓ, (1) the values $\mu_\ell^A(v)$ and $\mu_\ell^B(v)$ are known, and (2) it is known whether $(A \cap V_v) \leftrightarrow_\ell^{G_v} (B \cap V_v)$, then in polynomial time it can be decided whether $A \leftrightarrow_k^G B$.*

In particular, Theorem 4 implies that for any two independent sets A and B for a *cograph* G, it can be decided in polynomial time whether $A \leftrightarrow_k^G B$.

For all of our proofs, an essential (easy to see) fact is that for every node u, the vertex set V_u is a module of G. A *module* of a graph G is a set $M \subseteq V(G)$ such that for every $v \in V(G) \backslash M$, either $M \subseteq N(v)$ or $M \cap N(v) = \emptyset$. Note that we will also consider $V(G)$ to be a (trivial) module of G. Modules are very useful for independent set reconfiguration, since to some extent, we can reconfigure within the module and outside of the module independently; only the *number* of tokens on the module matters. The following two lemmas make this more precise, and present two useful properties for the proofs below.

Lemma 5. *(*) Let M be a module of a graph G, let k and ℓ be integers, and let A be an independent set of G, with $|A \cap M| \geq \max\{1, \ell\}$ and $|A| \geq k$. Denote $H = G[M]$. If there exists an independent set B of G with $A \leftrightarrow_k^G B$ and $|B \cap M| \leq \ell$, and if there exists an independent set C of H with $(A \cap M) \leftrightarrow_\ell^H C$, then there exists an independent set D of G with $A \leftrightarrow_k^G D$ and $D \cap M = C$.*

Lemma 6. *(*) Let M be a module of a graph G, such that M can be partitioned into two sets M_1 and M_2 with no edges between M_1 and M_2. Let A be an independent set of G, let B_1 be an independent set of G with $A \leftrightarrow_k^G B_1$, that maximizes $|B_1 \cap M_1|$ among all such sets, and let B_2 be an independent set of G with $A \leftrightarrow_k^G B_2$. Then there exists an independent set C of G with $A \leftrightarrow_k^G C$ and $C \cap M_i = B_i \cap M_i$ for $i \in \{1, 2\}$.*

4   Dynamic Programming Rules

Throughout this section, T denotes a generalized cotree of G and I denotes an independent set of G. We first show how to compute the values $\mu_\ell^I(u)$ for every type of node u. For trivial leaf nodes, this is easy.

Proposition 7. *Let $u \in V(T)$ be a trivial leaf node. Then $\mu_\ell^I(u) = 1$ for all ℓ.*

For join nodes u, the computation of $\mu_\ell^I(u)$ is still relatively straightforward. Note that for any independent set I, u has a child w with $V_w \cap I = \emptyset$.

Proposition 8. *Let* $u \in V(T)$ *be a join node. Let* w *be a child of* u *with* $I \cap V_v = \emptyset$, *and let* v *be the other child of* u. *Then* $\mu_\ell^I(u) = \mu_\ell^I(v)$ *for all* $\ell \geq 1$, *and* $\mu_0^I(u) = \max\{\mu_0^I(v), \mu_0^I(w)\}$.

Proof: Because all edges are present between G_v and G_w, a maximum independent set of G_u is either a maximum independent set of G_v or of G_w, so $\mu_0^I(u) = \alpha(G_u) = \max\{\alpha(G_v), \alpha(G_w)\} = \max\{\mu_0^I(v), \mu_0^I(w)\}$. Now consider the case $\ell \geq 1$, and thus $|I \cap V_u| \geq 1$. Then initially all tokens of I are on the child G_v. As long as there is at least one token on G_v, no tokens can be added to G_w. So essentially, G_w can be ignored, and thus $\mu_\ell^I(u) = \mu_\ell^I(v)$. $\square$

For union nodes u with children v and w, computing the values $\mu_\ell^I(u)$ is more complicated, and requires studying ℓ-TAR-sequences for G_u of the following type. Let $x_0 = I \cap V_v$. Observe that from the initial independent set $I \cap V_u$ we can reach an independent set with $\mu_{x_0}^I(v)$ tokens on V_v, and $y_0 := \max\{0, \ell - \mu_{x_0}^I(v)\}$ tokens on V_w, while keeping at least ℓ tokens on V_u throughout. Call this an independent set of *type* $(\mu_{x_0}^I(v), y_0)$. From this, we can subsequently reach an independent set of type $(x_1, \mu_{y_0}^I(w))$, with $x_1 := \max\{0, \ell - \mu_{y_0}^I(w)\}$. Next, an independent set of type $(\mu_{x_1}^I(v), y_1)$ with $y_1 := \max\{0, \ell - \mu_{x_1}^I(v)\}$ can be reached, etc. This process continues with finding ever lower x- and y-values, until a 'stable tuple' (x, y) is obtained. This motivates the following definition.

Definition 9. *For a union node* $u \in V(T)$ *with left child* v *and right child* w, *and integer* $\ell \leq |I \cap V_u|$, *call a tuple* (x, y) *of integers with* $x \leq |I \cap V_v|$ *and* $y \leq |I \cap V_w|$ ℓ-*stable if* $x = \max\{0, \ell - \mu_y^I(w)\}$ *and* $y = \max\{0, \ell - \mu_x^I(v)\}$. *Call an* ℓ-*stable tuple* (x, y) *maximum if there is no* ℓ-*stable tuple* (x', y') *with* $x' \geq x$, $y' \geq y$ *and* $(x, y) \neq (x', y')$.

It can be shown that there is a unique maximum ℓ-stable tuple, which can be characterized as follows. Using this characterization, Lemma 11 shows how the values $\mu_\ell^I(u)$ can be computed for a join node u.

Lemma 10. *(*) Let* $u \in V(T)$ *be a union node, with left child* v *and right child* w. *For* $\ell \in \{0, \ldots, |I \cap V_u|\}$, *let* $x = \min |J \cap V_v|$ *and* $y = \min |J \cap V_w|$, *where in both cases the minimum is taken over all independent sets* J *of* G_u *with* $(I \cap V_u) \leftrightarrow_\ell^{G_u} J$. *Then* (x, y) *is the unique maximum* ℓ-*stable tuple for* I *and* u.

Lemma 11. *(*) Let* $u \in V(T)$ *be a union node, with left child* v *and right child* w. *For* $\ell \in \{0, \ldots, |I \cap V_u|\}$, *let* (x, y) *be the unique maximum* ℓ-*stable tuple for* I *and* u. *Then* $\mu_\ell^I(u) = \mu_x^I(v) + \mu_y^I(w)$.

We will now show how the values $\lambda_k^I(v)$ can be computed for all nodes $v \in V(T)$. For the case that v is a union node, this requires knowledge of the unique maximum ℓ-stable tuple. For the root node of T, the value is trivial.

Proposition 12. *Let* r *be the root node of* T. *Then* $\lambda_k^I(r) = k$.

Proposition 13. *Let* $u \in V(T)$ *be a join node, with children* v *and* w *such that* $I \cap V_w = \emptyset$. *Then* $\lambda_k^I(v) = \lambda_k^I(u)$ *and* $\lambda_k^I(w) = 0$.

Proof: Considering I, $\lambda_k^I(w) = 0$ follows immediately. If $\lambda_k^I(u) = 0$, then obviously $\lambda_k^I(v) = 0$. Adding a token to G_w requires first reaching an independent set with no tokens on G_v, and thus requires $\lambda_k^I(u) = 0$. So if $\lambda_k^I(u) \geq 1$, then G_w can essentially be ignored, and therefore $\lambda_k^I(v) = \lambda_k^I(u)$ in that case.   □

Lemma 14. *Let $u \in V(T)$ be a union node, with left child v and right child w. Let $\ell = \lambda_k^I(u)$, and let (x, y) be the maximum ℓ-stable tuple for I and u. Then $\lambda_k^I(v) = x$ and $\lambda_k^I(w) = y$.*

Proof: Denote $I_u = I \cap V_u$. We first show that $\lambda_k^I(v) \geq x$ and $\lambda_k^I(w) \geq y$. Consider a k-TAR-sequence $I_0, \ldots, I_p$ for G with $I_0 = I$ and $|I_p \cap V_v| = \lambda_k^I(v)$. For every i, denote $I_i' = I_i \cap V_u$, and consider the sequence $I_0', \ldots, I_p'$. By definition of $\ell = \lambda_k^I(u)$, for every i it holds that $|I_i'| \geq \ell$, so $I_u \leftrightarrow_\ell^{G_u} I_p'$. Using Lemma 10 it then follows that $\lambda_k^I(v) = |I_p \cap V_v| \geq x$. Analogously, $\lambda_k^I(w) \geq y$ follows.

We will now prove that $\lambda_k^I(v) \leq x$ and $\lambda_k^I(w) \leq y$. The case $\ell = 0$ is obvious, so assume $\ell \geq 1$. By Lemma 10, there exist independent sets J_1 and J_2 of G_u with $I_u \leftrightarrow_\ell^{G_u} J_1$, $I_u \leftrightarrow_\ell^{G_u} J_2$, $|J_1 \cap V_v| = x$ and $|J_2 \cap V_w| = y$. By the definition of $\ell = \lambda_k^I(u)$, there exists an independent set B of G with $I \leftrightarrow_k^G B$ and $|B \cap V_u| = \ell$. We can now apply Lemma 5 twice, with V_u and I in the role of M and A, and J_1 or J_2 respectively in the role of C, to conclude that there exist independent sets D_1 and D_2 of G with $I \leftrightarrow_k^G D_1$, $I \leftrightarrow_k^G D_2$, $D_1 \cap V_u = J_1$ and $D_2 \cap V_u = J_2$. So $|D_1 \cap V_v| = x$ and $|D_2 \cap V_w| = y$, and thus $\lambda_k^I(v) \leq x$ and $\lambda_k^I(w) \leq y$.   □

5   Algorithm Summary and Main Theorems

We can now prove our two main theorems; first we prove our characterization of $A \leftrightarrow_k^G B$ in terms of the values $\lambda_k^I(u)$, and secondly we summarize how these values can be computed efficiently.

Proof of Theorem 2: The forward direction of the proof was given in Sect. 3. We now prove the backward direction. Assume that the two properties given in the theorem statement hold. So we may denote $\lambda_k(u) = \lambda_k^A(u) = \lambda_k^B(u)$ for all nodes u. We prove the following claim by induction over T:

Claim A: For all nodes $u \in V(T)$: $(A \cap V_u) \leftrightarrow_{\lambda_k(u)}^{G_u} (B \cap V_u)$.

For leaf nodes $u \in V(T)$ (induction base), the statement follows immediately from the second property. To prove the induction step, first consider a join node $u \in V(T)$ with children v and w. Suppose that $\lambda_k(v) \geq 1$. This implies $A \cap V_v \neq \emptyset$ and $B \cap V_v \neq \emptyset$. Therefore, since u is a join node, $A \cap V_u = A \cap V_v$ and $B \cap V_u = B \cap V_v$. In addition, $\lambda_k(u) = \lambda_k(v)$ (Proposition 13). From these facts, and the induction assumption $(A \cap V_v) \leftrightarrow_{\lambda_k(v)}^{G_v} (B \cap V_v)$, we conclude that $(A \cap V_u) \leftrightarrow_{\lambda_k(u)}^{G_u} (B \cap V_u)$. The case $\lambda_k(w) \geq 1$ is analog. On the other hand, if $\lambda_k(v) = \lambda_k(w) = 0$, then $\lambda_k(u) = 0$ (Proposition 13). Claim A follows for u since $(A \cap V_u) \leftrightarrow_0^{G_u} (B \cap V_u)$ trivially holds.

Next, consider the case that $u \in V(T)$ is a union node with left child v and right child w. Denote $\ell = \lambda_k(u)$, $x = \lambda_k(v)$ and $y = \lambda_k(w)$. By Lemma 14, (x, y) is the maximum ℓ-stable tuple for u, for both A and B. We define C_v to be an independent set of G_v with $(A \cap V_v) \leftrightarrow_x^{G_v} C_v$, with maximum size among all such sets, and define C_w to be an independent set of G_w with $(A \cap V_w) \leftrightarrow_y^{G_w} C_w$, with maximum size among all such sets. By induction, $(A \cap V_v) \leftrightarrow_x^{G_v} (B \cap V_v)$ holds, so it also holds that $(B \cap V_v) \leftrightarrow_x^{G_v} C_v$, and that C_v has maximum size among all such reachable sets. Analogously, $(B \cap V_w) \leftrightarrow_y^{G_w} C_w$ holds, and C_w has maximum size among all such reachable sets. Define $C_u = C_v \cup C_w$. We will now show that C_u is reachable from both $A \cap V_u$ and $B \cap V_u$, which proves Claim A for node u.

Lemma 10 shows that there exists an independent set J of G_u with $(A \cap V_u) \leftrightarrow_\ell^{G_u} J$ and $|J \cap V_v| = x$. Using this, we argue that there exists an independent set J_1 of G_u with $(A \cap V_u) \leftrightarrow_\ell^{G_u} J_1$ and $J_1 \cap V_v = C_v$. If $A \cap V_v = \emptyset$, then this claim is trivial. Otherwise, we can apply (module) Lemma 5 to draw this conclusion (using V_v, G_u, J and C_v in the roles of the module M, entire graph G, and independent sets B and C, respectively). Analogously, we may conclude that there exists an independent set J_2 of G_u with $(A \cap V_u) \leftrightarrow_\ell^{G_u} J_2$ and $J_2 \cap V_w = C_w$. Since $C_u = C_v \cup C_w$, we can now apply (module) Lemma 6 (with G_u in the role of the entire graph, V_v and V_w in the roles of disjoint modules M_1 and M_2, and J_1 and J_2 in the roles of B_1 and B_2), to conclude that $A \cap V_u \leftrightarrow_\ell^{G_u} C_u$. For this, we require the fact that C_v has maximum size among all independent sets of G_v that are reachable from $A \cap V_v$.

The argument from the previous paragraph also holds when replacing A by B, since C_v and C_w are also maximum reachable independent sets from $B \cap V_v$ and $B \cap V_w$. So $B \cap V_u \leftrightarrow_\ell^{G_u} C_u$ also holds. Hence $A \cap V_u \leftrightarrow_\ell^{G_u} C_u \leftrightarrow_\ell^{G_u} B \cap V_u$, which proves Claim A for u.

This concludes the induction proof of Claim A. Applying Claim A to the root node r of T shows that $A \leftrightarrow_k^G B$, since $\lambda_k(r) = k$ (Proposition 12), and $G = G_r$, and therefore concludes the proof of the theorem.   □

Proof of Theorem 4: First we use a *bottom up* dynamic programming algorithm, to compute the values $\mu_\ell^A(u)$ and $\mu_\ell^B(u)$ for every node u and relevant integer ℓ, and to compute the maximum ℓ-stable tuples for every union node and relevant integer ℓ. This can be done in polynomial time using the rules given in Proposition 7, Proposition 8 and Lemma 11. Note that maximum ℓ-stable tuples can easily be computed in polynomial time by testing a quadratic number of possible tuples (Definition 9).

Next, we start the *top down* phase of the dynamic programming algorithm, where we compute the values $\lambda_k^A(u)$ and $\lambda_k^B(u)$ for every node u. This can be done in polynomial time using the rules given in Propositions 12 and 13, and Lemma 14. Note that applying Lemma 14 to a union node u requires the previously computed maximum ℓ-stable tuple (x, y) for $I = A, B$, with $\ell = \lambda_k^I(u)$. This is why the bottom up phase is required. At this point, the characterization given in Theorem 2 can be used to conclude whether $A \leftrightarrow_k^G B$.   □

6   Examples of Suitable Graph Classes

Consider T, G, A, B and k as in Theorem 4. If $v \in V(T)$ is a trivial leaf, then for every relevant value ℓ, $\mu_\ell^A = 1$ and $\mu_\ell^B = 1$ hold (Proposition 7), and clearly $(A \cap V_v) \leftrightarrow_\ell^{G_v} (B \cap V_v)$ holds. So combined with the fact that a cotree of a cograph G can be found in linear time [12], Theorem 4 implies that the TAR-Reachability problem can be decided in polynomial time for a cograph G.

Theorem 4 is however much stronger, and implies that TAR-Reachability can be decided efficiently for much richer graph classes. Recall that a (simple) graph G is *chordal/even-hole-free/claw-free* if it does not contain as an *induced subgraph* a cycle of length at least four/a cycle of even length/a $K_{1,3}$, respectively. By applying independent set reconfiguration results from [22] for even-hole-free graphs, and from [8] for claw-free graphs, one can easily prove the following two theorems. Similar to Definition 3, for an independent set A of a graph G with $|A| \geq k$, we denote $\mu_k^A(G) = \max\{|J| : A \leftrightarrow_k^G J\}$.

Theorem 15. *(*) Let A and B be independent sets of an even-hole-free or claw-free graph G. Then in polynomial time, it can be decided whether $A \leftrightarrow_k^G B$.*

Theorem 16. *(*) Let A be an independent set of a graph G that is even-hole-free or claw-free. Then $\mu_k^A(G) = |A|$ if A is a dominating set of size k, and $\mu_k^A(G) = \alpha(G)$ otherwise.*

It follows that for an even-hole-free or claw-free graph G, $\mu_k^A(G)$ can be computed efficiently if $\alpha(G)$ can be computed efficiently. Unfortunately, for even-hole-free graphs G it is an open question whether this can be done (see [22,26]). Nevertheless, for the subclass of chordal graphs, an efficient algorithm to compute $\alpha(G)$ is known [15]. For claw-free graphs, $\alpha(G)$ can be computed efficiently as well [23,25]. Denote by $\mathcal{G}^*$ class of all graphs that are chordal or claw-free. We conclude that if for G, a cotree decomposition into $\mathcal{G}^*$-graphs is given, then the conditions of Theorem 4 are satisfied, and thus the TAR-Reachability problem can be solved efficiently for G. It only remains to find such a cotree decomposition efficiently. Recall that for a graph G, by $\overline{G}$ the *complement* of G is denoted, which is the graph $\overline{G} = (V(G), \{uv \mid uv \notin E(G)\})$.

Definition 17. *A* maximal cotree decomposition *is a generalized cotree decomposition T where for every leaf $u \in V(T)$, both G_u and $\overline{G_u}$ are connected.*

Proposition 18. *(*) For any graph G, a maximal cotree decomposition of G can be computed in polynomial time.*

A graph class $\mathcal{G}$ is called *hereditary* if for every $G \in \mathcal{G}$ and every induced subgraph H of G, $H \in \mathcal{G}$ holds. Clearly, the aforementioned class $\mathcal{G}^*$ is hereditary.

Lemma 19. *(*) Let $\mathcal{G}$ be a hereditary graph class, and let G be a graph that admits a cotree decomposition into $\mathcal{G}$-graphs. Then every maximal cotree decomposition of G is a cotree decomposition into $\mathcal{G}$-graphs.*

From Proposition 18 and Lemma 19 it follows that a cotree decomposition into $\mathcal{G}^*$-graphs can be computed in polynomial time. Together, these statements yield the main result of this section.

Theorem 20. *(*) Let G be a graph that admits a cotree decomposition into graphs that are chordal or claw-free, and let A and B be independent sets of G, both of size at least k. Then in polynomial time, we can decide whether $A \leftrightarrow_k^G B$.*

7 Discussion

In the full version of this paper [6], we show that our DP algorithm for cographs G can be implemented to run in time $O(n^2)$, where $n = |V(G)|$. The key to this is a more efficient computation of stable tuples, not based on Definition 9. Secondly, in [6] we show that components of $\text{TAR}_k(G)$ have diameter at most $4n - 2k$, if G is a cograph.

The following question related to independent set reconfiguration in cographs is still open: what is the complexity of deciding whether there exists a k-TAR-sequence of length at most ℓ between two independent sets of a cograph? (Recall that for general graphs, this is strongly NP-hard [22].) In [6], we also asked if it can be decided efficiently whether $\text{TAR}_k(G)$ is connected, using similar techniques. In subsequent research [2], this question has been answered affirmatively.

References

1. Bonamy, M., Bousquet, N.: Recoloring bounded treewidth graphs. Electron. Notes Discrete Math. **44**, 257–262 (2013)
2. Bonamy, M., Bousquet, N.: Reconfiguring independent sets in cographs, June 2014. arXiv:1406.1433
3. Bonamy, M., Johnson, M., Lignos, I., Patel, V., Paulusma, D.: Reconfiguration graphs for vertex colourings of chordal and chordal bipartite graphs. J. Comb. Optim. 1–12 (2012)
4. Bonsma, P.: Rerouting shortest paths in planar graphs. In: FSTTCS 2012. vol. 18, LIPIcs, pp. 337–349. Schloss Dagstuhl - Leibniz-Zentrum fuer Informatik (2012)
5. Bonsma, P.: The complexity of rerouting shortest paths. Theor. Comput. Sci. **510**, 1–12 (2013)
6. Bonsma, P.: Independent set reconfiguration in cographs, February 2014. arXiv:1402.1587
7. Bonsma, P., Cereceda, L.: Finding paths between graph colourings: PSPACE-completeness and superpolynomial distances. Theor. Comput. Sci. **410**(50), 5215–5226 (2009)
8. Bonsma, P., Kamiński, M., Wrochna, M.: Reconfiguring independent sets in claw-free graphs. In: Ravi, R., Gørtz, I.L. (eds.) SWAT 2014. LNCS, vol. 8503, pp. 86–97. Springer, Heidelberg (2014)
9. Cereceda, L., van den Heuvel, J., Johnson, M.: Connectedness of the graph of vertex-colourings. Discrete Appl. Mathe. **308**(5–6), 913–919 (2008)
10. Cereceda, L., van den Heuvel, J., Johnson, M.: Mixing 3-colourings in bipartite graphs. Eur. J. Comb. **30**(7), 1593–1606 (2009)

11. Cereceda, L., van den Heuvel, J., Johnson, M.: Finding paths between 3-colorings. J. Graph Theory **67**(1), 69–82 (2011)
12. Corneil, D., Perl, Y., Stewart, L.: A linear recognition algorithm for cographs. SIAM J. Comput. **14**(4), 926–934 (1985)
13. Courcelle, B., Olariu, S.: Upper bounds to the clique width of graphs. Discrete Appl. Math. **101**(13), 77–114 (2000)
14. Eggermont, C., Woeginger, G.J.: Motion planning with pulley, rope, and baskets. Theory Comput. Syst. **53**(4), 569–582 (2013)
15. Gavril, F.: Algorithms for minimum coloring, maximum clique, minimum covering by cliques, and maximum independent set of a chordal graph. SIAM J. Comput. **1**(2), 180–187 (1972)
16. Gopalan, P., Kolaitis, P.G., Maneva, E., Papadimitriou, C.H.: The connectivity of boolean satisfiability: computational and structural dichotomies. SIAM J. Comput. **38**(6), 1863–1920 (2009)
17. Hearn, R.A., Demaine, E.D.: PSPACE-completeness of sliding-block puzzles and other problems through the nondeterministic constraint logic model of computation. Theor. Comput. Sci. **343**(1–2), 72–96 (2005)
18. van den Heuvel, J.: The complexity of change. Surv. Comb. **2013**, 127–160 (2013)
19. Ito, T., Demaine, E.D.: Approximability of the subset sum reconfiguration problem. In: Ogihara, M., Tarui, J. (eds.) TAMC 2011. LNCS, vol. 6648, pp. 58–69. Springer, Heidelberg (2011)
20. Ito, T., Demaine, E.D., Harvey, N.J.A., Papadimitriou, C.H., Sideri, M., Uehara, R., Uno, Y.: On the complexity of reconfiguration problems. Theor. Comput. Sci. **412**(12–14), 1054–1065 (2011)
21. Kamiński, M., Medvedev, P., Milanič, M.: Shortest paths between shortest paths. Theor. Comput. Sci. **412**(39), 5205–5210 (2011)
22. Kamiński, M., Medvedev, P., Milanič, M.: Complexity of independent set reconfigurability problems. Theor. Comput. Sci. **439**, 9–15 (2012)
23. Minty, G.J.: On maximal independent sets of vertices in claw-free graphs. J. Comb. Theory Ser. B **28**(3), 284–304 (1980)
24. Mouawad, A.E., Nishimura, N., Raman, V., Simjour, N., Suzuki, A.: On the parameterized complexity of reconfiguration problems. In: Gutin, G., Szeider, S. (eds.) IPEC 2013. LNCS, vol. 8246, pp. 281–294. Springer, Heidelberg (2013)
25. Sbihi, N.: Algorithme de recherche d'un stable de cardinalité maximum dans un graphe sans étoile. Discrete Math. **29**(1), 53–76 (1980)
26. Vušković, K.: Even-hole-free graphs: a survey. Appl. Anal. Discrete Math. **4**(2), 219–240 (2010)
27. Wrochna, M.: Reconfiguration in bounded bandwidth and treedepth, May 2014. arXiv:1405.0847

Structural Parameterizations for Boxicity

Henning Bruhn[1]([✉]), Morgan Chopin[1], Felix Joos[1], and Oliver Schaudt[2]

[1] Institut Für Optimierung Und Operations Research, Universität Ulm,
HelmholtzstraßE 18, 89081 Ulm, Germany
{henning.bruhn,morgan.chopin,felix.joos}@uni-ulm.de
[2] Institut Für Informatik, Universität Zu Köln, Weyertal 80, 50931 Köln, Germany
schaudto@uni-koeln.de

Abstract. The boxicity of a graph G is the least integer d such that G has an intersection model of axis-aligned d-dimensional boxes. BOXICITY, the problem of deciding whether a given graph G has boxicity at most d, is NP-complete for every fixed $d \geq 2$. We show that BOXICITY is fixed-parameter tractable when parameterized by the cluster vertex deletion number of the input graph. This generalizes the result of Adiga et al. [4], that BOXICITY is fixed-parameter tractable in the vertex cover number. Moreover, we show that BOXICITY admits an additive 1-approximation when parameterized by the pathwidth of the input graph.

Finally, we provide evidence in favor of a conjecture of Adiga et al. [4] that BOXICITY remains NP-complete even on graphs of constant treewidth.

1 Introduction

Every graph G can be represented as an intersection graph of axis-aligned boxes in $\mathbb{R}^d$, provided d is large enough. The *boxicity* of G, denoted by $\mathrm{box}(G)$, introduced by Roberts [21], is the smallest dimension d for which this is possible. We denote the corresponding decision problem by BOXICITY: given G and $d \in \mathbb{N}$, determine whether G has boxicity at most d.

Boxicity has received a fair amount of attention. This is partially due to the wider context of graph representations, but also because graphs of low boxicity are interesting from an algorithmic point of view. While many hard problems remain so for graphs of bounded boxicity, some become solvable in polynomial time, notably max-weighted clique (as observed by Spinrad [23, p. 36]).

Cozzens [13] showed that BOXICITY is NP-complete. To cope with this hardness result, several authors [1,4,18] studied the parameterized complexity of BOXICITY. Since the problem remains NP-complete for constant $d \geq 2$ (Yannakakis [25] and Kratochvíl [20]), boxicity itself is ruled out as parameter. Instead more structural parameters have been considered. Our work follows this line. We prove:

Theorem 1. BOXICITY *is fixed-parameter tractable when parameterized by cluster vertex deletion number.*

© Springer International Publishing Switzerland 2014
D. Kratsch and I. Todinca (Eds.): WG 2014, LNCS 8747, pp. 117–128, 2014.
DOI: 10.1007/978-3-319-12340-0_10

The *cluster vertex deletion* number is the minimum number of vertices that have to be deleted to get a disjoint union of complete graphs or *cluster graph*. As discussed by Doucha and Kratochvíl [15] cluster vertex deletion is an intermediate parameterization between vertex cover and cliquewidth. A *d-box representation* of a graph G is a representation of G as intersection graph of axis-aligned boxes in $\mathbb{R}^d$.

Theorem 2. *Finding a d-box representation of G such that $d \le \text{box}(G) + 1$ can be done in $f(\text{pw}(G)) \cdot |V(G)|$ time where $\text{pw}(G)$ is the pathwidth of G.*

A natural parameter for BOXICITY is the treewidth $\text{tw}(G)$ of a graph G, in particular as Chandran and Sivadasan [11] proved that $\text{box}(G) \le \text{tw}(G) + 2$. However, Adiga, Chitnis and Saurabh [4] conjecture that BOXICITY is NP-complete on graphs of bounded treewidth. Our last result provides evidence in favor of this conjecture. For this, we mention the observation of Roberts [21] that a graph G has boxicity d if and only if G can be expressed as the intersection of d interval graphs.

Theorem 3. *There is an infinite family of graphs $\mathcal{G}$ of boxicity 2 and bandwidth $\mathcal{O}(1)$ such that, among any pair of interval graphs whose intersection is $G \in \mathcal{G}$, at least one has treewidth $\Omega(|V(G)|)$.*

Why do we see the result as evidence? An algorithm solving BOXICITY on graphs of bounded treewidth (or even stronger, of bounded bandwidth) is likely to exploit the local structure of the graph in order to make dynamic programming work. Yet, Theorem 3 implies that this locality may be lost in some dimensions, which constitutes a serious obstacle for any dynamic programming based approach. We discuss this in more detail in Sect. 5.

Figure 1 summarizes previously known parameterized complexity results on boxicity along with those obtained in this article. Adiga et al. [4] initiated this line of research when they parameterized BOXICITY by the minimal size k of a *vertex cover* in order to give an $2^{O(2^k k^2)} \cdot n$-time algorithm, where n denotes the number of vertices of the input graph, as usual. This result had already been observed earlier by Fellows et al. [17] in the context of well-quasi orders of certain graph classes. Adiga et al. [4] also described an approximation algorithm that, in time $2^{O(k^2 \log k)} \cdot n$, returns a box representation of at most $\text{box}(G) + 1$ dimensions. Both results were extended by Ganian [18] to the less restrictive parameter *twin cover*. Our Theorem 1 includes Ganian's.

Other structural parameters that were considered by Adiga et al. [4] for parameterized approximation algorithms are the size of a *feedback vertex set* – the minimum number of vertices that need to be deleted to obtain a forest – and *maximum leaf number* – the maximum number of leaves in a spanning tree of the graph. They proved that finding a d-box representation of a graph G such that $d \le 2\text{box}(G) + 2$ (resp. $d \le \text{box}(G) + 2$) can be done in $f(k) \cdot |V(G)|^{O(1)}$ time (resp. $2^{O(k^3 \log k)} \cdot |V(G)|^{O(1)}$ time) where k is the size of a feedback vertex set (resp. maximum leaf number). In [1], Adiga, Babu, and Chandran generalized these approximation algorithms to parameters of the type "distance to $\mathcal{C}$", where

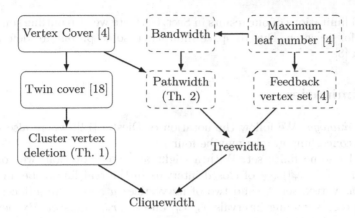

Fig. 1. Navigation map through our parameterized complexity results for BOXICITY. An arc from a parameter k_2 to a parameter k_1 means that there exists some function h such that $k_1 \leq h(k_2)$. A rectangle means fixed-parameter tractability for this parameter and a dashed rectangle means an approximation algorithm with running time $f(k) \cdot n^{O(1)}$ is known.

$\mathcal{C}$ is any graph class of bounded boxicity. More precisely, the parameter measures the minimum number of vertices whose deletion results in a graph that belongs $\mathcal{C}$.

The algorithm of Theorem 2 generalizes the approximation algorithm for the parameter vertex cover number, and improves the guarantee bound of the approximation algorithm for the parameter maximum leaf number.

There is merit in studying approximation algorithms from a parameterized perspective: not only is BOXICITY NP-complete, but the associated minimization problem cannot be approximated in polynomial time within a factor of $n^{1-\varepsilon}$ for any $\varepsilon > 0$ even when the input is restricted to bipartite, co-bipartite or split graphs (provided NP$\neq$ZPP). This is a result due to Chalermsook et al. [10] using the hardness reduction of Adiga, Bhowmick and Chandran [2]. There is, however, an approximation algorithm with factor $o(n)$ for general graphs; see Adiga et al. [1].

While Roberts [21] was the first to study the boxicity parameter, he was hardly the first to consider box representations of graphs. Already in 1948 Bielecki [6] asked, here phrased in modern terminology, whether triangle-free graphs of boxicity ≤ 2 had bounded chromatic number. This was answered affirmatively by Asplund and Grünbaum in [5]. Kostochka [19] treats this question in a much more general setting.

Following Roberts who proved that box$(G) \leq \frac{n}{2}$, other authors obtained bounds for boxicity. Adiga et al. [3], for instance, showed that box$(G) \leq \Delta(G) \log^2 \Delta(G)$, while Scheinerman [22] established that every outerplanar graph has boxicity at most two. This, in turn, was extended by Thomassen [24], who showed that planar graphs have boxicity at most three.

In the next section, we will give formal definitions of the necessary concepts for this article. We prove our main results in Sects. 3–5. Finally, we discuss the impact and limitations of our results in Sect. 6, where we also outline some future directions for research. Due to space limitation, some proofs are deferred to a full version [8].

2 Preliminaries

Graph terminology. We follow the notation of Diestel [14], where also all basic definitions concerning graphs may be found.

Let X be some finite set. With a slight abuse of notation, we consider a collection $I = ([\ell_v, r_v])_{v \in X}$ of closed intervals in the real line to be an *interval graph*: I has vertex set X, and two of its vertices u and v are adjacent if and only if the corresponding intervals $[\ell_u, r_u]$ and $[\ell_v, r_v]$ intersect. By perturbing the endpoints of the intervals we can ensure that no two intervals have a common endpoint, and that for every interval the left endpoint is distinct from the right endpoint. We always tacitly assume the intervals to be of that form.

The *bandwidth* of a graph G, say with vertex set $V(G) = \{v_1, v_2, \ldots, v_n\}$, is the least number k for which the vertices of G can be labeled with distinct integers $\ell(v_i)$ such that $k = \max\{|\ell(v_i) - \ell(v_j)| : v_i v_j \in E\}$. Equivalently, it is the least integer k for which the vertices of G can be placed at distinct integer points on the real line such that the length of the longest edge is at most k. We denote the bandwidth of a graph G by $\mathrm{bw}(G)$.

The *pathwidth* of a graph G, denoted $\mathrm{pw}(G)$, is the minimum size of the largest clique of any interval supergraph of G, minus 1.

The *treewidth* of a graph G, denoted $\mathrm{tw}(G)$, is the minimum size of the largest clique of any chordal supergraph of G, minus 1.

For the purpose of our paper it is important to remark that for every graph G we have $\mathrm{tw}(G) \leq \mathrm{pw}(G) \leq \mathrm{bw}(G)$.

Parameterized complexity. A decision problem parameterized by a problem-specific parameter k is called *fixed-parameter tractable* if there exists an algorithm that solves it in time $f(k) \cdot n^{O(1)}$, where n is the instance size. The function f is typically super-polynomial and only depends on k. One of the main tools to design such algorithms is the *kernelization* technique. A kernelization algorithm transforms in polynomial time an instance I of a given problem parameterized by k into an equivalent instance I' of the same problem parameterized by $k' \leq k$ such that the size of I' is bounded by $g(k)$ for some computable function g. The instance I' is called a *kernel* of size $g(k)$. The following folklore result is well known.

Theorem 4. *A parameterized problem P is fixed-parameter tractable if and only if P has a kernel.*

In the remainder of this paper, the kernel size is expressed in terms of the number of vertices.

For more background on parameterized complexity the reader is referred to Downey and Fellows [16].

Problem definition. We call an *axis-aligned d-dimensional box* (or *d-box*) a Cartesian product of d closed real intervals. A *d-box representation* of a graph G is a mapping that maps every vertex $v \in V(G)$ to a d-box B_v such that two vertices $u, v \in V(G)$ are adjacent if and only if their associated boxes have a non-empty intersection. The *boxicity* of G, denoted by $\text{box}(G)$, is the minimum integer d such that G admits a d-box representation. We consider the following problem.

BOXICITY
Input: A graph G and an integer d.
Question: Is $\text{box}(G) \leq d$?

Given a d-box representation of G, we denote by $[\ell_i(v), r_i(v)]$ the interval representing v in the i-th dimension.

Throughout the article, we make frequent use of the reformulation of boxicity in terms of interval graphs:

Theorem 5 (Roberts [21]). *The boxicity of a graph G is equal to the smallest integer d so that G can be expressed as the intersection of d interval graphs.*

3   Cluster Vertex Deletion

Theorem 1 follows immediately from the following lemma:

Lemma 1. BOXICITY *admits a kernel of at most $k^{2^{O(k)}}$ vertices, where k is the cluster vertex deletion number of the input graph.*

In the course of this section, we present the sequence of lemmas that are needed to prove the above kernelization result.

Two adjacent vertices u, v in a graph G are *true twins* if u and v have the same neighbourhoods in $G - \{u, v\}$. As observed by Ganian [18], deleting one of two true twins does not change the boxicity.

Lemma 2. *Let u, v be true twins of a graph G. Then $\text{box}(G) = \text{box}(G - u)$.*

We remark, without proof, that there is also a reduction for false twins (those that are non-adjacent): if there are at least three of them, then one may be deleted without changing the boxicity. We will not, however, make use of this observation.

Recall that a *cluster graph* is the disjoint union of complete graphs, called *clusters*. In what follows, we implicitly identify a cluster with its vertex set.

Let $G - X$ be a cluster graph for some $X \subseteq V(G)$. We call two clusters C, C' of $G - X$ *equivalent* if there is a bijection $C \to C'$, $v \mapsto v'$, such that $N_G(v) \cap X = N_G(v') \cap X$. Observe that, if $G - X$ has no true twins, then two clusters C and C' are equivalent if and only if $\{N_G(u) \cap X : u \in C\} = \{N_G(v) \cap X : v \in C'\}$.

Lemma 3. *Let G be a graph without true twins, and let X be a set of k vertices so that $G - X$ is a cluster graph. Then every cluster in $G - X$ contains at most 2^k vertices.*

We also need the following result.

Theorem 6 (Chandran and Sivadasan [11]). *It holds that* box$(G) \leq$ tw$(G)+$ 2 *for any graph* G.

In particular, box$(G) \leq$ pw$(G) + 2$ for any graph G.

Lemma 4. *Let G be a graph without true twins, and let X be a set of k vertices so that $G - X$ is a cluster graph. Moreover, let $\mathcal{D}$ be an equivalence class of clusters with $|\mathcal{D}| \geq 2(2k+2)^{2^{k+1}(2^k+k+1)}$. For every $C^* \in \mathcal{D}$,* box$(G) =$ box$(G - C^*)$.

Proof. As deleting vertices may only decrease the boxicity, it suffices to prove that box$(G) \leq$ box$(G - C^*)$.

Set $H = G - C^*$, $d = $ box(H), $k = |X|$ and $\mathcal{C} = \mathcal{D} \setminus \{C^*\}$. We claim that

$$d = \text{box}(H) \leq 2^k + k + 1. \tag{1}$$

Indeed, define a path decomposition with a bag W_C for every cluster C of $H - X$ such that $W_C = X \cup C$. This gives a path decomposition of H with width at most $k + 2^k - 1$, by Lemma 3. Theorem 6 now implies (1).

For the sake of simplicity, let us introduce the following notions. Fix a d-box representation of H. The set of *corners* of a box of a vertex is the Cartesian product $\times_{i=1}^{d}\{\ell_i(v), r_i(v)\}$. By rescaling every dimension, we can ensure that every endpoint of an interval of a vertex in X lies in $\{1, 2, \ldots, 2k\}$. Thus every corner of a box of X lies in the grid $\{1, 2, \ldots, 2k\}^d$. We may moreover assume that every other box of H is contained in $[0, 2k + 1]^d$. Points of $\{0, 1, \ldots, 2k+1\}^d$ are called *grid points*, and any set $[z_1, z_1+1] \times \ldots \times [z_d, z_d+1]$, where $z_i \in \{0, \ldots, 2k\}$, is a *grid cell*. In each dimension i we say that the grid induces the *grid intervals* $[0, 1], [1, 2], \ldots, [2k, 2k + 1]$. A box of a vertex in $H - X$ is a *cluster box*.

By perturbing the boxes slightly we may always assume that

> if s is a corner of a cluster box of a cluster C of $H - X$, and if
> t is a corner of the box of any vertex $z \in V(H - C)$ then $s_i \neq t_i$ (2)
> for all dimensions $i = 1, \ldots, d$.

Moreover, we may assume that any corner of a cluster box lies in the interior of a grid cell. A cluster box that does not contain any grid point is called a *thin box*.

We concentrate on *thin* clusters, that is, clusters that consist of thin boxes only. We claim that

$$\text{at least } (2k + 2)^{2^{k+1}(2^k+k+1)} \text{ clusters in } \mathcal{C} \text{ are thin.} \tag{3}$$

To prove this claim, observe that no grid point lies in a cluster box of two different clusters as then two vertices in distinct clusters would be adjacent. Thus, there is at most one cluster per grid point so that one of its cluster boxes contains the grid point. As, by (1), there are $(2k + 2)^d \leq (2k + 2)^{2^k+2k+1}$ grid points, it follows that $\mathcal{C}$ has at least $|\mathcal{C}| - (2k + 2)^{2^k+2k+1} \geq (2k + 2)^{2^{k+1}(2^k+k+1)}$ thin clusters.

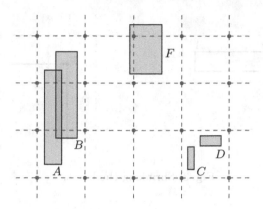

Fig. 2. Boxes A, B are in the same position, as are C and D; F is not thin.

We say that two cluster boxes B and B' are in the *same position* if every grid cell containing a corner of B also contains a corner of B' and vice versa (see Fig. 2). Note that if two vertices $v, v' \in V(H) - X$ have boxes in the same position then $N_H(v) \cap X = N_H(v') \cap X$. (Here we use the fact that cluster boxes have their corner strictly in the interior of grid cells).

For every cluster $C \in \mathcal{C}$ we fix a point $p(C)$ that lies in every cluster box of C: such a point exists by the Helly property for boxes in $\mathbb{R}^d$. We claim that, using this Helly point, we can modify our box representation of H so that

> for all thin clusters $C \in \mathcal{C}$ and for each dimension $i \in \{1, \ldots, d\}$
> holds the following: if $p(C)$ and a corner t of a box of C lie in
> the same grid interval in dimension i, that is, if there is a j so
> that $p_i(C), t_i \in [j, j+1]$, then $t_i = p_i(C)$. $\qquad$ (4)

To achieve (4), we proceed as follows. Let v be a vertex of any thin cluster $C \in \mathcal{C}$. Consider a dimension i where $\ell_i(v)$ or $r_i(v)$ lie in the same grid interval as $p_i(C)$. Note that $\ell_i(v) \leq p_i(C) \leq r_i(v)$. In dimension i, we shrink the box of v in the following way: if $\ell_i(v)$ lies in the same grid interval as $p_i(C)$, we replace $\ell_i(v)$ by $p_i(C)$. Similarly, if $r_i(v)$ lies in the same grid interval as $p_i(C)$, we replace $\ell_i(v)$ by $p_i(C)$. This procedure is illustrated in Fig. 3.

Since by shrinking a box we may only lose edges of the corresponding graph, it suffices to show that every edge is still present. Since the new box of v still contains $p(C)$, the vertex v is still adjacent to every other vertex in C. As we change the box of v only within a grid interval, the old and the new box of v are in the same position. Thus, we do not lose any edge from v to X. Performing this transformation iteratively for every box of C in every dimension, and for every thin cluster $C \in \mathcal{C}$, we obtain a box representation of H satisfying (4).

Next, we claim that

> there is a pair of distinct thin clusters $C, C' \in \mathcal{C}$ such that for
> every $v \in C$ and $v' \in C'$ with $N_H(v) \cap X = N_H(v') \cap X$, the
> boxes of v and v' are in the same position. $\qquad$ (5)

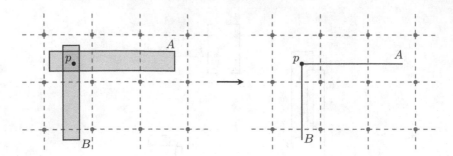

Fig. 3. Shrinking the boxes.

Note that, as C and C' are equivalent, there is indeed a bijection between the vertices of C and C' that maps a vertex v to $v' \in C'$ with $N_H(v) \cap X = N_H(v') \cap X$.

Observe that for the endpoints $\ell_i(v), r_i(v)$ of the interval representing a vertex $v \in V(H)$ in the i-th dimension, there are at most $(2k+1)^2$ many choices to select the grid intervals they lie in. Thus, any set of thin boxes, pairwise not in the same position, has size at most $(2k+1)^{2d}$. Because G is devoid of true twins, no cluster has two vertices whose boxes are in the same position.

Recall that every cluster has at most 2^k vertices. Thus, among any choice of more than $(2k+1)^{2d \cdot 2^k}$ thin clusters there are two thin clusters satisfying (5). As $(2k+1)^{2d \cdot 2^k} \leq (2k+1)^{2(2^k+k+1)) \cdot 2^k}$, by (1), and since $\mathcal{C}$ contains at least $(2k+2)^{2^{k+1}(2^k+k+1)}$ thin clusters, by (3), the claim follows.

Consider clusters C, C' as in (5). We now embed the deleted cluster C^* in the box representation of $H = G - C^*$. For this, choose $\epsilon > 0$ small enough so that

> *for all $v \in C$ and $w \in V(H - C)$ and all dimensions i it holds*   (6)
> *that $|s_i - t_i| > \epsilon$, when s is a corner of the box of v and t is a*
> *corner of the box of w.*

(If such an ϵ does not exist, we may again perturb the box representation slightly so as to guarantee (2) while keeping (4)).

Define $q \in \mathbb{R}^d$ by setting

$$q_i = \begin{cases} 1 & \text{if } p_i(C) < p_i(C') \\ -1 & \text{if } p_i(C) > p_i(C') \\ 0 & \text{if } p_i(C) = p_i(C'). \end{cases}$$

Let $v \mapsto v^*$ be the bijection between C and C^* with $N_G(v) \cap X = N_G(v^*) \cap X$. We define a box for every $v^* \in C^*$ by taking a copy of the box of v and shifting its coordinates by the vector $\epsilon \cdot q$, that is, for every dimension i we set

$$\ell_i(v^*) = \ell_i(v) + \epsilon q_i \text{ and } r_i(v^*) = r_i(v) + \epsilon q_i.$$

Note that, by choice of ϵ, the box of v^* and the box of v are in the same position.

Let $\tilde{G}$ be the graph defined by this new box representation. We claim that $\tilde{G} = G$, which then finishes the proof of the lemma.

To prove this, we first note that we only added edges between vertices in C^* and H, while all other adjacencies remain unchanged. Next, as $p(C) + \epsilon q$ is a point that lies in every box of C^*, it follows that $\tilde{G}[C^*]$ is a complete graph. Moreover, by choice of ϵ, we have

$$N_{\tilde{G}}(v^*) \setminus (C \cup C^*) = N_G(v) \setminus (C \cup C^*)$$

for any $v \in C$. In particular, $N_{\tilde{G}}(v^*) \cap C' = \emptyset$. It remains to show that also $N_{\tilde{G}}(v^*) \cap C = \emptyset$.

For this, let $w^* \in C^*$ and $v \in C$ be arbitrary, where we allow that $v = w$. Let us show that the boxes of v and w^* do not intersect.

Since v and w' are nonadjacent in H, there is a dimension i such that either $r_i(v) < \ell_i(w')$ or $r_i(w') < \ell_i(v)$. By symmetry, we may assume $r_i(v) < \ell_i(w')$. Let I be the grid interval such that $r_i(v) \in I$. If $\ell_i(w') \notin I$, then $r_i(v) < \ell_i(w^*)$, since by our construction $\ell_i(w^*)$ is in the same grid interval as $\ell_i(w')$. This means that the boxes of v and w^* do not intersect. Thus, we may assume that $\ell_i(w^*) \in I$. As v and w are in the same cluster and thus adjacent, it follows that $\ell_i(w) \le r_i(v)$, which implies that $p_i(C) \in [\ell_i(w), r_i(v)] \subseteq I$. Now, (4) implies that $r_i(v) = p_i(C) = \ell_i(w)$.

Since $p_i(C) = r_i(v) < \ell_i(w')$, it follows that $p_i(C) < p_i(C')$. Thus, $r_i(v) = \ell_i(w) < \ell_i(w) + \epsilon = \ell_i(w^*)$. Consequently, the boxes of v and w^* do not intersect. This completes the proof.   □

4   An Additive 1-Approximation Algorithm

Bounded pathwidth suggests a dynamic programming approach, and this is precisely what we do. There is a hitch, though. The standard approach would be to solve the BOXICITY problem on one bag after another of the path decomposition, so that the local solutions can be combined to a global one. BOXICITY, however, does not permit this: as we are constructing the box representation of the graph, we may have to completely rearrange the previous boxes to add a new one.

Thus, the key issue is to force the problem to become "localized". To this end, we introduce a special interval graph I^* that reflects the path structure of the graph: two vertices are adjacent if and only if they appear in the same bag of the path decomposition. Doing so, we can safely compute local box representations of the subgraphs induced by the bags without paying attention to how these representations overlap. Indeed, the interval graph I^* gets rid of any unwanted adjacency.

Theorem 7. *There is an algorithm that, for any graph G with a given path decomposition of width w, determines in $2^{O(w^2 \log w)} \cdot |V(G)|$ time a $d \in \mathbb{N}$ so that $d \le \mathrm{box}(G) \le d + 1$ together with a box representation of dimension $d + 1$.*

Together with the algorithm of Bodlaender [7] that computes a path-decomposition of a graph G of width $\mathrm{pw}(G)$ in $f(\mathrm{pw}(G)) \cdot |V(G)|$ time, we obtain Theorem 2. We note that the running time could conceivably be improved by using a faster approximation algorithm with, say, a constant approximation factor.

5   Bounded Bandwidth does not Help

It is an open problem whether boxicity is polynomial-time solvable on graphs of bounded treewidth. While we cannot solve the problem, we can offer an indication why we suspect boxicity to be hard.

The first approach to prove tractability is usually dynamic programming. Evidently, this is because Courcelle [12] proved that a vast number of problems, namely those expressible in monadic second order logic, can be solved in polynomial time by a generic dynamic programming algorithm, if the treewidth is bounded. However, nobody appears to know how to formulate "box$(G) \leq d$?" in monadic second order logic, and it is doubtful that this is possible at all. More generally, dynamic programming seems to fail. Why is that so? We think this is because the tree-like structure of the input graph does not translate to a tree-like structure in the interval representation: given an input graph G of bounded treewidth, it may very well be the case that at least one interval graph in any optimal interval representation of G has unbounded treewidth.

To illustrate this, consider a $K_{2,n}$, where the smaller bipartition class is comprised of two vertices x and y, and the larger consists of $v_1, \ldots, v_n$. Clearly, $K_{2,n}$ has pathwidth 2 and boxicity 2 as well: in fact, $K_{2,n} + xy$ and $K_{2,n} + \{v_i v_j : i, j\}$ are two interval graphs whose intersection is $K_{2,n}$. Now, let I_1, I_2 be any two interval graphs with $K_{2,n} = I_1 \cap I_2$. The vertices x and y are not adjacent in at least one of I_1 and I_2, say in I_1. Suppose that I_1 contains a pair of non-adjacent v_i, v_j: then $x v_i y v_j x$ is an induced 4-cycle, which is impossible in an interval graph. Thus, $\{v_i\}_{i=1}^n$ form a clique of size n in I_1, and I_1 has therefore pathwidth at least $n - 1$.

What about stronger width-parameters? We have found a similar, albeit more complicated, example for bounded bandwidth, a parameter even more restrictive than pathwidth. Theorem 3 is a direct consequence of the following lemma.

Lemma 5. *For every n there is a graph G^n of bandwidth at most 16 and boxicity 2, so that in any interval representation $G = I_1 \cap I_2$ one of I_1 and I_2 has treewidth $\geq |V(G^n)|/32$.*

In light of the lemma, we would like to strengthen the conjecture of Adiga et al. [4]: We believe that BOXICITY remains NP-complete even for graphs of bounded bandwidth.

6   Discussion

In some respect, the method of our first algorithm is a generalization of the true twin reduction. The key insight is that if there are many vertex sets (the clusters) that are identical in the graph then many of these sets will have essentially the same geometric realization. Deleting one of these many "geometric twins" is unlikely to change boxicity.

We believe this approach can exploited further. Indeed, we are convinced that with similar methods as developed in this article, we can also formulate a parameterized algorithm for BOXICITY when the parameter is *distance to stars* – the smallest number of vertices whose removal results in a disjoint union of stars. Like cluster vertex deletion, distance to stars provides a non-trivial parameterization for BOXICITY between vertex cover (solved) and feedback vertex set (open). Moreover, given a graph G, computing a minimum set $X \subseteq V(G)$ such that $G[V - X]$ is a disjoint union of stars can be done in $f(|X|) \cdot |V(G)|^{O(1)}$ time [9].

Our second algorithm yields an additive 1-approximation for BOXICITY on graphs of bounded pathwidth. Two questions that immediately arise are: can we get rid of the additive 1, such that the algorithm computes box(G) exactly? Can the algorithm be lifted to run on graphs of bounded treewidth?

We turn to the second question: why is it difficult to extend the algorithm to graphs of bounded treewidth? We rely heavily on the fact that the one extra dimension is sufficient to reflect the path decomposition of the whole graph. If we mimick this approach for bounded treewidth we have to describe the tree decomposition of the graph with as few extra dimensions as possible. How many extra dimensions would we need? As many as the boxicity of the chordal supergraph obtained by turning each bag of the decomposition into a clique. If we started with a path decomposition, the boxicity will be one. For a general tree decomposition, however, it could well be that the boxicity of this chordal graph is about the treewidth of the input graph [11]. This suggests that there might be input graphs G for which box(G) is much lower than the number of dimensions required to describe their tree decomposition, which makes it impossible to approximate using only the techniques of Sect. 4.

References

1. Adiga, A., Babu, J., Chandran, L.S.: Polynomial time and parameterized approximation algorithms for boxicity. In: Thilikos, D.M., Woeginger, G.J. (eds.) IPEC 2012. LNCS, vol. 7535, pp. 135–146. Springer, Heidelberg (2012)
2. Adiga, A., Bhowmick, D., Chandran, L.S.: The hardness of approximating the boxicity, cubicity and threshold dimension of a graph. Discrete Appl. Math. **158**(16), 1719–1726 (2010)
3. Adiga, A., Bhowmick, D., Chandran, L.S.: Boxicity and poset dimension. SIAM J. Discrete Math. **25**(4), 1687–1698 (2011)
4. Adiga, A., Chitnis, R., Saurabh, S.: Parameterized algorithms for boxicity. In: Cheong, O., Chwa, K.-Y., Park, K. (eds.) ISAAC 2010, Part I. LNCS, vol. 6506, pp. 366–377. Springer, Heidelberg (2010)
5. Asplund, E., Grünbaum, B.: On a coloring problem. Math. Scand. **8**, 181–188 (1960)
6. Bielecki, A.: Problem 56. Colloq. Math. **1**, 333 (1948)

7. Bodlaender, H.L.: A linear-time algorithm for finding tree-decompositions of small treewidth. SIAM J. Comput. **25**(6), 1305–1317 (1996)
8. Bruhn, H., Chopin, M., Joos, F., Schaudt, O.: Structural parameterizations for boxicity. CoRR (2014). abs/1402.4992
9. Cai, L.: Fixed-parameter tractability of graph modification problems for hereditary properties. Inf. Process. Lett. **58**(4), 171–176 (1996)
10. Chalermsook, P., Laekhanukit, B., Nanongkai, D.: Graph products revisited: tight approximation hardness of induced matching, poset dimension and more. In: Proceedings of the 24th Annual ACM-SIAM Symposium on Discrete Algorithms (SODA 2013), pp. 1557–1576 (2013)
11. Chandran, L.S., Sivadasan, N.: Boxicity and treewidth. J. Comb. Theor. Ser. B **97**(5), 733–744 (2007)
12. Courcelle, B.: The monadic second-order logic of graphs I. Recognizable sets of finite graphs. Inf. Comput. **85**(1), 12–75 (1990)
13. Cozzens, M.: Higher and multi-dimensional analogues of interval graphs. Ph.D. thesis, Department of Mathematics, Rutgers University, New Brunswick, NJ (1981)
14. Diestel, R.: Graph Theory, 4th edn. Springer, Heidelberg (2010)
15. Doucha, M., Kratochvíl, J.: Cluster vertex deletion: a parameterization between vertex cover and clique-width. In: Rovan, B., Sassone, V., Widmayer, P. (eds.) MFCS 2012. LNCS, vol. 7464, pp. 348–359. Springer, Heidelberg (2012)
16. Downey, R.G., Fellows, M.R.: Fundamentals of Parameterized Complexity. Springer, London (2013)
17. Fellows, M.R., Hermelin, D., Rosamond, F.A.: Well quasi orders in subclasses of bounded treewidth graphs and their algorithmic applications. Algorithmica **64**(1), 3–18 (2012)
18. Ganian, R.: Twin-cover: beyond vertex cover in parameterized algorithmics. In: Marx, D., Rossmanith, P. (eds.) IPEC 2011. LNCS, vol. 7112, pp. 259–271. Springer, Heidelberg (2012)
19. Kostochka, A.: Coloring intersection graphs of geometric figures with a given clique number. In: Pach, J. (ed.) Towards a Theory of Geometric Graphs of Contemp. Math., vol. 342, pp. 127–138. Amer. Math. Soc. (2004)
20. Kratochvíl, J.: A special planar satisfiability problem and a consequence of its NP-completeness. Discrete Appl. Math. **52**(3), 233–252 (1994)
21. Roberts, F.S.: On the boxicity and cubicity of a graph. In: Tutte, W.T. (ed.) Recent Progress in Combinatorics, pp. 301–310. Academic Press, New York (1969)
22. Scheinerman, E.: Intersection classes and multiple intersection parameters. Ph.D. thesis, Princeton University (1984)
23. Spinrad, J.: Efficient Graph Representations: Fields Institute monographs. American Mathematical Society, USA (2003)
24. Thomassen, C.: Interval representations of planar graphs. J. Comb. Theor. Ser. B **40**(1), 9–20 (1986)
25. Yannakakis, M.: The complexity of the partial order dimension problem. SIAM J. Algebraic Discrete Methods **3**(3), 351–358 (1982)

A New Characterization of P_k-free Graphs

Eglantine Camby[1]([✉]) and Oliver Schaudt[2]

[1] Département de Mathématique, Université Libre de Bruxelles, Boulevard du
Triomphe, 1050 Brussels, Belgium
ecamby@ulb.ac.be
[2] Institut für Informatik, Universität zu Köln, Weyertal 80, 50931 Köln, Germany
schaudto@uni-koeln.de

Abstract. The class of graphs that do not contain an induced path on
k vertices, P_k-free graphs, plays a prominent role in algorithmic graph
theory. This motivates the search for special structural properties of P_k-
free graphs, including alternative characterizations.

Let G be a connected P_k-free graph, $k \geq 4$. We show that G admits
a connected dominating set whose induced subgraph is either P_{k-2}-free,
or isomorphic to P_{k-2}. Surprisingly, it turns out that every minimum
connected dominating set of G has this property.

This yields a new characterization for P_k-free graphs: a graph G is P_k-
free if and only if each connected induced subgraph of G has a connected
dominating set whose induced subgraph is either P_{k-2}-free, or isomorphic
to C_k. This improves and generalizes several previous results; the particular case of $k = 7$ solves a problem posed by van 't Hof and Paulusma
[A new characterization of P_6-free graphs, COCOON 2008] [12].

In the second part of the paper, we present an efficient algorithm that,
given a connected graph G, computes a connected dominating set X of
G with the following property: for the minimum k such that G is P_k-free,
the subgraph induced by X is P_{k-2}-free or isomorphic to P_{k-2}.

As an application our results, we prove that HYPERGRAPH 2-COLORA-
BILITY, an NP-complete problem in general, can be solved in polynomial
time for hypergraphs whose vertex-hyperedge incidence graph is P_7-free.

Keywords: P_k-free graph · Connected domination · Computational
complexity

1   Introduction

A *dominating set* of a graph G is a vertex subset X such that every vertex not
in X has a neighbor in X. Dominating sets have been intensively studied in
the literature. The main interest in dominating sets is due to their relevance on
both theoretical and practical side. Moreover, there are interesting variants of
domination and many of them are well-studied.

A *connected dominating set* of a graph G is a dominating set X whose induced
subgraph, henceforth denoted $G[X]$, is connected. As usual, a connected dominating set such that every proper subset is not a connected dominating set

© Springer International Publishing Switzerland 2014
D. Kratsch and I. Todinca (Eds.): WG 2014, LNCS 8747, pp. 129–138, 2014.
DOI: 10.1007/978-3-319-12340-0_11

is called a *minimal connected dominating set*. A connected dominating set of minimum size is called a *minimum connected dominating set*.

We use the following standard notation. Let P_k be the induced path on k vertices and let C_k be the induced cycle on k vertices. If G and H are two graphs, we say that G is H-free if H does not appear as an induced subgraph of G. Furthermore, if G is H_1-free and H_2-free for some graphs H_1 and H_2, we say that G is (H_1, H_2)-free. If two graphs G and H are isomorphic, we write $G \cong H$.

The class of P_k-free graphs has received a fair amount of attention in the theory of graph algorithms. Given an NP-hard optimization problem, it is often fruitful to study its complexity when the instances are restricted to P_k-free graphs.

Let us mention two recent results in this direction: the polynomial time algorithm to compute a stable set of maximum weight, given by Lokshtanov *et al.* [10], and the result of Hoang *et al.* [6] showing that k-COLORABILITY is efficiently solvable on P_5-free graphs. The proof of the latter result relies on the fact that a connected P_5-free graph has a dominating clique or a dominating P_3.

Theorem 1 (Bácso and Tuza [1]**).** *Let G be a connected P_5-free graph. Then G has a dominating clique or a dominating induced P_3.*

An immediate implication of this result is the following.

Theorem 2 (Bácso and Tuza [1]**, Cozzens and Kelleher** [4]**).** *Let G be a graph. The following assertions are equivalent.*

(i) G is P_5-free.
(ii) Every induced subgraph H of G admits a connected dominating set X such that $H[X]$ is a clique or $H[X] \cong C_5$.

Later, van 't Hof and Paulusma [13] obtained a characterization for the class of P_6-free graphs in the flavour of Theorem 2. An earlier, slightly weaker result was given by Liu *et al.* [8], and the particular case of triangle free graphs was discussed before by Liu and Zhou [9].

Theorem 3 (van 't Hof and Paulusma [13]**).** *Let G be a graph. The following assertions are equivalent.*

(i) G is P_6-free.
(ii) Every connected induced subgraph H of G admits a connected dominating set X such that $H[X]$ has a complete bipartite spanning subgraph or $H[X] \cong C_6$.

Complementing Theorem 3, van 't Hof and Paulusma give a polynomial time algorithm that, given a connected P_6-free graph, computes a connected dominating set X such that $G[X]$ has a complete bipartite spanning subgraph or $G[X] \cong C_6$.

In view of Theorems 2 and 3, two questions arise. The first one is whether condition (ii) of Theorem 3 can be tightened, such that $H[X]$ is a P_4-free graph or $G[X] \cong C_6$. Note that if $H[X]$ is P_4-free, it is a connected cograph, and in particular has a complete bipartite spanning subgraph. This condition is the direct analogue of condition (ii) of Theorem 2 for P_6-free graphs. The advantage of the strengthened version is of course that the structure of cographs is well understood and more restricted compared to the class of graphs having a spanning complete bipartite graph.

The second question is whether similar characterizations can be given for the class of P_k-free graphs, for $k > 6$. In their paper, van 't Hof and Paulusma [13] explicitly ask for such a characterization in the case of $k = 7$.

1.1   Our Contribution

In this paper, we give an affirmative answer to these two questions. We show that every connected P_k-free graph, $k \geq 4$, admits a connected dominating set whose induced subgraph is either P_{k-2}-free, or isomorphic to P_{k-2}. Surprisingly, it turns out that every minimum connected dominating set has this property.

Theorem 4. *Let G be a connected P_k-free graph, $k \geq 4$, and let X be any minimum connected dominating set of G. Then $G[X]$ is P_{k-2}-free, or $G[X] \cong P_{k-2}$.*

From this result we derive the following characterization of P_k-free graphs.

Theorem 5. *Let G be a graph and $k \geq 4$. The following assertions are equivalent.*

(i) G is P_k-free.
(ii) Every connected induced subgraph H of G admits a connected dominating set X such that $H[X]$ is P_{k-2}-free or $H[X] \cong C_k$.

We now come to the algorithmic dimension of the problem. The proof of Theorem 4 is constructive in the sense that it yields an algorithm to compute, given a P_k-free graph, a connected dominating set whose induced subgraph is either P_{k-2}-free, or isomorphic to P_{k-2}. However, recall that the computation of a longest induced path in a graph is an NP-hard problem, as shown in Garey and Johnson [5, p. 196]. In other words, there is little hope of computing in polynomial time the minimum k for which the input graph is P_k-free. To overcome this obstacle, our algorithm can only make implicit use of the absent induced P_k, which is the main difficulty here.

Theorem 6. *Given a connected graph G on n vertices and m edges, one can compute in time $\mathcal{O}(n^5(n+m))$ a connected dominating set X with the following property: for the minimum $k \geq 3$ such that G is P_k-free, $G[X]$ is P_{k-2}-free or $G[X] \cong P_{k-2}$.*

Our last result is an application of the previous theorems. A *2-coloring* of a hypergraph assigns to each vertex one of two colors, such that each hyperedge contains vertices of both colors. The problem HYPERGRAPH 2-COLORABILITY is to decide whether a given hypergraph admits a 2-coloring. Garey and Johnson

[5, p. 221] explain that it is NP-complete in general. One successful approach to deal with this hardness is to put restrictions on the bipartite vertex-hyperedge incidence graph[1] of the input hypergraph.

As an application of Theorem 3, van 't Hof and Paulusma [13] show that HYPERGRAPH 2-COLORABILITY is solvable in polynomial time for hypergraphs with P_6-free incidence graph. Using our results, we settle the case of hypergraphs with P_7-free incidence graph.

Theorem 7. HYPERGRAPH 2-COLORABILITY *can be solved in polynomial time for hypergraphs with P_7-free incidence graph. If it exists, a 2-coloring can be computed in polynomial time.*

The proof of Theorems 4, 5 and 6 we give in the next section. Due to space limitations, the proof of Theorem 7 is omitted. We close the paper with a short discussion of our contribution.

2   Proofs

2.1   Proof of Theorems 4 and 5

We need the following lemma from an earlier paper of ours [3].

Lemma 1 (Camby and Schaudt [3]). *Let G be a connected graph that is (P_k, C_k)-free, for some $k \geq 4$, and let X be a minimal connected dominating set of G. Then $G[X]$ is P_{k-2}-free.*

When applied to P_k-free graphs, which are in particular (P_{k+1}, C_{k+1})-free, the above lemma implies that any minimal connected dominating set induces a P_{k-1}-free graph, for $k \geq 3$. We next prove a simple but useful lemma, which plays a key role also in the proof of Theorem 6. Let X be a connected dominating set of a graph G, and $x \in X$. Assuming that X is a minimal connected dominating set and $|X| \geq 2$, x is a cut-vertex of $G[X]$ or x has a *private neighbor*: a vertex $y \in V(G) \backslash X$ with $N_G(y) \cap X = \{x\}$.

Lemma 2. *Let G be a P_k-free graph, for some $k \geq 4$, and let X be a minimal connected dominating set of G. Assume that there is an induced P_{k-2} in $G[X]$, say on the vertices $x_1, x_2, \ldots, x_{k-2}$. Then any private neighbor y of x_1 is such that $(X \cup \{y\}) \backslash \{x_{k-2}\}$ is a connected dominating set of G.*

Proof. Note that G is in particular (P_{k+1}, C_{k+1})-free and thus, by Lemma 1, $G[X]$ is P_{k-1}-free.

Let $X' := \{x_1, x_2, \ldots, x_{k-2}\}$. Moreover, let y be any private neighbor of x_1, and let $Y := (X \cup \{y\}) \backslash \{x_{k-2}\}$. We have to prove that Y is a connected dominating set of G.

---

[1] Recall that for a hypergraph $H = (V, E)$ we define the bipartite vertex-hyperedge incidence graph as the bipartite graph on the set of vertices $V \cup E$ with the edges vY such that $v \in V$, $Y \in E$ and $v \in Y$. In the following, we just say the *incidence graph*.

Suppose for a contradiction that $G[Y]$ is not connected. Hence, x_{k-2} is a cut-vertex of $G[X]$. In particular, there is some vertex $y' \in X$ such that $N_G(y') \cap X' = \{x_{k-2}\}$. But then $G[X' \cup \{y'\}] \cong P_{k-1}$, a contradiction.

It remains to show that Y is a dominating set. Suppose the contrary, that is, there is some vertex x' with $N_G[x'] \cap Y = \emptyset$. As X is a dominating set, $N_G[x'] \cap X = \{x_{k-2}\}$. Because x_{k-2} is adjacent to Y and x' is not adjacent to Y, $x' \neq x_{k-2}$. But this means that $G[X' \cup \{y, x'\}] \cong P_k$, a contradiction.   □

Now we can state the proof of Theorem 4.

Proof (Proof of Theorem 4). Let X be a minimum connected dominating set of G. As G is in particular (P_{k+1}, C_{k+1})-free, $G[X]$ is P_{k-1}-free, by Lemma 1. We have to show that $G[X]$ is P_{k-2}-free or isomorphic to P_{k-2}.

To see this, assume there is an induced P_{k-2} in $G[X]$, say on the vertices $x_1, x_2, \ldots, x_{k-2}$. Let $X' := \{x_1, x_2, \ldots, x_{k-2}\}$. Note that x_1 is not a cut-vertex of $G[X]$: otherwise there is some vertex $y' \in X$ such that $N_G(y') \cap X' = \{x_1\}$, and hence $G[X' \cup \{y'\}] \cong P_{k-1}$. This is a contradiction. Thus, x_1 is not a cut-vertex of $G[X]$ and therefore has a private neighbor w.r.t. X, say y_1. By Lemma 2, $Y_1 := (X \cup \{y_1\}) \setminus \{x_{k-2}\}$ is a connected dominating set of G. As X is a minimum connected dominating set, Y_1 is a minimum connected dominating set, too. Moreover, y_1 has no neighbor in $X \setminus \{x_1\}$, in particular in $X \setminus X'$.

By reapplying the argumentation to Y_1 and the induced P_{k-2} on $y_1, x_1, x_2, \ldots, x_{k-3}$, We obtain a vertex $y_2 \in V(G) \setminus Y_1$ such that $Y_2 := (Y_1 \cup \{y_2\}) \setminus \{x_{k-3}\}$ is a minimum connected dominating set of G and $G[Y_2]$ contains an induced P_{k-2} on the vertices $y_2, y_1, x_1, x_2, \ldots, x_{k-4}$. Moreover, y_2 has no neighbor in $Y_1 \setminus \{y_1\}$, in particular in $X \setminus X'$.

Iteratively, we end up with a minimum connected dominating set Y_{k-2}, which is exactly $(X \setminus X') \cup \{y_1, \ldots, y_{k-2}\}$. Since, for $i = 1, 2, \ldots, k-2$, y_i is not adjacent to $X \setminus X'$ and $G[Y_{k-2}]$ is connected, $X \setminus X'$ must be empty, hence $X = X'$. Thus, $G[X] = G[X'] \cong P_{k-2}$. This completes the proof.   □

Proof (Proof of Theorem 5). Clearly P_k does not have a connected dominating set satisfying (ii). Hence, (ii) implies (i).

Conversely, let H be any connected induced subgraph of G, and let X be a minimum connected dominating set of H. By Theorem 4, $H[X]$ is P_{k-2}-free or $H[X] \cong P_{k-2}$. If $H[X]$ is P_{k-2}-free, the assertion of (ii) is satisfied. Otherwise, let $x_1, x_2, \ldots, x_{k-2}$ be a consecutive ordering of the induced path $H[X]$. In particular, x_1 and x_{k-2} are not cut-vertices of $H[X]$. As X is minimum, there exists a private neighbor y_i of x_i, for $i \in \{1, k-2\}$. It must be that $y_1 y_{k-2} \in E(H)$, since otherwise $H[X \cup \{y_1, y_{k-2}\}] \cong P_k$. Hence, $H[X \cup \{y_1, y_{k-2}\}] \cong C_k$, as desired. So, (i) implies (ii).   □

2.2 Proof of Theorem 6

Before we state our algorithm, we need to introduce some notation and definitions. For this, let us assume we are given a connected input graph G on n vertices and m edges. Let X be an arbitrary connected dominating set of G.

By $NC(X)$ we denote the set of vertices in X that are non-cutting in $G[X]$, i.e. for every $x \in NC(X), G[X \setminus \{x\}]$ is connected. Let x be a degree-1 vertex of $G[X]$. We define the *half-path* starting in x to be the maximal path $(x, x_1, x_2, \ldots, x_s)$ in X such that $|N_{G[X]}(x_i)| = 2$ for each $i \in \{1, 2, \ldots, s-1\}$. For example, if the neighbor $y \in X$ of x has degree at least 3, the half-path is simply (x, y). The *length* of the half-path is then s. To each $x \in X$ we assign a weight $w_X(x)$ as follows:

1. if $|N_{G[X]}(x)| \geq 2$, put $w_X(x) = 0$, and
2. if $|N_{G[X]}(x)| = 1$, put $w_X(x) = s$, where s is the length of the half-path starting in x.

Finally, the weight $w(X)$ of the set X given by

$$w(X) = \sum_{x \in X} (w_X(x))^2.$$

See Fig. 1 for an illustration of these definitions.

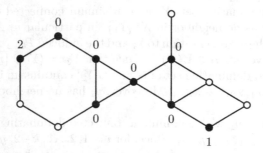

Fig. 1. A graph G. The black vertices form a connected dominating set X of G, with weights w_X as shown. We have $w(X) = 5$.

Let $\mathcal{X}$ be the family of all connected dominating sets of G. We next define a strict partial order $\prec$ on $\mathcal{X}$ as follows. For any two sets $X, Y \in \mathcal{X}$, we put $X \prec Y$ if

1. $|X| > |Y|$, or
2. $|X| = |Y|$ and $w(X) < w(Y)$.

The *height* of the strict poset $(\mathcal{X}, \prec)$ is the maximum set of mutually comparable elements of $\mathcal{X}$.

Lemma 3. *For a connected n-vertex graph G, the height of $(\mathcal{X}, \prec)$ is in $\mathcal{O}(n^3)$.*

Proof. If $G[X]$ is not an induced path, every vertex in X of degree at most 2 in $G[X]$ is contained in at most one half-path. Hence, $\sum_{x \in X} w_X(x) \leq |X|$. If $G[X]$

is an induced path, every vertex appears in at most two half-paths, implying $\sum_{x \in X} w_X(x) \leq 2|X|$. Thus

$$w(X) = \sum_{x \in X} (w_X(x))^2 \leq (\sum_{x \in X} w_X(x))^2 \leq 4|X|^2,$$

and so the weight of a connected dominating set is in $\mathcal{O}(n^2)$. Since there are at most n different possible sizes of connected dominating sets of G, the height of $(\mathcal{X}, \prec)$ is in $\mathcal{O}(n^3)$. $\qquad\square$

Proof (Proof of Theorem 6). Assume we are given a connected graph G on n vertices and m edges as input. Our algorithm works as follows, starting with the connected dominating set $Y := V(G)$. Its output is a connected dominating set X with the properties stated in Theorem 6.

1. Compute a minimal connected dominating set $X \subseteq Y$.
2. If $G[X]$ is an induced path, return X and terminate the algorithm.
3. Compute the set $NC(X)$ and the weight $w_X(x)$ for every $x \in NC(X)$.
4. Order the vertices of $NC(X)$ with non-increasing weight w_X, breaking ties arbitrarily. Let that order be $v_1, v_2, \ldots, v_{|NC(X)|}$.
5. For i from 1 to $|NC(X)|$ do the following:
   (a) Compute a private neighbor y_i of v_i w.r.t. X.
   (b) For j from $i + 1$ to $|NC(X)|$ do the following:
       i. Check whether $Y_{ij} := (X \cup \{y_i\}) \setminus \{v_j\}$ is a connected dominating set.
       ii. If yes, put $X \leftarrow Y_{ij}$ and go to Step 1.
6. Return X and terminate the algorithm.

We remark that the computation of y_i in Step 5a is always possible, since x_i is non-cutting in $G[X]$ and X is a minimal connected dominating set. The proof is completed by the following sequence of claims.

Claim 1. *When the algorithm terminates, the output X is a connected dominating set and $G[X]$ is P_{k-2}-free or $G[X] \cong P_{k-2}$.*

Since Step 1 is applied before the return is called, X is a minimal connected dominating set. If the algorithm terminates with Step 2, $G[X]$ is P_{k-1}-free by Lemma 1. Hence, either $G[X] \cong P_{k-2}$ or $G[X]$ is P_{k-2}-free.

Now assume that the algorithm terminates in Step 6. In particular, $G[X]$ is not an induced path. Suppose for a contradiction that $G[X]$ contains an induced P_{k-2}, say on the vertices $x_1, x_2, \ldots, x_{k-2}$. Like in the proof of Lemma 2, both x_1 and x_{k-2} cannot be cut-vertices of $G[X]$. Thus, $x_1, x_{k-2} \in NC(X)$.

After Step 4, the vertices of $NC(X)$ are ordered $v_1, v_2, \ldots, v_{|NC(X)|}$ with non-increasing weight. W.l.o.g. $x_1 = v_i$, $x_{k-2} = v_j$, and $i < j$. As X is returned, the set $Y_{ij} := (X \cup \{y_i\}) \setminus \{v_j\}$ is not a connected dominating set, in contradiction to Lemma 2. This proves our claim.

Claim 2. *Let X be a minimal connected dominating set considered in some iteration of the algorithm. Assume that the 'go to' is called in Step 5(b)ii because $Y_{ij} := (X \cup \{y_i\}) \setminus \{v_j\}$ is a connected dominating set. Let X' be the minimal connected dominating set computed in the subsequent Step 1. Then $X \prec X'$.*

Clearly $|X'| \leq |X|$. If $|X'| < |X|$, $X \prec X'$ by definition. So we may assume that $|X'| = |X|$, and hence $X' = Y_{ij}$. It remains to show that $w(X) < w(X')$.

Let $z \in X \backslash \{v_i, v_j\}$ be a degree-1 vertex of $G[X]$, and let $(z, x_1, x_2, \ldots, x_s)$ be a half-path starting in z. As $G[X]$ is not a path, x_s is a cut-vertex of $G[X]$. In particular, $x_s \neq v_j$. Hence, in $G[Y_{ij}]$, $(z, x_1, x_2, \ldots, x_s)$ is the initial segment of a half-path starting in z. In particular, $w_{X'}(z) \geq w_X(z)$.

If v_i is not a degree-1 vertex of $G[X]$, $w_{X'}(v_i) = w_X(v_i) = 0$, and (y_i, v_i) is the initial segment of a half-path starting in y_i. Hence, $w_{X'}(y_i) \geq 1$, and thus

$$w_{X'}(v_i) = 0 \text{ and } w_{X'}(y_i) \geq w_X(v_i) + 1. \tag{1}$$

If the degree of v_i in $G[X]$ is 1, let $(v_i, x_1, x_2, \ldots, x_s)$ be a half-path starting in v_i. Again, x_s is a cut-vertex of $G[X]$, and so $x_s \neq v_j$. Hence, in $G[X']$, $(y_i, v_i, x_1, x_2, \ldots, x_s)$ is the initial segment of a half-path starting in y_i. Again (1) holds.

Summing up, we see that (1) holds, and

$$w_{X'}(z) \geq w_X(z) \text{ for every vertex } z \in X' \backslash \{y_i, v_i\}. \tag{2}$$

We now turn to the vertex v_j. First assume that the degree of v_j in $G[X]$ is at least 2, and thus $w_X(v_j) = 0$. Then, by (2),

$$w(X') - w(X) \geq w_{X'}(y)^2 - w_X(v_j)^2 > 0,$$

and so $w(X') - w(X) > 0$.

Now assume that v_j is a vertex of degree 1 in $G[X]$, and so $w_X(v_j) \geq 1$. Let $N_{G[X]}(v_j) = \{x\}$. As $G[X]$ is not a path, $|N_{G[X]}(x)| \geq 2$, and so $w_X(x) = 0$. Thus $w_{X'}(x) = w_X(v_j) - 1$. Recall that (2) holds, and $w_{X'}(z) \geq w_X(z)$ for every vertex $z \in X' \backslash \{y_i, v_i\}$. We obtain the following inequality.

$$\begin{aligned} w(X') - w(X) &\geq w_{X'}(y_i)^2 + w_{X'}(x)^2 - w_X(v_i)^2 - w_X(v_j)^2 \\ &= (w_{X'}(y_i)^2 - w_X(v_i)^2) - (w_X(v_j)^2 - w_{X'}(x)^2) \\ &\geq [(w_X(v_i) + 1)^2 - w_X(v_i)^2] - [w_X(v_j)^2 - (w_X(v_j) - 1)^2] \end{aligned}$$

But $w_X(v_i) \geq w_X(v_j)$ implies

$$(w_X(v_i) + 1)^2 - w_X(v_i)^2 > w_X(v_j)^2 - (w_X(v_j) - 1)^2,$$

and thus $w(X') - w(X) > 0$ holds as in the previous case.

Hence, $X \prec X'$, proving our claim.

See Fig. 2 for an illustration of Step 5(b)ii.

Claim 3. *The algorithm terminates in $\mathcal{O}(n^5(n+m))$ time.*

By Claim 2, each call of the 'go to'-step and the subsequent application of Step 1 result in a connected dominating set that is properly larger in the order $\prec$. By Lemma 3, the height of the poset $(\mathcal{X}, \prec)$, and hence the number of iterations the whole algorithm performs, is in $\mathcal{O}(n^3)$.

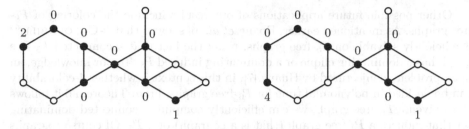

Fig. 2. Before (left) and after (right) an application of Step 5(b)ii. In the next iteration, the algorithm terminates with the right connected dominating set as output.

It remains to discuss the complexity of the particular steps. For this, recall that it can be checked in time $\mathcal{O}(n + m)$ whether a given vertex subset is a connected dominating set. Consequently, Step 1 can be performed in time $\mathcal{O}(n(n + m))$ by the immediate greedy procedure.

Step 2 and the computation of the weights in Step 3 can both be performed in linear time using the degree sequence of $G[X]$. The computation of the set $NC(X)$ in Step 3 can be done straightforwardly in time $\mathcal{O}(n(n + m))$.

It remains to discuss the complexity of the loop of Step 5. The computation of a private neighbor in Step 5a is clearly done in $\mathcal{O}(n + m)$ time. The inner loop of Step 5b consumes $\mathcal{O}(n)$ checks whether some vertex set is a connected dominating set, requiring $\mathcal{O}(n + m)$ time each. Hence, Step 5 can be done in $\mathcal{O}(n^2(n + m))$ time.

The overall running time computes to $\mathcal{O}(n^5(n + m))$, which completes the proof of both our claim and Theorem 6. □

3 Conclusion

In this paper we gave a description of the structure of connected dominating sets in P_k-free graphs. We have shown that any connected P_k-free graph admits a connected dominating set whose induced subgraph is P_{k-2}-free or isomorphic to P_{k-2}. In fact, any minimum connected dominating set has this property. Loosely speaking, this means that the restricted structure of connected P_k-free graphs results in an even more restricted structure of the induced subgraph of their minimum connected dominating sets.

Although we think that our results are of their own interest, our hope is that they might be useful in other contexts, too. One example we gave is the polynomial time solvability of HYPERGRAPH 2-COLORABILITY for hypergraphs with P_7-free incidence graph. It seems possible that, with more work, one could push this result to hypergraphs with P_8-free incidence graph. However, more interesting would be to know whether there is any k for which HYPERGRAPH 2-COLORABILITY for hypergraphs with P_k-free incidence graph is *not* solvable in polynomial time. So far, we do not have an opinion or an intelligent guess on this question.

Other possible future applications of our results include the coloring of P_k-free graphs. As mentioned earlier, Hoang *et al.* [6] showed that k-COLORABILITY is efficiently solvable on P_5-free graphs, using the fact that a connected P_5-free graph has a dominating clique or a dominating induced P_3. To our knowledge, an open problem, conjectured by Huang [7], in this context is whether 4-colorability can be decided in polynomial time for P_6-free graphs. From Theorem 6 it follows that, given a P_6-free graph, we can efficiently compute a connected dominating set that induces a P_4-free graph (that is a cograph) or a P_4. Of course cographs are less trivial than cliques, especially when it comes to coloring – but that does not rule out an approach similar to that of Hoáng *et al.* [6]. The fact that each vertex of the graph has some neighbor in this cograph leaves a 3-coloring problem for the rest of the graph, once the coloring of the cograph is fixed. Here, one might use the fact that 3-coloring is polynomial time solvable for P_6-free graphs, shown by Randerath and Schiermeyer [11], even in the pre-coloring extension version, proven by Broersma *et al.* [2].

References

1. Bácso, G., Tuza, Z.: Dominating cliques in P_5-free graphs. Period. Math. Hungar. **21**, 303–308 (1990)
2. Broersma, H., Fomin, F.V., Golovach, P.A., Paulusma, D.: Three complexity results on coloring P_k-free graphs. Europ. J. Combin. **34**, 609–619 (2013)
3. Camby, E., Schaudt, O.: The price of connectivity for dominating sets: upper bounds and complexity. Disc. Appl. Math. **177**, 53–59 (2014)
4. Cozzens, M.B., Kelleher, L.L.: Dominating cliques in graphs. Disc. Math. **86**, 101–116 (1990)
5. Garey, M.R., Johnson, D.S.: Computers and Intractability. W.H. Freeman and Co., New York (1979)
6. Hoáng, C.T., Kaminski, M., Lozin, V.V., Sawada, J., Shu, X.: Deciding k-colorability of P_5-free graphs in polynomial time. Algorithmica **57**, 74–81 (2010)
7. Huang, S.: Improved complexity results on k-coloring P_t-free graphs. In: Chatterjee, K., Sgall, J. (eds.) MFCS 2013. LNCS, vol. 8087, pp. 551–558. Springer, Heidelberg (2013)
8. Liu, J., Peng, Y., Zhao, C.: Characterization of P_6-free graphs. Disc. App. Math. **155**, 1038–1043 (2007)
9. Liu, J., Zhou, H.: Dominating subgraphs in graphs with some forbidden structures. Disc. Math. **135**, 163–168 (1994)
10. Lokshtanov, D., Vatshelle, M., Villanger, Y.: Independent set in P_5-free graphs in polynomial time. In: Proceedings of SODA, pp. 570–581. SIAM (2014)
11. Randerath, B., Schiermeyer, I.: 3-colorability $\in \mathcal{P}$ for P_6-free graphs. Disc. App. Math. **136**, 299–313 (2004)
12. van 't Hof, P., Paulusma, D.: A new characterization of P_6-free graphs. In: Proceedings of COCOON, pp. 415–424 (2008)
13. van 't Hof, P., Paulusma, D.: A new characterization of P_6-free graphs. Disc. App. Math. **158**, 731–740 (2010)

Contact Representations of Planar Graphs: Extending a Partial Representation is Hard

Steven Chaplick[1]([⊠]), Paul Dorbec[2], Jan Kratochvíl[3],
Mickael Montassier[4], and Juraj Stacho[5]

[1] Institut Für Mathematik, Technische Universität Berlin, Berlin, Germany
chaplick@math.tu-berlin.de
[2] University of Bordeaux, CNRS - LaBRI, UMR 5800, 33400 Talence, France
paul.dorbec@u-bordeaux.fr
[3] Department of Applied Mathematics, Faculty of Mathematics and Physics,
Charles University, Praha, Czech Republic
honza@kam.mff.cuni.cz
[4] Université Montpellier 2, CNRS - LIRMM, Montpellier, France
mickael.montassier@lirmm.fr
[5] Department of Industrial Engineering and Operations Research,
Columbia University, New York, NY, USA
stacho@cs.toronto.edu

Abstract. Planar graphs are known to have geometric representations of various types, e.g. as contacts of disks, triangles or - in the bipartite case - vertical and horizontal segments. It is known that such representations can be drawn in linear time, we here wonder whether it is as easy to decide whether a partial representation can be completed to a representation of the whole graph. We show that in each of the cases above, this problem becomes NP-hard. These are the first classes of geometric graphs where extending partial representations is provably harder than recognition, as opposed to e.g. interval graphs, circle graphs, permutation graphs or even standard representations of plane graphs.

On the positive side we give two polynomial time algorithms for the grid contact case. The first one is for the case when all vertical segments are pre-represented (note: the problem remains NP-complete when a subset of the vertical segments is specified, even if none of the horizontals are). Secondly, we show that the case when the vertical segments have only their x-coordinates specified (i.e., they are ordered horizontally) is polynomially equivalent to level planarity, which is known to be solvable in polynomial time.

1 Introduction

An intersection representation of a graph $G = (V, E)$ is a set family $\{S_v : v \in V\}$ such that uv is an edge of G iff $S_u \cap S_v \neq \emptyset$. Geometric representations

S. Chaplick: Supported by ESF Eurogiga project GraDR.
J. Kratochvíl: Supported by GAČR project GA14-14179S and partially by ESF Eurogiga project GraDR as GAČR GIG/11/E023.

D. Kratsch and I. Todinca (Eds.): WG 2014, LNCS 8747, pp. 139–151, 2014.
DOI: 10.1007/978-3-319-12340-0_12

(i.e., intersection representations where each set is a geometric object) of graphs have been intensively studied both for their practical motivations and interesting algorithmic properties. The motivations stem from VLSI designs, graphic layouts including the rectangular windows overlays, bioinformatics applications (including DNA sequencing), cellular description of reachability and interference in mobile networks, and many others. Geometric representations often also allow problems which are NP-hard in general to be solved in polynomial time.

The oldest and probably best understood class of intersection graphs are *interval graphs*, i.e., intersection graphs of intervals on a line [13]. They can be recognized in linear time and all basic optimization problems like independent set, clique or coloring can be solved on them in linear time as well. Generalizations of interval graphs include *circular arc graphs* [12,25], the intersection graphs of arcs on a circle. *Circle graphs* [4] are intersection graphs of chords of a circle and as such include *permutation graphs* [2,14], the intersection graphs of straight-line segments connecting points on two parallel lines. Intersection graphs of curves connecting points on two parallel lines, sometimes called the *function graphs* [14], are exactly the complements of comparability graphs (graphs admitting a transitive orientation). All these classes can be recognized in polynomial time and the independent set and clique problems can be solved in polynomial time on them as well. An overview of these and many other intersection-defined graph classes is given in many textbooks [5,22].

Geometric representations of graphs also help in visualizing the information grasped by the graph structure. Thus, the question of recognizing these classes and constructing a representation of a given type is rather important. Additionally, in some cases the polynomial time algorithms mentioned above exploit geometric representations. For the vast majority of the interesting classes of graphs the complexity of their recognition is well understood on the level of Polynomial versus NP-hard, with some cases where NP-membership is not known (for intersection graphs of straight-line segments or intersection graphs of convex sets in the plane, which are both known to be NP-hard to recognize, only PSPACE membership of the recognition problem is known, as their recognition is complete for the existential theory of the reals [24]). Recently, more attention has been paid to the question of extending partial representations of graphs. This setting corresponds to a situation where a part of the graph comes already represented from the applied instance or when the visualization task comes from a customer who does not want to see some part of the picture changed. Formally, we discuss the following decision problem, parameterized by an intersection-defined class $\mathcal{C}$:

REPEXT($\mathcal{C}$)

Instance: A graph G and a $\mathcal{C}$-representation R' of an induced subgraph of G.

Question: Does there exist a $\mathcal{C}$-representation R of G such that $R' \subseteq R$?
This question falls into a natural paradigm of extending a partial solution of a problem rather than building a solution from scratch, the latter approach being often easier. This common knowledge of architects and engineers can be observed in graph coloring problems where it is well known that every cubic

bipartite graph is 3-edge-colorable, but extending a partial edge-coloring is NP-complete [11], even for planar bipartite graphs [21]. Therefore it feels somewhat unexpected that for the resolved cases of geometric intersection graphs, extending partial solutions has not been harder than recognizing the particular classes[1]. In particular, REPEXT($\mathcal{C}$) is decidable in polynomial time when $\mathcal{C}$ is: interval graphs [16], proper interval graphs [16], unit interval graphs [19], circle graphs [6], permutation graphs [17], and function graphs [17]. These algorithms tend to extend the plain recognition ones in nontrivial ways through the use of special data structures which capture all representations. Interestingly, even though the classes of unit and proper interval graphs coincide, they are separated by the partial representation extension problem; i.e., there are instances of partial representations consisting of unit intervals that are extendible to a proper interval representation, but not to a unit one [16].

In this paper we consider REPEXT($\mathcal{C}$) when $\mathcal{C}$ is a *contact* graph class. This work is motivated by several elegant theorems that show that all planar graphs have geometric representations by contacts of various geometric objects. In general, an intersection representation is a contact one if the interiors of any two objects of the representation are disjoint. The classical example is Koebe's theorem, often referred to as the *kissing lemma* or the *coin representation*, which was rediscovered several times by several authors. It states that every planar graph is the contact graph of a collection of disks in the plane [20]. The proof of this theorem is nonconstructive but later Mohar [23] gave a polynomial time algorithm for producing an approximate representation (there are planar graphs that require irrational coordinates for some disk centers in any coin representation, and so approximate constructions are the best one can hope for, at least if we want to describe the coordinates and radii by rational numbers). De Fraysseix et al. [8] constructively proved that every planar graph is a contact graph of triangles in the plane. In 1991, Hartman et al. [3] showed that every bipartite planar graph has a *grid contact* representation, i.e., a contact representation in which vertices of one class of bipartition are represented by vertical segments and vertices of the other class by horizontal ones (this was also independently shown by de Fraysseix et al. [7]).

We prove that in all of these cases, deciding whether a partial contact representation of a planar (bipartite) graph can be extended to a contact representation of the entire graph is NP-hard. For geometric intersection graphs (i.e., for intersection graphs of planar objects defined by their shape or geometrical properties), this collection of results provides the first examples where extending partial representations is harder than deciding or constructing representations with no initial constraints. Note that for extending partial representations by triangles, convex sets, or disks, we only claim NP-hardness and not NP-completeness. This is because the membership in the class NP is not known

---

[1] The only exception is formed by partial subtree-in-tree or path-in-tree representations of chordal graphs, but there the NP-hardness follows from limited space issues, not any geometrical ones [18].

(similarly to recognizing intersection graphs of disks, convex sets or straight line segments, where only PSPACE membership is known).

In the last section of the paper we refocus on grid contact representations of planar graphs and show that the partial representation extension problem remains NP-hard if only some of the vertical segments are prerepresented. On the contrary, the problem becomes polynomially decidable if all of the vertical segments are given in the input, and also if only their x-coordinates are given (i.e., all horizontal segments and the vertical position of the vertical ones are unspecified). The last mentioned case is shown to be polynomially equivalent to testing level-planarity, a problem known to be decidable in linear time.

2   Grid Contact Graphs

For this section, let $G = (V \cup H, E)$ be a planar bipartite graph and let $n_1 = |V|, n_2 = |H|$. As already mentioned, de Fraysseix et al. [7] proved that G has a contact representation in which vertices of V are represented by vertical line-segments, vertices of H by horizontal line-segments, no parallel segments intersect, no two segments cross and any two segments u, v share a point (i.e., a point of contact) if and only if $uv \in E$ (for simplicity we use the same symbol for a vertex and the segment representing it). In particular, both V and H are independent sets of vertices in G. The proof [7] is based on bipolar orientations of planar graphs and their visibility representations. Such a representation can be constructed in polynomial time. We show that the task becomes harder if some of the vertices are pre-represented. The proof of the following theorem plays an important role in the rest of the paper. The NP-hardness reductions are all based on modifications of the gadgets constructed in it.

Theorem 1. *Given a planar bipartite graph G and some of its vertices represented by vertical or horizontal line-segments, it is NP-complete to decide if the partial representation can be extended to a grid contact representation of G.*

Proof. The NP-membership is straightforward, since a grid intersection representation can be described by the linear quasi-orders of n_1 coordinates for the vertical segments and n_2 coordinates for the horizontal ones.

For the NP-hardness proof we reduce from PLANAR-3-SAT. Given a Boolean formula Φ with a set C of clauses over a set X of variables such that the graph $G_\Phi = (C \cup X, \{xc : (x \in c \in C) \vee (\neg x \in c \in C)\})$ is planar, it is NP-complete to decide if Φ is satisfiable. This problem remains NP-complete even if every variable occurs in 3 clauses, once negated and twice positive, and every clause contains 2 or 3 literals [10] (in fact, Fellows et al. show NP-completeness even in a stronger way, for planar clause-linked formulas, i.e., for formulas whose incidence graphs remain planar after adding a cycle through all clause vertices, but we do not need this assumption). Given such a formula, we first draw the graph G_Φ in a rectilinear way so that edges are piece-wise linear curves with all segments either vertical or horizontal. We may further assume that the edges leaving each variable are positioned so that the edges corresponding to the two

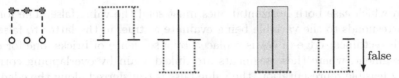

Fig. 1. The brick and its two possible representations.

Fig. 2. The brick and its two possible representations.

positive occurrences start with horizontal segments while the edge corresponding to the negative occurrence starts with a vertical one. The planarity of G_Φ can be tested in linear time, and a rectilinear drawing can be also constructed in linear time, even with a bounded number of bends per edge.

From this drawing we construct a graph G by a sequence of local replacements. Every variable is replaced by a copy of a variable gadget, every clause by a copy of a clause gadget, and the edges are replaced by chains of gadgets whose length depends on the number of bends on the edge. All gadgets are constructed from two building blocks. The basic one, the so called *brick*, is depicted in Fig. 1. The left part of the figure shows the subgraph, the right one a representation by contacts of segments. In all figures the black vertices and segments are those whose position is prescribed, and white vertices (dotted segments) are the flexible ones. The middle vertical black segment is an isolated vertex and thus cannot be crossed by any of the dotted ones. The dotted path connecting the other two black vertices can be represented either above or below the middle black segment. We use the schematic light grey rectangle depicted in Fig. 2 for the brick, and the side which bears the dotted segment encodes the value false. The bricks can also be rotated into a horizontal position, thus sending the false value to the left or to the right.

The variable gadget consists of three bricks whose vertices are pairwise non-adjacent. It is depicted in Fig. 3. From the overlapping corners of the bricks, it either sends the value false along the vertical edge, in which case both horizontal edges may transfer the value true, or it sends the value true along the vertical

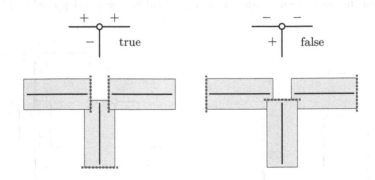

Fig. 3. The variable gadget.

edge, in which case both horizontal ones must send the value false. The former case corresponds to the variable being evaluate as true, in the latter to false.

Each rectilinear edge of G_Φ is replaced by a sequence of bricks, one for each linear segment, where these segments are linked again by overlapping corners. In every feasible representation, the value false is transferred along the edge, see Fig. 4. Note that it is possible for the edge gadget to transmit the value false even when its first brick is set to true, but this does not change the satisfiability of Φ.

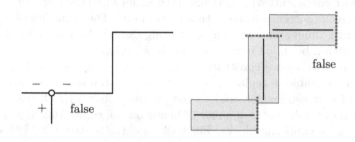

Fig. 4. The edge gadget.

For the clause gadget we use a *modified brick* depicted in Fig. 5. In any representation, at least one of the corners of the bounding rectangle must be used by a dotted segment. The clause gadget consists of two normal bricks and a modified one linked as depicted in Fig. 6. (For clauses containing 2 literals, we use the same gadget with one dummy variable represented by a single brick whose dotted path is pre-represented in the false position.) It is straightforward to see that if all three literals in the clause evaluate to false, all of the corners of the modified brick inside the clause gadget are blocked and the modified brick itself cannot be represented. Thus if the graph constructed as above has a representation, each clause must have at least one true literal and Φ is satisfiable. On the other hand, if Φ is satisfiable, we construct a representation following the lines and pictures above. Feasible representations of the clause gadget for the cases when the bottom or a side incoming literal is true are depicted in Fig. 7. ∎

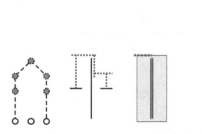

Fig. 5. The modified brick.

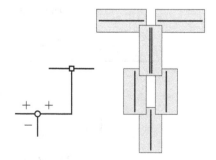

Fig. 6. The clause gadget.

3   Triangles, Disks, and Convex Sets

De Fraysseix et al. [8] proved that every planar graph is a contact graph of isosceles triangles with parallel bases. The construction is based on canonical ordering of the vertices in planar triangulations and can be performed in polynomial time. We show that again, given some vertices pre-represented, it is NP-hard to decide if the representation can be extended to a contact representation of the entire graph. We prove this in a stronger form, noting that triangles are convex sets.

Theorem 2. *Given a planar graph G and a partial representation R' by contacts of isosceles triangles, the following questions are NP-hard to decide*

1. if R' can be extended to an intersection representation of G by convex sets,
2. if R' can be extended to a contact representation of G by convex sets,
3. if R' can be extended to a contact representation of G by isosceles triangles.

Proof. We modify the proof of Theorem 1. First, note that the graph G constructed in the proof is a disjoint union of isolated vertices and paths and all *flexible* vertices (i.e., those whose segments are not prescribed) are of degree 2. Thus, any intersection representation by closed convex sets can be reduced to a contact representation by segments (take the sets representing flexible vertices one by one and replace each of them by the segment connecting the closest intersection points with its two neighbors). These segments, however, do not need to follow the vertical and horizontal directions. To force them to be "almost" bi-directional, adjust the bricks by predrawing auxiliary guiding segments, very close to each other, that leave a very narrow angle for the flexible (dotted) segments, as depicted in Fig. 8 (the guiding segments represent isolated vertices and so must not be crossed or touched by any other segments of the representation). If the width of the corridor between the guiding segments is small enough with respect to the length of the central black segment, the corners of the bounding rectangle are blocked by the dotted flexible segments as in the proof of Theorem 1. A similar modification is applied to the modified brick. To keep all

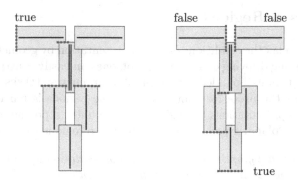

Fig. 7. The representations of the clause gadget.

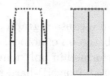

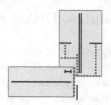

Fig. 8. The brick
for the convex set reduction.

Fig. 9. Extra blockage
for the convex set reduction.

flexible segments under control, we add extra blocking predrawn segments – all
loose corners will be blocked by an extra black segment and every corner of the
modified brick will be filled with three predrawn segments in an H-position as
shown in Fig. 9.

So if G has a representation by intersections of convex sets (extending the
given partial representation), then G has a contact representation by segments
with similar properties propagating the false assignment of variables as in the
proof of Theorem 1, and Φ is satisfied by the corresponding valuation. Moreover,
if Φ is satisfiable, a contact representation by vertical and horizontal segments
is a contact (i.e., also intersection) representation by convex sets. To achieve a
contact representation by isosceles triangles, it suffices to replace vertical segments by very thin triangles and the horizontal segments by very fat ones (with
very small height and whose base corresponds to a horizontal segment).   ∎

A similar modification of the gadgets by leaving a controlled space for the
flexible vertices shows the next result on disk contact (intersection) graphs (the
proof will be given in the full version of the paper).

Theorem 3. *Given a planar graph G and a partial representation R' by contacts
of disks, the following questions are NP-hard to decide*

1. *if R' can be extended to an intersection representation of G by disks,*
2. *if R' can be extended to a contact representation of G by disks.*

4   Contacts of Regions

One can further relax the conditions on the representation by geometrical objects.
From a non-crossing drawing of a planar graph one can easily construct a contact
graph of closed regions bounded by simple Jordan curves. (Disks are of course
such regions, but the proof for contacts of simple regions is much easier.) For
partial contact representations by regions we encounter a polynomially solvable
case. To maintain planarity, we insist that no three regions share a point.

Theorem 4. *Given a graph G and a partial representation by contacts of simple
regions, one can decide in linear time if the representation can be extended to a
contact region representation of the entire graph G.*

Proof. Add a master point M_u inside every region representing a vertex u and connect it by non-crossing curves to contact points on the boundary of its region. Consider this as a non-crossing drawing of a graph H and add vertices for the unrepresented vertices of G connected to the master points of their neighbors. Call the graph obtained in this way H'. Then G has a representation by contacts of regions if and only if H' has a planar drawing that extends the fixed drawing of H. This can be decided in linear time [1]. See an illustration in Fig. 10.   ∎

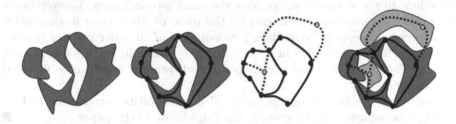

Fig. 10. Connecting region contact graphs and partially embedded planarity.

5   Grid Contact Graphs Revisited

In this section we modify the construction from the proof of Theorem 1 once more. We note that we may require that the pre-represented vertices belong to the same bipartition class.

Theorem 5. *Given a planar bipartite graph $G = (V \cup H, E)$ and some of the V vertices represented by vertical line-segments, it is NP-complete to decide if the partial representation can be extended to a grid contact representation of the entire graph.*

Proof. From the construction in the proof of Theorem 1, we replace each pre-represented horizontal segment as shown in Fig. 11. Specifically, we replace it with a flexible horizontal segment with three prescribed vertical neighbors. The new flexible horizontal segment has the same adjacency as the original prescribed segment and is locked in the same place by its new black vertical neighbors.   ∎
We conclude by observing that for the NP-hardness, it is important that some vertical vertices remain flexible:

Fig. 11. Lifting the pre-representation of horizontal segments.

Theorem 6. *Given a planar bipartite graph $G = (V \cup H, E)$ with all of the V vertices represented by vertical line-segments, it can be decided in polynomial time if the partial representation can be extended to a grid contact representation of the entire graph.*

Proof (Sketch). For every horizontal vertex $u \in H$, the x-coordinates of its (vertical) neighbors determine the x-coordinates of the left and right endpoints of the segment representing u. What remains to be determined is its height, i.e., the y-coordinate. The positions of its (vertical) neighbors determine a range $I(u)$ of possible y-coordinates of this segment. It can be shown that $I(u)$ is a union of at most $|V|$ real intervals, and hence can be described in polynomial time. Finally, one has to resolve conflicts among the horizontal segments as segments with equal y-coordinates must not intersect (i.e., segments with overlapping projections to the x axis must not have the same y-coordinate). The vertices u with infinite $I(u)$ can be disregarded for the moment, since their y-coordinates can always be chosen different than y-coordinates of all other vertices (we are processing a finite graph). It can be seen that if $I(u)$ is finite, it contains at most 2 values. The choice of the y-coordinates of such vertices can then be modeled by 2-SATISFIABILITY.

Pseudocode for this algorithm, a detailed proof of its correctness, and its running time analysis will be given in the full version of the paper. ∎

A further relaxation is when the vertical segments do not come with specified endpoints, but only their x-coordinates (or, equivalently, their left-to-right order) are given:

Theorem 7. *Given a planar bipartite graph $G = (V \cup H, E)$ and the order of x-coordinates of the vertical segments V, there is a polynomial-time algorithm to decide if there is a grid contact representation of G respecting this order.*

Proof. Here we sketch a polynomial reduction from this problem to *level-planarity testing* (a detailed proof will be given in the full version of the paper where we also show that level-planarity testing can be reduced to this problem). In level-planarity testing we are given a *leveled* graph G, i.e., a graph whose vertex set is partitioned into independent sets (*levels*) $S_1, \ldots, S_k$. The goal is to determine if there is a planar drawing of G where the vertices of each S_i are represented by points on the line $x = i$. (For convenience we represent levels vertically, rather than horizontally.)

Level-planarity testing is known to be solvable in linear time [9, 15]; the algorithm proceeds level-by-level and uses PQ-trees to record possible orderings on previous levels. Now to our problem.

Let $G = (V \cup H, E)$ be a given planar bipartite graph and let $v_1, v_2, \ldots, v_{n_1}$ be a given ordering of V by prescribed x-coordinates. For simplicity we use the same symbol for a vertex and the segment representing it.

We may assume that G is connected, otherwise we simply solve the corresponding problem on each connected component of G and put the representations one above the other. We may also assume that the degree of each vertex in H is at least two; all vertices of degree one in H can be safely removed and reattached later at arbitrary contact points.

The intuition regarding the connection between these problems comes from the special case when every horizontal segment has degree two. In this case we see the reduction immediately by respectively mapping vertical segments and horizontal segments to vertices and edges in the level-planarity instance. In particular, by giving each vertical segment its own level and ordering the levels by the x-coordinates of the verticals we are done.

We now turn to the more interesting case, when some horizontal segments have high (≥ 3) degree. In this case we replace each high degree horizontal segment h by a gadget which involves $O(\text{degree of } h)$ new segments where the horizontal segments have degree two and the vertical segments have degree three. This replacement is depicted in Fig. 12 and is formally described as Rule 1 below.

Rule 1. If H contains a vertex h of degree at least 3, then let $v_{i_0}, v_{i_1}, \ldots, v_{i_{k+1}}$ denote neighbors of h, where $i_0 < i_1 < \ldots < i_{k+1}$, and do the following:

(a) remove h from H and add a path

$$v_{i_0}, z_0, x_1^-, y_1^-, v_{i_1}, y_1^+, x_1^+, z_1, x_2^-, y_2^-, v_{i_2}, \ldots, v_{i_k}, y_k^+, x_k^+, z_k, v_{i_{k+1}}$$

where $x_j^-, x_j^+, y_j^-, y_j^+, z_j$ are new vertices such that:

x_j^-, x_j^+ are put in V and the rest in H.

(b) add new vertices $h_1, h_2, \ldots, h_k$ where each h_i is adjacent to x_i^- and x_i^+,
(c) modify the ordering of V by
  – inserting x_j^- right before and x_j^+ right after v_{i_j}, for all $j = 1, \ldots, k$.

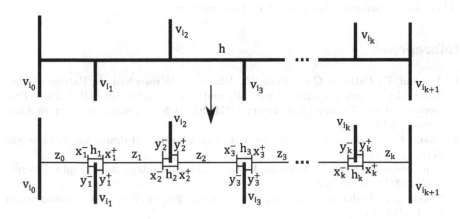

Fig. 12. Replacing a horizontal segment.

Moreover, from Fig. 12, it is easy to see that any solution to the original problem is preserved by applying Rule 1. So, we need to argue that any solution to the instance post-replacement must correspond to a solution pre-replacement. This amounts to two properties that we will need. The first is that the faces x_j^-, $h_j, x_j^+, y_j^+, v_{i_j}, y_j^-$ are empty in any solution and the second is that the path

$$z_0, x_1^-, h_1, x_1^+, z_1, x_2^-, h_2, x_2^+, z_2, \ldots, x_k^-, h_k, x_k^+, z_k$$

can be "straightened". Both of these conditions are easily observed, but require a bit of care to prove formally. Moreover, once they are attained we then simply reverse the replacement as in Fig. 12 to obtain a solution to our original instance.

Notice that the size of the instance of level-planarity we produced is linear with respect to our input graph. Thus, via Rule 1 and our argument regarding the degree two case, the reduction is complete.   ∎

6   Conclusion

In most of the cases we have encountered NP-hardness. One certainly wonders if additional assumptions may make the partial representation extension problems polynomially solvable. One possible direction is requiring the input graph to be highly connected (since the graph used in the proof of Theorem 1 is very sparse):

Problem 1. Is extendability of partial grid contact representations of planar quadrangulations decidable in polynomial time?

In view of Theorem 7 one may wonder what happens if only a part of the vertical segments is partially described:

Problem 2. Given a planar bipartite graph and a linear order of the x-coordinates of some of the vertical segments, can one decide in polynomial time if there is a grid contact representation respecting this order?

References

1. Angelini, P., Battista, G.D., Frati, F., Jelínek, V., Kratochvíl, J., Patrignani, M., Rutter, I.: Testing planarity of partially embedded graphs. In: SODA 2010: Proceedings of the Twenty-First Annual ACM-SIAM Symposium on Discrete Algorithms (2010)
2. Baker, K.A., Fishburn, P., Roberts, F.S.: Partial orders of dimension 2. Networks **2**(1), 11–28 (1972)
3. Ben-Arroyo Hartman, I., Newman, I., Ziv, R.: On grid intersection graphs. Discrete Math. **87**(1), 41–52 (1991)
4. Bouchet, A.: Reducing prime graphs and recognizing circle graphs. Combinatorica **7**, 243–254 (1987)
5. Brandstaedt, A., Le, V.B., Spinrad, J.P.: Graph Classes: A Survey. Society for Industrial and Applied Mathematics, Philadelphia (1999)
6. Chaplick, S., Fulek, R., Klavík, P.: Extending partial representations of circle graphs. In: Wismath, S., Wolff, A. (eds.) GD 2013. LNCS, vol. 8242, pp. 131–142. Springer, Heidelberg (2013)
7. de Fraysseix, H., Ossona de Mendez, P., Pach, J.: Representation of planar graphs by segments. In: Böröczky, K., Fejes Tóth, G. (eds.) Intuitive Geometry, Coll. Math. Soc. J. Bolyai **63**, 109–117 (1994)
8. de Fraysseix, H., Ossona de Mendez, P., Rosenstiehl, P.: On triangle contact graphs. Comb. Probab. Comput. **3**, 233–246 (1994)
9. di Battista, G., Nardelli, E.: Hierarchies and planarity theory. IEEE Trans. Syst. Man Cybern. **18**, 1035–1046 (1988)

10. Fellows, M.R., Kratochvíl, J., Middendorf, M., Pfeiffer, F.: The complexity of induced minors and related problems. Algorithmica **13**(3), 266–282 (1995)
11. Fiala, J.: NP completeness of the edge precoloring extension problem on bipartite graphs. J. Graph Theor. **43**(2), 156–160 (2003)
12. Gavril, F.: Algorithms on circular-arc graphs. Networks **4**(4), 357–369 (1974)
13. Gilmore, P.C., Hoffman, A.J.: A characterization of comparability graphs and of interval graphs. Canadian J. Math. **16**, 539–548 (1964)
14. Golumbic, M.C.: Algorithmic Graph Theory and Perfect Graphs. Computer Science and Applied Mathematics. Academic Press, New York (1980)
15. Jünger, M., Leipert, S., Mutzel, P.: Level planarity testing in linear time. In: Whitesides, S.H. (ed.) GD 1998. LNCS, vol. 1547, pp. 224–237. Springer, Heidelberg (1999)
16. Klavík, P., Kratochvíl, J., Vyskočil, T.: Extending partial representations of interval graphs. In: Ogihara, M., Tarui, J. (eds.) TAMC 2011. LNCS, vol. 6648, pp. 276–285. Springer, Heidelberg (2011)
17. Klavík, P., Kratochvíl, J., Krawczyk, T., Walczak, B.: Extending partial representations of function graphs and permutation graphs. In: Epstein, L., Ferragina, P. (eds.) ESA 2012. LNCS, vol. 7501, pp. 671–682. Springer, Heidelberg (2012)
18. Klavík, P., Kratochvíl, J., Otachi, Y., Saitoh, T.: Extending partial representations of subclasses of chordal graphs. In: Chao, K.-M., Hsu, T., Lee, D.-T. (eds.) ISAAC 2012. LNCS, vol. 7676, pp. 444–454. Springer, Heidelberg (2012)
19. Klavík, P., Kratochvíl, J., Otachi, Y., Rutter, I., Saitoh, T., Saumell, M., Vyskočil, T.: Extending partial representations of proper and unit interval graphs. In: Ravi, R., Gørtz, I.L. (eds.) SWAT 2014. LNCS, vol. 8503, pp. 253–264. Springer, Heidelberg (2014). http://arxiv.org/abs/1207.6960
20. Koebe, P.: Kontaktprobleme auf der konformen Abbildung. Ber. Verh. Saechs. Akad. Wiss. Leipzig Math.-Phys. Kl. **88**, 141–164 (1936)
21. Marx, D.: NP-completeness of list coloring and precoloring extension on the edges of planar graphs. J. Graph Theor. **49**(4), 313–324 (2005)
22. McKee, T.A., McMorris, F.R.: Topics in Intersection Graph Theory. SIAM Monographs on Discrete Mathematics and Applications. SIAM, Philadelphia (1999)
23. Mohar, B.: A polynomial time circle packing algorithm. Discrete Math. **117**(1–3), 257–263 (1993)
24. Schaefer, M.: Complexity of some geometric and topological problems. In: Eppstein, D., Gansner, E.R. (eds.) GD 2009. LNCS, vol. 5849, pp. 334–344. Springer, Heidelberg (2010)
25. Tucker, A.: An efficient test for circular-arc graphs. SIAM J. Comput. **9**(1), 1–17 (1980)

The Maximum Labeled Path Problem

Basile Couëtoux$^{(\boxtimes)}$, Elie Nakache, and Yann Vaxès

Aix-Marseille Université, CNRS, LIF UMR 7279, 13288 Marseille, France
{basile.couetoux,elie.nakache,yann.vaxes}@univ-amu.fr

Abstract. In this paper, we study the approximability of the Maximum Labeled Path problem: given a vertex-labeled directed acyclic graph D, find a path in D that collects a maximum number of distinct labels. Our main results are a $\sqrt{OPT}$-approximation algorithm for this problem and a self-reduction showing that any constant ratio approximation algorithm for this problem can be converted into a PTAS. This last result, combined with the APX-hardness of the problem, shows that the problem cannot be approximated within a constant ratio unless $P = NP$.

1 Introduction

Optimization network design problems over labeled graphs have been widely studied in the literature [2–8,10,11]. Since these problems are usually NP-hard, they have been mainly investigated toward the goal of finding efficiently approximate solutions. Most of these studies consider edge-labels that represent kinds of connections and the optimization concerns the number of different kinds of connections used. Our motivation is different, we consider vertex-labels that represent membership to different components. Our goal is then to maximize the number of components visited by a path in a directed graph. More precisely, the problem is defined on a directed graph with labels on the vertices and the objective is to find a path visiting a maximum number of distinct labels. We call this problem MAX-LABELED-PATH. Actually, the vertex-labeled and edge-labeled versions of this problem are equivalent but the vertex-labeled version is closer to our initial motivation. To our knowledge, there is no prior work on this simple and natural problem. A related problem is the Min LP $s-t$ problem that asks to find a path between s and t minimizing the number of different labels in this path. In [7] Hassin et al. achieves a $\sqrt{n}$ ratio for this problem and they show that it is hard to approximate within $O(\log n)$. We used a similar approach for our hardness result and the comparison is interesting since the maximization requires a much more precise analysis.

1.1 Contributions

In this paper we report both positive and negative results about the MAX-LABELED-PATH. Namely, we prove that this problem does not admit a constant factor approximation algorithm unless $P = NP$ and we propose an algorithm that returns a solution of value at least $\sqrt{OPT}$ where OPT is the value of an

© Springer International Publishing Switzerland 2014
D. Kratsch and I. Todinca (Eds.): WG 2014, LNCS 8747, pp. 152–163, 2014.
DOI: 10.1007/978-3-319-12340-0_13

optimal solution. In Sect. 2, the hardness proof starts with a reduction from MAX 3SAT preserving the approximation and therefore proving that MAX-LABELED-PATH is APX-hard. In Sect. 3, a polynomial self-reduction shows that finding a solution on a more complex graph enables us to find a solution with a better ratio on the initial graph. This, combined with the APX-hardness of the problem, shows that the problem cannot be approximated within a constant ratio unless $P = NP$. In Sect. 4, we describe a $\sqrt{OPT}$-approximation algorithm for MAX-LABELED-PATH. This algorithm requires a specific preprocessing and an inductive analysis that uses the poset structure of the problem.

1.2 Preliminaries

A vertex-labeled Directed Acyclic Graph $D = (V, A)$ is a DAG whose vertices are labeled by a function $l : V \to \mathcal{L}$. For each vertex $u \in V$, we denote by $\lambda(u)$ and call the *level* of u, the maximum number of vertices in a path having u as end-vertex. The *ith level set* L_i of D consists of all vertices $u \in V$ such that $\lambda(u) = i$. The vertices of L_1, i.e. having no ingoing arcs, are called the *sources* of D. The vertices having no outgoing arcs are called the sink. Let k be the largest integer such that $L_k \neq \emptyset$. L_k is a subset of the sinks. Let P be a (directed) path in D. P is maximal by inclusion if and only if it connects a source to a sink. The set of labels *collected* by P is the set $\{l(u) : u \in P\}$ of labels of vertices in P. Given a vertex-labeled DAG D, the problem MAX-LABELED-PATH consists in finding a path P in D maximizing the number of distinct labels collected by P. Any solution can be extended into a maximal path without decreasing its value, therefore we only consider solutions that connects a source to a sink. In this paper, we consider only maximization problem. Let D be an instance of a maximization problem, we denote by $OPT(D)$ its optimum. We say that an algorithm *achieves a constant performance ratio* α, if for every instance D, it returns a solution of value at least $\alpha\, OPT(D)$.

2 Maximum Labeled Path Is APX-Hard

In this section, we describe a reduction from MAX-3SAT establishing that MAX-LABELED-PATH is APX-hard even when restricted to instances satisfying the following conditions:

(C1) All maximal (by inclusion) paths of D contain the same number k of vertices.
(C2) D contains a path that collects all the labels, $OPT(D) = |\mathcal{L}|$.
(C3) D contains a path that collects each label exactly once, $OPT(D) = k = |\mathcal{L}|$.
(C4) $OPT(D) = k = |\mathcal{L}|$ is a power of two.

Note that (C4) is stronger than (C3) which is stronger than (C2). Applying our initial reduction to satisfiable instances of MAX-3SAT, we produce instances MAX-LABELED-PATH satisfying conditions (C1) with $k \leq 3|\mathcal{L}|$ and (C2) and proves Theorem 2. Then, we proceed in two steps: first we establish the APX-hardness for instances satisfying conditions (C1) and (C3) in Theorem 3 and then the APX-hardness for instances satisfying conditions (C1) and (C4) in Theorem 4. In the next section we use a self-reduction of MAX-LABELED-PATH to prove that MAX-LABELED-PATH does not belong to APX. This self-reduction is valid only for instances satisfying conditions (C1) and (C4).

Theorem 1. *(Håstad [9]) Assuming $P \neq NP$, no polynomial-time algorithm can achieve a performance ratio exceeding $\frac{7}{8}$ for MAX-3SAT even when restricted to satisfiable instances of the problem.*

Theorem 2. *Assuming $P \neq NP$, no polynomial-time algorithm can achieve a performance ratio exceeding $\frac{7}{8}$ for MAX-LABELED-PATH even when restricted to instances satisfying conditions (C1) with $k \leq 3|\mathcal{L}|$ and (C2).*

Before proving Theorem 2, we establish the following lemma showing that (C1) is not a strong requirement in the sense that each instance of MAX-LABELED-PATH can be converted into an equivalent instance satisfying (C1). The proof of Lemma 1 is omitted due to space limitation.

Lemma 1. *Given an instance D of MAX-LABELED-PATH, it is possible to construct an instance D' satisfying condition (C1) and such that there exists a mapping between the set of maximal paths in D and the set of maximal paths in D' preserving the number of labels collected.*

Proof (of Theorem 2). Given an instance F of MAX-3SAT, we define an instance $D_F = (V, A)$ of MAX-LABELED-PATH as follows. Let $\{w^1, w^2, ..., w^q\}$ be the set of variables of F. For all $j \in \{1, ..., q\}$, we denote by $|w^j|$ the number of occurrences of the literal w^j and by $|\neg w^j|$ the number of occurrences of its negation. We create $|w^j| + |\neg w^j|$ vertices and call them $w^j_1, w^j_2, ..., w^j_{|w^j|}$ and $\neg w^j_1, \neg w^j_2, ..., \neg w^j_{|\neg w^j|}$. We connect in a directed path $P(w^j)$ the vertices which represent the literal w^j, i.e. we create an arc (w^j_i, w^j_{i+1}) for all $i \in \{1, ..., |w^j| - 1\}$. In the same way, we connect in a directed path $P(\neg w^j)$ the vertices representing $\neg w^j$. For all $j \in \{1, ..., q-1\}$, we connect by an arc the last vertices of $P(w^j)$ and $P(\neg w^j)$ to the first vertices of $P(w^{j+1})$ and $P(\neg w^{j+1})$. Let us define the labeling function $l : V \rightarrow \mathcal{L} := \{1, ..., m\}$ where m is the cardinality of the set of clauses $\{C_1, C_2, ..., C_m\}$ of F. There is a one to one correspondence between the occurrences of the literals in the clauses and the vertices of D_F. A vertex u receives the label j if u corresponds to an occurrence of a literal in the clause C_j (see Fig. 1).

Applying the reduction to a satisfiable instance F of MAX-3SAT, we obtain an instance D_F of MAX-LABELED-PATH that contains a path collecting all the labels, i.e. that satisfies condition (C2). Moreover, since each clause contains at most three literals, the number k of vertices in a maximal path of D_F is at most thrice the number m of labels, i.e. $k \leq 3m$. In the resulting graph D_F, each maximal path P is a path from a vertex in $\{w^1_1, \neg w^1_1\}$ to a vertex in $\{w^q_{|w^q|}, \neg w^q_{|\neg w^q|}\}$ that contains for all $j \in \{1, ..., q\}$ either $P(w_j)$ or $P(\neg w_j)$ but not both. Therefore, it represents in an obvious way an assignment of the variables ($w_j = true \Leftrightarrow P(w_j) \subset P$). From the choice of the labeling of vertices in D_F, it is easy to verify that an assignment of the variables satisfying n clauses corresponds to a maximal path collecting n labels. This transformation produces in polynomial time an instance D_F satisfying the conditions (C2) with $k \leq 3|\mathcal{L}|$. It remains to ensure (C1), this can be done by applying the transformation of Lemma 1. Together with Theorem 1, this concludes the proof of Theorem 2.   □

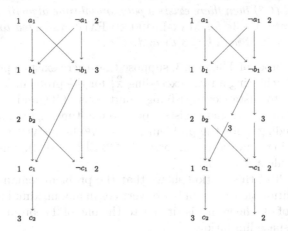

Fig. 1. The digraph D_F for the formula $F = (a \vee b \vee c) \wedge (\neg a \vee b \vee \neg c) \wedge (\neg b \vee c)$ before the transformation of Lemma 1 (to the left) and after (to the right).

The next step consists in showing that the problem MAX-LABELED-PATH remains APX-hard even when restricted to instances such that all maximal paths have the same number of vertices and contain a path collecting each label exactly once.

Theorem 3. *Assuming $P \neq NP$, no polynomial time algorithm can achieve a performance ratio exceeding $\frac{23}{24}$ for* MAX-LABELED-PATH *even when restricted to instances satisfying (C1) and (C3).*

Proof. Consider a DAG $D = (V, A)$ with a labeling function l that satisfies the conditions (C1) with $k \leq 3|\mathcal{L}|$ and (C2). Every maximal path in D contains the same number k of vertices. Let $m := |\mathcal{L}| \leq k$ be the number of labels of vertices in D. We construct a DAG D' by adding to D, for each vertex $v \in V$, a set $\{v^1, \ldots, v^r\}$ of $r := k - m$ copies of the vertex v. There is an arc between two vertices in D' if and only if there is an arc between their preimages in D (the preimage of a vertex $v \in V$ is v itself). Every maximal path in D' corresponds to a maximal path in D, in particular it contains exactly k vertices. The set of labels of D' is $\mathcal{L}' := \mathcal{L} \cup \{m + 1, m + 2, \ldots, m + r = k\}$. For each vertex v of D and each integer $j \in \{1, 2, \ldots, r\}$ the label of the vertex v^j is $m + j$. The labels in D' of the vertices that belong to D remain unchanged. We call the resulting instance D' the *extension* of the instance D.

The following two lemmata (whose proofs are omitted due to space limitation) establish a close relationship between the optimum of the instances D and D'.

Lemma 2. *If there is a path in D collecting n labels then there is a path in D' collecting $n + r$ labels. If there is a path in D' collecting n labels then there is a path in D collecting at least $n - r$ labels.*

Lemma 3. *If there exists a polynomial time algorithm that achieves a performance ratio $1 - \epsilon$ for MAX-LABELED-PATH restricted to instances satisfying conditions (C1) and (C3) then there exists a polynomial time algorithm that achieves a performance ratio $1 - 3\epsilon$ for MAX-LABELED-PATH restricted to instances satisfying conditions (C1) with $k \leq 3|\mathcal{L}|$ and (C2).*

To complete the proof of Theorem 3, suppose that there exists a polynomial time algorithm ALG′ achieving a ratio exceeding $\frac{23}{24}$ for the problem MAX-LABELED-PATH restricted to instances satisfying conditions (C1) and (C3). Then, by Lemma 3, we deduce that there exists a polynomial time algorithm ALG achieving a ratio exceeding $\frac{7}{8}$ for the problem MAX-LABELED-PATH restricted to the instances satisfying conditions (C1) with $k \leq 3|\mathcal{L}|$ and (C2), this cannot occur by Theorem 2, unless $P = NP$. □

The last result of this section shows that the problem remains APX-hard if we add the condition that the number of vertices in any maximal path is a power of two. The proof of Theorem 4 is similar to the one of Theorem 3 and has been omitted due to space limitation.

Theorem 4. *Assuming $P \neq NP$, no polynomial time algorithm can achieve a performance ratio exceeding $\frac{47}{48}$ for MAX-LABELED-PATH even when restricted to instances satisfying conditions (C1) and (C4).*

3   Maximum Labeled Path Does Not Belong to APX

In this section, using a self-reduction of the problem MAX-LABELED-PATH, we will prove the following result:

Theorem 5. *Assuming $P \neq NP$, no polynomial time algorithm can achieve a constant performance ratio for MAX-LABELED-PATH even when restricted to instances satisfying conditions (C1) and (C4).*

3.1   Self-reduction

In Sect. 3, we will consider only instances of MAX-LABELED-PATH satisfying conditions (C1) and (C4). Namely, a DAG $D = (V, A)$ whose vertices are labeled by a function $l : V \to \mathcal{L} = \{1, \ldots, k\}$ such that there exists a path collecting each label exactly once and the number $k = |\mathcal{L}|$ of vertices in any maximal path is a power of two. We will prove that such instances of the problem MAX-LABELED-PATH cannot be approximated in polynomial time within a constant factor. For the sake of simplicity, we also assume that there is only one source s and one sink t. Therefore, any maximal path is a path from s to t and all vertices of D belong to a path from s to t. Recall that, for each vertex $u \in V$, $\lambda(u)$ is the number of vertices in a path from s to u (all such paths have the same length because D satisfies (C4)). For all $u \in V$, $\lambda(s) = 1 \leq \lambda(u) \leq k = \lambda(t)$.

Pseudo Square and Pseudo Cubic Acyclic Digraph. The *pseudo square digraph* $\bar{D}$ of D is obtained from D by replacing each vertex $u \in V$ by a copy D_u of the digraph D. We denote by v_u the copy of the vertex $v \in V$ in the digraph D_u. There is an arc $v_u w_u$ in $\bar{D}$ if and only if there is an arc vw in D. In addition to the arcs of the subgraphs D_u, $u \in V$, we add to $\bar{D}$ an arc $t_u s_v$ for each arc from uv in D. The *pseudo cubic digraph* $\tilde{D}$ of D is obtained from $\bar{D}$ by replacing each vertex v_u of $\bar{D}$ by a path $P(v_u)$ with k vertices. Each arc entering a vertex v_u in $\bar{D}$ is replaced by an arc of $\tilde{D}$ entering the first vertex of $P(v_u)$. Analogously, each arc leaving the vertex v_u in $\bar{D}$ is replaced by an arc of $\tilde{D}$ leaving the last vertex of $P(v_u)$ (see Fig. 2). We define a new instance of MAX-LABELED-PATH on the digraph $\tilde{D}$ with the first vertex of $P(s_s)$ as a source and the last vertex of $P(t_t)$ as a sink and a labeling function $\tilde{l}$ defined as follows.

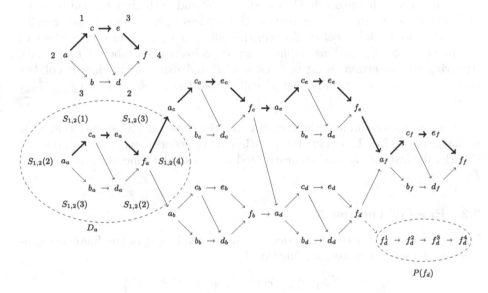

$P(f_d)$

Fig. 2. An example of pseudo square digraph $\bar{D}$ with $k = |\mathcal{L}| = 4$. An optimal path P in D and the corresponding optimal path $\bar{P}$ in $\bar{D}$ are drawn in bold. In the subgraph D_a, each vertex v of $\bar{D}$ is labeled by the subset of labels received by the vertices of the path $P(v)$ of $\tilde{D}$. In $\tilde{D}$, the vertex f_d of $\bar{D}$ is replaced by the path $P(f_d) = (f_d^1, f_d^2, f_d^3, f_d^4)$.

Labeling. Let v_u be a vertex of $\tilde{D}$, the set of labels of the vertices of $P(v_u)$ will depend on the labels of u and v in D and on the level of u in D. Since either all vertices of $P(v_u)$ are visited by a path from the source to the sink or none of them are, our labeling function assigns a set of labels to the path $P(v_u)$ and does not precise the order in which the labels appear on $P(v_u)$. The set of labels $\tilde{\mathcal{L}}$ used to define the labeling of $\tilde{D}$ consists of k disjoint subsets $\tilde{\mathcal{L}}_1, \ldots, \tilde{\mathcal{L}}_k$ such that $|\tilde{\mathcal{L}}_1| = \ldots = |\tilde{\mathcal{L}}_k| = k^2$. For each label $c \in \mathcal{L}$ and each level $i \in \{1, \ldots, k\}$, we construct a partition $\mathcal{S}_{i,c} := \{S_{i,c}(c') : c' \in \mathcal{L}\}$ of $\tilde{\mathcal{L}}_c$ into k subsets of size k such that any two subsets arising from different partitions intersect in exactly one label, i.e. if $i_1 \neq i_2$ for all $c', c'' \in \mathcal{L}$, $|S_{i_1,c}(c') \cap S_{i_2,c}(c'')| = 1$. Since k^2 is a power of two ($k^2 = 2^r$), such partitions can be easily constructed as classes of parallel lines of a finite affine plane (each class of parallel lines induces a partition in which the subsets are the lines). The construction of finite affine planes from finite fields is described for instance in [1]. This construction can be done in polynomial time in the size of D by first identifying an irreducible polynomial of degree r by brute force and then constructing the corresponding finite fields $GF(2^r)$. The labeling function $\tilde{l}$ assigns to the vertices of $P(v_u)$ the labels that belong to the subset $S_{\lambda(u),l(u)}(l(v))$ of the partition $\mathcal{S}_{\lambda(u),l(u)}$.

Claim. There is a path in $\tilde{D}$ that collects each label in $\tilde{\mathcal{L}}$ exactly once.

Proof. Let P be the path of D collecting all the labels in $\mathcal{L}$. Consider the path $\tilde{P}$ passing via each subgraph D_u for all $u \in P$ and such that the subpath $\tilde{P}_u$ of $\tilde{P}$ inside the subgraph D_u consists of the vertices v_u for all $v \in P$ (see Fig. 2). Since P collects each label in $\mathcal{L}$ once, the subpath $\tilde{P}_u$ collects every subset of the partition $\mathcal{S}_{\lambda(u),l(u)}$. This implies that $\tilde{P}_u$ collects each label of $\tilde{\mathcal{L}}_{l(u)}$ once. Applying this assertion to all vertices $u \in P$ and using again that P collects each label in $\mathcal{L}$, we conclude that $\tilde{P}$ collects all the labels of $\tilde{\mathcal{L}} = \bigcup_{u \in P} \tilde{\mathcal{L}}_{l(u)}$ once. $\square$

The previous claim and the fact that $|\tilde{\mathcal{L}}|$ is a power of two ensure that $\tilde{D}$ is an instance of MAX-LABELED-PATH satisfying the conditions of (C1) and (C4). Clearly, the instance $\tilde{D}$ can be constructed in polynomial time from the instance D.

3.2   Proof of Theorem 5

Let g denote the reciprocal function on the interval $[0, 1]$ of the following continuous and strictly increasing function h:

$$h(x) := \begin{cases} h_1(x) := x(x^2 - x + 1) & \text{if } 0 < x < \frac{1}{2}; \\ h_2(x) := x^2 - \frac{1}{4}x + \frac{1}{4} & \text{if } \frac{1}{2} \leq x \leq 1. \end{cases}$$

Lemma 4. *For each $0 < \beta < 1$, the sequence β_n defined by $\beta_0 = \beta$ and $\beta_{n+1} = g(\beta_n)$ has a limit of 1.*

In the next section, we show the following two results:

Lemma 5. *Given any path Q in $\tilde{D}$ that collects at least βk^3 labels, a path P in D that collects at least $g(\beta)k$ labels can be computed in polynomial time.*

Lemma 6. *If there is a polynomial-time algorithm with a ratio β for MAX-LABELED-PATH then there is a polynomial-time algorithm with a ratio $g(\beta)$ for MAX-LABELED-PATH.*

Proof. Suppose there exists a polynomial time algorithm ALG_β with a ratio at least β for MAX-LABELED-PATH. Let D be an instance of MAX-LABELED-PATH, we use the following algorithm:

Function ALG(D): a maximal path in D that collects $g(\beta)k$ labels

---

Construct the digraph $\tilde{D}$ from the digraph D;

Perform ALG$_\beta$ to obtain a path Q of $\tilde{D}$ that collects βk^3 labels;

Derive from Q a path P of D that collects at least $g(\beta)k$ labels;

Return P;

This algorithm is clearly polynomial because all the steps are, thus we have a polynomial time algorithm with a ratio $g(\beta)$ for MAX-LABELED-PATH. $\square$

Suppose there exists an approximation algorithm with a constant factor β for MAX-LABELED-PATH. By Lemma 4, there exists an integer n such that $\beta_n > \frac{47}{48}$. Applying n times Lemma 6, we derive a polynomial-time algorithm for the problem MAX-LABELED-PATH with a ratio exceeding $\frac{47}{48}$. A similar argument shows that any constant factor approximation algorithm for MAX-LABELED-PATH can be converted into a PTAS for this problem. Such an algorithm does not exist unless $P = NP$ by Theorem 4. Assuming Lemma 5, this concludes the proof of Theorem 5.

3.3   Proof of Lemma 5

We explain how to construct in polynomial time a path P in D that collects a set $\mathcal{L}^P \subseteq \mathcal{L}$ containing at least $g(\beta)k$ labels from a path Q in $\tilde{D}$ that collects a set $\tilde{\mathcal{L}}^Q \subseteq \tilde{\mathcal{L}}$ containing at least βk^3 labels. We denote by $V^Q \subseteq V$ the set of vertices u such that Q passes via D_u and by $\mathcal{L}^Q \subseteq \mathcal{L}$ the set of labels of the vertices in V^Q. For each vertex $u \in V^Q$, we define $W_u^Q \subseteq V$ the set of vertices v such that Q contains $P(v_u)$ as a subpath and by $\mathcal{L}_u^Q \subseteq \mathcal{L}$ the set of labels of the vertices in W_u^Q. Let $\alpha_u := |\mathcal{L}_u^Q|/k$. We will prove that either $|\mathcal{L}^Q| \geq g(\beta)k$ or there exists a vertex $u \in V^Q$ such that $|\mathcal{L}_u^Q| = \alpha_u k \geq g(\beta)k$. In the first case, the vertices of V^Q induce in D a path that collects $g(\beta)k$ labels. In the second case, the vertices of Q that belong to the subgraph D_u induce in D a path that collects $g(\beta)k$ labels. Therefore, if one of the two assertions hold, one can derive in polynomial time a path P of D collecting $g(\beta)k$ labels and we are done.

Suppose by way of contradiction that none of the two assertions hold. Namely, $|\mathcal{L}^Q| < g(\beta)k$ and for all $u \in V^Q$, $\alpha_u < g(\beta)$. Let c be a label in $\mathcal{L}^Q$. We denote by $V_c^Q \subseteq V^Q$ the set of vertices $u \in V^Q$ such that $l(u) = c$ and we define $\alpha_c := \max_{u \in V_c^Q} \alpha_u$ and $u_c := \arg\max_{u \in V_c^Q} \alpha_u$. By assumption, $\alpha_c < g(\beta)$. In D_{u_c}, Q collects $\sum_{c' \in \mathcal{L}_{u_c}^Q} |S_{c,\lambda(u)}(c')| = \sum_{c' \in \mathcal{L}_{u_c}^Q} k = \alpha_c k^2$ labels.

Let u be a vertex of $V_c^Q - \{u_c\}$. The number of labels collected by Q in D_u that are not collected by Q in D_{u_c} is the sum over all labels $c' \in \mathcal{L}_u^Q$ of

$$\left| S_{c,\lambda(u)}(c') - \bigcup_{c'' \in \mathcal{L}_{u_c}^Q} S_{c,\lambda(u_c)}(c'') \right| = k - \left| \bigcup_{c'' \in \mathcal{L}_{u_c}^Q} (S_{c,\lambda(u)}(c') \cap S_{c,\lambda(u_c)}(c'')) \right|$$

$$= k - \sum_{c'' \in \mathcal{L}_{u_c}^Q} \left| S_{c,\lambda(u)}(c') \cap S_{c,\lambda(u_c)}(c'') \right|$$

$$= k - \sum_{c'' \in \mathcal{L}_{u_c}^Q} 1$$

$$= k - \alpha_c k$$

The first equation follows $\left| S_{c,\lambda(u)}(c') \right| = k$ and trivial set properties. For the second equation, recall that the family $\{ S_{c,\lambda(u_c)}(c'') : c'' \in \mathcal{L}_{u_c}^Q \}$ is a partition of $\tilde{\mathcal{L}}_c$. The choice of the partitions used to define the labeling function of $\tilde{D}$ ensures that $\left| S_{c,\lambda(u)}(c') \cap S_{c,\lambda(u_c)}(c'') \right| = 1$ and yields the third equation. For the last equation, we use $|\mathcal{L}_{u_c}^Q| = \alpha_c k$. We conclude that the number of labels collected by Q in D_u and not collected by Q in D_{u_c} is $|\mathcal{L}_u^Q|(k - \alpha_c k)$. Since $(k - \alpha_c k) \geq 0$ and $|\mathcal{L}_u^Q| = \alpha_u k \leq \alpha_c k$, this number is at most $\alpha_c k (k - \alpha_c k)$.

Using this bound for all vertices $u \in V_c^Q - \{u_c\}$ and the fact that $\alpha_c k^2$ labels are collected by Q in D_{u_c}, we obtain that the following bound on the number of labels of $\tilde{\mathcal{L}}_c$ collected by Q:

$$\left| \tilde{\mathcal{L}}^Q \cap \tilde{\mathcal{L}}_c \right| \leq \alpha_c k^2 + (|V_c^Q| - 1)\alpha_c k(k - \alpha_c k)$$
$$\leq k^2(\alpha_c + \alpha_c(|V_c^Q| - 1)(1 - \alpha_c))$$

Summing over all labels $c \in \mathcal{L}^Q$, we obtain that the total number of labels collected by Q is upper bounded as follows:

$$\left| \tilde{\mathcal{L}}^Q \right| \leq k^2 \sum_{c \in \mathcal{L}^Q} (\alpha_c + \alpha_c(|V_c^Q| - 1)(1 - \alpha_c))$$
$$< k^2 \sum_{c \in \mathcal{L}^Q} (g(\beta) + \alpha_c(|V_c^Q| - 1)(1 - \alpha_c)) \qquad (*)$$

This last inequality is obtained using the initial assumption $\alpha_c < g(\beta)$.

We distinguish two cases depending on the value of $g(\beta)$. First, suppose that $g(\beta) \geq \frac{1}{2}$. Note that the maximum $\frac{1}{4}$ of the function $x(1-x)$ on the interval $[0,1]$ is realized for $x = \frac{1}{2}$. Therefore for all $c \in \mathcal{L}^Q$, $\alpha_c(1 - \alpha_c) \leq \frac{1}{4}$ and we derive from $(*)$:

$$\left| \tilde{\mathcal{L}}^Q \right| < k^2 \sum_{c \in \mathcal{L}^Q} (g(\beta) + \frac{1}{4}(|V_c^Q| - 1))$$
$$< k^2 \left((g(\beta) - \frac{1}{4}) \sum_{c \in \mathcal{L}^Q} 1 + \frac{1}{4} \sum_{c \in \mathcal{L}^Q} |V_c^Q| \right)$$
$$< k^2 \left((g(\beta) - \frac{1}{4}) g(\beta)k + \frac{1}{4}k \right)$$
$$< k^3 \left(g(\beta)^2 - \frac{1}{4}g(\beta) + \frac{1}{4} \right)$$
$$< k^3 (h(g(\beta)))$$
$$< k^3 \beta$$

In the third inequality, the upper bound on the left operand follows from the initial assumption $g(\beta)k > |\mathcal{L}^Q| = \sum_{c \in \mathcal{L}^Q} 1$ and $(g(\beta) - \frac{1}{4}) \geq 0$. The upper bound on the right operand follows from the fact that any path in D from s

to t contains exactly k vertices, therefore $\sum_{c \in \mathcal{L}^Q} |V_c^Q| = k$. The last equation contradicts the choice of Q and concludes the proof for the case $g(\beta) \geq \frac{1}{2}$.

Now, suppose that $g(\beta) < \frac{1}{2}$. Since the function $x(1-x)$ is a strictly increasing function on the interval $[0, \frac{1}{2}]$ and $|V_c^Q| - 1 \geq 0$ for all $c \in \mathcal{L}^Q$, we can replace α_c by $g(\beta)$ in the inequality $(*)$:

$$
\begin{aligned}
|\tilde{\mathcal{L}}^Q| &< k^2 \sum_{c \in \mathcal{L}^Q} \left(g(\beta) + g(\beta)(|V_c^Q| - 1)\,(1 - g(\beta)) \right) \\
&< k^2 g(\beta) \left(\sum_{c \in \mathcal{L}^Q} 1 - (1 - g(\beta)) + |V_c^Q|\,(1 - g(\beta)) \right) \\
&< k^2 g(\beta) \left(g(\beta) \sum_{c \in \mathcal{L}^Q} 1 + (1 - g(\beta)) \sum_{c \in \mathcal{L}^Q} |V_c^Q| \right) \\
&< k^2 g(\beta) \left(g(\beta)^2 k + (1 - g(\beta))\,k \right) \\
&< k^3 g(\beta) \left(g(\beta)^2 - g(\beta) + 1 \right) \\
&< k^3 h(g(\beta)) \\
&< k^3 \beta
\end{aligned}
$$

Again we use $\sum_{c \in \mathcal{L}^Q} 1 < g(\beta)k$ and $\sum_{c \in \mathcal{L}^Q} |V_c^Q| = k$ to derive the fourth inequality. In the two cases, we obtain a contradiction with the assumption that the path Q collects at least βk^3 labels. This concludes the proof of Lemma 5.

4 $\sqrt{OPT}$-Approximation for Max-Labeled-Path

4.1   Algorithm

In this section, we describe a polynomial algorithm that computes for each instance D of Max-Labeled-Path, a path of D collecting $\sqrt{OPT(D)}$ labels. Again, for the sake of simplicity, we assume that there is only one source s and one sink t. Our algorithm can be easily adapted to handle the case with several sources and several sinks. First, we define a function $F : V \to \mathbb{N}$ such that $F(u)$ can be computed for all vertices $u \in V$ in time $O(|V|^3)$. Then, we prove that, for any vertex $u \in V$, $F(u)$ is an upper bound on the number of labels collected by a path from s to u. Finally, we describe an algorithm that computes for any vertex $u \in V$ a path that collects at least $\lfloor \sqrt{F(u)} \rfloor$ labels. Applying this algorithm to t, we obtain a path from s to t that collects at least $\lfloor \sqrt{OPT} \rfloor$ labels.

For each pair of vertices $u, v \in V$, let $D_{u,v}$ be the subgraph of D consisting of all paths from u to v. We denote by $\Gamma(u, v)$ the number of labels in $D_{u,v}$. Let $F : V \to \mathbb{N}$ be the function recursively defined as follows :

$$
F(u) := \begin{cases} 1, & \text{if } u = s \text{ ;} \\ \max\limits_{P \in \mathcal{P}^u} \min\limits_{ww' \in P} F(w) + \Gamma(w', u), & \text{otherwise.} \end{cases}
$$

where $\mathcal{P}^u$ denotes the set of the paths from s to u. Let $P(u)$ be a path in $\mathcal{P}^u$ that realizes the maximum, i.e. such that $F(u) = \min\limits_{ww' \in P(u)} F(w) + \Gamma(w', u)$.

The following lemma shows that, for any vertex $u \in V$, $F(u)$ is an upper bound on the number of labels that can be collected by a path from s to u.

Lemma 7. *If $P = (s = u_0, u_1, ..., u_n = u)$ is a path between s and u that collects α labels then $F(u) \geq \alpha$.*

Proof. By induction on n. For $n = 0$, $F(u_0) = F(s) = 1$. For $n > 0$, consider a path $P = (s = u_0, u_1, ..., u_n = u)$ that collects α labels. For any $i = 1, ..., n$, let α_i be the number of labels collected by the path $(u_0, u_1, ..., u_i)$. The path $(u_i, ..., u_n)$ collects at least $\alpha - \alpha_{i-1}$ labels and belongs to $D_{u_i, u}$, therefore $\Gamma(u_i, u) \geq \alpha - \alpha_{i-1}$. Since, by induction, $F(u_{i-1}) \geq \alpha_{i-1}$, $F(u_{i-1}) + \Gamma(u_i, u) \geq \alpha$ for any $i = 1, ..., n$ yielding $F(u) \geq \alpha$. $\qquad\square$

Corollary 1. *If OPT is the maximum number of labels that can be collected by a path from s to t then $F(t) \geq OPT$.*

Suppose that $F(v)$ and $P(v)$ have been already computed for all $v \in V$, this can be done in $O(|V|^3)$ using standard data structures. Let u be a vertex in V. The algorithm `ComputePath` returns a path between s and u that collects at least $\lfloor\sqrt{F(u)}\rfloor$ labels. By Corollary 1, applying this procedure with $u = t$ we obtain a path from s to t that collects at least $\lfloor\sqrt{OPT}\rfloor$ labels.

Function ComputePath($u \in V$): a su-path that collects $\lfloor\sqrt{F(u)}\rfloor$ labels

---

if $u = s$ **then**
$\quad\vert$ **return** (s)
else
$\quad\vert\quad$ Let ww' be an arc of $P(u)$ with $F(w) \leq (\lfloor\sqrt{F(u)}\rfloor - 1)^2$ and
$\quad\vert\quad$ $F(w') \geq (\lfloor\sqrt{F(u)}\rfloor - 1)^2$;
$\quad\vert\quad$ $P' \leftarrow$ ComputePath(w') ;
$\quad\vert\quad$ **if** P' collects at least $\lfloor\sqrt{F(u)}\rfloor$ labels **then**
$\quad\vert\quad\quad\vert$ **return** $P'.Q$ where Q is any path from w' to u ;
$\quad\vert\quad$ **else**
$\quad\vert\quad\quad\vert\quad$ Perform a BFS in $D_{w', u}$ to find a vertex v with $l(v)$ not in P' ;
$\quad\vert\quad\quad\vert\quad$ **return** $P'.Q$ where Q is a $w'u$-path passing via v ;

The following lemma is useful to prove that the algorithm `ComputePath` is correct.

Lemma 8. *If $F(u) \geq 4$ then there is an arc ww' in $P(u)$ such that $F(w) \leq (\lfloor\sqrt{F(u)}\rfloor - 1)^2$ and $F(w') \geq (\lfloor\sqrt{F(u)}\rfloor - 1)^2$. Moreover, for any such arc, $\Gamma(w', u) \geq \lfloor\sqrt{F(u)}\rfloor + 1$.*

Proof. The first assertion is true because $F(s) = 1 \leq (\lfloor\sqrt{F(u)}\rfloor - 1)^2$ and $F(u) \geq (\lfloor\sqrt{F(u)}\rfloor - 1)^2$. To verify the second assertion, let ww' be an arc such that $F(w) \leq (\lfloor\sqrt{F(u)}\rfloor - 1)^2$ and $F(w') \geq (\lfloor\sqrt{F(u)}\rfloor - 1)^2$. Since $ww' \in P(u)$, $F(w) + \Gamma(w', u) \geq F(u)$. This implies $\Gamma(w', u) \geq F(u) - F(w) \geq \lfloor\sqrt{F(u)}\rfloor^2 - (\lfloor\sqrt{F(u)}\rfloor - 1)^2 = 2\lfloor\sqrt{F(u)}\rfloor - 1 \geq \lfloor\sqrt{F(u)}\rfloor + 1$, because $\lfloor\sqrt{F(u)}\rfloor \geq 2$. $\qquad\square$

Theorem 6. ComputePath(u) *computes a path P that collects at least $\lfloor\sqrt{F(u)}\rfloor$ labels.*

Proof. If $F(u) < 4$, any path from s to u collects at least $\lfloor\sqrt{F(u)}\rfloor = 1$ labels. Now suppose that $F(u) \geq 4$. We proceed by induction on the number of recursive calls. If $u = s$ the algorithm returns the path (s) that collects $F(s) = 1$ labels. Otherwise, the first assertion of Lemma 8 ensures that $P(u)$ contains an arc ww' such that $F(w) \leq (\lfloor\sqrt{F(u)}\rfloor - 1)^2$ and $F(w') \geq (\lfloor\sqrt{F(u)}\rfloor - 1)^2$. By induction hypothesis, ComputePath(w') returns a path P' collecting at least $\lfloor\sqrt{F(u)}\rfloor - 1$ labels. If P' collects at least $\lfloor\sqrt{F(u)}\rfloor$ labels, the path $P'.Q$ returned by the algorithm is a correct answer. Now, suppose that the path P' collects exactly $\lfloor\sqrt{F(u)}\rfloor - 1$ labels. By Lemma 8, $\Gamma(w', u) \geq \lfloor\sqrt{F(u)}\rfloor + 1$. This implies that $D_{w',u} - \{w'\}$ contains at least $\lfloor\sqrt{F(u)}\rfloor$ labels. Among them at least one is not collected by P'. A BFS traversal of $D_{w',u}$ will find a vertex v having this label together with a path Q from w' to u passing via v. Finally, the path $P'.Q$ that collects at least $\lfloor\sqrt{F(u)}\rfloor$ labels is a correct answer. $\qquad\square$

Using standard data structures, computing $F(u)$ and $P(u)$ for every vertex $u \in V$ can be done in time $O(|V|^3)$.

Acknowledgment. We are grateful to Jérôme Monnot for suggesting the use of a self-reduction to prove the hardness result of Sect. 3.

References

1. Batten, L.M.: Combinatorics of Finite Geometries. Cambridge University Press, New York (1997)
2. Broersma, H., Li, X.: Spanning trees with many or few colors in edge-colored graphs. Discuss. Math. Graph Theory **17**, 259–269 (1997)
3. Broersma, H., Li, X., Woeginger, G.J., Zhang, S.: Paths and cycles in colored graphs. Aust. J. Comb. **31**, 299–311 (2005)
4. Brüggemann, T., Monnot, J., Woeginger, G.J.: Local search for the minimum label spanning tree problem with bounded color classes. Oper. Res. Lett. **31**, 195–201 (2003)
5. Chang, R.-S., Leu, S.-J.: The minimum labeling spanning trees. IPL **31**, 195–201 (2003)
6. Couëtoux, B., Gourvès, L., Monnot, J., Telelis, O.: Labeled traveling salesman problems: complexity and approximation. Discrete Optim. **7**, 74–85 (2010)
7. Hassin, R., Monnot, J., Segev, D.: Approximation algorithms and hardness results for labeled connectivity problems. J. Comb. Optim. **14**, 437–453 (2007)
8. Hassin, R., Monnot, J., Segev, D.: The complexity of bottleneck labeled graph problems. In: Brandstädt, A., Kratsch, D., Müller, H. (eds.) WG 2007. LNCS, vol. 4769, pp. 328–340. Springer, Heidelberg (2007)
9. Håstad, J.: Some optimal inapproximability results. J. ACM **48**, 798–859 (2001)
10. Krumke, S.O., Wirth, H.-C.: Approximation algorithms and hardness results for labeled connectivity problems. IPL **66**, 81–85 (1998)
11. Monnot, J.: The labeled perfect matching in bipartite graphs. IPL **96**, 81–88 (2005)

Minimum Spanning Tree Verification
Under Uncertainty

Thomas Erlebach and Michael Hoffmann[✉]

Department of Computer Science, University of Leicester, Leicester, UK
{te17,mh55}@mcs.le.ac.uk

Abstract. In the *verification under uncertainty* setting, an algorithm is given, for each input item, an *uncertainty area* that is guaranteed to contain the exact input value, as well as an assumed input value. An *update* of an input item reveals its exact value. If the exact value is equal to the assumed value, we say that the update *verifies* the assumed value. We consider verification under uncertainty for the minimum spanning tree (MST) problem for undirected weighted graphs, where each edge is associated with an uncertainty area and an assumed edge weight. The objective of an algorithm is to compute the smallest set of updates with the property that, if the updates of all edges in the set verify their assumed weights, the edge set of an MST can be computed. We give a polynomial-time optimal algorithm for the MST verification problem by relating the choices of updates to vertex covers in a bipartite auxiliary graph. Furthermore, we consider an alternative uncertainty setting where the vertices are embedded in the plane, the weight of an edge is the Euclidean distance between the endpoints of the edge, and the uncertainty is about the location of the vertices. An update of a vertex yields the exact location of that vertex. We prove that the MST verification problem in this vertex uncertainty setting is NP-hard. This shows a surprising difference in complexity between the edge and vertex uncertainty settings of the MST verification problem.

1   Introduction

In this paper we consider settings where a solution to a combinatorial problem needs to be computed and where the input data of the problem might change over time. We assume that the data cannot change arbitrarily and thus the new data is guaranteed to be somewhat close to the old data, represented by an *uncertainty area* for each input data item. The operation of checking the current exact value of an input item, which we also refer to as an *update*, may be expensive, so we want to avoid applying it to all input data items. Moreover, it is possible that the input data is stable and has not changed. One would then like to *verify* for a small set of input data that their values have not changed so that a solution to the combinatorial problem can be calculated based on the verified input data and the given uncertainty areas. We refer to problems of this kind as *verification under uncertainty*.

© Springer International Publishing Switzerland 2014
D. Kratsch and I. Todinca (Eds.): WG 2014, LNCS 8747, pp. 164–175, 2014.
DOI: 10.1007/978-3-319-12340-0_14

In practice, such settings arise naturally, e.g., when maintaining an optimal routing structure in wireless networks with nodes that are generally static but may occasionally move within a limited area. If the exact node positions were known at some point in the recent past, the possible node positions at the current time are known to lie in uncertainty areas that are limited regions around the original positions of the nodes. One can also imagine scenarios in which nodes automatically send a notification message if their location changes by more than a certain threshold. In such scenarios, if none of the nodes has sent a notification since the last determination of exact positions, the area within the threshold distance of the previous location of a node becomes its uncertainty area. As the size of such uncertainty areas is independent of the time that has elapsed since the last determination of the exact position, frequent requests to compute a solution are better addressed in the verification setting. Finally, in a network setting where edge weights represent link congestion, we may again have scenarios where exact weights were known at a point in the past and the current edge weights are guaranteed to lie in certain intervals represented by uncertainty areas. These scenarios have in common that it is possible to obtain the exact current data (node positions or link congestion values) at some cost, and one is interested in being able to compute a solution after verifying only for a small subset of the input data that the data has not changed.

In this paper, we consider the minimum spanning tree (MST) verification problem under uncertainty. The MST is one of the most fundamental graph structures and relevant in many application areas, including routing in wireless ad-hoc networks. We study two uncertainty settings: In the *edge uncertainty* setting, each edge e has an uncertainty area A_e that is guaranteed to contain its current weight, and an update of the edge reveals its exact current weight. In the *vertex uncertainty* setting, the graph is a complete graph embedded in the plane and the weight of an edge is the Euclidean distance between its endpoints. The uncertainty is in the positions of the nodes, and an update of a node reveals its exact current position. In both settings, the goal is to compute a minimum set of updates such that, if these updates verify the expected input data, the edge set of an MST can be calculated.

Our Results. We obtain the following results for MST verification under uncertainty:

- For MST verification under edge uncertainty, provided that the uncertainty areas are open sets or trivial (i.e., contain only one value), we obtain a polynomial-time optimal algorithm by relating sets of updates to vertex covers in a bipartite auxiliary graph that is constructed by adapting a witness set algorithm.
- We show that MST verification under vertex uncertainty is NP-hard even if the uncertainty areas are trivial or open disks. The proof is by reduction from the vertex cover problem for planar graphs with maximum degree 3.
- As an auxiliary result used in the NP-hardness proof, we show that every planar graph of maximum degree 5 can be represented by a unit disk graph (after introducing degree-two vertices on each edge of the planar graph). Although

embeddings of planar graphs as unit disk graphs have been used in the past for NP-hardness proofs, our embedding of planar graphs of degree 5 may be of independent interest.

Our work contributes to the wider research area of computing under uncertainty that studies the problem of minimizing the cost of obtaining exact input values in settings where some of the input data is uncertain. Traditional research in optimization assumes that all input data is given precisely. In cases where the input data is not known precisely (e.g., only a probability distribution for the input data values is known), a substantial amount of research in areas such as stochastic programming or robust optimization has focussed on computing solutions that are good (e.g., in expectation, with high probability, or in the worst case) no matter what the exact values of the input data are. The area of computing under uncertainty approaches problems with uncertain input data from a different angle by assuming that an algorithm can obtain the exact value of an input data item at a certain cost (by performing an update), and aiming to minimize the cost of updates while guaranteeing that an exact solution can be computed.

Work in computing under uncertainty falls in three main categories: In the *adaptive online* setting an algorithm initially knows only the uncertainty areas and performs updates one by one (determining the next update based on the information from previous updates) until it has obtained sufficient information to determine a solution. Algorithms are typically evaluated by competitive analysis, comparing the number of updates they make with the minimum number of updates that, in hindsight, would have been sufficient to determine a solution (referred to as the offline optimum). In the *non-adaptive online* setting an algorithm is also given only the uncertainty areas initially, but it must determine a set U of updates such that after performing all updates in U it is guaranteed to have sufficient information to determine a solution. Finally, there is the *verification* setting that was already described above. It is worth noting that the optimal update set of the *verification* setting is also the offline optimum of the adaptive online setting. Therefore, algorithms solving the verification problem are also useful for the experimental evaluation of algorithms for the adaptive online setting.

Related Work. Kahan [7] presented a model for handling imprecise but update-able input data. He demonstrated his model on a set of real numbers where instead of the precise value of each number an interval was given. That interval when updated reveals that number. The aim is to determine the maximum, the median, or the minimal gap between any two numbers in the set, using as few updates as possible. His work included a competitive analysis for this type of online algorithm, where the number of updates is measured against the optimal number (OPT) of updates. For the problems considered, he presented online algorithms with optimal competitive ratio. Feder et al. [4] studied the problem of computing the value of the median of an uncertain set of numbers up to a certain tolerance. Applications of uncertainty settings can be found in many different areas including databases, geometry and structured data such as graphs. The work presented in this paper mainly concerns the latter two areas.

Bruce et al. [1] studied geometric uncertainty problems in the plane. Here, the input consists of points in the plane and the uncertainty information is for each point of the input an area that contains that point. They presented algorithms with optimal competitive ratio for the maximal point problem and the convex hull problem. Both algorithms are based on a more general technique called a *witness set algorithm* that was introduced in their paper.

Examples of uncertainty applications to graphs include [3], where Feder et al. investigated shortest paths on graphs with uncertain edge weights. They allowed a precision factor limiting the deviation from the actual shortest path and presented results on adaptive as well as non-adaptive updates.

In [2], Erlebach et al. studied the adaptive online setting for MST under two types of uncertainty: the edge uncertainty setting, which is the same as the one considered by Feder et al. [3], and the vertex uncertainty setting. In the latter setting, all vertices are points is the plane and the graph is a complete graph with the weight of an edge being the distance between the vertices it connects. The uncertainty is given by areas for the location of each vertex. For both settings, Erlebach et al. presented algorithms with optimal competitive ratio for the MST under uncertainty. The competitive ratios are 2 for edge uncertainty and 4 for vertex uncertainty, and the uncertainty areas must satisfy certain restrictions (which are satisfied by, e.g., open and trivial areas in the edge uncertainty case). A variant of computing under uncertainty where updates yield more refined estimates instead of exact values was studied by Gupta et al. [6].

A different setting of the MST under vertex uncertainty was studied by Kamousi et al. [8]. They assume that point locations are known exactly, but each point i is present only with a certain probability p_i. They show that it is #P-hard to compute the expected length of an MST even in 2-dimensional Euclidean space, and provide a fully polynomial randomized approximation scheme for metric spaces.

Paper Outline. In Sect. 2 we give formal definitions and preliminaries. Section 3 presents our optimal algorithm for MST verification under edge uncertainty. In Sect. 4 we give the proof of NP-hardness for MST verification under vertex uncertainty.

Some proofs of Sects. 3 and 4 are omitted.

2   Preliminaries

Within the wider field of problems under uncertainty we consider the minimum spanning tree problem for graphs under uncertainty. We consider undirected weighted graphs $G = (V, E)$ under two different types of uncertainty. In an *edge uncertainty* graph, the weight W_e of an edge e might not be known exactly, but instead a set A_e of possible values of W_e is given. We let W be the set that contains for each $e \in E$ the exact weight W_e, and $\mathcal{A}$ the set that contains for each $e \in E$ its uncertainty area A_e. We refer to $\mathcal{A}$ as the *areas of uncertainty*.

In this type of uncertainty, the update of an edge e reveals its exact weight W_e. This effectively changes the uncertainty area A_e to the singleton set containing just W_e (we call such a set *trivial*). An instance of the MST-EDGE-UNCERTAINTY problem is given by $(G, W, \mathcal{A})$.

In the second type of uncertainty, we consider *vertex uncertainty* graphs. Here the graph is a complete graph embedded in the plane. The weight of each edge is given by the distance of the vertices that it connects. For each vertex $v \in V$ the location of v might not be known exactly, but instead a set A_v of possible locations of v is given, and $\mathcal{A}$ is the family of all these uncertainty areas A_v. An update of a vertex v reveals the exact location P_v of v. An instance of the MST-VERTEX-UNCERTAINTY problem is given by $(G, P, \mathcal{A})$, where P is the set containing the precise vertex location P_v for each $v \in V$.

In the online setting, the precise information (W for edge uncertainty and P for vertex uncertainty graphs) is not known to the algorithm; the algorithm has to request updates until $\mathcal{A}$ is precise enough to allow the calculation of an MST of G. In the verification setting, the sets W or P respectively are given to the algorithm. This additional information is not used to calculate an MST of G directly, but it is used to determine which updates should be made so that an MST of G can be calculated based on the updated areas of uncertainty. A set of updates that reveals enough information so that an MST of G can be calculated is called an *update solution*, and the set of all update solutions is denoted by S. For a given instance of a problem, we denote the size of the smallest update solution by OPT. In the verification setting, the goal of an algorithm is to calculate an update solution of size OPT.

In the remainder of the paper we will use the following notion: $G = (V, E)$ is the weighted undirected graph for which an MST should be found; we say $\mathcal{U}$ is an *uncertainty graph of* G if it consists of the same edges and vertices as G, but only contains the uncertain information as specified by $\mathcal{A}$. S is the set of update solutions, and OPT is the size of the smallest element of S.

In Fig. 1, the left-hand side shows a graph G, and an uncertainty graph $\mathcal{U}$ of G is given on the right. G has two minimal spanning trees, with edge sets $\{b, c, d, f\}$ and $\{b, d, e, f\}$. None of them can be calculated based on the information of $\mathcal{U}$ alone, as for example the weight of the edge a could be smaller than that of b. The set of update solutions is $S = \{\{a, b, e\}, \{a, e\}, \{b, e\}\}$, and both $\{a, e\}$ and $\{b, e\}$ have minimum size. Thus, in this example OPT is 2.

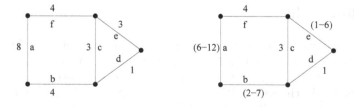

Fig. 1. Example of graph G (left) and edge uncertainty graph $\mathcal{U}$ (right)

3   Verification Under Edge Uncertainty

In this section we give a polynomial-time algorithm for MST-EDGE-UNCERTAINTY. We assume that the uncertainty areas of the edge weights are trivial or open areas. The first phase of the algorithm is based on the algorithm U-RED that was presented in [2] for the adaptive online version of the MST under edge uncertainty problem. Adopting the principles of this online algorithm to the verification setting is non trivial. Roughly speaking, the algorithm U-RED repeatedly identifies an edge e that, based on the current areas of uncertainty, may or may not be in an MST. It then identifies at most two edges such that without updating any of the two edges the edge e can neither be included nor excluded from any MST (these two edges thus form a so-called witness set), and updates both of the edges.

Whereas U-RED updates both of these edges, we utilize the additional information of the precise values that are available in the verification setting. It turns out that with this information we can always arrive at one of the following two cases: (1) There is a single edge f such that, without updating f, the edge e can neither be included nor excluded from any MST. In this case, we record that f is needed to be in any update solution and simulate for U-RED the update of f. (2) There is a choice of edge sets whose updates can determine whether e is in an MST or not. In this case, we record the choice. We can prove that this can only happen when e is not in any MST, so we remove that edge and continue the simulation of U-RED.

After the simulated run of U-RED, we have established a set of updates that are common among all update solutions and we have also recorded a set of choices. We show that each choice is between a single edge and a single set of edges. We also show that the set of choices has additional properties that allow us to model it as a bipartite graph in such a way that a minimum vertex cover of the bipartite graph yields a minimum set of updates to cover all choices. Together with the already established set of common updates this gives a minimum update solution for MST-EDGE-UNCERTAINTY.

Theorem 1. *There is a polynomial-time algorithm that computes an optimal update solution for instances of* MST-EDGE-UNCERTAINTY *where the uncertainty areas are trivial or open areas.*

In the remainder of this section we present the algorithm and prove its correctness, thus establishing Theorem 1. The algorithm runs in three phases. In the first phase, two sets A and R are constructed. The set $A \subseteq E$ is the set of edges that are common to all update solutions, i.e., $A = \bigcap_{s \in S} s$. (Recall that S denotes the set of all update solutions.) The set $R \subseteq E \times \mathcal{P}(E)$ consists of pairs (d, B) with $d \in E$ and $B \subseteq E$. Each pair $(d, B) \in R$ represents a choice with the property that every update solution must contain d or all elements of B. In addition, R is of the form that any combination of choosing either the single edge or the set of edges together with the set A is an update solution. As we will refer to these properties later on, we state them formally as follows.

Property 1. The sets A and $R = \{(d_1, B_1), \ldots, (d_n, B_n)\}$ satisfy the following properties:

- **p1:** $A = \bigcap_{s \in S} s$
- **p2:** If s is an update solution, then for all $1 \le i \le n$ we have $d_i \in s$ or $B_i \subseteq s$.
- **p3:** $S' = \{A \cup \{d_i | i \in I\} \cup \bigcup_{j \in J} B_j \mid I, J \text{ form a partition of } \{1, \ldots, n\}\}$ is a set of update solutions.

As a consequence of p1–p3, for every update solution s there exists $s' \in S'$ such that $s' \subseteq s$.

From the outset it is not clear that a set R satisfying p1–p3 exists. We will show that it does and how to construct it. In the second phase, redundant choices in R will be removed without altering the properties of R. In the third and final phase, we model the choice selection for R as a vertex cover problem in a bipartite graph. We will show that an optimal solution to the latter problem results in an update solution for the MST verification problem of minimum size.

Phase 1. The aim of this phase is to establish the sets A and R described above. The algorithm used in this phase is based on (a simulation of) the online algorithm U-RED presented in [2]. The significant changes include: (1) The online algorithm U-RED restarts after each update, whereas our algorithm avoids restarts and sorts the updated edges back into the running process. (2) When the online algorithm U-RED updates the edges in a witness set, we utilize the information of the exact weights of the edges involved and determine the appropriate contribution to the sets A and R instead. The resulting Algorithm PHASE1 is given in Fig. 2. It uses the notation of the following definitions.

Definition 1. *For an edge e in an edge-uncertainty graph, we denote the actual weight of the edge by W_e and the upper limit of A_e by $U_e = \lim \sup \{a \mid a \in A_e\}$ and the lower limit of A_e by $L_e = \lim \inf \{a \mid a \in A_e\}$.*

Note that, as edges are updated in the algorithm, the values for U_e and L_e for an edge e may change. In particular, after updating the edge e we have that $L_e = W_e = U_e$.

Definition 2. *The order by which the edges are sorted in Algorithm PHASE1 is as follows: Let $\mathcal{U}$ be an edge-uncertainty graph and let e, f be two edges of $\mathcal{U}$. We define $e < f$ if $L_e < L_f$ or ($L_e = L_f$ and $U_e < U_f$). Edges with the same upper and lower weight limit are ordered arbitrarily.*

Definition 3. *Let C be a cycle in $\mathcal{U}$ and $e \in C$. The edge e is said to be always maximal in C if for all possible weights that are consistent with the uncertainty areas given by $\mathcal{U}$ the weight of e is maximal among the weights of all edges in C.*

Note that updating edges in $\mathcal{U}$ only reduces the options for the edge weights. Hence, an always maximal edge in C remains an always maximal edge in C after updating arbitrary elements of $\mathcal{U}$.

```
01 Create a list L of all edges in the order of Definition 2 from low to high
02 Let Γ be U without any edge
03 while L is not empty do
04        add the head of L to Γ
05        remove the head of L
06        if Γ has a cycle C then
07              case (a): C contains an always maximal edge e.
08                    delete e from Γ
09              case (b): There exists e ∈ C whose update must be in any update solution.
10                    update e, remove e from Γ, and add e to A
11                    sort e back into L
12              case (c): There is a choice of updates that establish an edge as
13                          an always maximal edge in the cycle C.
14                    add the choice to R
15                    delete the always maximal edge from Γ
16        end if
17 end while
```

Fig. 2. Algorithm PHASE1

Before showing that there exists (in line 15) a unique always maximal edge, and that the sets A and R are built correctly, we establish the following lemma which gives a locality property: Updates required on the basis of just one cycle will never be made redundant by other updates or cycles in the graph.

Lemma 1. *During the run of the algorithm, when a cycle C is closed, let e be a non-trivial edge in C. If updating a set $U \subseteq E - \{e\}$ does not determine whether e is always maximal in C, then updating U will also not verify that e is always maximal in any other cycle.*

Once a cycle in Γ is formed during the run of the algorithm, different actions are taken. We will show that the cases listed in the algorithm cover all possibilities and that the sets A and R are built correctly.

The first check after a cycle C in Γ is formed is whether there exists, according to the current uncertainty information, an edge in C that is always maximal. If such an edge exists, the algorithm executes case (a) and the edge is deleted from Γ, no update is made, and the sets A and R stay unaltered.

If case (a) does not apply, let h be an edge in C with maximum upper limit U_h. Note that h must be non-trivial (otherwise case (a) would apply). There are four possible cases for how the actual weight W_h relates to the weights and limits of other edges in the cycle C. **Case 1.** If W_h is not maximal among the actual weights of all edges in C, then h needs to be updated in any update solution (the algorithm executes case (b)). **Case 2.** If W_h is maximal amongst the actual weights of edges in C and there exists an $f \in C$ with $U_f > W_h$, then f needs to be updated in any update solution (the algorithm executes case (b)). **Case 3.** If W_h is maximal amongst the actual weights of edges in C and there exists an $f \in C$ such that $W_f > L_h$, then h needs to be updated in any update solution (the algorithm executes case (b)). **Case 4.** If Cases 1–3 do not apply, every update solution must contain h or all edges of the set $B = \{c \in C - \{h\} \mid U_c > L_h\}$, so the algorithm executes case (c).

Remark 1. In the situation of Case 4, the edge h is greater in the order of edges used by the algorithm than any other edge in the cycle C (i.e., $h > c$ for all $c \in C - \{h\}$) and hence was the edge that closed the cycle.

From the above we can conclude that the set A only contains updates that are in any update solution, that for all $(d, B) \in R$ an update solution must include the edge d or all edges in B, and that any set of edges containing all elements of A and from every pair $(d, B) \in R$ at least d or all edges in B is an update solution. This shows that A and R satisfy properties p2 and p3.

Since for every pair $(d, B) \in R$ the edge d is not in B (see also Lemma 2 below), there exists for every $g \notin A$ an update solution not containing g. This shows that A is the intersection of all update solutions, establishing that p1 is satisfied as well. Before tidying up R in phase 2 (in a way that maintains p1–p3), we establish an additional property of R that will be used in phase 3 to build a bipartite graph.

Lemma 2. *Let (d, B) and (d', B') in R. Then $d \notin B'$.*

Proof. Assume there exist (d, B) and (d', B') in R with $d \in B'$. When a pair (d, B) is added to R, the edge d is deleted from Γ and hence will not be part of any pair that is added to R later. So for d being an element of B', the pair (d', B') must have been added to R before (d, B). By Remark 1, $d' \leq d$. Considering that $d \in B'$ we also have by the same remark that $d < d'$, which gives a contradiction.
□

Phase 2. As the sets A and R are built up simultaneously, it is possible that for a pair (d, B) in R some edges in B are added to A later on in the run of the Algorithm PHASE1. Since the edge d is deleted from Γ when a pair (d, B) is added to R, the edge d can never be added to A.

In this short phase, R is tidied up by the following steps: For every $(d, B) \in R$ all elements of B that are also in A will be removed from B. Where, as a result, B becomes empty, the entire pair (d, B) is removed from R. Formally R is replaced by $\{(d, B) \mid \exists (d, B') \in R, B = B' - A, B \neq \emptyset\}$. This does not affect the properties p1–p3 of Property 1.

Phase 3. In this final phase, an optimal update set is calculated from the sets A and R. As stated in p3 of Property 1, a set S' of update solutions can be formed from A and R. An update solution with minimum size amongst them can be established by modelling the choices as a vertex cover problem in a bipartite graph. We then show that there is no update solution with fewer updates. Recall the notation of p3: $R = \{d_1, B_1), \ldots, (d_n, B_n)\}$ and $S' = \{A \cup \{d_i \mid i \in I\} \cup \bigcup_{j \in J} B_j \mid I, J \text{ partition of } \{1, \ldots, n\}\}$. In phase 2, any overlap between elements of A and elements appearing in the pairs of R was removed from R. So, to find an element of S' with minimum size it is enough to find an element of minimum size in $R' = \{\{d_i | i \in I\} \cup \bigcup_{j \in J} B_j \mid I, J$ partition of $\{1, \ldots, n\}\}$, as A and the elements of R' are disjoint.

We now create a bipartite graph G' for which any element of R' is a vertex cover and any vertex cover must contain one element of R' as a subset. Loosely speaking, every edge of the uncertainty graph $\mathcal{U}$ occurring inside any element of R is a node in G'. For every choice $(d, B) \in R$, the node in G' corresponding to d is connected to the nodes in G' corresponding to the elements of B.

Let $G' = (V', E')$ be an undirected graph with $V' = \{d_1, \ldots, d_n\} \cup B_1 \cup \cdots \cup B_n$ and $E = \{(d_i, b) \mid b \in B_i, 1 \le i \le n\}$. By Lemma 2, the set $\{d_1, \ldots, d_n\}$ and $B_1 \cup \cdots \cup B_n$ are disjoint. As every edge in G' connects an element from $\{d_1, \ldots, d_n\}$ to an element of $B_1 \cup \cdots \cup B_n$, G' is a bipartite graph.

Every element of R' contains, for every i, the edge d_i or all elements of B_i. Therefore, every element of R' is a vertex cover for G'. Similarly if a vertex cover of G' does not include d_i for any i, then it must include all elements of B_i and hence it must contain an element of R' as a subset. Thus, a minimum vertex cover r^* of G' is also an element of R' with minimal size. Furthermore, $A \cup r^*$ is an element of minimum size in S'. By Property 1, for every update solution s there exists an $s' \in S'$ such that $s' \subseteq s$. So $A \cup r^*$ is of minimum size amongst all update solutions, and $OPT = |A \cup r^*|$.

Noting that the minimum vertex cover problem is polynomial for bipartite graphs, it is not difficult to show that the algorithm runs in polynomial time. Hence, a minimal update solution (of size OPT) for MST-EDGE-UNCERTAINTY under the restriction to open and trivial areas can be computed in polynomial time. This completes the proof of Theorem 1.

Furthermore, we note that the algorithm can be extended to a version of the problem where each edge e has an arbitrary update cost $c_e > 0$ and the goal is to minimize the total cost of the update solution. The approach is the same, except that the vertex cover problem in the bipartite auxiliary graph needs to be solved as a minimum-weight vertex cover problem.

Theorem 2. *For the* MST-EDGE-UNCERTAINTY *problem with arbitrary positive update costs and under the restriction to open or trivial areas, an optimal update solution can be computed in polynomial time.*

4   Verification Under Vertex Uncertainty

In this section we prove that MST-VERTEX-UNCERTAINTY is NP-hard. The proof uses a reduction from the vertex cover problem in planar graphs of maximum degree 3, which was shown to be NP-complete in [5]. In the reduction we use the following embedding result.

Theorem 3. *Let $G = (V, E)$ be a planar graph of maximum degree 5 with n vertices. Then there exists a value $s > 0$ and an embedding of G such that*

- *vertices are mapped to integer coordinates in an n by n grid,*
- *edges are mapped to non-crossing paths (consisting of straight line segments and circular arcs),*

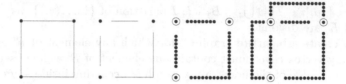

Fig. 3. Transformation of the graph G

- *the length of each path is polynomial in n,*
- *$1/s$ is polynomial in n,*
- *in the disk of radius s around each vertex, all edges are equally spaced straight lines,*
- *everywhere else the edges are at least s apart,*
- *the embedding can be constructed in polynomial time.*

We give an outline of the proof as the detailed proof is somewhat technical. The starting point is an arbitrary planar graph of maximum degree 3. Finding a minimum vertex cover for such graphs is NP-hard [5]. We transform this graph to an instance of MST-VERTEX-UNCERTAINTY by the following three steps (illustrated in Fig. 3). The steps are given here in an order that reflects the motivation for the steps, but for technical reasons the order will be different in the actual reduction.

Step 1: Create an uncertainty problem. After embedding the graph in the plane, we shorten each edge so that instead of connecting two vertices, it falls short at both ends. At one end it leaves a gap of $1 + 5\epsilon$ to the vertex, and at the other a gap of $1 + 8\epsilon$ to the other vertex. Finally, each vertex is replaced by an uncertainty area that is a disk of radius 2ϵ around the original location of the vertex. So, each edge has at one end a gap between $1 + 3\epsilon$ and $1 + 7\epsilon$ and at the other end a gap between $1 + 6\epsilon$ and $1 + 10\epsilon$. If one wants to know for each edge which end has the smaller gap, one has to update at least one of the vertices that it connected originally. If the precise location of each vertex lies at the center of its uncertainty disk, then updating either end vertex of an edge will determine at which end the edge has a smaller gap. Thus, finding the smallest set of vertices that needs to be updated to determine for each edge at which end it has the smaller gap is equivalent to finding a minimum vertex cover of the original graph.

Step 2: Create a vertex uncertainty graph. To convert the graph of step 1 to a vertex uncertainty graph, we replace each edge fragment by a dense sequence of new vertices. The position of these vertices is given exactly (i.e., their uncertainty areas are trivial). The distance of two neighboring vertices is less than $1/2$.

Step 3: Create a minimum spanning tree problem. To turn the question that asks at which side each former edge has the smaller gap into an MST problem, we place additional vertices such that all original vertices are connected via dense 'lines' made out of these new vertices, and the original vertices and the new 'lines' form a tree. The gap of such a 'line' to an original vertex is 1. If all lines can be placed in such a way that all vertices on one line are far away (at least distance 1) from any vertex on another line, solving the MST-VERTEX-UNCERTAINTY problem requires updating for each original edge at least one of its end points. The minimum set of such updates yields a minimum vertex cover of the original graph. In the actual proof we add the auxiliary 'lines' of Step 3 already before embedding the graph for Step 1, and the distances mentioned above are scaled down by an appropriate scaling factor.

Theorem 4. *Calculating OPT for* MST-VERTEX-UNCERTAINTY *is NP-hard.*

As the exact weight of any edge can be obtained by updating both of its vertices, the polynomial optimal algorithm for MST-EDGE-UNCERTAINTY can be used to obtain a 2-approximation of the MST-VERTEX-UNCERTAINTY problem.

Acknowledgements. The second author would like to thank the University of Leicester for supporting this research in granting him academic study leave.

References

1. Bruce, R., Hoffmann, M., Krizanc, D., Raman, R.: Efficient update strategies for geometric computing with uncertainty. Theor. Comput. Syst. **38**(4), 411–423 (2005)
2. Erlebach, T., Hoffmann, M., Krizanc, D., Mihalák, M., Raman, R.: Computing minimum spanning trees with uncertainty. In: Albers, S., Weil, P. (eds.) STACS. LIPIcs, vol. 1, pp. 277–288. Schloss Dagstuhl - Leibniz-Zentrum fuer Informatik, Germany (2008)
3. Feder, T., Motwani, R., O'Callaghan, L., Olston, C., Panigrahy, R.: Computing shortest paths with uncertainty. J. Algorithms **62**(1), 1–18 (2007)
4. Feder, T., Motwani, R., Panigrahy, R., Olston, C., Widom, J.: Computing the median with uncertainty. SIAM J. Comput. **32**(2), 538–547 (2003)
5. Garey, M., Johnson, D.: The rectilinear Steiner tree problem is NP-complete. SIAM J. Appl. Math. **32**(4), 826–834 (1977)
6. Gupta, M., Sabharwal, Y., Sen, S.: The update complexity of selection and related problems. In: IARCS Annual Conference on Foundations of Software Technology and Theoretical Computer Science (FSTTCS 2011). LIPIcs, vol. 13, pp. 325–338. Schloss Dagstuhl - Leibniz-Zentrum für Informatik (2011)
7. Kahan, S.: A model for data in motion. In: Proceedings of the 23rd Annual ACM Symposium on Theory of Computing (STOC'91), pp. 267–277 (1991)
8. Kamousi, P., Chan, T.M., Suri, S.: Stochastic minimum spanning trees in Euclidean spaces. In: Proceedings of the 27th Annual ACM Symposium on Computational Geometry (SoCG'11), pp. 65–74. ACM (2011)

Towards the Hanani-Tutte Theorem for Clustered Graphs

Radoslav Fulek[✉]

Department of Industrial Engineering
and Operations Research, Columbia University,
New York City, NY, USA
radoslav.fulek@gmail.com

Abstract. The weak variant of the Hanani–Tutte theorem says that a graph is planar, if it can be drawn in the plane so that every pair of edges cross an even number of times. Moreover, we can turn such a drawing into an embedding without changing the order in which edges leave the vertices. We prove a generalization of the weak Hanani–Tutte theorem that also easily implies the monotone variant of the weak Hanani–Tutte theorem by Pach and Tóth. Thus, our result can be thought of as a common generalization of these two neat results. In other words, we prove the weak Hanani-Tutte theorem for strip clustered graphs, whose clusters are linearly ordered vertical strips in the plane and edges join only vertices in the same cluster or in neighboring clusters with respect to this order.

Besides usual tools for proving Hanani-Tutte type results our proof combines Hall's marriage theorem, and a characterization of embedded upward planar digraphs due to Bertolazzi et al.

Keywords: Hanani–Tutte theorem · Hall's theorem · Upward planarity · C-planarity

1 Introduction

A *drawing* of G is a representation of G in the plane, where every vertex is represented by a unique point and every edge $e = uv$ is represented by a simple arc joining the two points that represent u and v. If it leads to no confusion, we do not distinguish between a vertex or an edge and its representation in the drawing and we use the words "vertex" and "edge" in both contexts. We assume that in a drawing no edge passes through a vertex, no two edges touch and every pair of edges cross in finitely many points. A drawing of a graph is an *embedding* if no two edges cross. A graph is *planar*, if it admits a planar embedding.

The author gratefully acknowledges support from the Swiss National Science Foundation Grant No. 200021-125287/1 GIG/11/E023.

© Springer International Publishing Switzerland 2014
D. Kratsch and I. Todinca (Eds.): WG 2014, LNCS 8747, pp. 176–188, 2014.
DOI: 10.1007/978-3-319-12340-0_15

1.1 Hanani–Tutte Theorem

The Hanani–Tutte theorem [13,22] is a classical result that provides an algebraic characterization of planarity with interesting algorithmic consequences. The (strong) Hanani–Tutte theorem says that a graph is planar as soon as it can be drawn in the plane so that no pair of independent edges crosses an odd number of times. Moreover, its variant known as the weak Hanani–Tutte theorem [3,14,17] states that if we have a drawing $\mathcal{D}$ of a graph G where every pair of edges cross an even number of times then G has an embedding that preserves the cyclic order of edges at vertices from $\mathcal{D}$. Note that the weak variant does not directly follow from the strong Hanani–Tutte theorem. For sub-cubic graphs, the weak variant implies the strong variant.

Other variants of the Hanani–Tutte theorem in the plane were proved for x-monotone drawings [10,15], partially embedded planar graphs, simultaneously embedded planar graphs [20], and two clustered graphs [9]. As for the closed surfaces of genus higher than zero, the weak variant is known to hold in all closed surfaces [18], and the strong variant was proved only for the projective plane [16]. It is an intriguing open problem to decide if the strong Hanani–Tutte theorem holds for closed surfaces other than the sphere and projective plane.

To prove a strong variant for a closed surface it is enough to prove it for all the minor minimal graphs (see e.g. [6] for the definition of a graph minor) not embeddable in the surface. Moreover, it is known that the list of such graphs is finite for every closed surface, see e.g. [6, Section 12]. Thus, proving or disproving the strong Hanani-Tutte theorem on a closed surface boils down to a search for a counterexample among a finite number of graphs. That sounds quite promising, since checking a particular graph is reducible to a finitely many, and not so many, drawings, see e.g. [21]. However, we do not have a complete list of such graphs for any surface besides sphere and projective plane.

On the positive side, the list of possible minimal counterexamples for each surface was recently narrowed down to vertex two-connected graphs [21]. See [19] for a recent survey on applications of the Hanani–Tutte theorem and related results.

1.2 Notation

In the present paper we assume that $G = (V, E)$ is a (multi)graph. We refer to an embedding of G as to a *plane* graph G. The *rotation* at a vertex v is the clockwise cyclic order of the end pieces of edges incident to v. The *rotation system* of a graph is the set of rotations at all its vertices. Two embeddings of a graph are the *same*, if they have the same rotation system up to switching the orientations of all the rotations simultaneously. A pair of edges in a graph is *adjacent* or *independent*, if they do not share a vertex. An edge in a drawing is (independently) *even*, if it crosses every other (non-adjacent) edge an even number of times. A drawing of a graph is (independently) *even*, if all edges are (independently) even. Note that an embedding is an even drawing. Let $x(v)$ (resp. $y(v)$) denote the x-coordinate (resp. y-coordinate) of a vertex in a drawing.

1.3 Hanani–Tutte for Strip Clustered Graphs

Borrowing the notation from [1] a *clustered graph*[1] is an ordered pair (G, T), where G is a graph, and $T = \{V_i | i = 1, \ldots, k\}$ is a partition of the vertex set of G into k parts. We call the sets V_i clusters. A *clustered graph* (G, T) is *strip clustered*, if $G = \left(V_1 \uplus \ldots \uplus V_k, E \subseteq \bigcup_i \binom{V_i \uplus V_{i+1}}{2} \right)$, i.e., the edges in G are either contained inside a part or join vertices in two consecutive parts. A drawing of a strip clustered graph (G, T) in the plane is *clustered*, if $i < x(v_i) < i + 1$ for all $v_i \in V_i$, and every vertical line $x = i$, $i \in \mathbb{Z}$, intersects every edge at most once. We use the term "cluster V_i" also, when referring to a vertical strip containing the vertices in V_i. A strip clustered graph (G, T) is *clustered planar* (or briefly *c-planar*) if (G, T) has a clustered embedding in the plane.

The notion of clustered planarity appeared for the first time in the literature in the work of Feng, Cohen and Eades [7,8] under the name of c-planarity. See, e.g., [5,7,8] for the general definition of c-planarity. Here, we consider only a special case of it. See, e.g., [5] for further references. We only remark that it has been an intriguing open problem for almost two decades to decide, if c-planarity is NP-hard, despite of considerable effort of many researchers and that already for strip clustered graphs the problem constitutes a challenge [1].

We show the following generalization of the weak Hanani–Tutte theorem for strip clustered graphs. See Fig. 1(a) and (b) for an illustration.

(a)   (b)

Fig. 1. (a) Even clustered drawing of a strip clustered graph; (b) Clustered embedding of the same clustered graph.

Theorem 1. *If a strip clustered graph* (G, T) *admits an even clustered drawing* $\mathcal{D}$ *then* (G, T) *is c-planar. Moreover, there exists a clustered embedding of* (G, T) *with the same rotation system as in* $\mathcal{D}$.

Due to the family of counterexamples in [9], Theorem 1 does not leave too much room for straightforward generalizations. Let (G, T) denote a clustered graph, and let $G' = G'(G, T)$ denote a graph obtained from (G, T) by contracting every cluster to a vertex and deleting all the loops and multiple edges. If (G, T) is a strip clustered graph, G' is a subgraph of a path. In this sense, the most general

---

[1] This type of clustered graphs is usually called flat clustered graph in the graph drawing literature. We chose this simplified notation in order not to overburden the reader with unnecessary notation.

variant of Hanani-Tutte, the weak or strong one, we can hope for, is the one for the class of clustered graphs (G, T), for which G' is an arbitrary tree.

By allowing G' to contain a cycle, c-planarity testing seems to be much harder than in the case, when it is acyclic. Already in the case of three clusters [4], if G (not G') is a cycle, the polynomial time algorithm for c-planarity is not trivial, while if G can be any graph, its existence is still open. For a comparison, if G is a cycle then a strip clustered graph (G, T) is trivially c-planar. We note that by an easy geometric argument a polynomial time algorithm for c-planarity in the case of three clusters would imply a polynomial time algorithm in the case of strip clustered graphs.

Our proof of Theorem 1 is slightly technical, and combines a characterization of upward planar digraphs from [2] and Hall's theorem [6, Sect. 2]. Using the result from [2] in our situation is quite natural, as was already observed in [1], where they solve an intimately related algorithmic question discussed below. The reason is that deciding the c-planarity for embedded strip clustered graphs is, essentially, a special case of the upward planarity testing. The technical part of our argument augments the even drawing with subdivided edges by using tricks from [10, 17] so that we are able to apply Hall's Theorem. Hence, the real novelty of our work lies in proving the marriage condition, which makes the characterization do the work for us. It took a considerable effort to make Hall's Theorem work here, and thus, we wonder if a more direct proof exists.

An edge e of a topological graph is *x-monotone*, if every vertical line intersects e at most once. Pach and Tóth [15] (see also [10] for a different proof of the same result) proved the following theorem.

Theorem 2. *Let G denote a graph, whose vertices are totally ordered. Suppose that there exists a drawing $\mathcal{D}$ of G, in which x-coordinates of vertices respect their order, edges are x-monotone and every pair of edges cross an even number of times. Then there exists an embedding of G, in which the vertices are drawn as in $\mathcal{D}$, the edges are x-monotone, and the rotation system is the same as in $\mathcal{D}$.*

We show that Theorem 1 easily implies Theorem 2. Our argument for showing that suggests a slightly different variant of Theorem 1 for not necessarily clustered drawings that directly implies Theorem 2 (see Sect. 2.1). The strong variant of Theorem 1, which we conjecture to hold, would imply the existence of a polynomial time algorithm for the corresponding variant of the c-planarity testing [9]. To the best of our knowledge, a polynomial time algorithm was given only in the case, when the underlying planar graph has a prescribed planar embedding [1]. Our weak variant gives a polynomial time algorithm if G is subcubic, and in the same case as in [1]. Nevertheless, we think that the weak variant is interesting in its own right. To support our conjecture we prove the strong variant of Theorem 1 under the condition that the underlying abstract graph G of a clustered graph is a subdivision of a vertex three-connected graph. In general, we only know that it is true for two clusters [9].

Theorem 3. *Let G denote a subdivision of a vertex three-connected graph. If a strip clustered graph (G, T) admits an independently even clustered drawing $\mathcal{D}$ then (G, T) is c-planar.*[2]

The proof of Theorem 3 reduces to Theorem 1 by correcting the rotations at the vertices of G so that the theorem becomes applicable. As we noted above, the weak Hanani-Tutte theorem fails already for three clusters. Moreover, the underlying graph in the counterexample is a cycle [9], and thus, the strong variant fails as well in general clustered graphs without imposing additional restrictions.

The paper is organized as follows.
In Sect. 2 we introduce terminology and tools for proving our results, where in Subsect. 2.2 we outline the proof of our main result Theorem 1. In Sect. 3 we give the proof of Theorem 3. In Sect. 4 we derive Theorem 2 from Theorem 1. Open problems are stated in Sect. 5.

2   Preliminaries

2.1   Even Drawings

We will use the following fact about closed curves in the plane. Let C denote a closed (possibly self-crossing) curve in the plane.

Lemma 1. *The regions in the complement of C can be two-colored so that two regions sharing a non-trivial part of the boundary receive opposite colors.*

Let us two-color the regions in the complement of C so that two regions sharing a non-trivial part of the boundary receive opposite colors. A point not lying on C is *outside* of C, if it is contained in the region with the same color as the unbounded region. Otherwise, such a point is *inside* of C. As a simple corollary of Lemma 1 we obtain a well-known fact that a pair of closed curves in the plane cross an even number of times. We use this fact tacitly throughout the paper.

Let G denote a planar graph. Since in the problem we study connected components of G can be treated separately, we can afford to assume that G is connected. A *face* in an embedding of G is a walk that corresponds to the boundary of the connected component of the complement of G in the plane. A vertex or an edge is *incident* to a face, if it appears on the corresponding walk.

Given a drawing of a graph G, where every pair of edges crosses an even number of times, by the weak Hanani-Tutte theorem [3, 14, 17], we can obtain an embedding of G with the same rotation system, and hence, the facial structure of an embedding of G is already present in an even drawing. This allows us to speak about faces in an even drawing of G. Hence, a face in an even drawing of

---

[2] The argument in the proof of Theorem 3 proves, in fact, a strong variant even in the case, when we require the vertices participating in a cut or two-cut to have the maximum degree three. Hence, we obtained a polynomial time algorithm even in the case of sub-cubic cuts and two-cuts.

G is the walk bounding the corresponding face in the embedding of G with the same rotation system.

Let $\gamma : V \to \mathbb{N}$ be a labeling of the vertices of G by integers. Given a face f in an even drawing of G, a vertex v incident to f is the *local minimum* (resp. *maximum*) of f, if in the corresponding facial walk W of f the value of $\gamma(v)$ is not smaller (not bigger) than the value of its successor and predecessor on W. The minimal (resp. maximal) local minimum (resp. maximum) of f is called *global minimum* (resp. *maximum*) of f. The face f is *simple* with respect to γ, if f has exactly one local minimum and one local maximum. The face f is *semi-simple* (with respect to γ), if f has exactly two local minima and these minima have the same value, and two local maxima and these maxima have the same value. A path P is *(strictly) monotone with respect to* γ, if the labels of the vertices on P form a (strictly) monotone sequence if ordered in the correspondence with their appearance on P.

Given a strip clustered graph (G, T) we naturally associate with it a labeling γ that for each vertex v returns the number of the cluster v belongs to. We refer to the cluster, whose vertices get label k, as to the k-th cluster. Let $(\vec{G}, T)$ denote the directed strip clustered graph obtained from (G, T) by orienting every edge uv from the vertex with the smaller label to the vertex with the bigger label, and in case of a tie orienting uv arbitrarily. A *sink* (resp. *source*) of $\vec{G}$ is a vertex with no outgoing (resp. incoming) edges.

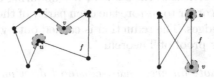

Fig. 2. Assignment of angles at u and v to f corresponding to an upward embedding (on the left), and assignment of angles that is not admissible in an upward embedding (on the right).

In our arguments we use a continuous deformation in order to transform a given drawing into a drawing with desired properties. Observe that during such transformation of a drawing of a graph the parity of crossings between a pair of edges is affected only when an edge e passes over a vertex v, in which case we change the parity of crossings of e with all the edges adjacent to v. Let us call such an event an *edge-vertex switch*.

Edge contraction and vertex split. A *contraction* of an edge $e = uv$ in a topological graph is an operation that turns e into a vertex by moving v along e towards u while dragging all the other edges incident to v along e. Note that by contracting an edge in an even drawing, we obtain again an even drawing.

We will also often use the following operation which can be thought of as the inverse operation of the edge contraction in a topological graph. A *vertex split* in a drawing of a graph G is the operation that replaces a vertex v by two vertices v' and v'' drawn in a small neighborhood of v joined by a short crossing free edge so that the neighbors of v are partitioned into two parts according to whether they are joined with v' or v'' in the resulting drawing, the rotations at v' and v'', resp., is inherited from the rotation at v, and the new edges are drawn in the small neighborhood of the edges they correspond to in G.

Bounded Edges. Theorem 1 can be extended to more general clustered graphs (G, T) that are not necessarily strip clustered, and drawings that are not necessarily clustered. The clusters $V_1, \ldots, V_k$ of (G, T) in our drawing $\mathcal{D}$ are still linearly ordered and drawn as vertical strips respecting this order. An edge $uv \in E(G)$, where $u \in V_i, v \in V_j$, can join any two vertices of G, but it must be drawn so that it intersects only clusters V_l such that $i \leq l \leq j$. We say that the edge uv is *bounded*, and the drawing *quasi-clustered*.

A similar extension of a variant of Hanani-Tutte theorem is also possible in the case of x-monotone drawings [10]. In the x-monotone setting instead of the x-monotonicity of edges in an (independently) even drawing it is only required that the vertical projection of each edge is bounded by the vertical projections of its vertices. Thus, each edge stays between its end vertices.

In the same vein as for x-monotone drawing the extension of our result to drawings $\mathcal{D}$ of clustered graphs with bounded edges can be proved by a reduction to the original claim, Theorem 1. To this end we just need to subdivide every edge e of (G, T) violating conditions of strip clustered drawings so that newly created edges join the vertices in the same or neighbouring clusters, and perform edge-vertex switches in order to restore the even parity of the number of crossings between every pair of edges. The reduction is carried out by the following lemma that is also used in the proof of Theorem 1.

Lemma 2. *Let $\mathcal{D}$ denote an even quasi-clustered drawing of a clustered graph (G, T). Let $e = uv$, where $u \in V_i, v \in V_j$ denote an edge of G. Let G' denote a graph obtained from G by subdiving e by $|i - j| - 1$ vertices. Let (G', T') denote the clustered graph, where T' is inherited from T so that the subdivided edge e is turned into a strictly monotone path w.r.t. γ. Then there exists an even quasi-clustered drawing $\mathcal{D}'$ of (G', T'), in which each new edge crosses the boundary of a cluster exactly once and in which no new intersections of edges with boundaries of the clusters are introduced.*

Proof. Refer to Fig. 3(a) and (b). First, we continuously deform e so that e crosses the boundary of every cluster it visits at most twice. During the deformation we could change the parity of the number of crossings between e and some edges of G. This happens when e passes over a vertex w. We remind the reader that we call this event an edge-vertex switch. Note that we can further deform e so that it performs another edge-vertex switch with each such vertex w, while introducing new crossings with edges "far" from w only in pairs. Thus, by performing the appropriate edge-vertex switches of e with vertices of G we maintain the parity of the number of crossings of e with the edges of G and we do not introduce intersections of e with the boundaries of the clusters.

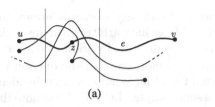 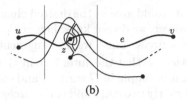

(a)   (b)

Fig. 3. (a) Subdivision of the edge e by the vertex z resulting into odd crossing pairs; (b) Restoration of the evenness by performing edge-vertex switches with z.

Second, if e crosses the boundary of a cluster twice, we subdivide e by a vertex z inside the cluster thereby turning e into two edges, the edge joining u with z and the edge joining z with v. After we subdivide e by z, the resulting drawing is not necessarily even. However, it cannot happen that an edge crosses an odd number of times exactly one edge incident to z, since prior to subdividing the edge e the drawing was even. Thus, by performing edge-vertex switches of z with edges that cross both edges incident to z an odd number of times we restore the even parity of crossings between all pairs of edges. By repeating the second step until we have no edge that crosses the boundary of a cluster twice we obtain a desired drawing of G'.   □

2.2 From Strip Clustered Graphs to the Marriage Condition

The main tool for proving Theorem 1 is [2, Theorem 3] of Bertolazzi et al. that characterizes embedded directed planar graphs, whose embedding can be straightened (the edges turned into straight line segments) so that all the edges are directed upward, i.e., every edge is directed towards the vertex with a higher y-coordinate. Here, it is not crucial that the edges are drawn as straight line segments, since we can straighten them as soon as they are y-monotone [15]. The theorem says that an embedded directed planar graph $\vec{G}$ admits such an embedding, if there exists an assignment of the sources and sinks of $\vec{G}$ to the faces of $\vec{G}$ that is easily seen to be necessary for such a drawing to exist (see Fig. 2 for an illustration).

Intuitively, a sink or source v is assigned to a face f, if and only if a pair of edges vw and vz, incident to f form in f a concave angle, i.e., an angle bigger than π in an upward embedding. Thus, a vertex can be assigned to a face only if it is incident to it. First, note that the number of sinks incident to a face f is the same as the number of sources incident to f. The mentioned easy necessary condition for the existence of an upward embedding is that an internal (resp. external) face with $2k$ sinks and sources have precisely $k-1$ (resp. $k+1$) of them assigned to it, and that the rotation at each vertex can be split into two parts consisting of incoming and outgoing, resp., edges. The embeddings satisfying the latter are dubbed *candidate embeddings* by [2].

Assuming that in (G, T) each cluster forms an independent set, we would like to prove that $(\vec{G}, T)$ satisfy this condition, if (G, T) admits an even clustered drawing.

That would give us the desired clustered drawing by an easy geometric argument. However, we do not know how to do it directly, if faces have arbitrarily many sinks and sources. Thus, we first augment the given even drawing by adding edges and vertices so that the outer face in $\overrightarrow{G}$ is incident to at most one sink and one source, i.e., it is simple in G w.r.t. γ, and each internal face, that is not simple, is incident to exactly two sinks and two sources, i.e., it is semi-simple.[3] Let (G', T') denote the resulting strip clustered graph. This reduces the proof to showing that there exists a bijection between the set of internal semi-simple faces, and the set of sinks and sources in $\overrightarrow{G}'$ without the source and sink incident to the outer face.

By [2, Lemma 5] the total number of sinks and sources is exactly the total demand by all the faces in a candidate embedding, which is also a direct consequence of the discretized version [11] of the Poincaré-Hopf index theorem [12]. Hence, by Hall's Theorem the bijection exists, if every subset of internal semi-simple faces of size l is incident to at least l sinks and sources.

2.3 Crossing Paths

Just to give a glimpse of the proof of our main result we present two observations, whose combination plays an important role in the proof of the required marriage condition that we need in order to apply [2, Theorem 3]. The first one is a simple parity variant of the Pigeon hole principle.

Observation 1. *Let $C = v_1 v_2 \ldots v_{2a}$, $a \geq 2$, denote an even cycle. Let V' denote a subset of the vertices of C of size at least $a + 2$. Then V' contains four vertices v_i, v_j, v_k and v_l, where $i < j < k < l$, such that i, k is odd and j, l is even (or vice versa).*

Proof. For the sake of contradiction we assume that V' does not contain four such vertices. Let V_0 and V_1, resp., denote the vertices of V with even and odd index. Similarly, let V_0' and V_1', resp., denote the vertices of V' with even and odd index. Suppose that $2 \leq |V_0'| \geq |V_1'|$ and fix a direction in which we traverse C. Between every two consecutive vertices of V_0' along C except for at most one pair of consecutive vertices we have a vertex in $V_1 - V_1'$. Thus, $|V_1 - V_1'| \geq |V_0'| - 1$. On the other hand, $|V_1 - V_1'| = a - |V_1'| \leq a - (a + 2 - |V_0'|) = |V_0'| - 2$ (contradiction). $\qquad \square$

Let G denote a graph with a rotation system. We define the *crossing index* of paths P_1 and P_2 in G as follows. Let us orient all the edges of P_1 and P_2, resp., so that P_1 and P_2 has only one sink and one source. Let P denote the subgraph of G which is the union of P_1 and P_2. We define $cr(v) = +1$ (resp. $cr(v) = -1$), if v is a vertex of degree four in P such that the paths P_1 and P_2 alternate in the rotation at v and at v the path P_2 crosses P_1 from left to right (resp. right to left) in the direction of P_1. We define $cr(v) = +1/2$ (resp. $cr(v) = -1/2$), if v is a vertex of degree three in P such that at v the path P_2 is oriented towards P_1

---

[3] We would not have to do anything that follows, if we could turn all the faces into simple ones. However, this seems to be a difficult task.

from left, or from P_1 to right (resp. towards P_1 from right, or from P_1 to left) in the direction of P_1. The crossing index of P_1 and P_2 is then the absolute value of the sum of $cr(v)$ over all vertices of degree three and four in P. Let $G' \subseteq G$. Let $\max(G')$ and $\min(G')$, resp., denote the maximal and minimal value of $\gamma(v)$, $v \in V(G')$.

A path P in G is an i-*cap* and i-*cup*, resp., if for the end vertices u and v of P we have $\min(P) = \gamma(u) = \gamma(v) = i$ and $\max(P) = \gamma(u) = \gamma(v) = i$. Note that we can define the crossing index for any two subgraphs of maximum degree two. Then two cycles in an even drawing of G cannot have an odd crossing index, since they would correspond to two curves in the plane crossing an odd number of times. This observation can be easily extended to special pairs of an i-cap and j-cup.

An i-cap P_1 and j-cup P_2 *cross*, if

(A) their crossing index is odd; and
(B) $\min(P_1) < \min(P_2) \leq \max(P_1) < \max(P_2)$.

Observation 2. *The clustered graph (G, T) admitting an even clustered drawing does not contain a pair P_1 and P_2 of an i-cap and j-cup, $i + 1 < j$, that cross.*

3 The Proof of Theorem 3

First, we prove a lemma that allows us to get rid of odd crossing pairs by doing only local redrawings and vertex splits.

A drawing of a graph G is obtained from the given drawing of G by *redrawing edges locally at vertices* if the resulting drawing of G differs from the given one only in small pairwise disjoint neighborhoods of vertices not containing any other vertex. The proof of the following lemma is inspired by the proof of [17, Theorem 3.1].

Lemma 3. *Let G denote a subdivision of a vertex three-connected graph drawn in the plane so that every pair of non-adjacent edges cross an even number of times. We can turn the drawing of G into an even drawing by a finite sequence of local redrawings of edges at vertices and vertex splits.*

Proof. We process cycles in G containing an edge crossed by one of its adjacent edges an odd number of times one by one until no such cycle exists. Let C denote a cycle of G. By local redrawings at the vertices of C we obtain a drawing of G, where every edge of C crosses every other edge an even number of times. Let v denote a vertex of C.

First, suppose that every edge incident to v and starting inside of C crosses every edge incident to v and starting outside of C an even number of times. In this case we perform at most two subsequent vertex splits. If there exists at least two edges starting at v inside (resp. outside) of C, we split v into two vertices v' and v'' joined by a very short crossing free edge so that v' is incident to the neighbors of v formerly joined with v by edges starting inside (resp. outside) of C, and v''

is incident to the rest of the neighbors of v. Thus, v'' replaces v on C. Notice that by splitting we maintain the property of the drawing to be independently even, and the property of our graph to be three-connected. Moreover, all the edges incident to the resulting vertex v'' of degree three or four cross one another an even number of times. Hence, no edge of C will ever be crossed by another edge an odd number of times, after we apply appropriate vertex splits at every vertex of C.

Second, we show that there does not exist a vertex v incident to C so that an edge vu starting inside of C crosses an edge vw starting outside of C an odd number of times. Since G is a subdivision of a vertex three-connected graph, there exist two distinct vertices u' and w' of C different from v such that u' and w', resp., is connected with u and w by a path internally disjoint from C. Let uP_1u' and wP_2w', resp., denote this path. Note that u can coincide with u' and w can coincide with w'. Let vP_3u' denote the path contained in C no passing through w'. Let C' denote the cycle obtained by concatenation of P_1, P_3, and vu. Let C'' denote the cycle obtained by concatenating P_2 and the portion of C between w' and v not containing u'. Since vw and vu cross an odd number of times and all the other pairs of edges $e \in E(C')$ and $f \in E(C'')$ cross an even number of times, the edges of C' and C'' cross an odd number of times. It follows that their corresponding curves cross an odd number of times (contradiction).

Notice that by vertex splits we decrease the value of the function $\sum_{v \in V(G)} deg^3(v)$ whose value is always non-negative. Hence, after a finite number of vertex splits we turn G into an even drawing of a new graph G'.   □

We turn to the actual proof of Theorem 3.
We apply Lemma 3 to the graph G thereby obtaining a clustered graph (G', T'), where each vertex obtained by a vertex split, belongs to the cluster of its parental vertex and the membership of other vertices to clusters is unchanged. By applying Theorem 1 to (G', T') we obtain a clustered embedding of (G', T'). Finally, we contract the pairs of vertices obtained by vertex splits in order to obtain a clustered embedding of (G, T).

4   Monotone Variant of the Weak Hanani–Tutte Theorem

In the present section we derive Theorem 2 from Theorem 1.

Given a graph G with a fixed order of vertices let $\mathcal{D}$ denote its drawing such that x-coordinates of the vertices of G respect their order, edges are drawn as x-monotone curves and every pair of edges cross an even number of times. We turn our drawing $\mathcal{D}$ of G into a clustered drawing $\mathcal{D}'$ of a strip clustered graph (G', T') which is still even.

We divide the plane by vertical lines such that each resulting strip contains exactly one vertex of G in its interior. Let (G, T) denote the clustered graph, in which every cluster consists of a single vertex, such that the clusters are ordered according to x-coordinates of the vertices. Thus, every vertical strip corresponds to a cluster of (G, T). Note that all the edges in the drawing of (G, T) are bounded, and hence, by Lemma 2 can be turned into paths so that the resulting clustered graph is strip clustered, and even drawing clustered. We denote the resulting strip clustered graph by (G', T') and drawing by $\mathcal{D}'$.

Applying Theorem 1 to $\mathcal{D}'$, we obtain an embedding of (G', T') that can be turned into an embedding of (G, T) by converting the subdivided edges in G' back to the edges of G. The obtained embedding is turned into an x-monotone embedding by replacing each edge e with a polygonal path whose bends are intersections of e with vertical lines separating clusters in (G, T).

5 Open Problems

We proved the weak variant of Hanani-Tutte theorem for strip clustered graphs, and verified the corresponding strong variant for three-connected graphs. Naturally, the main open problem we left open is to prove or disprove the strong variant, if the underlying abstract graph G is not a subdivision of a three-connected graph. We find the case, when G is guaranteed to be only two-connected, already quite challenging. A possible approach to prove the strong variant is to adapt the technique of "untangling" pairs of edges crossing an odd number times from [10, Sect. 3]. Another direction for further research would be the weak variant of Conjecture 1 from [9].

Acknowledgment. We would like to express our special thanks of gratitude to the organizers and participants of the 11th GWOP workshop, where we could discuss the research problems treated in the present paper. In particular, we especially benefited from the discussions with Bettina Speckmann, Edgardo Roldán-Pensado and Sebastian Stich. Furthermore, we would like to thank Ján Kynčl for useful discussions at the initial stage, and Gábor Tardos for comments at the final stage of this work.

References

1. Angelini, P., Da Lozzo, G., Di Battista, G., Frati, F.: Strip planarity testing. In: Wismath, S., Wolff, A. (eds.) GD 2013. LNCS, vol. 8242, pp. 37–48. Springer, Heidelberg (2013)
2. Bertolazzi, P., Di Battista, G., Liotta, G., Mannino, C.: Upward drawings of tri-connected digraphs. Algorithmica **12**(6), 476–497 (1994)
3. Cairns, G., Nikolayevsky, Y.: Bounds for generalized thrackles. Discrete Comput. Geom. **23**(2), 191–206 (2000)
4. Cortese, P.F., Di Battista, G., Patrignani, M., Pizzonia, M.: Clustering cycles into cycles of clusters. J. Graph Algorithms Appl. **9**(3), 391–413 (2005)
5. Cortese, P.F., Di Battista, G.: Clustered planarity (invited lecture). In: Twenty-first Annual Symposium on Computational Geometry (proc. SoCG 05), pp. 30–32. ACM (2005)
6. Diestel, R.: Graph Theory. Springer, New York (2010)
7. Feng, Q.-W., Cohen, R.F., Eades, R.: How to draw a planar clustered graph. In: Li, M., Du, D.-Z. (eds.) COCOON 1995. LNCS, vol. 959, pp. 21–30. Springer, Heidelberg (1995)

8. Feng, Q.-W., Cohen, R.F., Eades, P.: Planarity for clustered graphs. In: Spirakis, P.G. (ed.) ESA 1995. LNCS, vol. 979, pp. 213–226. Springer, Heidelberg (1995)
9. Fulek, R., Kynčl, J., Malinović, I., Pálvölgyi, D.: Efficient c-planarity testing algebraically. arXiv:1305.4519
10. Fulek, R., Pelsmajer, M., Schaefer, M., Štefankovič, D.: Hanani-Tutte, monotone drawings and level-planarity. In: Pach, J. (ed.) Thirty Essays in Geometric Graph Theory, pp. 263–288. Springer, New York (2012)
11. Gortler, S.J., Gotsman, C., Thurston, D.: Discrete one-forms on meshes and applications to 3D mesh parameterization. J. CAGD **23**, 83–112 (2006)
12. Guillemin, V., Pollack, A.: Differential Topology. Prentice-Hall (1974)
13. Hanani, H.: Über wesentlich unplättbare Kurven im drei-dimensionalen Raume. Fundam. Math. **23**, 135–142 (1934)
14. Pach, J., Tóth, G.: Which crossing number is it anyway? J. Combin. Theory Ser. B **80**(2), 225–246 (2000)
15. Pach, J., Tóth, J.: Monotone drawings of planar graphs. J. Graph Theory **46**(1), 39–47 (2004). http://arxiv.org/abs/1101.0967(Updated version)
16. Pelsmajer, M.J., Schaefer, M., Stasi, D.: Strong Hanani-Tutte on the projective plane. SIAM J. Discrete Math. **23**(3), 1317–1323 (2009)
17. Pelsmajer, M.J., Schaefer, M., Štefankovič, D.: Removing even crossings. J. Combin. Theory Ser. B **97**(4), 489–500 (2007)
18. Pelsmajer, M.J., Schaefer, M., Štefankovič, D.: Removing even crossings on surfaces. Eur. J. Comb. **30**(7), 1704–1717 (2009)
19. Schaefer, M.: Hanani-Tutte and related results. Bolyai Memorial Volume (2011)
20. Schaefer, M.: Toward a theory of planarity: Hanani-Tutte and planarity variants. In: Didimo, W., Patrignani, M. (eds.) GD 2012. LNCS, vol. 7704, pp. 162–173. Springer, Heidelberg (2013)
21. Schaefer, M., Štefankovič, D.: Block additivity of $\mathbb{Z}_2$-embeddings. In: Wismath, S., Wolff, A. (eds.) GD 2013. LNCS, vol. 8242, pp. 185–195. Springer, Heidelberg (2013)
22. Tutte, W.T.: Toward a theory of crossing numbers. J. Combin. Theory **8**, 45–53 (1970)

On Set Expansion Problems and the Small Set Expansion Conjecture

Rajiv Gandhi[✉] and Guy Kortsarz

Department of Computer Science,
Rutgers University-Camden,
Camden, NJ 08102, USA
{rajivg,guyk}@camden.rutgers.edu

Abstract. We study two problems related to the Small Set Expansion Conjecture [14]: the Maximum weight m'-edge cover (MWEC) problem and the Fixed cost minimum edge cover (FCEC) problem. In the MWEC problem, we are given an undirected simple graph $G = (V, E)$ with integral vertex weights. The goal is to select a set $U \subseteq V$ of maximum weight so that the number of edges with at least one endpoint in U is at most m'. Goldschmidt and Hochbaum [8] show that the problem is NP-hard and they give a 3-approximation algorithm for the problem. The approximation guarantee was improved to $2 + \epsilon$, for any fixed $\epsilon > 0$ [12]. We present an approximation algorithm that achieves a guarantee of 2. Interestingly, we also show that for any constant $\epsilon > 0$, a $(2 - \epsilon)$-ratio for MWEC implies that the Small Set Expansion Conjecture [14] does not hold. Thus, assuming the Small Set Expansion Conjecture, the bound of 2 is tight. In the FCEC problem, we are given a vertex weighted graph, a bound k, and our goal is to find a subset of vertices U of total weight at least k such that the number of edges with at least one edges in U is minimized. A $2(1 + \epsilon)$-approximation for the problem follows from the work of Carnes and Shmoys [3]. We improve the approximation ratio by giving a 2-approximation algorithm for the problem and show a $(2 - \epsilon)$-inapproximability under Small Set Expansion Conjecture conjecture. Only the NP-hardness result was known for this problem [8]. We show that a natural linear program for FCEC has an integrality gap of $2 - o(1)$. We also show that for any constant $\rho > 1$, an approximation guarantee of ρ for the FCEC problem implies a $\rho(1 + o(1))$ approximation for MWEC. Finally, we define the Degrees density augmentation problem which is the density version of the FCEC problem. In this problem we are given an undirected graph $G = (V, E)$ and a set $U \subseteq V$. The objective is to find a set W so that $(e(W) + e(U, W))/deg(W)$ is maximum. This problem admits an LP-based exact solution [4]. We give a combinatorial algorithm for this problem.

Rajiv Gandhi: Supported in part by NSF awards 1050968 and 1218620. This work was started when the author was visiting IBM Research, New Delhi. India.
Guy Kortsarz: Supported in part by NSF award 1218620.

D. Kratsch and I. Todinca (Eds.): WG 2014, LNCS 8747, pp. 189–200, 2014.
DOI: 10.1007/978-3-319-12340-0_16

1   Introduction

Given a graph $G = (V, E)$ and a subset $S \subseteq V$, let $deg(S)$ denote sum of degrees of all vertices in S and let $e(S, \overline{S})$ denote the number of edges that have one endpoint in S and the other in $V \setminus S$. Then the edge expansion, $\phi_G(S)$ is given by $\phi_G(S) = e(S, \overline{S})/deg(S)$. Given some δ, $0 < \delta \leq 1/2$ and a d-regular graph F, let $\mathcal{S}$ denote all subsets of V of size $\delta|V|$. Let

$$\phi_G(\delta) = \min_{S \in \mathcal{S}} \frac{e(S, \overline{S})}{deg(S)}$$

The Small Set Expansion Conjecture states that for any constant η, it is NP-hard to distinguish whether $\phi_G(\delta) \geq 1 - \eta$ or $\phi_G(\delta) \leq \eta$. In [14], Raghavendra and Steurer showed that proving the Small Set Expansion Conjecture implies a proof for the Unique Games Conjecture and an algorithm that refutes the Unique Games Conjecture refutes the Small Set Expansion Conjecture.

In this paper we relate the Small Set Expansion Conjecture to two other edge expansion problems. We say that an edge e is *touched* by a set of vertices U or that e *touches* the set of vertices U, if at least one of e's endpoints is in U. Specifically, the problems that we study are as follows. The Maximum weight m'-edge cover (MWEC) problem that we study was first introduced by Goldschmidt and Hochbaum [8]. In this problem, we are given an undirected simple graph $G = (V, E)$ with integral vertex weights. The goal is to select a subset $U \subseteq V$ of maximum weight so that the number of edges touching U is at most m'. This problem is motivated by application in loading of semi-conductor components to be assembled into products [8].

We also study the closely related Fixed cost minimum edge cover (FCEC) problem in which given a graph $G = (V, E)$ with vertex weights and a number W, our goal is to find $U \subseteq V$ of weight at least W such that the number of edges touching U is minimized.

Finally, we study the Degrees density augmentation problem which is the density version of the FCEC problem. In the Degrees density augmentation problem, we are given an undirected graph $G = (V, E)$ and a set $U \subseteq V$ and our goal is to find a set W with maximum augmenting density i.e., a set W that maximizes $(e(W) + e(U, W))/deg(W)$.

1.1   Related Work

Goldschmidt and Hochbaum [8] introduced the MWEC problem. They show that the problem is NP-complete and give algorithms that yield 2-approximate and 3-approximate algorithm for the unweighted and the weighted versions of the problem, respectively. Their NP-hardness proof applies to FCEC as well. Liang [12] improved the bound of 3 to $2 + \epsilon$, for any fixed $\epsilon > 0$.

A class of related problems are the density problems – problems in which we are to find a subgraph and the objective function considers the ratio of the total number or weight of edges in the subgraph to the number of vertices in the subgraph. A well known problem in this class is the Dense k-subgraph problem (DkS)

in which we want to find a subset of vertices U of size k such that the total number of edges in the subgraph induced by U is maximized. The best ratio known for the problem is $n^{1/4+\epsilon}$ [2,5], which is an improvement over the bound of $O(n^{1/3-\epsilon})$, for ϵ close to $1/60$ [5]. The Dense k-subgraph problem is APX-hard under the assumption that NP problems can not be solved in subexponential time [9]. Interestingly, if there is no bound on the size of U then the problem can be solved in polynomial time [7,11].

Consider an objective function in which we minimize $deg(U)$. One can associate a cost $c_u = deg(u)$ with each vertex u and a size $s_u = w(u)$ for each vertex u, and then the objective is just to minimize $deg(U)$ subject to $\sum s_u x_u \geq k$. Carnes and Shmoys [3] give a $(1+\epsilon)$-approximation for the problem. Using this result and the observation that the objective function is at most a factor of 2 away from the objective function for the FCEC problem, a $2(1+\epsilon)$-approximation follows for the FCEC problem.

Variations of the Dense k-subgraph problem in which the size of U is at least k ($Dalk$) and the size of U is at most k ($Damk$) have been studied [1,10]. In [1,10], they give evidence that $Damk$ is just as hard as DkS. They also give 2-approximate solutions to the $Dalk$ problem. In [10], they also consider the density versions of the problems in directed graphs. Gajewar and Sarma [6] consider a generalization in which we are give a partition of vertices $U_1, U_2, \ldots, U_t$, and non-negative integers $r_1, r_2, \ldots, r_t$. the goal is to find a densest subgraph such that partition U_i contributes at least r_i vertices to the densest subgraph. They give a 3-approximation for the problem, which was improved to 2 by Chakravarthy et al. [4], who also consider other generalizations. They also show using linear programming that the Degrees density augmentation problem can be solved optimally.

A problem parameterized by k is Fixed Parameter Tractable [13], if it admits an exact algorithm with running time of $f(k) \cdot n^{O(1)}$. The function f can be exponential in k or larger. Proving that a problem is W[1]-hard (with respect to parameter k) is a strong indication that it has no FPT algorithm with parameter k (similar to NP-hardness implying the likelihood of no polynomial time algorithm). The FCEC problem parameterized by k is W[1] hard but admits a $f(k, \epsilon) \cdot n^{O(1)}$ time, $(1+\epsilon)$-approximation, for any constant $\epsilon > 0$ [13]. This is in contrast to our result that shows that it is highly unlikely that FCEC admits a polynomial time approximation scheme (PTAS), if the running time is bounded by a polynomial in k.

1.2 Preliminaries

The input is an undirected simple graph $G = (V, E)$ and vertex weights are given by $w(\cdot)$. Let $n = |V|$ and $m = |E|$. For any subset $S \subseteq V$, let $\overline{S} = V \setminus S$. Let $e(P, Q)$ be the set of edges with one endpoint in P and the other in Q. Let $deg(S)$ denote the sum of degrees of all vertices in S, i.e., $deg(S) = \sum_{v \in S} deg(v)$. Let $deg_H(v)$ denote the number of neighbors of v among the vertices in H. Let $deg_H(S)$ denote the quantity $\sum_{v \in S} deg_H(v)$. We use OPT to denote an optimal solution as well as the cost of an optimal solution. The meaning will be clear from the context in which it is used.

For set $U \subseteq V$, let $T(U)$ be the collection of all edges with at least one endpoint in U. Namely, is the set of edges touching U. We denote $t(U) = |T(U)|$. The set of edges with both endpoints in U, also called *internal* edges of U, is denoted by $E(U)$. We denote $e(U) = |E(U)|$. We denote by $e(X,Y)$ the number of edges with one endpoint in X and one in Y. Let $e_U(X,Y)$ be the number of edges between $X \cap U$ and $Y \cap U$ in the graph $G(U)$ induced by U. The following lemma is known [12]; we present it here for completeness.

Lemma 1. *The* FCEC *problem admits a simple 2-approximate solution in case of uniform vertex weights.*

Proof. Let Z be the set of k lowest degree vertices in G. The set Z is a 2-approximate solution by the following argument. Let b be the average degree of vertices in Z. Thus $t(Z) \leq bk$. The claim follows since $t(OPT) \geq deg(OPT)/2 \geq bk/2$.

Claim. For every set U, $t(U) = deg(U) - e(U)$.

Proof. Consider separately the edges $E(U, V \setminus U)$ and $E(U)$. Note that the edges $E(U, V \setminus U)$ are counted once in the sum of degrees, but edges in $E(U)$ are counted twice. Thus in order to get the number of edges touching U, we need to subtract $e(U)$ from $deg(U)$.

1.3   Our Results

Our contributions in this paper are as follows.

- For the MWEC problem we give an algorithm that yields an approximation guarantee of 2. This improves the approximation guarantee of 3 given by Goldschmidt and Hochbaum [8] and the ratio of $2 + \epsilon$, for fixed $\epsilon >$, given by Liang [12]. We also give a $(2 - \epsilon)$-inapproximability for the problem under the Small Set Expansion Conjecture. Note that only a NP-hardness result was known for this problem [8].
- We give a 2-approximate solution to the FCEC problem. This improves the $2(1 + \epsilon)$-ratio that follows from the work of Carnes and Shmoys [3]. We also give a $2 - \epsilon$ inapproximability for the problem, where $\epsilon > 0$ is any constant.
- We also show that a natural LP for FCEC has an integrality gap of $2(1 - o(1))$, even for the unweighted case.
- For any constant $\rho > 1$, we show that if FCEC admits a ρ-approximation algorithm then MWEC admits a $\rho(1 + o(1))$-approximation algorithm.
- We give a combinatorial algorithm that solves the Degrees density augmentation problem optimally.

2   Tight 2-Approximation for MWEC and FCEC

In this section we present 2-approximation algorithms for MWEC and FCEC, and show that the results are tight under the Small Set Expansion Conjecture.

2.1 A 2-Approximation for Maximum Weight m'-Edge Cover

We give a dynamic programming based solution for the MWEC problem. The idea of using dynamic programming in this context was first proposed by Goldschimdt and Hochbaum [8]. Recall that in the MWEC problem, we are given an undirected simple graph $G = (V, E)$ with integral vertex weights. The goal is to select a subset $U \subseteq V$ of maximum weight so that the number of edges touching U is at most m'.

We will guess the following entities (by trying all possibilities) and for each guess, we use dynamic programming to solve the problem.

1. $H^* = \{v_h\}$, where v_h is the heaviest vertex in an optimal solution.
2. $P_{H^*} = e(H^*, OPT \setminus H^*)$ – the number of neighbors of v_h in the optimal solution. There are at most n possibilities.
3. $D_{H^*} = deg_{\overline{H}^*}(OPT \setminus H^*)$: total degree of vertices in $OPT \setminus H^*$ in the graph induced by vertices in $V \setminus H^*$. There are at most n^2 possibilities.

We will try all combinations of the above entities. Since there are at most polynomial number of possibilities for each entity, we have at most polynomial number of possibilities in total. We define the following subproblems as part of our dynamic programming solution. For a guess H for H^*, let $\{v_1, v_2, \ldots, v_{|\overline{H}|}\}$ be the vertices in $\overline{H}$. Then, any H, we solve the following subproblems.

$A[H, i, P_H, D_H]$ denote the maximum weighted subset $Q \subseteq \{v_1, v_2, \ldots, v_i\}$ such that $e(H, Q) \geq P_H$ and $deg_{\overline{H}}(Q) \leq D_H/2$.

Note that while the natural bound on $deg_{\overline{H}}(Q)$ is D_H, using such a bound will lead to an infeasible solution. For fixed parameters H, P_H, and D_H, we are interested in $A[H, |\overline{H}|, P_H, D_H]$. We use the following recurrence as the basis for our dynamic programming solution: the value of $A[H, i, P_H, D_H] = -\infty$ in any of the following three cases – (i) $i = 0$ and $P_H > 0$, (ii) $i = 0$ and $D_H/2 < 0$, and (iii) $D_H/2 > m' - e(H, \overline{H})$. When $i = 0$, $P_H \leq 0$ and $D_H/2 \geq 0$, the value of $A[H, i, P_H, D_H] = 0$. Otherwise, we have

$$A[H, i, P_H, D_H] = \max\{A[H, i-1, P_H, D_H], w(v_i) + A[H, i-1, P'_H, D'_H]\}$$

where, $P'_H = P_H - deg_H(v_i)$ and $D'_H = D_H - 2(deg_{\overline{H}}(v_i))$. Our solution is given by $\max_{H, P_H, D_H}\{w(H) + A[H, |\overline{H}|, P_H, D_H]\}$.

Analysis

Lemma 2. *Our algorithm yields a feasible solution.*

Proof. Let $H' \cup Q'$, where $Q' \subseteq V \setminus H'$, be the set of vertices returned by our solution. The number of edges with at least one endpoint in $H' \cup Q'$, is

$$= e(H', \overline{H}') + e(Q', \overline{H}') \leq e(H', \overline{H}') + deg_{\overline{H}'}(Q') \leq e(H', \overline{H}') + \frac{D_{H'}}{2}$$
$$\leq e(H', \overline{H}') + (m' - e(H', \overline{H}')) = m'$$

Lemma 3. *The above algorithm results in a 2-approximate solution.*

Proof. Recall that H^* consists of the highest degree vertex in the optimal solution. Let Q^* be the remaining vertices in the optimal solution. Consider the scenario when our algorithm makes the correct guess for H^*. Let $Q \subseteq \overline{H^*}$ be the solution returned by the dynamic program in this setting. We know that

$$deg_{\overline{H}^*}(Q) \leq \frac{deg_{\overline{H}^*}(Q^*)}{2}$$

We now use ideas from [8] to show that $w(H^* \cup Q) \geq 2w(H^* \cup Q^*)$. Recall that $H' \cup Q'$ be the output of our algorithm. Since $w(H' \cup Q') \geq w(H^* \cup Q^*)$, it follows that our solution is a factor of at most 2 away from OPT.

Consider any arbitrary ordering of vertices $v_1, v_2, \ldots$ in Q^*. Note that the weight of each vertex in Q^* is at most $w(H^*)$. Let Q_r^* denote the first r vertices in the above ordering of vertices of Q^*. Let p be the first index such that $deg_{\overline{H}^*}(Q_p^*) > deg_{\overline{H}^*}(Q^*)/2$. This implies the following – (i) $deg_{\overline{H}^*}(Q_{p-1}^*) \leq deg_{\overline{H}^*}(Q^*)/2$, and (ii) $deg_{\overline{H}^*}(Q^* \setminus Q_p^*) < deg_{\overline{H}^*}(Q^*)/2$. Note that both the sets Q_{p-1}^* and $Q^* \setminus Q_p^*$ (neither set contains v_p) are feasible candidates for the set Q, the solution returned by our algorithm when the heaviest vertex set was chosen to be H^*. Since $w(Q) \geq w(Q_{p-1}^*)$, $w(Q) \geq w(Q^* \setminus Q_p^*)$, and $w(v_p) \leq w(H^*)$, we have

$$w(OPT) \leq w(H^* \cup Q^*) \leq w(H^*) + w(Q^*) \leq w(H^*) + w(Q_{p-1}^*) + w(v_p) + w(Q^* \setminus Q_p^*)$$
$$\leq w(H^*) + w(Q) + w(H^*) + w(Q) = 2w(H^* \cup Q) \leq 2w(H' \cup Q')$$

2.2   A 2-Approximation for Fixed Weight Minimum Edge Cover

Recall the FCEC problem: Given a graph $G = (V, E)$ with arbitrary vertex weights and a positive integer W, our objective is to choose a set $S \subseteq V$ of vertices of total weight at least W such that the number of edges with at least one end point in S is minimized.

We will solve the following related problem optimally and then show that an optimal solution to the problem is a 2-approximation to FCEC: we want to find a subset S of vertices such that $deg(S)$ is smallest and $w(S)$ is at least W.

We use the dynamic programming algorithm of the well-known Knapsack problem to find a solution to the above problem. For completeness, we restate the dynamic programming formulation below.

$P[i, D]$: maximum weight of set $Q \subseteq \{v_1, v_2, \ldots, v_i\}$ such that $deg(Q)$ is at most D.

Note that $P[0, D] = 0$, for all values of D is the base case. For all other case, we invoke the following recurrence.

$$P[i, D] = \max\{P[i - 1, D], w(v_i) + P[i - 1, D - w(v_i)]\}$$

After filling the table P using dynamic programming, we scan all entries of the form $P[|V|, D]$ to find the smallest value of D for which $P[|V|, D] \geq W$. Let S be the corresponding set.

Lemma 4. *The is a 2-approximate solution to the Fixed Cost Minimum Edge Cover Problem as follows.*

$$t(S) \leq deg(S) \leq deg(OPT) = 2(deg(OPT)/2) \leq 2OPT$$

2.3 A $(2 - \epsilon)$-inapproximability for FCEC Under the Small Set Expansion Conjecture

The Small Set Expansion Conjecture is the following. The expansion of a set $S \subseteq V$ is defined as

$$\phi_G(S) = \frac{e(S, V - S)}{deg(S)}.$$

Here $e(S, V - S)$ are the number of edges with one vertex in S and one in $V - S$ and $deg(S)$ is the sum of degrees of the vertices in S.

The conjecture is about the expansion of sets of small size. Given some $\delta \leq 1/2$ and a d-regular graph F, consider all subsets S of V of size $\delta \cdot |V|$. Let

$$\phi_G(\delta) = \min_{S \in \mathcal{S}} \frac{\phi_G(S)}{deg(S)}.$$

The Small Set Expansion Conjecture states that for any constant η, it is NP-hard to distinguish whether $\phi_G(\delta) \geq 1 - \eta$ or $\phi_G(\delta) \leq \eta$.

Theorem 1. *If the Fixed Cost Minimum Edge Cover problem with uniform vertex weights admits better than a $2 - 6\eta$ approximation, then the Small Set Expansion Conjecture does not hold.*

Proof. Consider first the case that $\phi_G(\delta) \geq 1 - \eta$, and let S be the set achieving this bound. Let k be the size of S.

As the graph is d regular, we have

$$e(S, V - S) \geq (1 - \eta)deg(S) = (1 - \eta)k \cdot d.$$

As $t(S) \geq e(S, V - S)$ we get that $t(S) \geq (1 - \eta) \cdot k \cdot d$.

We now consider the second case in which

$$e(S, V - S) \leq \eta \cdot deg(S) = \eta \cdot k \cdot d.$$

Thus $deg(S) = d \cdot k \leq \eta \cdot k \cdot d + 2e(S)$. Therefore, $e(S) \geq dk(1 - \eta)/2$ and $t(S) = deg(S) - e(S) \leq k \cdot d(1 + \eta)/2$.

The ratio between a yes and a no instance is:

$$\frac{(1 - \eta)k \cdot d}{(1 + \eta)/2 \cdot k \cdot d} = 2 \cdot \frac{1 - \eta}{1 + \eta} \geq 2 \cdot (1 - 3\eta).$$

This means that if FCEC admits an approximation ratio smaller than $2(1 - 3\eta)$ the Small Set Expansion Conjecture is disproved. We can pick $\epsilon = \eta/6$ and get the $(2 - \epsilon)$-inapproximability. As we can choose any small ϵ, (the conjecture allows us to chose η as small constant as we want) the inapproximability can be made close to 2 almost matching the lower bound.

2.4   A Hardness of $2 - \epsilon$ for Maximum Weight m'-Edge Cover

Recall that in our hardness example for FCEC for hardness $2 - \epsilon$, the graph is d regular.

For clarity in exposition, we ignore the small constant ϵ in this section. We will assume 2-inapproximability for FCEC and show that better than 2-approximation for MWEC implies a better than 2-approximation for FCEC.

In the reduction from FCEC to MWEC, we use the same d-regular graph that is part of the FCEC instance and set $m' = kd/2$. Thus for a "yes" instance of FCEC we assume that $t(OPT_{\text{FCEC}}) = kd/2$. And that for a "no" instance it is kd.

Since $OPT_{\text{MWEC}} = k$, an approximation ratio better than 2 for MWEC will give a set S of more than $k/2$ vertices. The number of vertices still required to be added to transform S to a legal FCEC output is strictly less than $k - k/2 = k/2$. We can complete the set S to size k by *any* set S' of $k/2$ vertices. Thus $t(S \cup S') < kd/2 + kd/2 = kd$ and thus we can distinguish between a no and a yes instance of FCEC, refuting the Small Set Expansion Conjecture.

This implies $(2 - \epsilon)$-inapproximability for MWEC.

3   Integrality Gap for Fixed Cost Minimum Edge Cover

Consider the following natural integer linear program for the problem: min $\sum_e y_e$, subject to (i) $\sum_{v \in V} x_v \geq k$, (ii) $\forall e = (u, v), y_e \geq x_u$, (iii) $\forall e = (u, v), y_e \geq x_v$, (iv) $\forall v \in V, x_v \in \{0, 1\}$, (v) $\forall e \in E, y_e \in \{0, 1\}$.

The LP relaxation can be obtained by relaxing the integrality constraints on x_v and y_e to $x_v \geq 0, \forall v \in V$ and $y_e \geq 0, \forall e \in E$.

Theorem 2. *The above LP has an integrality gap of $2(1 - o(1))$.*

Let $k = \lfloor \sqrt{n} \rfloor$. Construct a graph G on n vertices as follows. For each pair of vertices, include an edge between the pair with a probability $1/\lfloor \sqrt{n} \rfloor$. For any vertex v, $\mathbf{E}[deg(v)] = n(1/\lfloor \sqrt{n} \rfloor) \leq \lceil \sqrt{n} \rceil$. Using Chernoff bounds, for $0 < \delta < 1$, we have

$$\sqrt{n}(1 - o(1)) \leq deg(v) \leq \sqrt{n}(1 + o(1))$$

Consider any subset Q of vertices in G such that $|Q| = \lfloor \sqrt{n} \rfloor$. Then we have

$$\mathbf{E}[e(Q)] = \frac{1}{\lfloor \sqrt{n} \rfloor} \binom{Q}{2} = \frac{\lfloor \sqrt{n} \rfloor(\lfloor \sqrt{n} \rfloor - 1)}{2\lfloor \sqrt{n} \rfloor} = \frac{\lfloor \sqrt{n} \rfloor - 1}{2}$$

Thus, $n \geq 4$, we have $\sqrt{n}/4 \leq \mathbf{E}[e(Q)] < \sqrt{n}/2$. We use the following Chernoff bound to obtain the probability that $e(Q) \geq n^{1-\epsilon}$, for a constant ϵ.

$$\Pr[e(Q) \geq (1 + \delta)\mathbf{E}[e(Q)]] \leq \left(\frac{\exp(\delta)}{(1 + \delta)^{(1+\delta)}}\right)^{\mathbf{E}[e(Q)]}$$

In our case, $2n^{1/2-\epsilon} \leq 1 + \delta \leq 4n^{1/2-\epsilon}$, thus we get

$$\Pr[e(Q) \geq n^{1-\epsilon}] \leq \left(\frac{\exp(4n^{1/2-\epsilon})}{(2n^{1/2-\epsilon})^{2n^{1/2-\epsilon}}}\right)^{\sqrt{n}/4}$$

Let $f(n,\epsilon) = \left(\frac{\exp(n^{1/2-\epsilon})}{(2n^{1/2-\epsilon})(n^{1/2-\epsilon}/2)}\right)^{\sqrt{n}}$. The number of sets of size $\lfloor\sqrt{n}\rfloor$ is given by $\binom{n}{\sqrt{n}} \leq (ne/\lfloor\sqrt{n}\rfloor)^{\sqrt{n}} = (\lceil\sqrt{n}\rceil e)^{\sqrt{n}}$. The probability that there is no subset of size $\lfloor\sqrt{n}\rfloor$ that has at least $n^{1-\epsilon}$ edges is given by the union-bound as follows

$$f(n,\epsilon)\binom{n}{\sqrt{n}} \ll 1$$

The number of edges with at least one end point in Q is given by

$$t(Q) = deg(Q) - e(Q) \geq \lfloor\sqrt{n}\rfloor \cdot \sqrt{n}(1 - o(1)) - n^{1-\epsilon} = n(1 - o(1))$$

On the other hand, consider the fractional solution in which $x_v = 1/\sqrt{n}$, for each v and $y_e = 1/\sqrt{n}$, for each $e \in E$. This LP solution is feasible and has a cost of $|E|/\sqrt{n}$. The number of edges $|E| = n\sqrt{n}/2(1 + o(1))$. Thus the cost of the LP solution is at most $n(1 + o(1))/2$, which results in a gap of $2(1 - o(1))$.

4   An Approximation for Fixed Cost Minimum Edge Cover Implies the Same Approximation for Maximum Weight m'-Edge Cover

We first transform the input instance for the MWEC problem to one in which the optimum value of the objective function is at most n^5 by paying a very small penalty in the approximation ratio.

Lemma 5. *For the Maximum weight m'-subgraph problem, we can convert the input instance $\langle G, w, m'\rangle$, with an optimal solution denoted by OPT into an instance $\langle G', w', m'\rangle$, with optimal solution OPT'', such that $OPT'' \leq n^5$. Furthermore, if OPT' is the total weight of the vertices in OPT'' under the weight function w, then*

$$OPT' \geq OPT(1 - 1/n)(1 - 1/n^2)$$

Proof. Let $v_1, v_2, \ldots, v_n$ be the vertices in G such that $w(v_1) \geq w(v_2) \geq \cdots \geq w(v_n)$. Let v_p be the last vertex in the ordering such that $w(v_p) \geq w(v_1)/n^2$. In other words, for each j, $p < j \leq n$, $w(v_1) > n^2 w(v_j)$. Let G' is the graph induced on vertices $v_1, v_2, \ldots, v_p$. Let OPT_1 be the optimal solution for the instance $\langle G', w, m'\rangle$. Note that OPT may choose some vertices from the set $\{v_{p+1}, v_{p+1}, \ldots, v_n\}$. The error incurred in not considering these vertices is at most $n(w(v_1)/n^2) \leq OPT/n$. Thus we get $OPT_1 \geq OPT(1 - 1/n)$. We now scale the weights of vertices in G' to create an instance $\langle G', w', m'\rangle$, where

$$w'(v_j) = \left\lfloor\left(\frac{w(v_j)}{w(v_p)}\right)n^2\right\rfloor$$

Let OPT'' be an optimal solution to $\langle G', w', m'\rangle$. Clearly, $OPT'' \leq n^5$. Let OPT' be the cost of the solution OPT'' under the weight function w, i.e., $OPT' = \sum_{v\in OPT''} w(v)$. Thus we have

$$OPT' \geq OPT_1\left(1 - \frac{1}{n^2}\right) \geq OPT\left(1 - \frac{1}{n}\right)\left(1 - \frac{1}{n^2}\right)$$

Theorem 3. *For some constant α, an α approximation guarantee for* FCEC *implies an $\alpha(1 + o(1))$ approximation guarantee for* MWEC.

Proof. Suppose that we have an $\alpha > 1$ approximation algorithm for FCEC, for some constant α. Using Lemma 5, we transform the MWEC instance (G, m') with an optimal weight W^* to an instance in which the optimum weight $W^* \leq n^4$. This increase the approximation ratio by a factor of only $(1 + o(1))$. We now consider the modified instance (G', m') as an input to FCEC. We guess the value of W^* by trying all possible integral values between 1 and n^4. For each guess of W^*, we apply the α-approximation algorithm for FCEC to the new instance. When our guess W^* is correct and we apply the algorithm, we obtain a set U of vertices of cost at least W^* and that touch at most $\alpha \cdot m'$ edges.

Create a new set B in which every vertex from U is chosen with a probability $1/\alpha$. We say that an edge e is *deleted* if $e \notin E(B)$. Let τ be a constant.

We consider the following "bad" events: (i) $w(B) \leq W^*/((1 + \tau)\alpha)$, (ii) $t(B) > m'$.

We first bound the probability that $w(B) \leq W^*/((1+\tau)\alpha)$. The expected cost of B is $w(U)/\alpha = W^*/\alpha$. Consider the expected cost of $U \setminus B$. The expected cost is $W^* - W^*/\alpha$. The event that $w(B) \leq W^*/(\alpha(1+\tau)))$ is equivalent to the event $w(U) - w(B) \geq W^* - W^*/(\alpha(1 + \tau)) = W^*(1 - 1/(\rho(1 + \tau)))$. By the Markov's inequality, the last event has probability at most $(1 - 1/\alpha)/(1 - 1/(\alpha(1 + \tau))) = 1 - \tau/(\alpha + \alpha \cdot \tau - 1)$.

We now bound the probability of the second bad event. The expected number of edges in $E(B)$ is at most $m'(1 - (1 - \frac{1}{\alpha})^2)$. Note that the events that edges are deleted are positively correlated because given that an edge (v, u) is deleted, one of the possibilities that can cause this event, is that v is deleted, and in that case all edges of v are deleted with probability 1. Clearly, we can assume that $m' \geq c$ for any constant c. Otherwise, we can solve the MWEC problem in polynomial time by checking all subsets of edges. By the Chernoff bound, the probability that the number of edges is more than m' is bounded by $exp(-c\delta^2/2)$, for some $\delta < 1$. We can choose a large enough c so that the above probability is at most $\tau/(2(\alpha + \alpha \cdot \tau - 1))$. This would mean that the sum of probabilities of bad events is strictly smaller than 1. This construction can be derandomized by the method of conditional expectations.

5   Exact Algorithm for the Degrees Density Augmentation Problem

The Degrees density augmentation problem is as follows: Given a graph $G = (V, E)$ and a subset $U \subseteq V$, the objective is to find a subset $W \subseteq V \setminus U$ such that

$$\rho = \frac{e(W) + e(U, W)}{deg(W)} \text{ is maximized}$$

The Degrees density augmentation problem is related to the FCEC problem in the same way as the Densest subgraph problem is related to the Dense k-subgraph problem. A natural heuristic for the FCEC problem would be to iteratively find a set W with good augmentation degrees density. A polynomial time exact solution for the problem using linear programming is given in [4]. Here we present a combinatorial algorithm.

We solve the **Degrees density augmentation** problem exactly by finding minimum s-t cut in the flow network constructed as follows. Let $\overline{U}$ denote the set $V \setminus U$. In addition to the source s and the sink t, the vertex set contains $V_{E'} \cup \overline{U}$, where $V_{E'} = \{v_e \mid e \in E$ and both end points of e are in $\overline{U}\}$. There is an edge from s to every vertex in $V_{E'} \cup \overline{U}$. If a is a vertex in $V_{E'}$ then the capacity of the edge (s, a) is 1, otherwise, the capacity of the edge is $deg_U(a)$. For each vertex $v_e \in V_{E'}$, where $e = (p, q)$, there are edges (v_e, p) and (v_e, q). Each such edge has a large capacity of $M = \infty$ (any capacity of at least n^5 would work). Finally, each vertex $p \in \overline{U}$ is connected to t and has a capacity of $\rho \cdot deg(p)$.

5.1 Algorithm

For a particular value of ρ, let $W_s \subseteq \overline{U}$ be the vertices that are on the $s(t)$ side of a minimum s-t cut. Let $V_{E'}^s \subseteq V_{E'} (V_{E'}^t \subseteq V_{E'})$ be the vertices in $V_{E'}$ that are on the $s(t)$ side of the minimum s-t cut. We now state the algorithm.

1. Construct the flow network as shown above.
2. For each value of ρ, compute a minimum s-t cut and find the resulting value of $e(W_s) + e(U, W_s) - \rho deg(W_s)$. Find the largest value of ρ for which the expression is at least 0.
3. Return W_s corresponding to the largest value of ρ.

Due to space constraints, for the analysis of the above algorithm, the reader may refer to the full version of the paper.

Acknowledgements. We thank V. Chakravarthy for introducing the **FCEC** problem to us. We also thank V. Chakravarthy and S. Roy for useful discussions. Thanks also to U. Feige for bringing the Small Set Expansion Conjecture to our attention.

References

1. Andersen, R., Chellapilla, K.: Finding dense subgraphs with size bounds. In: Avrachenkov, K., Donato, D., Litvak, N. (eds.) WAW 2009. LNCS, vol. 5427, pp. 25–37. Springer, Heidelberg (2009)
2. Bhaskara, A., Charikar, M., Chlamtac, E., Feige, U., Vijayaraghavan, A: Detecting high log-densities: an $O(n^{1/4})$ approximation for densest k-subgraph. In: STOC, pp. 201–210 (2010)
3. Carnes, T., Shmoys, D.: Primal-dual schema for capacitated covering problems. In: Lodi, A., Panconesi, A., Rinaldi, G. (eds.) IPCO 2008. LNCS, vol. 5035, pp. 288–302. Springer, Heidelberg (2008)
4. Chakravarthy, V.T., Modani, N., Natarajan, S.R, Roy, S., Sabharwal, Y.: Density functions subject to co-matroid constraint. In: Foundations of Software Technology and Theoretical Computer Science, pp. 236–248 (2012)
5. Feige, U., Kortsarz, G., Peleg, D.: The dense k-subgraph problem. Algorithmica **29**(3), 410–421 (2001)

6. Gajewar, A., Sarma, A.: Multi-skill collaborative teams based on densest subgraphs. In: SDM, pp. 165–176 (2012)
7. Goldberg, A.V.: Finding a maximum density subgraph. Technical report UCB/CSD-84-171, EECS Department, University of California, Berkeley (1984)
8. Goldschmidt, O., Hochbaum, D.S.: k-edge subgraph problems. Discrete Appl. Math. **74**(2), 159–169 (1997)
9. Khot, S.: Ruling out ptas for graph min-bisection, dense k-subgraph, and bipartite clique. SIAM J. Comput. **36**(4), 1025–1071 (2006)
10. Khuller, S., Saha, B.: On finding dense subgraphs. In: Albers, S., Marchetti-Spaccamela, A., Matias, Y., Nikoletseas, S., Thomas, W. (eds.) ICALP 2009, Part I. LNCS, vol. 5555, pp. 597–608. Springer, Heidelberg (2009)
11. Lawler, Eugene L.: Combinatorial optimization: networks and matroids. Holt, Rinehart and Winston, New York (1976)
12. Liang, H.: On the k-edge-incident subgraph problem and its variants. Discrete Appl. Math. **161**(18), 2985–2991 (2013)
13. Marx, D.: Parameterized complexity and approximation algorithms. Comput. J. **51**(1), 60–78 (2008)
14. Raghavendra, P., Steurer, D.: Graph expansion and the unique games conjecture. In: STOC, pp. 755–764 (2010)

Hadwiger Number of Graphs with Small Chordality

Petr A. Golovach[1]([⊠]), Pinar Heggernes[1], Pim van 't Hof[1],
and Christophe Paul[2]

[1] Department of Informatics, University of Bergen, Bergen, Norway
{petr.golovach,pinar.heggernes,pim.vanthof}@ii.uib.no
[2] CNRS, LIRMM, Montpellier, France
paul@lirmm.fr

Abstract. The Hadwiger number of a graph G is the largest integer h such that G has the complete graph K_h as a minor. We show that the problem of determining the Hadwiger number of a graph is NP-hard on co-bipartite graphs, but can be solved in polynomial time on cographs and on bipartite permutation graphs. We also consider a natural generalization of this problem that asks for the largest integer h such that G has a minor with h vertices and diameter at most s. We show that this problem can be solved in polynomial time on AT-free graphs when $s \geq 2$, but is NP-hard on chordal graphs for every fixed $s \geq 2$.

1   Introduction

The Hadwiger number of a graph G, denoted by $h(G)$, is the largest integer h such that the complete graph K_h is a minor of G. The Hadwiger number has been the subject of intensive study, not in the least due to a famous conjecture by Hugo Hadwiger from 1943 [8] stating that the Hadwiger number of any graph is greater than or equal to its chromatic number. In a 1980 paper, Bollobás et al. [2] called Hadwiger's conjecture "one of the deepest unsolved problems in graph theory." Despite many partial results the conjecture remains wide open more than 70 years after it first appeared in the literature.

Given the vast amount of graph-theoretic results involving the Hadwiger number, it is natural to study the computational complexity of the HADWIGER NUMBER problem, which is to decide, given an n-vertex graph G and an integer h, whether the Hadwiger number of G is greater than or equal to h (or, equivalently, whether G has K_h as a minor). Rather surprisingly, it was not until 2009 that this problem was shown to be NP-complete by Eppstein [6]. Two years earlier, Alon et al. [1] observed that the problem is fixed-parameter tractable when parameterized by h due to deep results by Robertson and Seymour [10]. This shows that the problem of determining the Hadwiger number of a graph is in

The research leading to these results has received funding from the Research Council of Norway and the European Research Council under the European Union's Seventh Framework Programme (FP/2007–2013)/ERC Grant Agreement n. 267959.

© Springer International Publishing Switzerland 2014
D. Kratsch and I. Todinca (Eds.): WG 2014, LNCS 8747, pp. 201–213, 2014.
DOI: 10.1007/978-3-319-12340-0_17

some sense easier than the closely related problem of determining the clique number of a graph, as the decision version of the latter problem is W[1]-hard when parameterized by the size of the clique. Alon et al. [1] showed that the same holds from an approximation point of view: they provided a polynomial-time approximation algorithm for the HADWIGER NUMBER problem with approximation ratio $O(\sqrt{n})$, contrasting the fact that it is NP-hard to approximate the clique number of an n-vertex graph in polynomial time to within a factor better than $n^{1-\epsilon}$ for any $\epsilon > 0$ [13].

Bollobás et al. [2] referred to the Hadwiger number as the *contraction clique number*. This is motivated by the observation that for any integer h, a connected graph G has K_h as a minor if and only if G has K_h as a contraction. In this context, it is worth mentioning another problem that has recently attracted some attention from the parameterized complexity community. The CLIQUE CONTRACTION problem takes as input an n-vertex graph G and an integer k, and asks whether G can be modified into a complete graph by a sequence of at most k edge contractions. Since every edge contraction reduces the number of vertices by exactly 1, it holds that (G, k) is a yes-instance of the CLIQUE CONTRACTION problem if and only if G has the complete graph K_{n-k} as a contraction (or, equivalently, as a minor). Therefore, the CLIQUE CONTRACTION problem can be seen as the parametric dual of the HADWIGER NUMBER problem, and is NP-complete on general graphs. When parameterized by k, the CLIQUE CONTRACTION problem was recently shown to be fixed-parameter tractable [4,9], but the problem does not admit a polynomial kernel unless NP $\subseteq$ coNP/poly [4].

In this paper, we study the computational complexity of the HADWIGER NUMBER problem on several graph classes of bounded chordality. For chordal graphs, which form an important subclass of 4-chordal graphs, the HADWIGER NUMBER problem is easily seen to be equivalent to the problem of finding a maximum clique, and can therefore be solved in linear time on this class [12]. In Sect. 3, we present polynomial-time algorithms for solving the HADWIGER NUMBER problem on two other well-known subclasses of 4-chordal graphs: cographs and bipartite permutation graphs. We also prove that the problem remains NP-complete on co-bipartite graphs, and hence on 4-chordal graphs. The latter result implies that the problem is also NP-complete on AT-free graphs, a common superclass of cographs and bipartite permutation graphs.

In Sect. 4, we consider a natural generalization of the HADWIGER NUMBER problem, and provide additional results about finding large minors of bounded diameter. We show that the problem of determining the largest integer h such that a graph G has a minor with h vertices and diameter at most s can be solved in polynomial time on AT-free graphs if $s \geq 2$. In contrast, we show that this problem is NP-hard on chordal graphs for every fixed $s \geq 2$, and remains NP-hard for $s = 2$ even when restricted to split graphs. Observe that when $s = 1$, the problem is equivalent to the HADWIGER NUMBER problem and thus NP-hard on AT-free graphs and linear-time solvable on chordal graphs due to our aforementioned results.

Due to space restrictions, proofs are either omitted or just sketched in this extended abstract. The full version of the paper is available at [5].

2 Preliminaries

We consider finite undirected graphs without loops or multiple edges. For each of the graph problems considered in this paper, we let $n = |V(G)|$ and $m = |E(G)|$ denote the number of vertices and edges, respectively, of the input graph G. For a graph G and a subset $U \subseteq V(G)$ of vertices, we write $G[U]$ to denote the subgraph of G induced by U. We write $G - U$ to denote the subgraph of G induced by $V(G) \setminus U$, and $G - u$ if $U = \{u\}$. For a vertex v, we denote by $N_G(v)$ the set of vertices that are adjacent to v in G. The *distance* $\text{dist}_G(u, v)$ between vertices u and v of G is the number of edges on a shortest path between them. The *diameter* $\text{diam}(G)$ of G is $\max\{\text{dist}_G(u, v) \mid u, v \in V(G)\}$. The *complement* of G is the graph $\overline{G}$ with vertex set $V(G)$, where two distinct vertices are adjacent in $\overline{G}$ if and only if they are not adjacent in G. For two disjoint vertex sets $X, Y \subseteq V(G)$, we say that X and Y are *adjacent* if there are $x \in X$ and $y \in Y$ that are adjacent in G.

We say that P is a (u, v)-*path* if P is a path that joins u and v. The vertices of P different from u and v are the *inner* vertices of P. We denote by P_n and C_n the path and the cycle on n vertices respectively. The *length* of a path is the number of edges in the path. A set of pairwise adjacent vertices is a *clique*. A *matching* is a set M of edges such that no two edges in M share an end-vertex. A vertex incident to an edge of a matching M is said to be *saturated* by M. We write K_n to denote the *complete* graph on n vertices, i.e., graph whose vertex set is a clique. For two integers $a \leq b$, the *(integer) interval* $[a, b]$ is defined as $[a, b] = \{i \in \mathbb{Z} \mid a \leq i \leq b\}$. If $a > b$, then $[a, b] = \emptyset$.

The *chordality* $\text{chord}(G)$ of a graph G is the length of a longest induced cycle in G; if G has no cycles, then $\text{chord}(G) = 0$. For a non-negative integer k, a graph G is k-*chordal* if $\text{chord}(G) \leq k$. A graph is *chordal* if it is 3-chordal. A graph is *chordal bipartite* if it is both 4-chordal and bipartite. A graph is a *split* graph if its vertex set can be partitioned in an independent set and a clique. For a graph F, we say that a graph G is F-*free* if G does not contain F as an induced subgraph. A graph is a *cograph* if it is P_4-free. Let σ be a permutation of $\{1, \ldots, n\}$. A graph G is said to be a *permutation graph for* σ if G has vertex set $\{1, \ldots, n\}$ and two vertices i, j are adjacent if and only if i, j are reversed by the permutation. A graph G is a *permutation* graph if G is a permutation graph for some σ. A graph is a *bipartite permutation* graph if it is bipartite and permutation. An *asteroidal triple (AT)* is a set of three non-adjacent vertices such that between each pair of them there is a path that does not contain a neighbor of the third. A graph is *AT-free* if it contains no AT. Each of the above-mentioned graph classes can be recognized in polynomial (in most cases linear) time, and they are closed under taking induced subgraphs [3,7]. See the monographs by Brandstädt et al. [3] and Golumbic [7] for more properties and characterizations of these classes and their inclusion relationships.

Minors, Induced Minors, and Contractions. Let G be a graph and let $e \in E(G)$. The *contraction* of e removes both end-vertices of e and replaces them by a new vertex adjacent to precisely those vertices to which the two end-vertices were adjacent. We denote by G/e the graph obtained from G be the contraction of e. For a set of edges S, G/S is the graph obtained from G by the contraction of all edges of S. A graph H is a *contraction* of G if $H = G/S$ for some $S \subseteq E(G)$. We say that G is *k-contractible* to H if $H = G/S$ for some set $S \subseteq E(G)$ with $|S| \leq k$. A graph H is an *induced minor* of G if a H is a contraction of an induced subgraph of G. Equivalently, H is an induced minor of G if H can be obtained from G by a sequence of vertex deletions and edge contractions. A graph H is a *minor* of a graph G if H is a contraction of a subgraph of G. Equivalently, H is a minor of G if H can be obtained from G by a sequence of vertex deletions, edge deletions, and edge contractions.

Let G and H be two graphs. An *H-witness structure* $\mathcal{W}$ of G is a partition $\{W(x) \mid x \in V(H)\}$ of the vertex set of a (not necessarily proper) subgraph of G into $|V(H)|$ sets called *bags*, such that the following two conditions hold:

(i) each bag $W(x)$ induces a connected subgraph of G;
(ii) for all $x, y \in V(H)$ with $xy \in E(H)$, bags $W(x)$ and $W(y)$ are adjacent in G.

In addition, we may require an H-witness structure to satisfy one or both of the following additional conditions:

(iii) for all $x, y \in V(H)$ with $xy \notin E(H)$, bags $W(x)$ and $W(y)$ are not adjacent in G;
(iv) every vertex of G belongs to some bag.

By contracting each of the bags into a single vertex we observe that H is a contraction, an induced minor, or a minor of G if and only if G has an H-witness structure $\mathcal{W}$ that satisfies conditions (i)–(iv), (i)–(iii), or (i)–(ii), respectively. We will refer to such a structure $\mathcal{W}$ as an *H-contraction structure*, an *H-induced minor structure*, and an *H-minor structure*, respectively. Observe that, in general, such a structure $\mathcal{W}$ is not uniquely defined.

Let $\mathcal{W}$ be an H-witness structure of G, and let $W(x)$ be a bag of $\mathcal{W}$. We say that $W(x)$ is a *singleton* if $|W(x)| = 1$ and $W(x)$ is an *edge-bag* if $|W(x)| = 2$. We say that $W(x)$ is a *big bag* if $|W(x)| \geq 2$.

We conclude this section by presenting three structural lemmas that will be used in the polynomial-time algorithms presented in Sect. 3. The first lemma readily follows from the definitions of a minor, an induced minor, and a contraction.

Lemma 1. *For every connected graph G and non-negative integer p, the following statements are equivalent:*

- *G has K_p as a contraction;*
- *G has K_p as an induced minor;*
- *G has K_p as a minor.*

We say that an H-induced minor structure $\mathcal{W} = \{W(x) \mid x \in V(H)\}$ is *minimal* if there is no H-induced minor structure $\mathcal{W}' = \{W'(x) \mid x \in V(H)\}$ with $W'(x) \subseteq W(x)$ for every $x \in V(H)$ such that at least one inclusion is proper.

Lemma 2. *For any minimal K_p-induced minor structure of a graph G, each bag induces a subgraph of diameter at most $\max\{\text{chord}(G) - 3, 0\}$.*

Note that Lemma 2 immediately implies the aforementioned equivalence on chordal graphs between the HADWIGER NUMBER problem and the problem of finding a maximum clique. Lemma 2 also implies the following result.

Corollary 1. *If G is a graph of chordality at most 4, then for any minimal K_p-induced minor structure in G, each bag is a clique.*

We say that a K_p-induced minor structure is *nice* if each bag is either a singleton or an edge-bag.

Lemma 3. *Let G be a $\overline{C}_6$-free graph of chordality at most 4. If K_p is an induced minor of G, then G has a nice K_p-induced minor structure.*

3   Computing the Hadwiger Number

First, we show that HADWIGER NUMBER problem can be solved in polynomial time on bipartite permutation graphs.

Let us for a moment consider the class of chordal bipartite graphs. Recall that these are exactly the bipartite graphs that have chordality at most 4. It is well-known that chordal bipartite graphs form a proper superclass of the class of bipartite permutation graphs. Since chordal bipartite graphs have chordality at most 4 and are $\overline{C}_6$-free due to the absence of triangles, we can apply Lemma 3 to this class. Let us additionally observe that the number of singletons in any K_p-induced minor structure of a bipartite graph is at most 2.

The above observations allow us to reduce the HADWIGER NUMBER problem on chordal bipartite graphs to a special matching problem as follows. We say that a matching M in a graph G is a *clique-matching* if for any two distinct edges $e_1, e_2 \in M$, there is an edge in G between an end-vertex of e_1 and an end-vertex of e_2. Now consider the following decision problem:

CLIQUE-MATCHING
Instance: A graph G and a positive integer k.
Question: Is there a clique-matching of size at least k in G?

Lemma 4. *If the CLIQUE-MATCHING problem can be solved in $f(n, m)$ time on chordal bipartite graphs, then the HADWIGER NUMBER problem can be solved in $O((n + m) \cdot f(n, m))$ time on this graph class.*

We will use the following characterization of bipartite permutation graphs given by Spinrad et al. [11] (see also [3]). Let G be a bipartite graph and let V_1, V_2 be a bipartition of $V(G)$. An ordering of vertices of V_2 has the *adjacency property* if for every $u \in V_1$, $N_G(u)$ consists of vertices which are consecutive in the ordering of V_2. An ordering of vertices of V_2 has the *enclosure property* if for every pair of vertices $u, v \in V_1$ such that $N_G(u) \subseteq N_G(v)$, vertices in $N_G(v) \setminus N_G(u)$ occur consecutively in the ordering of V_2.

Lemma 5 [11]. *Let G be a bipartite graph with bipartition V_1, V_2. The graph G is a bipartite permutation graph if and only there is an ordering of V_2 that has the adjacency and enclosure properties. Moreover, bipartite permutation graphs can be recognized and the corresponding ordering of V_2 can be constructed in linear time.*

Theorem 1. *The CLIQUE-MATCHING problem can be solved in $O(mn^4)$ time on bipartite permutation graphs.*

Proof. Let G be a bipartite permutation graph and let V_1, V_2 be a bipartition of the vertex set. We assume without loss of generality that G has no isolated vertices. Let $n_1 = |V_1|$ and $n_2 = |V_2|$. We present a dynamic programming algorithm for the problem. For simplicity, the algorithm we describe only finds the size of a maximum clique-matching M in G, but the algorithm can be modified to find a corresponding clique-matching as well.

Our algorithm starts by constructing an ordering σ_2 of V_2 that has the adjacency and enclosure properties, which can be done in linear time due to Lemma 5. From now on, we denote the vertices of V_2 by their respective rank in σ_2, that is $V_2 = \{1, \ldots, n_2\}$. Observe that for every vertex $u \in V_1$, $N_G(u)$ forms an interval of σ_2. The *rightmost* (resp. *leftmost*) neighbor of u in σ_2 is the vertex of $N_G(u)$ which is the largest (resp. smallest) in σ_2.

Let $uv \in E(G)$ with $u \in V_1$ and $v \in V_2$ be an edge in G such that uv belongs to some maximum clique-matching in G and there is no $v' \in V_2$ with $v' < v$ such that v' is saturated by a maximum clique-matching in G. Our algorithm guesses the edge uv by trying all different edges of G. For each guess of uv, it does as follows.

By the definition of uv, we can safely delete all vertices $v' \in V_2$ with $v' < v$. To simplify notation, we assume without loss of generality that $v = 1$, so $uv = u1$. Denote by r the rightmost neighbor of u. Then, by the adjacency property of σ_2, we have that $N_G(u) = [1, r]$.

The algorithm now performs the following preprocessing procedure.

- Find the vertices $v_1, \ldots, v_l \in V_1 \setminus \{u\}$ (decreasingly ordered with respect to their rightmost neighbor) such that $[1, r] \subseteq N_G(v_i)$. By consecutively checking the intervals $N_G(v_1), \ldots, N_G(v_l)$ and selecting the rightmost available (i.e., not selected before) vertex in the considered interval, find the maximum set $S = \{j_1, \ldots, j_h\}$ of integers such that $j_1 > \ldots > j_h > r$ and $j_i \in N_G(v_i)$ for $i \in \{1, \ldots, h\}$. Delete $v_1, \ldots, v_h$ from G.

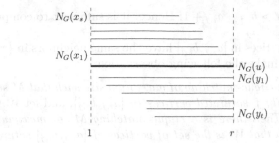

Fig. 1. Structure of the neighborhoods of u, $x_1, \ldots, x_s$ and $y_1, \ldots, y_t$ after the pre-processing procedure.

- Find the vertices $x_1, \ldots, x_s \in V_1 \setminus \{u\}$ (decreasingly ordered with respect to their rightmost neighbor) such that $[1, 2] \subseteq N_G(x_i)$.
- Find the vertices $y_1, \ldots, y_t \in V_1$ (increasingly ordered with respect to their leftmost neighbor) such that $1 \notin N_G(y_i)$ and $r \in N_G(y_i)$.
- Delete the vertices $r + 1, \ldots, n_2$ from V_2.

The structure of the neighborhoods of u, $x_1, \ldots, x_s$ and $y_1, \ldots, y_t$ after this preprocessing procedure is shown in Fig 1.

We prove that the preprocessing procedure is safe in the following claim.

Claim 1. *Let M be a clique-matching of maximum size in G such that $u1 \in M$. Then there is a clique-matching M' of maximum size such that $u1 \in M'$ and*

(i) $v_1 j_1, \ldots, v_h j_h \in M'$,
(ii) for any $vj \in M'$ such that $vj \neq u1$ and $v \notin \{v_1, \ldots v_h\}$, it holds that $v \in \{x_1, \ldots, x_s\} \cup \{y_1, \ldots, y_t\}$ and $j \in [2, r]$.

In the next stage of the algorithm we apply dynamic programming. For every $i \in \{0, \ldots, s\}$, $j \in \{0, \ldots, t\}$ and non-negative integer ℓ, let $c(i, j, \ell)$ denote the size of a maximum clique-matching M such that

(a) $u1 \in M$,
(b) for any $vp \in M$ such that $vp \neq u1$, it holds that $v \in \{x_1, \ldots, x_i\} \cup \{y_1, \ldots, y_j\}$, and
(c) there are at most ℓ vertices in $[a_{i,j}, b_{i,j}] = (\bigcap_{p=1}^{i} N_G(x_p)) \cap (\bigcap_{q=1}^{j} N_G(y_q))$ saturated by M.

Recall that the vertices of X and Y are ordered with respect to their rightmost and leftmost neighbors, respectively. Hence, for any $1 \leq p < q \leq i$, we have $1 \in N_G(x_q) \subseteq N_G(x_p) \subseteq [1, r]$, and for any $1 \leq p < q \leq j$, we have $1 \notin N_G(y_q) \subseteq N_G(y_p) \subseteq [2, r]$. In particular, $[a_{i,j}, b_{i,j}] = N_G(x_i) \cap N_G(y_j)$ for $i, j > 0$. In other words, if $[a_{i,j}, b_{i,j}] \neq \emptyset$, then $a_{i,j}$ is the left end-point of the interval $N_G(y_j)$ and $b_{i,j}$ is the right end-point of the interval $N_G(x_j)$. Observe that it can happen that $[a_{i,j}, b_{i,j}] = \emptyset$. Observe also that $c(i, j, \ell) = c(i, j, b_{i,j} - a_{i,j} + 1)$

if $[a_{i,j}, b_{i,j}] \neq \emptyset$ and $\ell > b_{i,j} - a_{i,j} + 1$. Hence, it is sufficient to compute $c(i, j, \ell)$ for $\ell \leq b_{i,j} - a_{i,j} + 1 \leq n_2$.

Because all the vertices in $[a_{i,j}, b_{i,j}]$ have the same neighbors in $\{x_1, \ldots, x_i\} \cup \{y_1, \ldots, y_j\}$, we can make the following observation.

Claim 2. *Let M be a clique-matching of maximum size such that M satisfies (a)–(c) and M has exactly f saturated vertices in $[a_{i,j}, b_{i,j}]$, and let $W \subseteq [a_{i,j}, b_{i,j}]$ be a set of size f. Then there is a clique-matching M' of maximum size that satisfies (a)–(c) such that W is the set of vertices of $[a_{i,j}, b_{i,j}]$ saturated by M'.*

If $i = j = 0$, then we set $c(i, j, \ell) = 1$ taking into account the matching with the unique edge $u1$. For other values of i, j, $c(i, j, \ell)$ is computed as follows. To simplify notation, we assume that $x_0 = y_0 = u$.

Computation of $c(i, j, \ell)$ for $i > 0, j = 0$. Because $1 \in N_G(x_q) \subseteq N_G(x_p) \subseteq [1, r]$ for every $1 \leq p < q \leq i$, any matching with edges incident to $x_1, \ldots, x_i$ is a clique-matching. This observation also implies that a maximum matching can be obtained in greedy way. Notice that $[a_{i,0}, b_{i,0}] = N_G(x_i)$. By consecutively checking the intervals $N_G(x_1), \ldots, N_G(x_i)$ and selecting the rightmost available (i.e., not selected before) vertex in the considered interval, we find the maximum set $\{p_1, \ldots, p_q\}$ of integers such that $t \geq p_1 > \ldots > p_q > 1$, $p_f \in N_G(x_f)$ for $f \in \{1, \ldots, q\}$, and $|\{p_1, \ldots, p_q\} \cap [a_{i,0}, b_{i,0}]| \leq \ell - 1$. Taking into account the edge $u1$, we observe that $M = \{u1, x_1p_1, \ldots, x_qp_q\}$ is a required matching, and we have that $c(i, j, \ell) = q + 1$.

Computation of $c(i, j, \ell)$ for $i = 0, j > 0$. Now we have that $r \in N_G(y_q) \subseteq N_G(y_p) \subseteq [2, r]$ for every $1 \leq p < q \leq j$. Hence, any matching with edges incident to $y_1, \ldots, y_j$ is a clique-matching and a maximum matching can be obtained in greedy way. Notice that $[a_{0,j}, b_{0,j}] = N_G(y_j)$. By consecutively checking the intervals $N_G(y_1), \ldots, N_G(y_j)$ and selecting the leftmost available (i.e., not selected before) vertex in the considered interval, we find the maximum set $\{p_1, \ldots, p_q\}$ of integers such that $1 < p_1 < \ldots < p_q \leq r$, $p_f \in N_G(y_f)$ for $f \in \{1, \ldots, q\}$, and $|\{p_1, \ldots, p_q\} \cap [a_{0,j}, b_{0,j}]| \leq \ell$. It is straightforward to see that $M = \{u1, y_1p_1, \ldots, y_qp_q\}$ is a required matching, and we have that $c(i, j, \ell) = q + 1$.

Computation of $c(i, j, \ell)$ for $i > 0, j > 0$. We compute $c(i, j, \ell)$ using the tables of already computed values $c(i - 1, j', \ell')$ for $j' \leq j$. We find the size of a maximum clique-matching M by considering all possible choices for the vertex x_i and then take the maximum among the obtained values. We distinguish three cases. Recall that $[a_{i,j}, b_{i,j}] = N_G(x_i) \cap N_G(y_j)$.

Case 1. The vertex x_i is not saturated by M. We have that $[a_{i-1,j}, b_{i-1,j}] = N_G(x_{i-1}) \cap N_G(y_j) \subseteq [a_{i,j}, b_{i,j}]$ and $|[a_{i,j}, b_{i,j}] \setminus [a_{i-1,j}, b_{i-1,j}]| = b_{i-1,j} - b_{i,j}$. By Claim 2 implies that for any maximum clique-matching M that satisfies (a)–(c) and has no edge incident to x_i, it holds that a clique-matching M' of maximum size that satisfies (a)–(b), has no edge incident to x_i, and has at most $\ell' = \ell + b_{i-1,j} - b_{i,j}$ saturated vertices in $[a_{i-1,j}, b_{i-1,j}]$ has the same size as M. Hence $c(i, j, \ell) = c(i - 1, j, \ell')$.

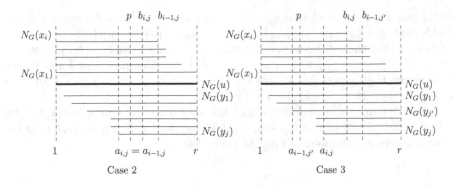

Fig. 2. Structure of the neighborhoods of u, $x_1, \ldots, x_i$ and $y_1, \ldots, y_j$ in Cases 2 and 3.

Now we consider the cases when x_i is saturated by M. Denote by $p \in N_G(x_i)$ the vertex such that $x_i p \in M$.

Case 2. Vertex $p \in [a_{i,j}, b_{i,j}]$ (see Fig. 2). Observe that p is adjacent to every vertex in $\{x_1, \ldots, x_{i-1}\} \cup \{y_1, \ldots, y_j\}$. Hence, for any edge vq such that $v \in \{u\} \cup \{x_1, \ldots, x_{i-1}\} \cup \{y_1, \ldots, y_j\}$ and $q \neq p$, $x_i p$ and vq have adjacent end-vertices, i.e., this choice of p does not influence the selection of other edges of M except that we can have at most $\ell - 1$ other saturated vertices in $[a_{i,j}, b_{i,j}]$. We have that $[a_{i-1,j}, b_{i-1,j}] = N_G(x_{i-1}) \cap N_G(y_j) \subseteq [a_{i,j}, b_{i,j}]$ and $|[a_{i,j}, b_{i,j}] \setminus [a_{i-1,j}, b_{i-1,j}]| = b_{i-1,j} - b_{i,j}$. By Claim 2, we obtain that for any maximum clique-matching M that satisfies (a)–(c) and $x_i p \in M$, a clique-matching M' of maximum size that satisfies (a)–(b), has no edge incident to x_i and has at most $\ell' = \ell + b_{i-1,j} - b_{i,j} - 1$ saturated vertices in $[a_{i-1,j}, b_{i-1,j}]$ has the same size as M. Hence $c(i, j, \ell) = c(i - 1, j, \ell')$.

Case 3. Vertex $p \notin [a_{i,j}, b_{i,j}]$, i.e., $p < a_{i,j}$ (see Fig. 2). Let $j' = \max\{f \mid p \in N_G(y_f), 0 \leq f \leq j\}$. As $p < a_{i,j}$, it holds that $j' < j$.

Let $f \in \{j' + 1, \ldots, j\}$, $g \in N_G(y_f)$ and $g > b_{i,j}$. Recall that $b_{i,j}$ is the right end-point of $N_G(x_i)$. Hence, $x_i g \notin E(G)$. Because $f > j'$, $x_f p \notin E(G)$. We conclude that such edges cannot be in M. Similarly, let $f \in \{j' + 1, \ldots, j\}$, $g \in N_G(y_f)$ and $g \leq b_{i,j}$. Then for any $v \in \{x_1, \ldots, x_i\} \cup \{y_1, \ldots, y_{j'}\}$, it holds that $vg \in E(G)$. Also if $j' + 1 \leq f < f' \leq j$, then for any $g \in N_G(x_{f'})$, $x_f g \in E(G)$. We have that it is safe to include in a clique-matching edges $x_f q$ for $f \in \{j' + 1, \ldots, j\}$, $g \in N_G(y_f)$ and $g \leq b_{i,j}$. We select such edges in a greedy way. By consecutively checking the intervals $N_G(y_{j'+1}), \ldots, N_G(y_j)$ and selecting the leftmost available (i.e., not selected before) vertex in the considered interval, we find the maximum set $\{g_1, \ldots, g_q\}$ of integers such that $p < g_1 < \ldots < g_q \leq b_{i,j}$, $g_f \in N_G(y_{f+j'})$ for $f \in \{1, \ldots, q\}$ and $|\{g_1, \ldots, g_q\} \cap [a_{i,j}, b_{i,j}]| \leq \ell$.

Claim 3. *Let M be a clique-matching of maximum size that satisfies (a)–(c) and $x_i p \in M$. Then there is a clique-matching M' of maximum size that satisfies* (a)–(c) and $x_i p \in M'$ such that $y_{j'+1} g_1, \ldots, y_{j'+q} g_q \in M'$ and for any $vf \in M'$, it holds that $v \in \{y_{j'+1}, \ldots, y_{j'+q}\} \cup \{x_1, \ldots, x_i\} \cup \{y_1, \ldots, y_{j'}\}$.

Observe that the total number of saturated vertices in $[a_{i-1,j'}, b_{i-1,j'}]$ should be at most $(a_{i,j} - a_{i-1,j'}) + (b_{i-1,j'} - b_{i,j}) + \ell$. Using Claims 2 and 3 and taking into account that $x_i p \in M$, we obtain that $c(i, j, \ell) = c(i - 1, j', \ell')$ for $\ell' = (a_{i,j} - a_{i-1,j'}) + (b_{i-1,j'} - b_{i,j}) + \ell - (q + 1)$.

By our dynamic programming algorithm we eventually compute $c(s, t, \ell)$ for $\ell = 0$ if $[a_{i,j}, b_{i,j}] = \emptyset$ or $\ell = b_{i,j} - a_{i,j} + 1$ if $[a_{i,j}, b_{i,j}] \neq \emptyset$. Then $c(s, t, \ell)$ is the size of a maximum clique-matching M such that

(a) $u1 \in M$,
(b) for any $yp \in M$ such that $vp \neq u1$, it holds that $v \in \{x_1, \ldots, x_i\} \cup \{y_1, \ldots, y_j\}$.

By Claim 1, the size of a maximum clique-matching M in G such that $u1 \in M$ is $c(s, t, \ell) + |S|$, where S is the set of vertices constructed during the preprocessing procedure. Recall that the algorithm tries all possible choices for the edge uv, implying that our algorithm indeed computes the size of a maximum clique-matching in G.

It remains to evaluate the running time to complete the proof. Constructing the ordering σ_2 of V_2 can be done in $O(n + m)$ time by Lemma 5. The algorithm considers m choices for the edge uv. For each of these choices, the preprocessing procedure can be performed in $O(n)$ time given the orderings of V_1 and V_2 (notice that Lemma 5 is symmetric with respect to V_1, V_2, so we can obtain an ordering of V_1 with the adjacency and enclosure properties, too). Each step of the dynamic programming can be done in $O(n^2)$ time using the orderings of V_1, V_2. Observe that in this time we can compute $c(i, j, \ell)$ for all values of ℓ. Hence, the dynamic programming algorithm runs in time $O(n^4)$. We conclude that the total running time is $O(mn^4)$. $\qquad \Box$

Combining Lemma 4 and Theorem 1 yields the following result.

Corollary 2. *The* HADWIGER NUMBER *problem can be solved in* $O((n + m) \cdot mn^4)$ *time on bipartite permutation graphs.*

We also show that the Hadwiger number of a cograph can be determined in polynomial time.

Theorem 2. *The* HADWIGER NUMBER *problem can be solved in* $O(n^3)$ *time on cographs.*

We complement the aforementioned algorithmic results by showing that the HADWIGER NUMBER problem is NP-complete on co-bipartite graphs, another well-known subclass of the class of 4-chordal graphs.

Theorem 3. *The* HADWIGER NUMBER *problem is* NP-*complete on co-bipartite graphs.*

4   Minors of Bounded Diameter

In this section, we consider a generalization of the HADWIGER NUMBER problem where the aim is to obtain a minor of bounded diameter. Let s be a positive integer. An s-club is a graph that has diameter at most s. We consider the following problem:

MAXIMUM s-CLUB MINOR

  Instance: A graph G and a non-negative integer h.
  Question: Does G have a minor with h vertices and diameter at most s?

When $s = 1$, the above problem is equivalent to the HADWIGER NUMBER problem. Recall that, due to Lemma 1, the HADWIGER NUMBER problem can be seen as the parametric dual of the CLIQUE CONTRACTION problem. The following straightforward lemma, which generalizes Lemma 1, will allow us to formulate the parametric dual of the MAXIMUM s-CLUB MINOR problem in a similar way.

Lemma 6. *For every connected graph G and non-negative integers p and s, the following statements are equivalent:*

- *G has a graph with p vertices and diameter at most s as a contraction;*
- *G has a graph with p vertices and diameter at most s as an induced minor;*
- *G has a graph with p vertices and diameter at most s as a minor.*

Lemma 6 implies that for any non-negative integer s, the parameteric dual of the MAXIMUM s-CLUB MINOR problem can be formulated as follows:

s-CLUB CONTRACTION

  Instance: A graph G and a positive integer k.
  Question: Does there exist a graph H with diameter at most s such that G
    is k-contractible to H?

Observe that 1-CLUB CONTRACTION is NP-complete on AT-free graphs as a result of Theorem 3. This is in stark contrast with our next result.

Theorem 4. *For any $s \geq 2$, the s-CLUB CONTRACTION problem can be solved in $O(m^4 n^3)$ time on AT-free graphs, even if s is given as a part of the input.*

On chordal graphs, the situation turns out to be opposite. Recall that the HADWIGER NUMBER problem, and hence the 1-CLUB CONTRACTION problem, can be solved in linear time on chordal graphs.

Theorem 5. *For any $s \geq 2$, the s-CLUB CONTRACTION problem on chordal graphs is NP-complete as well as W[2]-hard when parameterized by k. Moreover, 2-CLUB CONTRACTION is NP-complete and W[2]-hard when parameterized by k even on split graphs.*

5   Concluding Remarks

In Sect. 3, we showed that the HADWIGER NUMBER problem can be solved in polynomial time on cographs and on bipartite permutation graphs, respectively. A natural question is how far the results in those two sections can be extended to larger graph classes. An easy reduction from the HADWIGER NUMBER problem on general graphs, involving subdividing every edge of the input graph exactly once, implies that the problem is NP-complete on bipartite graphs. Since bipartite permutation graphs form exactly the intersection of bipartite graphs and permutation graphs, and the class of permutation graphs properly contains the class of cographs, our results naturally raise the question whether the HADWIGER NUMBER problem can be solved in polynomial time on permutation graphs. We leave this as an open question. We point out that the problem is NP-complete on co-comparability graphs, a well-known superclass permutation graphs, due to Theorem 3 and the fact that co-bipartite graphs form a subclass of co-comparability graphs.

In Sect. 4, we proved that the s-CLUB CONTRACTION problem is polynomial on AT-free graphs for $s \geq 2$. An interesting direction for further research is to identify other non-trivial graph classes for which the s-CLUB CONTRACTION problem is polynomial-time solvable (or fixed-parameter tractable when parameterized by k) for all values of $s \geq 2$.

References

1. Alon, N., Lingas, A., Wahlén, M.: Approximating the maximum clique minor and some subgraph homeomorphism problems. Theor. Comput. Sci. **374**, 149–158 (2007)
2. Bollobás, B., Catlin, P.A., Erdös, P.: Hadwiger's conjecture is true for almost every graph. Eur. J. Comb. **1**, 195–199 (1980)
3. Brandstädt, A., Le, V., Spinrad, J.: Graph Classes: A Survey. SIAM Monographs on Discrete Mathematics and Applications. SIAM, Philadelphia (1999)
4. Cai, L., Guo, C.: Contracting few edges to remove forbidden induced subgraphs. In: Gutin, G., Szeider, S. (eds.) IPEC 2013. LNCS, vol. 8246, pp. 97–109. Springer, Heidelberg (2013)
5. Golovach, P.A., Heggernes, P., van 't Hof, P., Paul C.: Hadwiger number of graphs with small chordality. CoRR abs/1406.3812 (2014)
6. Eppstein, D.: Finding large clique minors is hard. J. Graph Algorithms Appl. **13**(2), 197–204 (2009)
7. Golumbic, M.C.: Algorithmic Graph Theory and Perfect Graphs. Annals of Discrete Mathematics, vol. 57, 2nd edn. Elsevier, Amsterdam (2004)
8. Hadwiger, H.: Über eine klassifikation der streckenkomplexe. Vierteljschr. Naturforsch. Ges. Zürich **88**, 133–143 (1943)
9. Lokshtanov, D., Misra, N., Saurabh, S.: On the hardness of eliminating small induced subgraphs by contracting edges. In: Gutin, G., Szeider, S. (eds.) IPEC 2013. LNCS, vol. 8246, pp. 243–254. Springer, Heidelberg (2013)
10. Robertson, N., Seymour, P.: Graph minors. XIII. The disjoint paths problem. J. Comb. Theory Ser. B **63**, 65–110 (1995)

11. Spinrad, J., Brandstädt, A., Stewart, L.: Bipartite permutation graphs. Discrete Appl. Math. **18**(3), 279–292 (1987)
12. Tarjan, R.E., Yannakakis, M.: Simple linear-time algorithms to test chordality of graphs, test acyclicity of hypergraphs, and selectively reduce acyclic hypergraphs. SIAM J. Comput. **13**(3), 566–579 (1984)
13. Zuckerman, D.: Linear degree extractors and the inapproximability of max clique and chromatic number. In: Proceedings of the STOC 2006, pp. 681–690 (2006)

Recognizing Threshold Tolerance Graphs in $O(n^2)$ Time

Petr A. Golovach[1], Pinar Heggernes[1], Nathan Lindzey[2(✉)],
Ross M. McConnell[3], Vinícius Fernandes dos Santos[4], and Jeremy P. Spinrad[5]

[1] Department of Informatics, University of Bergen,
P.O.Box 7803, 5020 Bergen, Norway
{Petr.Golovach,pinar}@ii.uib.no
[2] Mathematics Department, Colorado State University,
Fort Collins, CO 80523-1873, USA
lindzey@cs.colostate.edu
[3] Computer Science Department, Colorado State University,
Fort Collins, CO 80523-1873, USA
rmm@cs.colostate.edu
[4] Departamento de Matematica Aplicada, Universidade do Estado do
Rio de Janeiro - UERJ, Rio de Janeiro, Brazil
vinicius@ime.uerj.br
[5] Department of Electrical Engineering and Computer Science,
Vanderbilt University, Nashville, TN, USA
spin@vuse.vanderbilt.edu

Abstract. A graph $G = (V, E)$ is a *threshold tolerance* graph if each vertex $v \in V$ can be assigned a weight w_v and a tolerance t_v such that two vertices $x, y \in V$ are adjacent if $w_x + w_y \geq \min(t_x, t_y)$. Currently, the most efficient recognition algorithm for threshold tolerance graphs is the algorithm of Monma, Reed, and Trotter which has an $O(n^4)$ runtime. We give an $O(n^2)$ algorithm for recognizing threshold tolerance and their complements, the co-threshold tolerance (co-TT) graphs, resolving an open question of Golumbic, Weingarten, and Limouzy.

1 Introduction

Tolerance graphs are an important subclass of perfect graphs that generalizes both interval graphs and permutation graphs [7]. They have been written about extensively and they model constraints in various combinatorial optimization and decision problems [7–9]. They have a rich structure and history, and interesting relationships to other graph classes. For a detailed overview of the class, see [9].

A graph $G = (V, E)$ is *threshold tolerance* if each vertex $v \in V$ can be assigned a weight w_v and a tolerance t_v such that two vertices $x, y \in V$ are adjacent when $w_x + w_y \geq \min(t_x, t_y)$ [12]. When the tolerances of the vertices are all the same, we obtain the subclass of *threshold graphs* [3].

© Springer International Publishing Switzerland 2014
D. Kratsch and I. Todinca (Eds.): WG 2014, LNCS 8747, pp. 214–224, 2014.
DOI: 10.1007/978-3-319-12340-0_18

Their complements, the *co-threshold tolerance graphs* (*co-TT graphs*), have also received attention as they have an interesting interpretation as a generalization of *interval graphs*. They are a special case of the tolerance graphs.

A graph $G = (V, E)$ is an *interval graph* if and only if each vertex $v \in V$ can be assigned an interval $I_v = [a(v), b(v)]$ on the real line such that two vertices $x, y \in V$ are adjacent exactly when their corresponding intervals intersect, in which case $\mathcal{I} = \{[a(v), b(v)] : v \in V\}$ forms an *interval model* of G. See [2,5,15] for surveys of the properties of this class and its relationship to other graph classes. To illustrate the relationship of the interval graphs to the threshold graphs, the definition can be rephrased:

Definition 1. *A graph $G = (V, E)$ is an interval graph if and only if there exist functions $a, b : V \mapsto \mathbb{R}$ such that:*

- *$a(x) \le b(x)$ for all $x \in V$;*
- *$xy \in E \Leftrightarrow a(x) \le b(y) \wedge a(y) \le b(x)$ for all $x, y \in V$.*

By this definition, $[a(x), b(x)]$ is the interval that represents x in the model. Relaxing the requirement that $a(x) \le b(x)$, gives the class of co-TT graphs:

Definition 2. *[12] A graph $G = (V, E)$ is a co-TT graph if and only if there exist functions $a, b : V \mapsto \mathbb{R}$ such that:*

- *$xy \in E \Leftrightarrow a(x) \le b(y) \wedge a(y) \le b(x)$ for all $x, y \in V$.*

The definition also works if the inequalities are strict. It is easy to see that this class is the complement of threshold tolerance graphs by setting $a(x) = w_x$ and $b(x) = t_x - w_x$ for all $x \in V$. Moreover, given a co-TT model, this gives a way of finding weights and tolerances that realize the complement, by assigning $w_x = a(x)$ and $t_x = b(x) + w_x$.

Following the notation in Golumbic, Weingarten and Limouzy [10], let the *blue-red partition* of V given by a co-TT model be (B, R), where $B = \{x | x \in V$ and $a(x) \le b(x)\}$ and $R = \{x | x \in V$ and $b(x) < a(x)\}$. Given such a partition, let B be the *blue vertices* and R be the *red vertices*. The *red intervals* are the intervals $[b(x), a(x)]$ corresponding to red vertices and the *blue intervals* are the intervals $[a(x), b(x)]$ corresponding to blue vertices.

The following is easily verified:

Lemma 1. *[10] Given a co-TT model of a co-TT graph $G = (V, E)$, let (B, R) be its blue-red partition. Then:*

- *If $\{x, y\} \subseteq B$, then $xy \in E \Leftrightarrow [a(x), b(x)]$ and $[a(y), b(y)]$ intersect;*
- *If $\{x, y\} \subseteq R$, then $xy \notin E$;*
- *If $x \in B$ and $y \in R$, then $xy \in E \Leftrightarrow [b(y), a(y)]$ is contained in $[a(x), b(x)]$.*

A *chord* on a cycle C in a graph is an edge not on the cycle but whose endpoints are on the cycle. A graph is *chordal* if every cycle on four or more vertices has a chord, see, for example [5]. A chord xy in an even cycle C is *odd* when the distance in C between x and y is odd. A graph is *strongly chordal* if it is chordal and every even-length cycle of size at least six has an odd chord [4].

The following illustrates an interesting relationship between chordal graphs, strongly chordal graphs, co-TT graphs and interval graphs. A graph is chordal if and only if there is a *perfect elimination ordering*, which is an ordering $(v_1, v_2, \ldots, v_n)$ of its vertices such that for every vertex v_i, v_i and its neighbors in $\{v_{i+1}, v_{i+2}, \ldots, v_n\}$ induce a complete subgraph. A graph is strongly chordal if and only if it has a *simple elimination ordering*, which is a special case of a perfect elimination ordering. A graph is a co-TT graph if and only if it has a *proper elimination ordering* [12], which is a special case of a simple elimination ordering. To complete this taxonomy, a graph is an interval graph if and only if it admits a proper elimination ordering $(v_1, v_2, \ldots, v_n)$, such that, whenever $v_i v_j \notin E$ and $i < j$, then v_i is to the left of all members of $N[v_j]$. Ordering the vertices in left-to-right order right endpoint in an interval model gives such an ordering. Conversely, it is easy to obtain an interval model, given such an ordering.

Note that R is an independent set, and that the blue intervals are an interval model of $G[B]$. Hence $G[B]$ is an interval graph. A vertex v of a graph is *simplicial* if its neighbors induce a complete subgraph. For each $r \in R$, the intervals corresponding to neighbors of r contain r's interval, so they have a common intersection point. Since the neighbors of r are blue, r is simplicial.

Henceforth, we will denote a co-TT model as $\mathcal{I}(B, R)$, which is a set of intervals on the line, together with an implied bijection from vertices to the intervals, and where (B, R) is the blue-red partition. If x maps to an interval $[l, r]$, then if $x \in B$, $a(x)$ is implicitly l and $b(x)$ is implicitly r, whereas if $x \in R$, $a(x)$ is implicitly r and $b(x)$ is implicitly l.

Despite the similarities between co-TT models and interval models, the best time bound for recognition of threshold tolerance and co-TT graphs until now has been $O(n^4)$ [12], whereas linear-time recognition of interval graphs has been known for some time [1].

A graph is a *split graph* if its vertices can be partitioned into a complete subgraph and an independent set. Golumbic, Weingarten and Limouzy [10] showed that *split co-TT* graphs, that is, those graphs that are both split graphs and co-TT graphs, can be recognized in $O(n^2)$ time and a forbidden subgraph characterization for split co-TT graphs was given. We generalize this bound to recognition of arbitrary co-TT graphs. The structural insight of Sect. 4, developed in [10] is essential to our approach. This gives an $O(n^2)$ bound for recognition of threshold tolerance graphs also, since it now takes $O(n^2)$ time to recognize whether the complement of a graph is a co-TT graph.

2   Preliminaries

Two sets *overlap* if they intersect and neither is a subset of the other. Given a binary $(0, 1)$ matrix, we treat the rows and columns as bit-vector representations of sets. A row is the set of columns where the row has a 1, and similarly for columns. This allows us to apply set operations to rows or to columns, such as evaluating whether one row is a subset of another.

Given a set A of directed edges, let A^T denote the *transpose* $\{(y,x)|(x,y) \in A\}$. An undirected graph $G = (V,E)$ is a special case of a symmetric directed graph, so we may refer to the directed edges of an undirected graph. For $\emptyset \subset V' \subseteq V$, let $G[V']$ denote the subgraph of G induced by V'. For $v \in V$, let the *open neighborhood* of v, denoted $N_G(v)$, be the set of neighbors of v in G, and let its *closed neighborhood*, denoted $N_G[v]$, be $N_G(v) \cup \{v\}$. When G is understood, we may denote these $N(v)$ and $N[v]$.

A *clique* of a graph is a complete subgraph that is not properly contained in any other complete subgraph. Two vertices u and v are *false twins* if $N(u) = N(v)$. Note that this implies that they are nonadjacent. They are *true twins* if $N[u] = N[v]$, which implies that they are adjacent. The simplicial vertices that have true twins can be found in $O(n+m)$ time by radix sorting.

3   Reduction to the Case where G is a Co-TT Graph

We give an $O(n^2)$ algorithm that has the precondition that its input graph G is a co-TT graph and the postcondition that it has returned a valid co-TT model. The reason that this suffices for recognition is that such an algorithm must fail to return a valid co-TT model if and only if its input graph is not a co-TT graph, since no valid co-TT model exists. (Our algorithm sometimes returns an invalid model, and sometimes halts when it recognizes that G lacks a property that co-TT graphs have.) Given a graph $G = (V,E)$ and co-TT model $\mathcal{I}(B,R)$ on V, it trivially takes $O(n^2)$ time to determine whether $\mathcal{I}(B,R)$ is a valid co-TT model of G, by applying Lemma 1 to each pair of intervals and comparing the result with the corresponding adjacency-matrix entry for G. We show how to implement the algorithm so that it halts in $O(n^2)$ time, whether or not G meets its precondition.

In the rest of this paper, we assume that the precondition to the algorithm is met, that is, that G is a co-TT graph, except when we analyze the running time in the case where G is not a co-TT graph.

4   Inferring a Blue-Red Partition

A key element of our approach is the following insight, which is given by Golumbic, Weingarten and Limouzy in [10].

Lemma 2. *If G is a co-TT graph, then there exists a co-TT model where the red vertices are the simplicial vertices that have no true twins in G, and the blue vertices are all others.*

By this, the authors implied an obvious linear-time algorithm for finding a blue-red partition in an arbitrary co-TT graph. However, since they only addressed split co-TT graphs, they did not give it explicitly. For completeness, we give it here:

Lemma 3. *If G is a co-TT graph, it takes $O(n + m)$ time to find a blue-red partition (B, R).*

Proof. It takes $O(n + m)$ time to recognize whether a graph is chordal, the sum of cardinalities of the cliques of a chordal graph is $O(n + m)$ and they take $O(n + m)$ time to find [13]. Since a co-TT graph is chordal, find its cliques, in $O(n+m)$ time [13]. A vertex is simplicial if and only if it is a member of exactly one clique. It takes time proportional to the sum of cardinalities of the cliques to test this on all vertices, by traversing the vertices in each clique, marking them, and marking each vertex as non-simplicial if it has already been marked during traversal of another clique. It takes $O(n + m)$ time to identify all equivalence classes of true twins by radix sorting.   □

This gives a reduction of the problem of recognizing co-TT graphs to that of finding whether a graph G' has a co-TT model with a given partition (B', R'), where B' has no true twins and the vertices in R' have no false twins. The reduction is given in Algorithm 1.

Data: A co-TT graph G
Result: A co-TT model of G
1 Find the blue-red partition (B, R) for some co-TT model of G (Lemma 3);
2 $B' \longleftarrow$ one representative from each equivalence class of true twins of B ;
3 Remove any vertices from R that are isolated in G;
4 $R' \longleftarrow$ one representative from each equivalence class of false twins of R ;
5 $G' \longleftarrow G[B' \cup R']$;
6 Find a co-TT model $\mathcal{I}'(B', R')$ of G' (Algorithm 2);
7 **for** $b \in B \setminus B'$ **do**
8   Insert a blue interval for b to $\mathcal{I}'$ equal to that of the representative of b's true-twin class;
9 **for** $r \in R \setminus R'$ **do**
10   **if** *r is an isolated vertex* **then**
11   Insert a red interval for r that contains all blue intervals;
12   **else**
13   Insert a red interval for r to $\mathcal{I}'$ equal to that of the representative of r's false-twin class;
14 Return the resulting model $\mathcal{I}(B, R)$;

Algorithm 1. Co-TT-Model(G)

Lemma 4. *The reduction of Algorithm 1 is correct, and can be implemented to run in $O(n + m)$ time whether or not the input graph is a co-TT graph.*

Proof. If $\mathcal{I}(B, R)$ is a co-TT model of G, then $\mathcal{I}(B, R) \cap (B' \cup R') = \mathcal{I}(B', R')$ is a co-TT model of G'. Therefore, G' has a co-TT model with blue-red partition (B', R'). Given $\mathcal{I}'(B', R')$, correctness of the construction of a co-TT model of G is then immediate from Lemmas 1 and 3. The time bound given in the proof of Lemma 3 depends only on the input graph G being chordal, which takes $O(n+m)$ time to determine [13]. Since all co-TT graphs are chordal, G can be rejected as a co-TT graph if it is not chordal.   □

The motivation for the reduction to the case where there are no true blue twins and no false red twins is given by the following lemma, which we use to simplify the analysis:

Lemma 5. *Let G', B', R' be as in Algorithm 1. For any pair $\{x, y\}$ of distinct vertices, $N[x] \neq N[y]$, and for any pair $\{r, r'\} \subseteq R'$ $N(r) \neq N(r')$.*

Proof. The red vertices of G remain simplicial in G'. If b and b' are two members of B', then since each true-twin equivalence class of B' has only one member, $N[b] \neq N[b']$. If r is a red vertex, it has no true twins in G, hence it has no true twins in G', and $N[r] \neq N[y]$ for any other vertex y. Since r is the only red member of its false-twin equivalence class in G', $N(r) \neq N(r')$ for any other red vertex r'. $\qquad\square$

- Henceforth in the paper, we will let G', B' and R' denote these elements of the reduction of Algorithm 2, and let $V' = B' \cup R'$ denote the vertices of G'.

5 Strongly Chordal Graphs and Chordal Bipartite Graphs

An *edge-vertex incidence matrix* for a graph has one row for each vertex, one column for each edge, and a 1 in row i, column j if edge j is incident to vertex i. A binary matrix is *totally balanced* if and only if it does not have as a submatrix the edge-vertex incidence matrix of a cycle of length at least three. The *augmented adjacency matrix* of a graph on vertex set $\{v_1, v_2, \ldots, v_n\}$ is the binary matrix that has a 1 in row i, column j if $v_j \in N[v_i]$. That is, it is the result of adding 1's on the diagonal to the adjacency matrix. The *bipartite adjacency matrix* for a bipartite graph $G = (\{v_1, v_2, \ldots, v_j\}, \{w_1, w_2, \ldots, w_k\}, E)$ is the binary matrix that has a 1 in row i, column j if w_j is a neighbor of v_i.

Theorem 1. *[4] A graph is strongly chordal if and only if its augmented adjacency matrix is totally balanced.*

A bipartite graph is *chordal bipartite* if every cycle of length greater than or equal to six has a chord. (See [15] for a survey.)

Theorem 2. *[6] A bipartite graph is chordal bipartite if and only if its bipartite adjacency matrix is totally balanced.*

The elimination orderings discussed in the introduction establish the following:

Theorem 3. *[12] Every co-TT graph is strongly chordal.*

Lemma 6. *Let $\{V_1, V_2\}$ be a partition of vertices of a strongly chordal graph, $G = (V, E)$. Then the bipartite graph $H = (V_1, V_2, \{xy | x \in V_1, y \in V_2, and xy \in E\})$ is chordal bipartite.*

Proof. Any cycle C in H is a cycle of G. If $|C| \geq 6$, then it is a cycle that has an odd chord in G. Since the vertices on C alternate between V_1 and V_2 around C, the odd chord has one end in V_1 and the other in V_2, hence it is a chord of C in H. □

Theorem 4. *[14] Given a $p \times q$ binary matrix, it takes $O(pq)$ time to determine whether it is totally balanced, and, if so, to determine for each ordered pair (i, j) of rows whether row i is a subset of row j.*

The main consequence of these results for this paper is summarized as follows:

Lemma 7. *In G', it takes $O(n^2)$ time to find whether $N[b_1] \subset N[b_2]$ for each ordered pair (b_1, b_2) of distinct vertices of B' and whether $N(r_1) \subset N(r_2)$ for each ordered pair of distinct vertices in R'.*

Proof. The bound for closed neighborhood containments follows from Theorems 1, 3, 4 and Lemma 5. Since R' is an independent set of G', the bound for open neighborhood containments follows from Theorems 2, 3, 4, Lemma 5, and an application of Lemma 6 to the bipartite graph $H = (R', B', E \cap (R' \times B'))$. □

6   Finding a Co-TT Model $\mathcal{I}(B', R')$ of G'

In this section, we give an algorithm to find a co-TT model $\mathcal{I}(B', R')$ of G', where (G', B', R') are as defined in the reduction of Algorithm 2.

Definition 3. *A set $\mathcal{I}$ of n intervals on the line is in* standard form *if all endpoints are distinct and elements of $\{1, 2, \ldots, 2n\}$.*

Definition 4. *Let $G = (V, E)$ be a graph. Let A_V denote $\{(x, y) | x, y \in V$ and $x \neq y\}$. Let $\mathcal{I}$ be a set of intervals in standard form. For each vertex x, assign a member $I_x \in \mathcal{I}$ to x, so that the assignment is a bijection from V to $\mathcal{I}$.*

For $(x, y) \in A_V$, let its intersection type be an overlap *if I_x and I_y overlap, a* non-intersection *if I_x and I_y are disjoint, a* containment *if I_x contains I_y, and a* subset relationship *if I_x is a subset of I_y. These labels are* intersection labels *of elements of A_V. Let us say that $\mathcal{I}$* realizes *the assignment of these labels. Let $E_n(\mathcal{I})$ denote the set of elements of A_V that are labeled as non-intersections, $E_o(\mathcal{I})$ those that are labeled as overlaps, $A_s(\mathcal{I})$ those that are labeled as subset relationships, and $A_c(\mathcal{I})$ be those that are labeled as containments. If $\mathcal{I}$ is understood, we may denote them E_n, E_o, A_s and A_c, respectively.*

Lemma 8. *[11] Given a set V and an arbitrary assignment of intersection labels to the elements of A_V, it takes $O(n^2)$ time to find a set of intervals in standard form that realizes the labeling, or else to determine that no such set of intervals exists.*

Lemma 8 further reduces the problem of finding a co-TT model $\mathcal{I}(B', R')$ to that of assigning intersection labels to elements of A_V that are consistent with some co-TT model $\mathcal{I}'(B', R')$ of G'. In the rest of this section, we show how to assign such a labeling in $O(n^2)$ time.

Definition 5. *Let $\mathcal{I}$ be an interval model of $G'[B']$ in standard form. The* red extension *of $\mathcal{I}$ is constructed as follows. For $r \in R'$, let its interval, I_r be the intersection of intervals of $\mathcal{I}$ that are neighbors. This is the maximal interval that r could have without misrepresenting a neighbor as a non-neighbor. The resulting model is not in standard form, since sets of right endpoints and sets of left endpoints coincide. Each maximal set $\mathcal{I}'$ of intervals whose right endpoints coincide contains exactly one blue interval. Move the right endpoints of red members of $\mathcal{I}'$ by epsilon amounts so that are ordered from left to right in ascending order of the length of the interval they belong to, and so that the blue member's endpoint is last. This has no effect on the represented graph, and causes all of these intervals to form a chain in the containment relation. Perform the symmetric operation on maximal sets of coinciding left endpoints. Put the model in standard form by listing the endpoints in left-to-right order.*

Lemma 9. *Let $\mathcal{I}$ be as in Definition 5. If the preconditions of Algorithm 2 are met, a red extension of $\mathcal{I}$ exists.*

Proof. Suppose the preconditions are met, but that a red extension fails to exist. Then $r_r < l_r$ for some $r \in R'$. Since $\mathcal{I}$ is an interval model of $G'[B']$, this implies that r has two blue neighbors that are nonadjacent to each other, and r fails to be simplicial, a contradiction. □

Proposition 1. *Suppose $\mathcal{I}(B', R')$ is a co-TT model of G'. Then the red extension $\mathcal{I}'(B', R')$ of $\mathcal{I}(B', R')[B']$ is a co-TT model of G'.*

Definition 6. *If G' is a co-TT graph, then a co-TT model $\mathcal{I}(B', R')$ of G' is* normalized *if $\mathcal{I}(B', R')$ is the red extension of $\mathcal{I}(B', R')[B']$.*

By Proposition 1, any normalized co-TT model can be turned into a co-TT model by replacing the model with the red extension of its blue intervals:

The algorithm for assigning the intersection labels to A_V is given as Algorithm 2, and gives a summary of the roles of the lemmas that follow in this section.

Lemma 10. *There exists a co-TT model $\mathcal{I}(B', R')$ of G' such that for each $(b_1, b_2) \in A_{B'}$:*

- *$(b_1, b_2) \in E_n$ if b_1 and b_2 are nonadjacent;*
- *$(b_1, b_2) \in A_s$ and $(b_2, b_1) \in A_c$ if $N[b_1] \subset N[b_2]$;*
- *$(b_1, b_2) \in E_o$ if b_1 and b_2 are adjacent but neither $N[b_1] \subset N[b_2]$ nor $N[b_2] \subset N[b_1]$.*

By Lemmas 8 and 10, we may now find an assignment $\mathcal{I}_B$ of intervals to B' such that the intersection types are the same as those of $\mathcal{I}(B', R')[B']$ for some co-TT model $\mathcal{I}(B', R')$ of G' in $O(n^2)$ time. Note that $\mathcal{I}_B$ and $\mathcal{I}(B', R')[B']$ are both interval models of $G'[B']$. Since there may be many interval models satisfying these intersection types, it is not necessarily the case that $\mathcal{I}_B = \mathcal{I}'(B', R')[B']$ for any co-TT model $\mathcal{I}'(B', R')$ with blue-red partition.

Data: A co-TT graph with blue-red partition (B', R'), B' has no true twins, and R' has no isolated vertices or false twins

Result: A labeling of elements of A_V with their intersection types in a co-TT model $\mathcal{I}(B', R')$ of G'

1 **for** $(x, y) \in A_{V'}$ **do**
2 $\quad$ Find whether $N[x] \subset N[y]$ (Lemma 7);
3 **for** $(r_1, r_2) \in A_{R'}$ **do**
4 $\quad$ Find whether $N(x) \subset N(y)$ (Lemma 7);
5 **for** $(b_1, b_2) \in A_{B'}$ *(Lemma 10)* **do**
6 $\quad$ **if** $b_1 b_2 \notin E'$ **then**
7 $\quad\quad$ Assign (b_1, b_2) to E_n ;
8 $\quad$ **else if** $N[b_1] \subset N[b_2]$ **then**
9 $\quad\quad$ Assign (b_1, b_2) to A_s and (b_2, b_1) to A_c ;
10 **for** $(b_1, b_2) \in A_{B'}$ **do**
11 $\quad$ **if** (b_1, b_2) *has not already been assigned* **then**
12 $\quad\quad$ Assign (b_1, b_2) to E_o ;
13 Construct an interval model $\mathcal{I}_B$ of $G[B']$ realizing these labels (Lemma 8) ;
14 Let $\mathcal{I}'$ be the red extension of $\mathcal{I}_B$;
15 **for** $(r_1, r_2) \in A_{R'}$ **do**
16 $\quad$ **if** $N(r_1) \subset N(r_2)$ **then**
17 $\quad\quad$ Assign (r_1, r_2) to A_c and (r_2, r_1) to A_s (Lemma 11);
18 $\quad$ **else if** $(r_1, r_2) \in E_n(\mathcal{I}')$ **then**
19 $\quad\quad$ Assign (r_1, r_2) to E_n (Lemma 12);
20 **for** $(b, r) \in B' \times R'$ **do**
21 $\quad$ **if** $br \in E'$ **then**
22 $\quad\quad$ Assign (b, r) to A_c and (r, b) to A_s (Lemma 1);
23 $\quad$ **else if** $(b, r) \in A_s(\mathcal{I}')$ **then**
24 $\quad\quad$ Assign (b, r) to A_s and (r, b) to A_c (Lemma 12);
25 $\quad$ **else if** $(b, r) \in E_n(\mathcal{I}')$ **then**
26 $\quad\quad$ Assign $(b, r), (r, b) \in E_n$ (Lemma 12);
27 **for** $(x, y) \in A_{V'}$ **do**
28 $\quad$ **if** (x, y) *has not been assigned* **then**
29 $\quad\quad$ Assign (x, y) to E_o ;
30 Apply Lemma 8 to find a co-TT model $\mathcal{I}(B', R')$ of G';

Algorithm 2. Co-TT-Model(G', B', R')

– Henceforth, as in Algorithm 2, we will let $\mathcal{I}_B$ denote this interval model of $G'[B']$.

Lemma 11. *If $\mathcal{I}(B', R')$ is a normalized co-TT model of G', then for distinct members $r_1, r_2 \in R'$, $(r_1, r_2) \in A_s(\mathcal{I}(B', R'))$ if and only if $N(r_2) \subset N(r_1)$.*

Though $\mathcal{I}_B$ realizes the intersection types of $\mathcal{I}(B', R')[B']$ for some co-TT model $\mathcal{I}(B', R')$, it may not be the case that $\mathcal{I}_B = \mathcal{I}'(B', R')[B']$ for any co-TT model $\mathcal{I}'(B', R')$ of G'. There may be many interval models of $G'[B']$ that realize these intersection types. Therefore, the red extension of $\mathcal{I}_B$ might not be a co-TT model. We can nevertheless use it to derive some of the intersection types in $\mathcal{I}(B', R')$:

Lemma 12. *Let $\mathcal{I}(B', R')$ be a normalized co-TT model of G' whose intersection types among the blue vertices are the same as those given by $\mathcal{I}_B$. Let $\mathcal{I}'$ be the red extension of $\mathcal{I}_B$.*

1. *For $r_1, r_2 \in R'$, $(r_1, r_2) \in E_n(\mathcal{I}')$ if and only if $(r_1, r_2) \in E_n(\mathcal{I}(B', R'))$;*
2. *For $r \in R', b \in B'$, $(b, r) \in E_n(\mathcal{I}')$ if and only if $(b, r) \in E_n(\mathcal{I}(B', R'))$, and $(b, r) \in A_s(\mathcal{I}')$ if and only if $(b, r) \in A_s(\mathcal{I}(B', R'))$.*

Lemma 13. *Algorithm 2 is correct.*

Proof. Since every red vertex in a co-TT model is simplicial, the preconditions imply that every vertex in R' is simplicial.

By Lemma 10, the labeling of intersection types conducted by the for loop on blue pairs is consistent with those in a co-TT model $\mathcal{I}(B', R')$. Therefore, $\mathcal{I}_B$ gives the same intersection labels as $\mathcal{I}_1(B', R')[B']$ does. This is true also for the red extension $\mathcal{I}_2(B', R') = \mathcal{I}(B', R')[B']$, which, by Proposition 1 and Definition 6 is a normalized co-TT model of G'. By Lemmas 11 and 12, those members (x, y) of $A_{V'}$ such that at least one of x and y is a member of R' are assigned to E_n, A_s or $A_{V'}$ in the next two loops if and only if they have those intersection types in $\mathcal{I}_2(B', R')$. Since $\{E_n(\mathcal{I}_2(B', R')), A_s(\mathcal{I}_2(B', R')), A_c(\mathcal{I}_2(B', R')), E_o(\mathcal{I}_2(B', R'))\}$ is a partition of $A_{V'}$, any elements not yet assigned must belong to $E_o(\mathcal{I}_2(B', R'))$, and the final loop correctly assigns these.

The intersection labels assigned to $A_{V'}$ are those of $\mathcal{I}_2(B', R')$. The set $\mathcal{I}_3$ of intervals given by Lemma 8 has these intersection types. Since $\mathcal{I}_2(B', R')$ is a co-TT model, so is $\mathcal{I}_3(B', R')$, and this is the model returned by the algorithm. $\square$

Lemma 14. *Algorithm 2 can be implemented to take $O(n^2)$ time even when (G', B', R') does not meet the preconditions.*

Proof. The application of Lemma 7 requires that G' be strongly chordal. This can be checked before the lemma is applied, by Theorems 1 and 4, and G' can be rejected as a co-TT graph if it fails this test, by Theorem 3.

Otherwise, the lemma gives the required neighborhood containments, whether or not G is a co-TT graph, in $O(n^2)$ time. The loop at Line 7 takes $O(n^2)$ time, whether or not G is a co-TT graph. Lemma 8 either gives a set $\mathcal{I}_B$ of intervals that realizes this labeling, or determines that none exists, in $O(n^2)$ time. By Lemma 10, such a set exists if (G', B', R') meets the precondition, so (G', B', R') can be rejected as failing to meet the preconditions. By Lemma 10, the intersection types of $\mathcal{I}_B$ are the same as they are for some co-TT model $\mathcal{I}(B', R')$ if (G', B', R') meets the preconditions.

Constructing a red extension $\mathcal{I}'$ of $\mathcal{I}_B$ or determining that none exists takes $O(n^2)$ time by elementary methods. By Lemma 9, there is a red extension of $\mathcal{I}_B$ if the preconditions are met, so if there is no red extension, (G', B', R') can be rejected as not meeting the precondition.

Otherwise, the time required for the remaining loops do not depend on any additional assumptions about the inputs, and they take $O(n^2)$ time. The final application of Lemma 8 takes $O(n^2)$ time whether or not it succeeds in producing a set of intervals that realizes the labeling. $\square$

Theorem 5. *Recognition of threshold tolerance and co-TT graphs takes* $O(n^2)$ *time.*

Proof. The problems reduce to each other in $O(n^2)$ time, so we show the result for co-TT graphs. Let G be a graph passed to Algorithm 1. Whether or not G is a co-TT graph, the algorithm halts in $O(n^2)$ time, by Lemmas 4 and 14. If G is a co-TT graph, it returns a co-TT model of G by Lemmas 4 and 13. If G is not a co-TT graph, it produces an incorrect co-TT model, since no co-TT model of G exists, or else it halts without producing one. If it halts without producing one, G can be rejected as a co-TT graph. If the algorithm produces a co-TT model, it takes $O(n^2)$ time to check whether it is a valid co-TT model for G, and if it is, G can be accepted, and if it is not, it can be rejected, since this only happens when G is not a co-TT graph.   □

References

1. Booth, K.S., Lueker, G.S.: Testing for the consecutive ones property, interval graphs, and graph planarity using pq-tree algorithms. J. Comput. Syst. Sci. **13**(3), 335–379 (1976)
2. Brandstaedt, A., Le, V.B., Spinrad, J.P.: Graph Classes: A Survey. SIAM Monographs on Discrete Mathematics. SIAM, Philadelphia (1999)
3. Chvatal, V., Hammer, P.L.: Aggregation of inequalities in integer programming. In: Korte, B.H., Hammer, P.L., Johnson, E.L., Nemhauser, G.L. (eds.) Studies in Integer Programming. Annals of Discrete Mathematics, vol. 1, pp. 145–162. North-Holland (Elsevier), Amsterdam (1977)
4. Farber, M.: Characterizations of strongly chordal graphs. Discrete Math. **43**, 173–189 (1983)
5. Golumbic, M.C.: Algorithmic Graph Theory and Perfect Graphs. Academic Press, New York (1980)
6. Golumbic, M.C., Goss, C.F.: Perfect elimination and chordal bipartite graphs. J. Graph Theory **2**(2), 155–163 (1978)
7. Golumbic, M.C., Monma, C.L., Trotter Jr., W.T.: Tolerance graphs. Discrete Appl. Math. **9**(2), 157–170 (1984)
8. Golumbic, M.C., Siani, A.: Coloring algorithms for tolerance graphs: reasoning and scheduling with interval constraints. In: Calmet, J., Benhamou, B., Caprotti, O., Henocque, L., Sorge, V. (eds.) AISC 2002 and Calculemus 2002. LNCS (LNAI), vol. 2385, pp. 196–207. Springer, Heidelberg (2002)
9. Golumbic, M.C., Trenk, A.N.: Tolerance Graphs. Cambridge Studies in Advanced Mathematics. Cambridge University Press, New York (2004)
10. Golumbic, M.C., Weingarten, N.L., Limouzy, V.: Co-TT graphs and a characterization of split co-TT graphs. Discrete Appl. Math. **165**, 168–174 (2014)
11. McConnell, R.M.: Linear-time recognition of circular-arc graphs. Algorithmica **37**(2), 93–147 (2003)
12. Monma, C.L., Reed, B., Trotter, W.T.: Threshold tolerance graphs. J. Graph Theory **12**(3), 343–362 (1988)
13. Rose, D.J., Tarjan, R.E., Lueker, G.S.: Algorithmic aspects of vertex elimination on graphs. SIAM J. Comput. **5**(2), 266–283 (1976)
14. Spinrad, J.P.: Doubly lexical ordering of dense 0 - 1 matrices. Inf. Process. Lett. **45**(5), 229–235 (1993)
15. Spinrad, J.P.: Efficient Graph Representations. American Mathematical Society, Providence, RI (2003)

Induced Disjoint Paths in Circular-Arc Graphs in Linear Time

Petr A. Golovach[1]([✉]), Daniël Paulusma[2], and Erik Jan van Leeuwen[3]

[1] Department of Informatics, University of Bergen, Bergen, Norway
petr.golovach@ii.uib.no
[2] School of Engineering and Computer Science, Durham University, Durham, UK
daniel.paulusma@durham.ac.uk
[3] Max-Planck Institut für Informatik, Saarbrücken, Germany
erikjan@mpi-inf.mpg.de

Abstract. The INDUCED DISJOINT PATHS problem is to test whether a graph G with k distinct pairs of vertices (s_i, t_i) contains paths $P_1, \ldots, P_k$ such that P_i connects s_i and t_i for $i = 1, \ldots, k$, and P_i and P_j have neither common vertices nor adjacent vertices (except perhaps their ends) for $1 \leq i < j \leq k$. We present a linear-time algorithm that solves INDUCED DISJOINT PATHS and finds the corresponding paths (if they exist) on circular-arc graphs. For interval graphs, we exhibit a linear-time algorithm for the generalization of INDUCED DISJOINT PATHS where the pairs (s_i, t_i) are not necessarily distinct.

1 Introduction

A classic algorithmic problem on a graph G with k distinct pairs of vertices (s_i, t_i) is to find vertex-disjoint paths $P_1, \ldots, P_k$ such that P_i connects s_i and t_i for $i = 1, \ldots k$. Known as the DISJOINT PATHS problem, it is NP-complete on general graphs [15], but can be solved in $O(n^3)$ time for any fixed integer k [24] (i.e. it is fixed-parameter tractable). The INDUCED DISJOINT PATHS problem also takes as input a graph G with k distinct pairs of vertices (s_i, t_i) and also asks whether there are paths $P_1, \ldots, P_k$ such that P_i connects s_i and t_i for $i = 1, \ldots, k$, but with the extra condition that $P_1, \ldots, P_k$ must be *mutually induced*, that is, no two paths P_i, P_j have common or adjacent vertices (except perhaps their end-vertices). Notice that the DISJOINT PATHS problem can be reduced to INDUCED DISJOINT PATHS by subdividing every edge of the graph. The INDUCED DISJOINT PATHS problem is NP-complete even for instances with $k = 2$ [2,5], and thus in particular is not fixed-parameter tractable unless P=NP.

The hardness of both DISJOINT PATHS and INDUCED DISJOINT PATHS on general graphs inspired research on their complexity on structured graph classes.

This work is supported by EPSRC (EP/K025090/1) and Royal Society (JP100692). The research leading to these results has also received funding from the European Research Council under the European Union's Seventh Framework Programme (FP/2007-2013)/ERC Grant Agreement n. 267959.

© Springer International Publishing Switzerland 2014
D. Kratsch and I. Todinca (Eds.): WG 2014, LNCS 8747, pp. 225–237, 2014.
DOI: 10.1007/978-3-319-12340-0_19

On the negative side, DISJOINT PATHS remains NP-complete on line graphs [19] and split graphs [14]. INDUCED DISJOINT PATHS remains NP-complete on claw-free graphs [6] (in fact, even on line graphs). Both problems remain NP-complete on planar graphs [8,18]. In these cases, however, fixed-parameter algorithms are known [9,14,16,23,24]. On the positive side, polynomial-time algorithms for DISJOINT PATHS exist on graphs of bounded treewidth [22] and graphs of cliquewidth at most 2 [12], and for INDUCED DISJOINT PATHS on AT-free graphs [8] and chordal graphs [1].

We focus on the complexity of INDUCED DISJOINT PATHS on circular-arc graphs. Recall that a *circular-arc graph* G has a *representation* in which each vertex of G corresponds to an arc of a circle, and two vertices of G are adjacent if and only if their corresponding arcs intersect. Circular-arc graphs generalize *interval graphs*, which have a representation in which each vertex corresponds to an interval of the line, and two vertices are adjacent if and only if their corresponding intervals intersect. The complexity of DISJOINT PATHS is known: it is NP-complete on interval graphs [21]. In contrast, for INDUCED DISJOINT PATHS, the authors of the present work recently showed a polynomial-time algorithm on circular-arc graphs [9] (for a weaker problem variant, such an algorithm is also implied by a general framework [7]). This work, as well as the polynomial-time algorithms on AT-free graphs [8] and chordal graphs [1], imply a polynomial-time algorithm on interval graphs. These algorithms, however, do not settle the complexity of INDUCED DISJOINT PATHS on circular-arc graphs (and interval graphs), as the question whether a linear-time algorithm exists is left open.

In this paper, we exhibit a linear-time algorithm for INDUCED DISJOINT PATHS on circular-arc graphs. This improves on the known algorithm for circular-arc graphs as well as the known algorithms for interval graphs. We also introduce a generalization of INDUCED DISJOINT PATHS called REQUIREMENT INDUCED DISJOINT PATHS, which is to find r_i paths that connect s_i and t_i for $i = 1, \ldots, k$, such that all paths are mutually induced. We present a linear-time algorithm for REQUIREMENT INDUCED DISJOINT PATHS on interval graphs. To solve these problems, our algorithms first preprocesses the instance. Some of the preprocessing rules build on our earlier work on INDUCED DISJOINT PATHS [8,9], but care is required to adapt them for REQUIREMENT INDUCED DISJOINT PATHS and to execute them in linear time. Most preprocessing rules, however, are novel. After the preprocessing stage, the algorithms identify a set of candidate paths for each pair (s_i, t_i). For each candidate path for a pair (s_i, t_i), we add an arc with color i that corresponds to the path to an auxiliary graph H. Finally, we show that it suffices to find an independent set in H that contains r_i arcs of each color. We show that the algorithms perform all stages in linear time.

2   Preliminaries

We only consider finite undirected graphs that have no loops and no multiple edges. We refer to the textbook of Diestel [4] for any standard graph terminology not defined here. Let $G = (V, E)$ be a graph. For a set $S \subseteq V$, the graph $G[S]$

denotes the subgraph of G *induced by* S; that is, the graph with vertex set S and edge set $\{uv \in E \mid u, v \in S\}$. We write $G - S = G[V \setminus S]$. We denote the (open) neighborhood of a vertex u by $N_G(u) = \{v \mid uv \in E\}$ and its closed neighborhood by $N_G[u] = N_G(u) \cup \{u\}$. We denote the neighborhood of a set $U \subseteq V$ by $N_G(U) = \{v \in V \setminus U \mid uv \in E \text{ for some } u \in U\}$ and $N_G[U] = U \cup N_G(U)$. We denote the degree of a vertex u by $\deg_G(u) = |N_G(u)|$.

We denote an unordered pair of elements x, y by $\{x, y\}$ (i.e. $\{x, y\} = \{y, x\}$).

Problem Definition. Let $P = v_1 \cdots v_r$ be a path (we call such a path a $v_1 v_r$-*path*). The vertices v_1 and v_r are the *ends* or *end-vertices* of P, and the vertices $v_2, \ldots, v_{r-1}$ are the *inner vertices* of P. We say that an edge $v_i v_j$, $i + 1 < j$, is an *inner chord* of P if v_i or v_j is an inner vertex of P. Distinct paths $P_1, \ldots, P_\ell$ in a graph G are *mutually induced* if:

(i) each P_i has no inner chords;
(ii) any distinct P_i, P_j may only share vertices that are ends of both paths;
(iii) no inner vertex u of any P_i is adjacent to a vertex v of some P_j for $j \neq i$, except when v is an end-vertex of both P_i and P_j.

Notice that condition (i) may be assumed without loss of generality. This definition is more general than the definition in Sect. 1, as it allows the end-vertices of distinct paths to be the same or adjacent. We can now formally state our decision problem (where a *terminal* is some specified vertex).

REQUIREMENT INDUCED DISJOINT PATHS

Instance: a graph G, k pairs of distinct terminals $(s_1, t_1), \ldots, (s_k, t_k)$ such that $\{s_i, t_i\} \neq \{s_j, t_j\}$ for $0 \leq i < j \leq k$, and k positive integers $r_1, \ldots, r_k$.
Question: does G have $\ell = r_1 + \ldots + r_k$ mutually induced paths $P_1, \ldots, P_\ell$ such that exactly r_i of these paths join s_i and t_i for $1 \leq i \leq k$?

If $r_1 = \ldots = r_k = 1$, then the problem is called INDUCED DISJOINT PATHS. The paths $P_1, \ldots, P_\ell$ are said to form a *solution* for a given instance, and we call every such path a *solution path*.

The problem definition allows a vertex v to be a terminal in two or more pairs (s_i, t_i) and (s_j, t_j). For instance, $v = s_i = s_j$ is possible. This corresponds to property (ii) of our definition of "being mutually induced". In order to avoid any confusion, we will view s_i and s_j as two different terminals "placed on" vertex v. Formally, we call v a *terminal vertex* that *represents* a terminal s_i or t_i if $v = s_i$ or $v = t_i$, respectively. We let T_v denote the set of terminals represented by v. If $T_v = \emptyset$, we call v a *non-terminal* vertex. We say that the two terminals s_i and t_i of a terminal pair (s_i, t_i) are *partners* of each other. If s_i is represented by u and t_i by v, then we also call a uv-path an $s_i t_i$-*path*. By our problem definition, each terminal pair (s_i, t_i) consists of two distinct terminals. Hence, two partners are never represented by the same vertex.

By Property (i), each solution path P has no inner chords and P is an induced path if and only if its ends are non-adjacent. If two adjacent vertices u and v

represent terminals vertices belonging to the same pair (s_i, t_i), then the path uv is called a *terminal path* for s_i, t_i. We need the following observation.

Observation 1. *Any yes-instance of* REQUIREMENT INDUCED DISJOINT PATHS *has a solution that contains all possible terminal paths. In particular, a terminal path for a pair (s_i, t_i) is the unique $s_i t_i$-path in this solution if $r_i = 1$.*

Graph Classes. Recall the definition of circular-arc and interval graphs from the introduction. Both graph types can be recognized in linear time and a corresponding representation can be found in linear time:

Theorem 1 ([3], see also [13,17]). *An interval graph G with n vertices and m edges can be recognized in $O(n + m)$ time. In the same time, a representation of G can be constructed with interval end-points $1, \ldots, 2n$.*

Theorem 2 [20]. *A circular-arc graph G with n vertices and m edges can be recognized in $O(n + m)$ time. In the same time, a representation of G can be constructed with arc end-points clockwise enumerated as $1, \ldots, 2n$.*

By Theorems 1 and 2, we always assume that an interval or circular-arc graph is given both by its adjacency list and its representation. Moreover, we assume that all the end-points of the intervals/arcs in the representation are distinct integers $1, \ldots, 2n$. Notice that using a representation we can check adjacency in $O(1)$ time. By slight abuse of notation, we often do not distinguish between the vertices and their corresponding intervals/arcs; e.g., we may speak of terminal intervals/arcs instead of terminal vertices.

For a vertex u of an interval graph, l_u and r_u denote the left and right end-point of u, respectively. Note that the degree of u is at least $(r_u - l_u - 1)/2$. For circular-arc graphs, we equate "left" to "counterclockwise" and "right" to "clockwise". Then, in the same way as for interval graphs, we let l_u and r_u denote the left and right end-point of a vertex u, respectively. In this way we are able to define similar terminology for both interval and circular-arc graphs. For two points x, y on the line, we write $x \leq y$ if y lies to the right with respect to x, and $x < y$ if $x \leq y$ and $x \neq y$, and we say that a point z *lies between* points x and y, if $x \leq z \leq y$. If x, y, z are points on a circle we write $x \leq z \leq y$ (or $x \leq z$ and $z \leq y$) to indicate that z is in the interval with the left end-point x and the right end-point y. We say that a vertex u *lies between* points x and y if $x \leq l_u < r_u \leq y$ (recall that l_u and r_u are distinct integers). Finally, a vertex u *lies between* two other vertices v, w if it lies between r_v and l_w; note that in that case we have in fact that $r_v < l_u < r_u < l_w$ by our assumption on the interval representation.

An *independent set* in a graph G is a set of vertices that are pairwise non-adjacent. At some stage, our algorithm for INDUCED DISJOINT PATHS on circular-arc graphs needs to compute a largest independent set of a circular-arc graph. This takes linear time:

Theorem 3 [11]. *If the arc end-points of a circular-arc graph G are sorted, then a largest independent set of G can be found in $O(n)$ time.*

3   Interval Graphs

In this section we develop a linear-time algorithm that solves REQUIREMENT INDUCED DISJOINT PATHS on interval graphs.[1] A possible approach would be the following greedy algorithm: find a terminal vertex with the leftmost right end-point and trace path(s) for the corresponding terminal pairs by a greedy procedure that iteratively chooses the non-terminal vertex with the leftmost right end-point that does not conflict with vertices already chosen. However, we do not elaborate on this approach for two reasons. First, this approach would require a thorough case analysis (just like our algorithm, and thus not be substantially simpler). Second, and more importantly, the goal of this paper is to design a linear-time algorithm for INDUCED DISJOINT PATHS on circular-arc graphs, where we have no natural starting point for a similar greedy approach and guessing such a starting point would irrevocably lead to a quadratic-time algorithm.

We describe the main constructs of our algorithm. Consider an instance of REQUIREMENT INDUCED DISJOINT PATHS. Let P be an $s_i t_i$-path that is not a terminal path, i.e. that has at least one inner vertex. Let I_P be the interval on the line obtained by taking the union of the intervals that correspond to the inner vertices of P. We say that P *covers* the interval I_P. Because P is an $s_i t_i$-path, we say that I_P has color i.

Lemma 1. *Let $P_1, \ldots, P_\ell$ form a solution. The following statements hold:*

(i) *For $1 \leq i \leq k$, any interval I_{P_a} with color i intersects the intervals that represent s_i and t_i and does not intersect any other terminal interval;*

(ii) *For $1 \leq a < b \leq \ell$, $I_{P_a} \cap I_{P_b} = \emptyset$;*

(iii) *For $1 \leq i < j \leq k$, there is no interval with color j that lies between two intervals with color i, or vice versa.*

We now outline our algorithm. Following Observation 1, we take all terminal paths into the solution. This might reduce the requirement r_i by 1 for some i. To find the remaining paths for all i, we determine a set of "candidate paths" that might or might not be used in the solution that we are constructing. The set of candidate paths is constructed such that for any $s_i t_i$ solution path P there is a candidate path P' such that P' is also an $s_i t_i$-path and $I_{P'} \subseteq I_P$. We guarantee that the set of candidate paths has size $O(n)$. By Lemma 1, the paths that are selected in a solution must cover distinct parts of the line. Therefore, we create an auxiliary interval graph H that consists of all intervals covered by the candidate paths. The intervals covered by candidate $s_i t_i$-paths all receive color i, for $i = 1, \ldots, k$. It then suffices to find an independent set with the required number of vertices of each color in H.

In the remainder of this section, we describe all steps of the algorithm in detail. We say that a step is *safe* if it runs in time $O(n + m + k)$ and is correct in the following sense:

---

[1] Due to space restrictions some proofs in this section and in the next ones are omitted or sketched. The full paper, with complete proofs, can be found in [10].

(i) a No-answer is given for no-instances only;
(ii) if a new instance is obtained, then it has a solution if and only if the original instance has a solution.
(iii) if a set of intervals that are all colored with color i is added to H, then this set has size $O(n)$ and corresponds to a *candidate set* of candidate paths.

The algorithm assumes that an interval representation of G is known, as given by Theorem 1. It also maintains an auxiliary interval graph H, initially empty. Recall that any vertex that we add to H will correspond to a candidate path for a solution. While adding vertices to H, we maintain an interval representation of H. Finally, the algorithm maintains a set $\mathcal{P}$ of paths, initially empty, which will form a solution for the instance (should it be a yes-instance). We let $T = \{s_1, t_1, \ldots, s_k, t_k\}$ be the set of all terminals. A terminal pair (s_i, t_i) is a *multi-pair* if $r_i \geq 2$, and a *simple pair* otherwise. The algorithm roughly consists of three stages: preprocess, construct H, and find an independent set.

3.1   Stage I: Preprocess

The only operations performed on G by our algorithm are vertex deletions. Hence, the graph that we obtain after each step is still interval. For simplicity, we denote this graph by G as well.

Step 1. Delete all non-terminal vertices that are adjacent to at least three terminal vertices.

Step 2. Check if there is a multi-pair that is represented by two non-adjacent terminal vertices. If so, then return a No-answer.

Lemma 2. *Steps 1 and 2 are safe.*

Suppose that we have not returned a No-answer after performing Step 2. In the next step, for each multi-pair, we identify a set of paths that together with the terminal paths form all candidate paths.

Step 3. For each non-terminal vertex u adjacent to terminal vertices v and w representing multi-pair terminals s_i and t_i, add I_{vuw} with color i to V_H, and delete u from G.

Lemma 3. *Step 3 is safe. Moreover, for any multi-pair (s_i, t_i), if P is a solution $s_i t_i$-path with at least one inner vertex, then there is a candidate $s_i t_i$-path P' with $I_{P'} \subseteq I_P$.*

In the next two steps, which are inspired by our earlier work on INDUCED DISJOINT PATHS [8,9], we get rid of all adjacent terminal vertices that represent the same terminal pair. This includes (but is not limited to) all multi-pairs.

Step 4. Find the set Z of all terminal vertices v such that v only represents terminals whose partners are in $N_G(v)$. Delete the vertices of Z and all non-terminal vertices of $N_G(Z)$ from G. Delete from T the terminals of all terminal pairs (s_i, t_i) with $s_i \in T_v$ or $t_i \in T_v$ for some $v \in Z$. Put all terminal paths corresponding to deleted terminal pairs in $\mathcal{P}$.

After Step 4, each terminal vertex represents at least one terminal whose partner is at distance at least 2. There may still be terminal pairs whose terminals are represented by adjacent vertices. We deal with such pairs in the next step.

Step 5. Delete all terminals s_i and t_i represented by adjacent terminal vertices from the terminal list, and delete all common non-terminal neighbors of the terminal vertices that represent s_i and t_i. Put all terminal paths corresponding to deleted terminals in $\mathcal{P}$.

Call a terminal pair *long* if its two terminals are represented by vertices of distance at least 2. After Step 5, all terminal pairs are long. Therefore, by Step 2, there are no multi-pairs anymore. Assume that there are $k' \leq k$ terminal pairs left; note that $k' = 0$ is possible.

Step 6. Check if there exists a terminal vertex that represents three or more terminals. If so, then return a No-answer.

After Step 6, a terminal vertex may represent at most two terminals (which must belong to different terminal pairs). We now observe that terminals should be ordered, and we let our algorithm find this ordering.

Step 7. Check if there exist three terminal vertices u, v, w such that u and w represent terminals from the same pair such that $l_u \leq l_v < l_w$. If so, then return a No-answer. Otherwise, order and rename the terminals such that $r_{u_i} < l_{v_i}$ and $l_{v_i} \leq l_{u_{i+1}}$ for $i = 1, \ldots, k' - 1$, where u_i, v_i are the vertices representing s_i, t_i, respectively.

Step 8. For $i \in \{1, \ldots, k' - 1\}$, if t_i and s_{i+1} are represented by distinct vertices u and v, delete all non-terminal vertices adjacent to both u and v.

Lemma 4. *Steps 4–8 are safe.*

3.2   Stage II: Construct H

We now construct the auxiliary H. Note that some intervals were already added to H as part of our preprocessing stage (see Step 3).

Step 9. For each $i \in \{1, \ldots, k'\}$, perform steps 9a–9d (where u and v are terminal vertices that represent s_i and t_i, respectively).

9a. For every common neighbor w of u and v, add the interval I_{uwv} to H with color i, and delete w from G.

9b. For each neighbor x of u not adjacent to v, determine whether there exists a neighbor y of v adjacent to x. If so, then choose y such that the right end-point of y is leftmost amongst all such neighbours of v. Add the interval I_{uxyv} to H with color i.

9c. Determine the connected components $C_1, \ldots, C_p$ of $G - (N[u] \cup N[v])$ whose vertices lie between r_u and l_v. For each C_j, determine the vertex $l(C_j)$ with the leftmost left end-point and the vertex $r(C_j)$ with the rightmost right end-point. Then among the neighbors that $l(C_j)$ and u have in common, let $s_i(C_j)$ be the one with the rightmost left end-point (if it exists). Similarly, let $t_i(C_j)$ be the neighbor that $r(C_j)$ and v have in common and that has the leftmost right end-point (if it exists). Add the interval between the left end-point of $s_i(C_j)$ and the right end-point of $t_i(C_j)$ to H with color i, if it has not been added already in Step 9b (which might be the case if $s_i(C_j)$ and $t_i(C_j)$ intersect).

Lemma 5. *Step 9 is safe. Moreover, for $i = 1, \ldots, k'$, if P is a solution s_it_i-path, then there is a candidate s_it_i-path P' with $I_{P'} \subseteq I_P$.*

Proof. We first prove that Step 9 is correct. Let $i \in \{1, \ldots, k'\}$. Let u and v be the (non-adjacent) vertices of G representing s_i and t_i, respectively. Let P be a solution path for (s_i, t_i).

Suppose that P has length 2. Then P has exactly one inner vertex w, which is adjacent to both u and v. By Step 9a, H contains the interval I_P.

Suppose that P has length 3. Then P has exactly two inner vertices x and y' that are adjacent to u and v, respectively. Let y be the neighbor of v that is adjacent to x and has the leftmost right end-point among all such vertices. Then $P' = uxyv$ is an s_it_i-path. Notice that $I_{P'} \subseteq I_P$ by the choice of y and by the fact that u and v have no common neighbors after Step 9a. Therefore, in any solution that contains P, P can be replaced P'. By Step 9b, H contains $I_{P'}$.

Finally, suppose that P has length at least 4. Because P is an induced path, there is a connected component C_j of $G - (N[u] \cup N[v])$ whose vertices all lie between r_u and l_v, such that all inner vertices of P except two neighbors of u and v are in C_j. Let x' and y' be the neighbors of u and v on P, respectively. Let $x = s_i(C_j)$ and $y = t_i(C_j)$. Then from P we can construct an s_it_i-path P' by replacing x' and y' with x and y, respectively. Notice that $I_{P'} \subseteq I_P$ by the choice of y and by the fact that u and v have no common neighbors after Step 9a. Therefore, in any solution that contains P, P can be replaced P'. By Step 9c, H contains $I_{P'}$.

Observe that the above arguments prove that for $i = 1, \ldots, k'$, if P is a solution s_it_i-path, then there is a candidate s_it_i-path P' with $I_{P'} \subseteq I_P$.

We now show how to perform Step 9 in $O(n+m)$ time. In Step 9a, we add all the intervals that correspond to common neighbors of s_i and t_i for $i = 1, \ldots, k'$, and delete these common neighbors from G. Common neighbors of s_i and t_i are not common neighbors of terminals of any other pair by Step 8. Therefore, Step 9a takes $O(n + m)$ time in total, and $O(n)$ intervals are added to H. In Step 9b, for $i = 1, \ldots, k'$, we find for each neighbor x of s_i (recall that x is not adjacent to t_i after Step 9a), the neighbor y of t_i such that x and y are adjacent and the right end-point of y is leftmost. By using the adjacency lists for the neighbors of u, Step 9b takes $O(n + m)$ time in total, and $O(n)$ intervals are added to H. In Step 9c, we first find the connected components $C_1, \ldots, C_\ell$. This can be done by performing a breadth-first search. Because the connected components that we consider (and their vertices) are unique to a terminal pair, Step 9c takes $O(n + m)$ time in total. Again, $O(n)$ intervals are added to H. $\square$

3.3   Stage III: Find Independent Set

It remains to find a particular independent set in H.

Step 10. Find an independent set in H that, for $i = 1, \ldots, k$, contains exactly $r_i - 1$ or r_i vertices colored i depending on whether (s_i, t_i) is a multi-pair or not. If such a set exists, add the corresponding candidate paths to $\mathcal{P}$ and return $\mathcal{P}$. Otherwise, return a No-answer.

Lemma 6. *Step 10 is safe.*

Proof. We first prove that Step 10 is correct. We do this by proving that our instance is a yes-instance if and only if H has an independent set as described in Step 10. First, suppose that H has such an independent set $\mathcal{I}$. For each interval u of color i, we can find an $s_i t_i$-path in G with inner vertices that are used to construct u. Taking into account the terminal paths that are already included in $\mathcal{P}$, we obtain r_i $s_i t_i$-paths for each $i \in \{1, \ldots, k\}$. We have to show that these paths are mutually induced. Because $\mathcal{I}$ is an independent set, distinct paths have no adjacent inner vertices. It remains to show that each $u \in \mathcal{I}$ does not intersect any terminal vertex (interval) of G except the vertices representing s_i, t_i. If u is added to H in Step 3, then it follows immediately from the fact that all non-terminal vertices that are adjacent to at least three terminals are deleted in Step 1 and from the description of Step 3. If u is added to H in Step 9, then notice u does not intersect any terminal vertex deleted in Step 4, because we delete them together with adjacent non-terminal vertices. Similarly, it does not interfere with any terminal deleted in Step 5, as proved in Lemma 4. Moreover, each interval added in Step 9 intersects exactly two remaining terminal vertices that are partners by Step 8. Hence, the instance is a yes-instance.

Now suppose that our instance is a yes-instance. Let $\ell_i = r_i - 1$ if (s_i, t_i) is a multi-pair, and let $\ell_i = r_i$ otherwise. By Observation 1, we can assume that the solution includes all terminal paths. Therefore, the solution contains exactly ℓ_i $s_i t_i$-path with inner vertices. By Lemma 3 and Lemma 5, for each such solution $s_i t_i$-path P, there is a candidate $s_i t_i$ path P' such that $I_{P'} \subseteq I_P$. Therefore, we can replace each solution path by a candidate path, and obtain a solution that uses only candidate paths. Let $\mathcal{I}$ denote the set of intervals covered by these paths. By Lemma 1, the intervals of $\mathcal{I}$ do not intersect each other. Moreover, by construction, $\mathcal{I}$ contains ℓ_i intervals with color i. Therefore, H has an independent set as described in Step 10.

We now show how to perform Step 10 in $O(n + m)$ time. We do this by performing the following procedure, which is a modification of the well-known greedy algorithm for finding a largest independent set in an interval graph.

1. Construct $2n$ buckets $L_1, \ldots, L_{2n}$ and $2n$ buckets $R_1, \ldots, R_{2n}$.
2. For each vertex u of H, put u in the buckets L_{l_u} and R_{r_u}.
3. Set $\mathcal{I} = \emptyset$ and $h = 2n$. For $i = 1, \ldots, k$, set $\ell_i = r_i - 1$ if (s_i, t_i) is a multi-pair, and set $\ell_i = r_i$ otherwise.
4. Scan the buckets $L_h, \ldots, L_1$ until we find a bucket L_j that contains a vertex u of H of some color i such that $\ell_i > 0$. Then u is included in $\mathcal{I}$. Find the set

of vertices X from the buckets $R_j, \ldots, R_h$, and delete them from H. Then set $\ell_i = \ell_i - 1$, $h = j$, and repeat the procedure. We stop as soon as we cannot find the next bucket L_j.

If $\mathcal{I}$ contains less than ℓ_i vertices of color i for some $i \in \{1, \ldots, k\}$, then stop and return a No-answer. Otherwise, return $\mathcal{I}$. This procedure takes $O(|V(H)|) = O(n)$ time, and the corresponding paths can be found in $O(n + m)$ time. Hence, it remains to show that the procedure is correct. We need the following claim (proof omitted).

Claim 1. *Let U_i, U_j be the set of vertices (intervals) of H colored by distinct colors i and j respectively. Then for any $u \in U_i$ and $v \in U_j$, $l_u \neq l_v$. Moreover, if $l_u < l_v$ for some $u \in U_i$ and $v \in U_j$, then $l_x < l_y$ for any $x \in U_i$ and $y \in U_j$.*

Claim 1 implies that between the left endpoints of two intervals with a color i there can be no left endpoint of an interval with color $j \neq i$. Then, similar as the correctness of the well-known greedy algorithm for finding a largest independent set in an interval graphs, we can argue that the above procedure outputs the required independent set.   □

As each step in our algorithm is safe, we obtain the following result.

Theorem 4. *The* REQUIREMENT INDUCED DISJOINT PATHS *problem can be solved in time $O(n + m + k)$ for interval graphs on n vertices and m edges with k terminal pairs.*

4   Circular-Arc Graphs

In this section, we modify the algorithm of the previous section to work for the INDUCED DISJOINT PATHS problem on circular-arc graphs. The general idea of the approach remains the same, but some preprocessing steps are no longer needed, and some steps need modification. In particular, we do not need colors here. We will again show that each step of the algorithm is *safe*, where the definition of a safe step remains the same, mutatis mutandis. The algorithm assumes that an arc representation of G is known, as given by Theorem 2. It maintains an auxiliary circular-arc graph H, initially empty, in a similar manner and function as before. It also maintains a set $\mathcal{P}$ of paths, initially empty.

The algorithm first performs Step 1. Note that Steps 2 and 3 are not necessary, as there are no multi-pairs now, and thus we do not apply them. We then continue with Steps 4 and 5.

Lemma 7. *Steps 1, 4, and 5 are safe.*

After Step 5, for each remaining terminal pairs (s_i, t_i), s_i and t_i are represented by vertices at distance at least two, and as before, we call such pairs *long*. Let k' be the number of remaining terminal pairs. Notice that it can happen that $k' \leq 1$ after Step 5. It is convenient to handle this case separately.

Step 5⁺. If $k' = 0$, then stop and return the solution $\mathcal{P}$. If $k' = 1$, then consider the terminal vertices u and v representing the terminals of the unique pair of T. Find a shortest uv-path P if it exists. If P exists, then add P to $\mathcal{P}$, and return the solution $\mathcal{P}$. Otherwise, stop and return a No-answer.

Lemma 8. *Step 5⁺ is safe.*

Now we can assume that $k' \geq 2$. Since all pairs are long and $k' \geq 2$, there is only one direction around the circle that a solution path can go, and therefore, intuitively, the problem starts to behave roughly as it does on interval graphs. We perform Steps 6, 7, 8, and 9, where in Step 9 we do not color the vertices.

Lemma 9. *Steps 6, 7, 8, and 9 are safe. Moreover, for $i = 1, \ldots, k'$, if P is a solution $s_i t_i$-path, then there is a candidate $s_i t_i$-path P' with $I_{P'} \subseteq I_P$.*

Finally, we execute the following simplified version of Step 10.

Step 10*. Find a largest independent set in H using Theorem 3. If such a set exists, add the corresponding candidate paths to $\mathcal{P}$ and return $\mathcal{P}$. Otherwise, return a No-answer.

Lemma 10. *Step 10* is safe.*

As each step in our algorithm is safe, we obtain the following result.

Theorem 5. *The* INDUCED DISJOINT PATHS *problem can be solved in time $O(n + m + k)$ for circular-arc graphs on n vertices and m edges with k terminal pairs.*

5 Conclusion

We gave a linear-time algorithm for REQUIREMENT INDUCED DISJOINT PATHS on interval graphs, and for INDUCED DISJOINT PATHS on circular-arc graphs. By the application of the same ideas, we can solve REQUIREMENT INDUCED DISJOINT PATHS on n-vertex circular-arc graphs in time $O(n^2)$. The increase in running time is because to solve the auxiliary problem of finding a multicolored independent set we must "guess" a starting point for the greedy selection of such a set. As an aside, we can prove that finding a multicolored independent set is NP-complete when no order on the colors is given, even on interval graphs [10].

References

1. Belmonte, R., Golovach, P.A., Heggernes, P., van 't Hof, P., Kaminski, M., Paulusma, D.: Detecting fixed patterns in chordal graphs in polynomial time. Algorithmica **69**, 501–521 (2014)

2. Bienstock, D.: On the complexity of testing for odd holes and induced odd paths. Discrete Math. **90**, 85–92 (1991). See also Corrigendum, Discrete Mathematics **102** (1992), 109
3. Booth, K.S., Lueker, G.S.: Testing for the consecutive ones property, interval graphs, and graph planarity using PQ-tree algorithms. J. Comput. Syst. Sci. **13**, 335–379 (1976)
4. Diestel, R.: Graph Theory. Springer, New York (2005). Electronic Edition
5. Fellows, M.R.: The Robertson-Seymour theorems: a survey of applications. In: Proceedings of the AMS-IMS-SIAM Joint Summer Research Conference, Contemporary Mathematics, vol. 89, pp. 1–18. American Mathematical Society, Providence (1989)
6. Fiala, J., Kamiński, M., Lidicky, B., Paulusma, D.: The k-in-a-path problem for claw-free graphs. Algorithmica **62**, 499–519 (2012)
7. Fomin, F.V., Todinca, I., Villanger, Y.: Large induced subgraphs via triangulations and CMSO. In: Proceedings of SODA 2014, pp. 582–593. SIAM (2014)
8. Golovach, P.A., Paulusma, D., van Leeuwen, E.J.: Induced disjoint paths in AT-Free graphs. In: Fomin, F.V., Kaski, P. (eds.) SWAT 2012. LNCS, vol. 7357, pp. 153–164. Springer, Heidelberg (2012)
9. Golovach, P.A., Paulusma, D., van Leeuwen, E.J.: Induced disjoint paths in claw-free graphs. In: Epstein, L., Ferragina, P. (eds.) ESA 2012. LNCS, vol. 7501, pp. 515–526. Springer, Heidelberg (2012)
10. Golovach, P.A., Paulusma, D., van Leeuwen, E.J.: Induced disjoint paths in circular-arc graphs in linear time. CoRR abs/1403.0789 (2014)
11. Golumbic, M.C., Hammer, P.L.: Stability in circular arc graphs. J. Algorithms **9**, 56–63 (1988)
12. Gurski, F., Wanke, E.: Vertex disjoint paths on clique-width bounded graphs. Theor. Comput. Sci. **359**, 188–199 (2006)
13. Habib, M., McConnell, R.M., Paul, C., Viennot, L.: Lex-BFS and partition refinement, with applications to transitive orientation, interval graph recognition and consecutive ones testing. Theor. Comput. Sci. **234**, 59–84 (2000)
14. Heggernes, P., van 't Hof, P., van Leeuwen, E.J., Saei, R.: Finding disjoint paths in split graphs. In: Geffert, V., Preneel, B., Rovan, B., Štuller, J., Tjoa, A.M. (eds.) SOFSEM 2014. LNCS, vol. 8327, pp. 315–326. Springer, Heidelberg (2014)
15. Karp, R.M.: On the complexity of combinatorial problems. Networks **5**, 45–68 (1975)
16. Kobayashi, Y., Kawarabayashi, K.: A linear time algorithm for the induced disjoint paths problem in planar graphs. J. Comput. Syst. Sci. **78**, 670–680 (2012)
17. Korte, N., Möhring, R.H.: An incremental linear-time algorithm for recognizing interval graphs SIAM. J. Comput. **18**, 68–81 (1989)
18. Kramer, M., van Leeuwen, J.: The complexity of wirerouting and finding minimum area layouts for arbitrary VLSI circuits. Adv. Comput. Res. **2**, 129–146 (1984)
19. Lynch, J.F.: The equivalence of theorem proving and the interconnection problem. SIGDA Newsl. **5**, 31–36 (1975)
20. McConnell, R.M.: Linear-time recognition of circular-arc graphs. Algorithmica **37**, 93–147 (2003)
21. Natarajan, S., Sprague, A.P.: Disjoint paths in circular arc graphs. Nordic J. Comput. **3**, 256–270 (1996)
22. Reed, B.A.: Tree width and tangles: A new connectivity measure and some applications. In: Bailey, R.A. (ed.) Surveys in Combinatorics, pp. 87–162. Cambridge University Press, Cambridge (1997)

23. Reed, B.A., Robertson, N., Schrijver, A., Seymour, P.D.: Finding disjoint trees in planar graphs in linear time. In: Robertson, N., Seymour, P.D. (eds.) Contemporary Mathematics, vol. 147, pp. 295–301. American Mathematical Society, Robertson (1993)
24. Robertson, N., Seymour, P.D.: Graph minors. XIII. The disjoint paths problem. J. Comb. Theory, Ser. B **63**, 65–110 (1995)

Near-Linear Time Constant-Factor
Approximation Algorithm
for Branch-Decomposition of Planar Graphs

Qian-Ping Gu$^{(\boxtimes)}$ and Gengchun Xu

School of Computing Science, Simon Fraser University, Burnaby, BC, Canada
qgu@cs.sfu.ca, gxa2@sfu.ca

Abstract. We give constant-factor approximation algorithms for branch-decomposition of planar graphs. Our main result is an algorithm which for an input planar graph G of n vertices and integer k, in $O(n \log^4 n)$ time either constructs a branch-decomposition of G with width at most $(2 + \delta)k$, $\delta > 0$ is a constant, or a $(k + 1) \times \lceil \frac{k+1}{2} \rceil$ cylinder minor of G implying $\mathrm{bw}(G) > k$, $\mathrm{bw}(G)$ is the branchwidth of G. This is the first $\tilde{O}(n)$ time constant-factor approximation for branchwidth/treewidth and largest grid/cylinder minors of planar graphs and improves the previous $\min\{O(n^{1+\epsilon}), O(nk^3)\}$ ($\epsilon > 0$ is a constant) time constant-factor approximations. For a planar graph G and $k = \mathrm{bw}(G)$, a branch-decomposition of width at most $(2+\delta)k$ and a $g \times \frac{g}{2}$ cylinder/grid minor with $g = \frac{k}{\beta}$, $\beta > 2$ is constant, can be computed by our algorithm in $O(n \log^4 n \log k)$ time.

Keywords: Branch-/tree-decompositions · Grid minor · Planar graphs · Approximation algorithm

1 Introduction

The notions of branchwidth and branch-decomposition introduced by Robertson and Seymour [27] in relation to the notions of treewidth and tree-decomposition have important algorithmic applications. The branchwidth $\mathrm{bw}(G)$ and the treewidth $\mathrm{tw}(G)$ of graph G are linearly related: $\mathrm{bw}(G) \leq \mathrm{tw}(G) + 1 \leq \lfloor \frac{3}{2}\mathrm{bw}(G) \rfloor$ for every G with more than one edge, and there are simple translations between branch-decompositions and tree-decompositions that meet the linear relations [27]. A graph G of small branchwidth (treewidth) admits efficient algorithms for many NP-hard problems [2,6]. These algorithms first compute a branch-/tree-decomposition of G and then apply a dynamic programming algorithm based on the decomposition to solve the problem. The dynamic programming step usually runs in polynomial time in the size of G and exponential time in the width of the branch-/tree-decomposition computed.

Deciding the branchwidth/treewidth and computing a branch-/tree-decomposition of minimum width have been extensively studied. For an arbitrary graph G of n vertices, the following results have been known: Given an integer k, it

© Springer International Publishing Switzerland 2014
D. Kratsch and I. Todinca (Eds.): WG 2014, LNCS 8747, pp. 238–249, 2014.
DOI: 10.1007/978-3-319-12340-0_20

is NP-complete to decide whether $\mathrm{bw}(G) \leq k$ [30] ($\mathrm{tw}(G) \leq k$ [1]). If $\mathrm{bw}(G)$ ($\mathrm{tw}(G)$) is upper-bounded by a constant then both the decision problem and the optimal decomposition problem can be solved in $O(n)$ time [7,9]. However, the linear time algorithms are mainly of theoretical importance because the constant behind the Big-Oh is huge. The best known polynomial time approximation factor is $O(\sqrt{\mathrm{bw}(G)})$ for branchwidth and $O(\sqrt{\log \mathrm{tw}(G)})$ for treewidth [15]. The best known exponential time approximation factors are as follows: an algorithm giving a branch-decomposition of width at most $3\mathrm{bw}(G)$ in $2^{O(\mathrm{bw}(G))}n^2$ time [28]; an algorithm giving a tree-decomposition of width at most $3\mathrm{tw}(G) + 4$ in $2^{O(\mathrm{bw}(G))}n \log n$ time [5]; and an algorithm giving a tree-decomposition of width at most $5\mathrm{tw}(G) + 4$ in $2^{O(\mathrm{tw}(G))}n$ time [5]. By the linear relation between the branchwidth and treewidth, the algorithms for tree-decompositions are also algorithms of same approximation factors for branch-decompositions, while from a branchwidth approximation α, a treewidth approximation 1.5α can be obtained.

Better results have been known for planar graphs G. Seymour and Thomas show that whether $\mathrm{bw}(G) \leq k$ can be decided in $O(n^2)$ time and an optimal branch-decomposition of G can be computed in $O(n^4)$ time [30]. Gu and Tamaki improve the $O(n^4)$ time for optimal branch-decomposition to $O(n^3)$ [17]. By the linear relation between the branchwidth and treewidth, the above results imply polynomial time 1.5-approximation algorithms for the treewidth and optimal tree-decomposition of planar graphs. It is open whether deciding $\mathrm{tw}(G) \leq k$ is NP-complete or polynomial time solvable for planar graphs G.

Fast algorithms for computing small width branch-/tree-decompositions of planar graphs have received much attention as well. Tamaki gives an $O(n)$ time heuristic algorithm for branch-decomposition [32]. Gu and Tamaki give an algorithm which for an input planar graph G of n vertices and integer k, either constructs a branch-decomposition of G with width at most $(c + 1 + \delta)k$ or outputs $\mathrm{bw}(G) > k$ in $O(n^{1+\frac{1}{c}})$ time, where c is any fixed positive integer and $\delta > 0$ is a constant [18]. By this algorithm and a binary search, a branch-decomposition of width at most $(c+1+\delta)k$ can be computed in $O(n^{1+\frac{1}{c}} \log k)$ time, $k = \mathrm{bw}(G)$. Recently, Kammer and Tholey give an algorithm which for input G and k, either constructs a tree-decomposition of G with width at most $(9 + \delta)k$, $\delta > 0$ is a constant, or outputs $\mathrm{tw}(G) > k$ in $O(nk^3)$ time [24,25]. This implies that a tree-decomposition of width at most $(9 + \delta)k$ can be computed in $O(nk^3 \log k)$ time, $k = \mathrm{tw}(G)$. Computational studies on branch-decomposition can be found in [3,4,21,22,31,32].

Grid and cylinder minors of graphs are notions closely related to branch-/tree-decompositions [12,13,19,29]. A $k \times h$ cylinder is a Cartesian product of a cycle on k vertices and a path on h vertices. For a graph G, let $\mathrm{cm}(G)$ be the largest integer k such that G has a $k \times \lceil \frac{k}{2} \rceil$ cylinder as a minor. It is shown in [19] that $\mathrm{cm}(G) \leq \mathrm{bw}(G) \leq 2\mathrm{cm}(G)$ for planar graphs. The $O(n^{1+\frac{1}{c}})$ time algorithm in [18] actually constructs a branch-decomposition of G with width at most $(c + 1 + \delta)k$ or a $(k + 1) \times \lceil \frac{k+1}{2} \rceil$ cylinder minor. Other work on the lower bound for the branchwidth/treewidth of planar graphs can be found in [8,16].

We propose an $\tilde{O}(n)$ time (the $\tilde{O}$ notation disregards poly-logarithmic terms) constant-factor approximation algorithm for branch-/tree-decompositions of planar graphs. This result is stated as follows.

Theorem 1. *There is an algorithm which given a planar graph G of n vertices and an integer k, in $O(n \log^4 n)$ time either constructs a branch-decomposition of G with width at most $(2 + \delta)k$, $\delta > 0$ is a constant, or a $(k + 1) \times \lceil \frac{k+1}{2} \rceil$ cylinder minor of G.*

Since a $(k + 1) \times \lceil \frac{k+1}{2} \rceil$ cylinder has branchwidth at least $k + 1$ [19], a cylinder minor given in Theorem 1 implies $\mathrm{bw}(G) > k$.

By the linear relation between the branchwidth and treewidth, Theorem 1 implies an algorithm which for an input planar graph G and integer k, in $O(n \log^4 n)$ time constructs a tree-decomposition of G with width at most $(3+\delta)k$ or outputs $\mathrm{tw}(G) > k$. For a planar graph G and $k = \mathrm{bw}(G)$, by Theorem 1 and a binary search, a branch-decomposition of width at most $(2 + \delta)k$ can be computed in $O(n \log^4 n \log k)$ time. This improves the previous result of a branch-decomposition of width at most $(c + 1 + \delta)k$ in $O(n^{1+\frac{1}{c}} \log k)$ time [18]. Similarly, for a planar graph G and $k = \mathrm{tw}(G)$, a tree-decomposition of width at most $(3+\delta)k$ can be computed in $O(n \log^4 n \log k)$ time, improving the previous result of a tree-decomposition of width at most $(9 + \delta)k$ in $O(nk^3 \log k)$ time [24,25] when $k > c'(\log n)^{\frac{4}{3}}$ for some constant $c' > 0$. Our algorithm can also be used to compute a $g \times \lceil \frac{g}{2} \rceil$ cylinder (grid) minor with $g = \frac{\mathrm{bw}(G)}{\beta}$, $\beta > 2$ is a constant, and a $g \times g$ cylinder (grid) minor with $g = \frac{\mathrm{bw}(G)}{\beta}$, $\beta > 3$ is a constant, of G in $O(n \log^4 n \log k)$ time. This improves the previous results of $g \times \lceil \frac{g}{2} \rceil$ with $g \geq \frac{\mathrm{bw}(G)}{\beta}$, $\beta > (c+1)$, and $g \times g$ with $g \geq \frac{\mathrm{bw}(G)}{\beta}$, $\beta > (2c+1)$, in $O(n^{1+\frac{1}{c}} \log k)$ time. As an application, our algorithm removes a bottleneck in work of [26] for computing a shortest path oracle and reduces its preprocessing time complexity in Theorem 6.1 from $O(n^{1+\frac{1}{c}} \log k \log n + S \log^2 n)$ to $O(n \log^5 n \log k + S \log^2 n)$.

Our algorithm for Theorem 1 uses the approach in the previous work of [18] described below. Given a planar graph G and integer k, let $\mathcal{Z}$ be the set of biconnected components of G with a normal distance (a definition is given in the next section) $h = ak$, $a > 0$ is a constant, from a selected edge e_0 of G. For each $Z \in \mathcal{Z}$, a minimum vertex-cut set $\partial(A_Z)$ which partitions $E(G)$ into edge subsets A_Z and $\overline{A}_Z = E(G) \setminus A_Z$ is computed such that $Z \subseteq A_Z$ and $e_0 \in \overline{A}_Z$ ($\partial(A_Z)$ separates Z and e_0). If $|\partial(A_Z)| > k$ for some $Z \in \mathcal{Z}$ then $\mathrm{bw}(G) > k$ is concluded. Otherwise, a branch-decomposition of graph H obtained from G by removing all A_Z is constructed. For each subgraph $G[A_Z]$ induced by A_Z, a branch-decomposition is constructed or $\mathrm{bw}(G[A_Z]) > k$ is concluded recursively. Finally, a branch-decomposition of G with width $O(k)$ is constructed from the branch-decomposition of H and those of $G[A_Z]$ or $\mathrm{bw}(G) > k$ is concluded.

Our algorithm uses a recent result in computing minimum face separating cycles in planar graphs to find $\partial(A_Z)$ for every $Z \in \mathcal{Z}$. Borradaile et al. give an algorithm which in $O(n \log^4 n)$ time computes an oracle for the all pairs minimum face separating cycle problem in a planar graph G [10,11]. For any pair of faces f

and g in G, the oracle in $O(|C|)$ time returns a minimum (f, g)-separating cycle C (C cuts the sphere on which G is embedded into two regions, one contains f and the other contains g). By this result, we show that a minimum vertex-cut set $\partial(A_Z)$ for every $Z \in \mathcal{Z}$ in all recursive steps can be computed in $O(n \log^4 n)$ time and get an algorithm for Theorem 1.

When $|\mathcal{Z}|$ is small, we show that $\partial(A_Z)$ for every $Z \in \mathcal{Z}$ can be computed more efficiently. Let n_z be the total number of components to be separated in all recursive steps, we have the following results.

Theorem 2. *There is an algorithm which given a planar graph G of n vertices and integer k, in $O((n + n_z\sqrt{n}) \log^3 n)$ time either constructs a branch-decomposition of G with width at most $(2 + \delta)k$ or a $(k + 1) \times \lceil \frac{k+1}{2} \rceil$ cylinder minor of G, where $\delta > 0$ is a constant.*

Theorem 3. *There is an algorithm which given a planar graph G of n vertices and integer k, in $O(nk + n_z k^3)$ time, either construct a branch-decomposition of G with width at most $(2 + \delta)k$ or a $(k + 1) \times \lceil \frac{k+1}{2} \rceil$ cylinder minor of G, where $\delta > 0$ is a constant.*

The next section gives the preliminaries of the paper. We prove Theorem 1 in Sect. 3, and briefly introduce the algorithms used to prove Theorems 2 and 3 in Sect. 4. Full proofs of these Theorems can be found in [20]. The final section concludes the paper.

2 Preliminaries

It is convenient to view a vertex-cut set $\partial(A_Z)$ in a graph as an edge in a hypergraph in some cases. A hypergraph G consists of a set $V(G)$ of vertices and a set $E(G)$ of edges, each edge is a subset of $V(G)$ with at least two elements. A hypergraph G is a graph if for every $e \in E(G)$, e has two elements. For a subset $A \subseteq E(G)$, we denote $\cup_{e \in A} e$ by $V(A)$ and denote $E(G) \setminus A$ by $\overline{A}$. For $A \subseteq E(G)$, the pair $(A, \overline{A})$ is a *separation* of G and we denote by $\partial(A)$ the vertex set $V(A) \cap V(\overline{A})$. The *order* of separation $(A, \overline{A})$ is $|\partial(A)|$. A hypergraph H is a subgraph of G if $V(H) \subseteq V(G)$ and $E(H) \subseteq E(G)$. For $A \subseteq E(G)$ and $W \subseteq V(G)$, we denote by $G[A]$ and $G[W]$ the subgraphs of G induced by A and W, respectively.

The notions of branchwidth and branch-decomposition are introduced by Robertson and Seymour [27]. A *branch-decomposition* of hypergraph G is a pair (ϕ, T) where T is a ternary tree and ϕ is a bijection from the set of leaves of T to $E(G)$. We refer the edges of T as links and the vertices of T as nodes. Consider a link e of T and let L_1 and L_2 denote the sets of leaves of T in the two respective subtrees of T obtained by removing e. We say that the separation $(\phi(L_1), \phi(L_2))$ is induced by this link e of T. We define the width of the branch-decomposition (ϕ, T) to be the largest order of the separations induced by links of T. The *branchwidth* of G, denoted by $\mathrm{bw}(G)$, is the minimum width of all branch-decompositions of G. In the rest of this paper, we identify a branch-decomposition (ϕ, T) with the tree T, leaving the bijection implicit and regarding each leaf of T as an edge of G.

A walk in graph G is a sequence of edges $e_1, e_2, ..., e_k$, where $e_i = \{v_{i-1}, v_i\}$. We call v_0 and v_k the end vertices and other vertices the internal vertices of the walk. A walk is a path if all vertices in the walk are distinct. A walk is a cycle if it has at least three vertices, $v_0 = v_k$ and $v_1, ..., v_k$ are distinct.

Let Σ be a sphere. For an open segment s homeomorphic to $\{x | 0 < x < 1\}$ in Σ, we denote by $\text{cl}(s)$ the closure of s. A planar embedding of a graph G is a mapping $\rho : V(G) \cup E(G) \to \Sigma \cup 2^{\Sigma}$ such that

- for $u \in V(G)$, $\rho(u)$ is a point of Σ, and for distinct $u, v \in V(G)$, $\rho(u) \neq \rho(v)$;
- for each edge $e = \{u, v\} \in E(G)$, $\rho(e)$ is an open segment in Σ with $\rho(u)$ and $\rho(v)$ the two end points in $\text{cl}(\rho(e)) \setminus \rho(e)$; and
- for distinct $e_1, e_2 \in E(G)$, $\text{cl}(\rho(e_1)) \cap \text{cl}(\rho(e_2)) = \{\rho(u) | u \in e_1 \cap e_2\}$.

A graph G is planar if it has a planar embedding ρ, and (G, ρ) is called a plane graph. We may simply use G to denote the plane graph (G, ρ), leaving the embedding ρ implicit. For a plane graph G, each connected component of $\Sigma \setminus (\cup_{e \in E(G)} \text{cl}(\rho(e)))$ is a face of G. We denote by $V(f)$ and $E(f)$ the set of vertices and the set of edges incident to face f, respectively. We say face f is bounded by the edges of $E(f)$.

A plane graph G is *biconnected* if for any distinct vertices $u, v, w \in V(G)$, there is a path of G between u and v that does not contain w. It suffices to prove Theorems 1, 2 and 3 for a biconnected G because if G is not biconnected, the problems of finding branch-decompositions and cylinder minors of G can be solved individually for each biconnected component.

For a plane graph G, a curve μ on Σ is *normal* if μ does not intersect any edge of G. The length of a normal curve μ is the number of connected components of $\mu \setminus \cup_{v \in V(G)} \{\rho(v)\}$. For vertices $u, v \in V(G)$, the *normal distance* $\text{nd}_G(u, v)$ is defined as the length of the shortest normal curve between $\rho(u)$ and $\rho(v)$. The *normal distance* between two vertex-subsets $U, W \subseteq V(G)$ is defined as $\text{nd}_G(U, W) = \min_{u \in U, v \in W} \text{nd}_G(u, v)$. We also use $\text{nd}_G(U, v)$ for $\text{nd}_G(U, \{v\})$ and $\text{nd}_G(u, W)$ for $\text{nd}_G(\{u\}, W)$.

A *noose* of G is a closed normal curve on Σ that does not intersect with itself. A noose ν of G separates Σ into two open regions R_1 and R_2 and induces a separation $(A, \overline{A})$ of G with $A = \{e \in E(G) \mid \rho(e) \subseteq R_1\}$ and $\overline{A} = \{e \in E(G) \mid \rho(e) \subseteq R_2\}$. We also say ν induces edge subset A $(\overline{A})$. A separation (resp. an edge subset) of G is called *noose-induced* if there is a noose which induces the separation (resp. edge subset). A noose ν separates two edge subsets A_1 and A_2 if ν induces a separation $(A, \overline{A})$ with $A_1 \subseteq A$ and $A_2 \subseteq \overline{A}$. We also say that the noose induced subset A separates A_1 and A_2.

For plane graph G and a noose ν induced $A \subseteq E(G)$, we denote by $G|A$ the plane hypergraph obtained by replacing all edges of A with edge $\partial(A)$ (i.e., $V(G|A) = (V(G) \setminus V(A)) \cup \partial(A)$ and $E(G|A) = (E(G) \setminus A) \cup \{\partial(A)\}$). An embedding of $G|A$ can be obtained from G with $\rho(\partial(A))$ an open disk (homeomorphic to $\{(x, y) | x^2 + y^2 < 1\}$) which is the open region separated by ν and contains A. For a collection $\mathcal{A} = \{A_1, .., A_r\}$ of mutually disjoint edge-subsets of G, $(..(G|A_1)|..)|A_r$ is denoted by $G|\mathcal{A}$.

3 $O(n \log^4 n)$ Time Algorithm

We give an algorithm to prove Theorem 1. Our algorithm follows the approach of the work in [18]. Given a plane graph G of n vertices, an edge e_0 of G and integers $k, h > 0$, let $\mathcal{Z}$ be the set of biconnected components of G such that for each $Z \in \mathcal{Z}$, $\mathrm{nd}_G(e_0, V(Z)) = h$ (notice that the subgraph of G induced by the vertices with normal distance at least h from e_0 may not be biconnected, and we handle every biconnected component of the subgraph). For each $Z \in \mathcal{Z}$, our algorithm computes a minimum noose induced subset A_Z separating Z and e_0. If for some $Z \in \mathcal{Z}$, $|\partial(A_Z)| > k$ then the algorithm constructs a $(k+1) \times h$ cylinder minor of G in $O(n)$ time by Lemma 1 proved in [18]. Otherwise, a set $\mathcal{A}$ of noose induced subsets with the following properties is computed: (1) for every $A_Z \in \mathcal{A}$, $|\partial(A_Z)| \leq k$, (2) for every $Z \in \mathcal{Z}$, there is an $A_Z \in \mathcal{A}$ which separates Z and e_0 and (3) for distinct $A_Z, A_{Z'} \in \mathcal{A}$, $A_Z \cap A'_Z = \emptyset$. Such an $\mathcal{A}$ is called a *good-separator* for $\mathcal{Z}$ and e_0.

Lemma 1. *[18] Given a plane graph G and integers $k, h > 0$, let A_1 and A_2 be edge subsets of G satisfying the following conditions: (1) each of separations $(A_1, \overline{A_1})$ and $(A_2, \overline{A_2})$ is noose-induced; (2) $G[A_2]$ is biconnected; (3) $\mathrm{nd}_G(V(\overline{A_1}), V(A_2)) \geq h$; and (4) every noose of G that separates $\overline{A_1}$ and A_2 has length $> k$. Then G has a $(k+1) \times h$ cylinder minor and given $(G|\overline{A_1})|A_2$, such a minor can be constructed in $O(|V(A_1 \cap \overline{A_2})|)$ time.*

Given a good-separator $\mathcal{A}$ for $\mathcal{Z}$ and e_0, our algorithm constructs a branch-decomposition of plane hypergraph $G|\mathcal{A}$ with width at most $k + 2h$ by Lemma 2 shown in [19,32]. For each $A_Z \in \mathcal{A}$, the algorithm computes a cylinder minor or a branch-decomposition for the plane hypergraph $G|\overline{A_Z}$ recursively. If a branch-decomposition of $G|\overline{A_Z}$ is found for every $A_Z \in \mathcal{A}$, the algorithm constructs a branch-decomposition of G with width at most $k + 2h$ from the branch-decomposition of $G|\mathcal{A}$ and those of $G|\overline{A_Z}$ by Lemma 3 which is straightforward from the definitions of branch-decompositions.

Lemma 2. *[19,32] Let $k > 0$ and $h > 0$ be integers. Let G be a plane hypergraph with each edge of G incident to at most k vertices. If there is an edge e_0 such that for any vertex v of G, $\mathrm{nd}_G(e_0, v) \leq h$ then given e_0, a branch-decomposition of G with width at most $k + 2h$ can be constructed in $O(|V(G)| + |E(G)|)$ time.*

The upper bound $k + 2h$ is shown in Theorem 3.1 in [19]. The normal distance in [19] between a pair of vertices is twice of the normal distance in this paper between the same pair of vertices. Tamaki gives a linear time algorithm to construct a branch-decomposition of width at most $k + 2h$ [32].

Lemma 3. *Given a plane hypergraph G and a noose-induced separation $(A, \overline{A})$ of G, let T_A and $T_{\overline{A}}$ be branch-decompositions of $G|\overline{A}$ and $G|A$ respectively.*

Let $T_A + T_{\overline{A}}$ to be the tree obtained from T_A and $T_{\overline{A}}$ by joining the link incident to the leaf $\partial(A)$ in T_A and the link incident to the leaf $\partial(A)$ in $T_{\overline{A}}$ into one link and removing the leaves $\partial(A)$. Then $T_A + T_{\overline{A}}$ is a branch-decomposition of G with width $\max\{|\partial(A)|, k_A, k_{\overline{A}}\}$ where k_A is the width of T_A and $k_{\overline{A}}$ is the width of $T_{\overline{A}}$.

To make a concrete progress in each recursive step, the following technique in [18] is used to compute $\mathcal{A}$. For a plane hypergraph G, a vertex subset e_0 of G and an integer $d \geq 0$, let

$$\mathrm{reach}_G(e_0, d) = \bigcup \{v \in V(G) | \mathrm{nd}_G(e_0, v) \leq d\}$$

denote the set of vertices of G with normal distance at most d from set e_0. Let $\alpha > 0$ be an arbitrary constant. For integer $k \geq 2$, let $d_1 = \lceil \frac{\alpha k}{2} \rceil$ and $d_2 = d_1 + \lceil \frac{k+1}{2} \rceil$. The *layer tree* $\mathrm{LT}(G, e_0)$ is defined as follows:

1. the root of the tree is G;
2. each biconnected component X of $G[V(G) \setminus \mathrm{reach}_G(e_0, d_1 - 1)]$ is a node in level 1 of the tree and is a child of the root; and
3. each biconnected component Z of $G[V(G) \setminus \mathrm{reach}_G(e_0, d_2 - 1)]$ is a node in level 2 of the tree and is a child of the biconnected component X in level 1 that contains Z.

For $h = d_2$, $\mathcal{Z}$ is the set of leaf nodes of $\mathrm{LT}(G, e_0)$ in level 2. For a node X of $\mathrm{LT}(G, e_0)$ in level 1, let $\mathcal{Z}_X$ be the set of child nodes of X. It is shown in [18] that for any $Z \in \mathcal{Z}_X$, if a minimum noose in the plane hypergraph $(G|\overline{X})|\mathcal{Z}_X$ separating Z and $\overline{X}$ has length $> k$ then G has a $(k+1) \times \lceil \frac{k+1}{2} \rceil$ cylinder minor. From this, a good-separator $\mathcal{A}_X$ for $\mathcal{Z}_X$ and $\overline{X}$ can be computed in hypergraph $(G|\overline{X})|\mathcal{Z}_X$, and the union of $\mathcal{A}_X$ for every X gives a good-separator $\mathcal{A}$ for $\mathcal{Z}$ and e_0.

To compute $\mathcal{A}_X$, we convert $(G|\overline{X})|\mathcal{Z}_X$ to a weighted plane graph and compute a minimum noose induced subset A_Z separating $Z \in \mathcal{Z}_X$ and $\overline{X}$ by finding a minimum face separating cycle in the weighted plane graph. We use the algorithm by Borradaile et al. [10,11] to compute the face separating cycles.

For an open disk D in Σ, let $\mathrm{cl}(D)$ be the closure of D and $\mathrm{bd}(D) = \mathrm{cl}(D) \setminus D$ be the boundary of D. For edge $\partial(\overline{X})$ in $(G|\overline{X})|\mathcal{Z}_X$, the embedding $\rho(\partial(\overline{X}))$ is an open disk and $E_{\overline{X}} = \mathrm{bd}(\rho(\partial(\overline{X}))) \setminus \{\rho(u)|u \in \partial(\overline{X})\}$ is a set of open segments. Similarly, for each edge $\partial(Z)$ in $(G|\overline{X})|\mathcal{Z}_X$, $E_Z = \mathrm{bd}(\rho(\partial(Z))) \setminus \{\rho(u)|u \in \partial(Z)\}$ is a set of open segments. We convert hypergraph $(G|\overline{X})|\mathcal{Z}_X$ to a plane graph G_X as follows: edge $\partial(\overline{X})$ is replaced by the set of edges which are segments in $E_{\overline{X}}$ and for each $Z \in \mathcal{Z}_X$, edge $\partial(Z)$ is replaced by the set of edges which are segments in E_Z.

We denote the face in G_X bounded by edges of $E_{\overline{X}}$ by $f_{\overline{X}}$. For each $Z \in \mathcal{Z}_X$, we denote the face in G_X bounded by edges of E_Z by f_Z. A face in G_X which is not $f_{\overline{X}}$ or any of f_Z is called a *natural face* in G_X. We convert G_X to a weighted plane graph H_X as follows: For each natural face f in G_X with $|V(f)| > 3$, we add a new vertex u_f and new edges $\{u_f, v\}$ in f for every vertex v in $V(f)$.

Each new edge $\{u_f, v\}$ is assigned the weight $1/2$. Each edge of G_X is assigned the weight 1. The length of a cycle (resp. path) in H_X is the sum of the weights assigned to the edges in the cycle (resp. path). For $Z \in \mathcal{Z}_X$, a minimum $(f_Z, f_{\overline{X}})$-separating cycle is a cycle separating f_Z and $f_{\overline{X}}$ with the minimum length. A noose in G_X is called a *natural noose* if it intersects only natural faces in G_X. It is shown in [18] that for each $Z \in \mathcal{Z}_X$, a minimum natural noose in G_X separating E_Z and $E_{\overline{X}}$ in G_X is a minimum noose separating Z and $\overline{X}$ in $(G|\overline{X})|\mathcal{Z}_X$. By Lemma 4, such a natural noose ν can be computed by finding a minimum $(f_Z, f_{\overline{X}})$-separating cycle C in H_X. The subset A_Z induced by ν in $(G|\overline{X})|\mathcal{Z}_X$ is also called cycle C induced subset.

Lemma 4. *[20] Let H_X be the weighted plane graph obtained from G_X. For any $(f_Z, f_{\overline{X}})$-separating cycle C in H_X, there is a natural noose ν which separates E_Z and $E_{\overline{X}}$ in G_X with the same length as that of C. For any minimum natural noose ν in G_X separating E_Z and $E_{\overline{X}}$, there is a $(f_Z, f_{\overline{X}})$-separating cycle C in H_X with the same length as that of ν.*

We assume that for every pair of vertices u, v in H_X, there is a unique shortest path between u and v. This can be realized by perturbating the edge weight $w(e)$ of each edge e in H_X as follows. Assume that the edges in H_X are $e_1, ... e_m$. For each edge e_i, let $w'(e_i) = w(e_i) + \frac{1}{2^{i+1}}$. Then it is easy to check that for any pair of vertices u and v in H_X, there is a unique shortest path between u and v w.r.t. to w'; and the shortest path between u and v w.r.t. w' is a shortest path between u and v w.r.t. w.

For a plane graph G, a *minimum cycle base tree* (MCB tree) introduced in [10, 11] is an edge-weighted tree $\tilde{T}$ such that

- There is a bijection from the faces of G to the nodes of $\tilde{T}$;
- removing each edge e from $\tilde{T}$ partitions $\tilde{T}$ into two subtrees $\tilde{T}_1$ and $\tilde{T}_2$; this edge e corresponds to a cycle which separates every pair of faces f and g with f in $\tilde{T}_1$ and g in $\tilde{T}_2$; and
- for any distinct faces f and g, the minimum-weight edge on the unique path between f and g in $\tilde{T}$ has weight equal to the length of a minimum (f, g)-separating cycle.

The next lemma gives the running time for computing a MCB tree of a plane graph and that for obtaining a cycle from the MCB tree.

Lemma 5. *[10, 11] Given a plane graph G of n vertices with positive edge weights, a MCB tree of G can be computed in $O(n \log^4 n)$ time. Further, for any distinct faces f and g in G, given a minimum weight edge in the path between f and g in the MCB tree, a minimum (f, g)-separating cycle C can be obtained in $O(|C|)$ time, $|C|$ is the number of edges in C.*

Using Lemma 5 for computing a MCB tree $\tilde{T}$ of H_X and $\mathcal{A}_X$, our algorithm is summarized in Procedure Branch-Minor below. In the procedure, U is a noose induced edge subset and initially $U = \{e_0\}$.

Procedure: Branch-Minor($G|U$)
Input: A biconnected plane hypergraph $G|U$ with $\partial(U)$ specified, $|\partial(U)| \leq k$ and every other edge has two vertices.
Output: Either a branch-decomposition of $G|U$ of width at most $k+2h$, $h = d_2$, or a $(k+1) \times \lceil \frac{k+1}{2} \rceil$ cylinder minor of G.

1. If $\text{nd}_{G|U}(\partial(U), v) \leq h$ for every $v \in V(G|U)$ then apply Lemma 2 to find a branch-decomposition of $G|U$. Otherwise, proceed to the next step.
2. Compute the layer tree $LT(G|U, \partial(U))$.
   For every node X of $LT(G|U, \partial(U))$ in level 1, compute $\mathcal{A}_X$ as follows:
   (a) Compute H_X from $(G|\overline{X})|\mathcal{Z}_X$.
   (b) Compute a MCB tree $\tilde{T}$ of H_X by Lemma 5.
   (c) Find a face f_Z, $Z \in \mathcal{Z}_X$, in $\tilde{T}$ by a breadth first search such that the path between f_Z and $f_{\overline{X}}$ in $\tilde{T}$ does not contain $f_{Z'}$ for any $Z' \in \mathcal{Z}_X$ with $Z' \neq Z$. Find the minimum weight edge e_Z in the path between f_Z and $f_{\overline{X}}$, and the cycle C from edge e_Z.

   If C has length $> k$ then compute a $(k+1) \times \lceil \frac{k+1}{2} \rceil$ cylinder minor by Lemma 1 and terminate.

   Otherwise, compute the cycle C induced subset A_Z and include A_Z to $\mathcal{A}_X$. For each node f of $\tilde{T}$, if edge e_Z is in the path between f and $f_{\overline{X}}$ in $\tilde{T}$ then delete f from $\tilde{T}$.

   Repeat the above until $\tilde{T}$ does not contain any f_Z for $Z \in \mathcal{Z}_X$.

   Let $\mathcal{A} = \cup_{X:\text{level 1 node}} \mathcal{A}_X$ and proceed to the next step.
3. For each $A \in \mathcal{A}$, call Branch-Minor($G|\overline{A}$) to construct a branch-decomposition T_A or a cylinder minor of $G|\overline{A}$.

   If a branch-decomposition T_A is found for every $A \in \mathcal{A}$, Lemma 2 is applied to $(G|U)|\mathcal{A}$ to construct a branch-decomposition T_0 of $(G|U)|\mathcal{A}$ and Lemma 3 is used to combine these branch-decompositions T_A, $A \in \mathcal{A}$, and T_0 into a branch-decomposition T of $G|U$ and return T.

Proof of Theorem 1: The input hypergraph $G|\overline{A}$ of our algorithm in each recursive step for $A \in \mathcal{A}$ is biconnected. For the $\mathcal{A}_X$ computed in Step 2, obviously (1) for every $A_Z \in \mathcal{A}_X$, $|\partial(A_Z)| \leq k$; (2) due to the way we find the cycles from the MCB tree, for every $Z \in \mathcal{Z}_X$, there is exactly one subset $A_Z \in \mathcal{A}_X$ separating Z and $\overline{X}$; and (3) from the unique shortest path in H_X, for distinct $A_Z, A_{Z'} \in \mathcal{A}_X$, $A_Z \cap A_{Z'} = \emptyset$. Therefore, $\mathcal{A}_X$ is a good-separator for $\mathcal{Z}_X$ and $\overline{X}$. From this, $\mathcal{A}$ is a good separator for $\mathcal{Z}$ and U and our algorithm computes a branch-decomposition or a $(k+1) \times \lceil \frac{k+1}{2} \rceil$ cylinder minor of G. The width of the branch-decomposition computed is at most

$$k + 2h = k + 2(d_1 + \lceil \frac{k+1}{2} \rceil) \leq k + 2(\lceil \frac{\alpha k}{2} \rceil) + (k+1) \leq (2 + \delta)k,$$

where δ is the smallest constant with $\delta k \geq \alpha k + 3$.

Let M, m_x, m be the numbers of edges in $G[\text{reach}_{G|U}(\partial(U), d_2)], (G|\overline{X})|\mathcal{Z}_X$, H_X, respectively. Then $m = O(m_x)$. In Step 2, the layer tree $LT(G|U, \partial(U))$ can be computed in $O(M)$ time. For each level 1 node X, it takes $O(m)$ time to compute H_X and by Lemma 5, it takes $O(m \log^4 m)$ time to compute a MCB tree $\tilde{T}$ of H_X. In Step 2(c), it takes $O(m)$ time to compute a cylinder minor by Lemma 1. From Property (3) of a good-separator, each edge of H_X appears in at most two cycles which induce the subsets in $\mathcal{A}_X$. So Step 2(c) takes $O(m)$ time to compute $\mathcal{A}_X$. Therefore, the total time for Steps 2(a)-(c) is $O(m \log^4 m)$. For distinct level 1 nodes X and X', the edge sets of subgraphs $(G|\overline{X})|\mathcal{Z}_X$ and $(G|\overline{X'})|\mathcal{Z}_{X'}$ are disjoint. From this, $\sum_{X:\text{level 1 node}} m_x = O(M)$. Therefore, Step 2 takes $\sum_{X:\text{level 1 node}} O(m_x \log^4 m_x) = O(M \log^4 M)$ time.

The time for other steps in Procedure Branch-Minor($G|U$) is $O(M)$. The number of recursive calls in which each vertex of $G|U$ is involved in the computation of Step 2 is $O(\frac{1}{\alpha}) = O(1)$. Therefore, the running time of the algorithm is $O(n \log^4 n)$. □

4   Algorithms for Theorems 2 and 3

A connected subgraph of a plane graph G is called a *piece* of G. For a piece P of G, the vertex-cut set ∂P partitioning $E(G)$ into $E(P)$ and $E(G) \setminus E(P)$ is called the *boundary* of P. To prove Theorem 2, we decompose H_X into pieces which form a *recursive* $\mathbf{r}$-*division* of H_X [23], compute $int\text{DDG}(P)$ and $ext\text{DDG}(P)$ (see [10,11] for definitions) for every piece in the recursive $\mathbf{r}$-division, and then compute minimum face separating cycles using the techniques in [10,11]. More precisely, we replace Steps 2(b)(c) in Procedure Branch-Minor with the following steps to get the algorithm.

– Compute a recursive $\mathbf{r}$-division R_H of H_X, where $\mathbf{r} = (r_1, r_2, .., r_l)$ with $r_1 = n/2$, $r_i = \frac{r_{i-1}}{2}$ for $1 < i \leq l$ and $r_l = \theta(\sqrt{m})$.
– For every piece P in R_H, compute $int\text{DDG}(P)$ and $ext\text{DDG}(P)$.
– For every $Z \in \mathcal{Z}_X$, compute a minimum $(f_Z, f_{\overline{X}})$-separating cycle C using R_H and $int\text{DDG}(P)/ext\text{DDG}(P)$.
  If the length of C is greater than k then compute a $(k+1) \times \lceil \frac{k+1}{2} \rceil$ cylinder minor by Lemma 1 and terminate. Otherwise, keep this cycle.
– Compute $\mathcal{A}_X$ from the minimum face separating cycles.

To prove Theorem 3, we decompose H_X into pieces by crest separators introduced in [24,25], compute the *good mountain structure tree* [24] GMST($H_X, \mathcal{S}_X$, $\mathcal{W}_X$) and $up\text{DDG}(S)$ and $low\text{DDG}(S)$ (see [20] for definitions) for every crest separator $S \in \mathcal{S}$, and find the minimum face separating cycles using the GMST and $up\text{DDG}(S)/low\text{DDG}(S)$. More specifically, we replace Steps 2(b)(c) in Procedure Branch-Minor with the following steps to get an algorithm for Theorem 3.

– Decompose H_X by crest separators into a good mountain structure tree GMST $(H_X, \mathcal{S}_X, \mathcal{W}_X)$.
– Compute $up\text{DDG}(S)$ and $low\text{DDG}(S)$ for every crest separator $S \in \mathcal{S}$.
– Mark every crest in $\mathcal{W}_X$ as non-separated, repeat the following until all crests are marked as separated. If there exist a non-separated crest $Z \in \mathcal{W}_X$, compute a minimum $(f_Z, f_{\overline{X}})$-separating cycle C using the GMST and

upDDG$(S)/low$DDG(S). We call C the cycle computed for crest Z. If the length of C is greater than k then compute a $(k+1) \times \lceil \frac{k+1}{2} \rceil$ cylinder minor by Lemma 1 and terminate. Otherwise, keep this cycle and mark every crest in ins(C) as separated.
- Compute $\mathcal{A}_X$ from the minimum face separating cycles.

5   Concluding Remarks

If we modify the definition for d_2 in Sect. 3 from $d_2 = d_1 + \lceil \frac{k+1}{2} \rceil$ to $d_2 = d_1 + (k+1)$, we get an algorithm which given a planar graph G and integer $k > 0$, in $O(n \log^4 n)$ time either computes a branch-decomposition of G with width at most $(3 + \delta)k$, where $\delta > 0$ is a constant, or a $(k+1) \times (k+1)$ cylinder minor (or grid minor). It is interesting to develop an $O(n)$ time constant factor approximation algorithm for the branchwidth and largest grid (cylinder) minors.

References

1. Arnborg, S., Cornell, D., Proskurowski, A.: Complexity of finding embedding in a k-tree. SIAM J. Discrete Math. **8**, 277–284 (1987)
2. Arnborg, S., Lagergren, J., Seese, D.: Easy problems for tree-decomposable graphs. J. Algorithms **12**, 308–340 (1991)
3. Bian, Z., Gu, Q.P., Marjan, M., Tamaki, H., Yoshitake, Y.: Empirical study on branchwidth and branch decomposition of planar graphs. In: Proceedings of Algorithm Engineering and Experimentation (ALENEX2008), pp. 152–165 (2008)
4. Bian, Z., Gu, Q.-P.: Computing branch decomposition of large planar graphs. In: McGeoch, C.C. (ed.) WEA 2008. LNCS, vol. 5038, pp. 87–100. Springer, Heidelberg (2008)
5. Bodlaender, H.L., Drange, P.G., Dreg, M.S., Fomin, F.V., Lokshtanov, D., Pilipczuk, M. : An $O(c^k n)$ 5-approximation algorithm for treewidth. In: Proceedings of the 2013 Annual Symposium on Foundation of Computer Science, (FOCS2013), pp. 499–508 (2013)
6. Bodlaender, H.L.: A tourist guide through treewidth. Acta Cybern. **11**, 1–21 (1993)
7. Bodlaender, H.L.: A linear time algorithm for finding tree-decomposition of small treewidth. SIAM J. Comput. **25**, 1305–1317 (1996)
8. Bodlaender, H.L., Grigoriev, A., Koster, A.M.C.A.: Treewidth lower bounds with brambles. Algorithmica **51**(1), 81–98 (2008)
9. Bodlaender, H.L., Thilikos, D.M.: Constructive linear time algorithm for branch-width. In: Degano, P., Gorrieri, R., Marchetti-Spaccamela, A. (eds.) ICALP 1997. LNCS, vol. 1256, pp. 627–637. Springer, Heidelberg (1997)
10. Borradaile, G., Sankowski, P., Wulff-Nilsen, C.: Min st-cut oracle for planar graphs with near-linear time preprocessing time. In: Proceedings of the 51st Annual Symposium on Foundations of Computer Science (FOCS2010), pp. 601–610 (2010) (also arXiv:1003.1320v2, April 2010)
11. Borradaile, G., Sankowski, P., Wulff-Nilsen, C.: Min st-cut oracle for planar graphs with near-linear time preprocessing time. arXiv:1003.1320v4 Oct 2013 (to appear in ACM TALG)

12. Demaine, E.D., Hajiaghayi, M.T.: Graphs excluding a fixed minor have grids as large as treewidth, with combinatorial and algorithmic applications through bidimensionality. In: Proceedings of the 2005 Symposium on Discrete Algorithms (SODA 2005), pp. 682–689 (2005)
13. Dorn, F., Fomin, F.V., Hajiaghayi, M.T., Thilikos, D.M.: Subexponential parameterized algorithms on bounded-genus graphs and H-minor-free graphs. J. of ACM 52(6), 866–893 (2005)
14. Fakcharoenphol, J., Rao, S.: Planar graphs, negative weight edges, shortest paths, and near linear time. J. Comput. Syst. Sci. 72, 868–889 (2006)
15. Feige, U., Hajiaghayi, M.T., Lee, J.R.: Improved approximation algorithms for minimum weight vertex separators. SIAM J. Comput. 38(2), 629–657 (2008)
16. Grigoriev, A.: Tree-width and large grid minors in planar graphs. Discrete Math. Theor. Comput. Sci. 13(1), 13–20 (2011)
17. Gu, Q.P., Tamaki, H.: Optimal branch decomposition of planar graphs in $O(n^3)$ time. ACM Trans. Algorithms 4(3), Article No. 30, 1–13 (2008)
18. Gu, Q.P., Tamaki, H.: Constant-factor approximations of branch-decomposition and largest grid minor of planar graphs in $O(n^{1+\epsilon})$ time. Theor. Comput. Sci. 412, 4100–4109 (2011)
19. Gu, Q.P., Tamaki, H.: Improved bound on the planar branchwidth with respect to the largest grid minor size. Algorithmica 64, 416–453 (2012)
20. Gu, Q.P., Xu, G.: Near-linear time constant-factor approximation algorithm for branch-decomposition of planar graphs. arXiv:1407.6761, July 2014
21. Hicks, I.V.: Planar branch decompositions I: the ratcatcher. INFORMS J. Comput. 17(4), 402–412 (2005)
22. Hicks, I.V.: Planar branch decompositions II: the cycle method. INFORMS J. Comput. 17(4), 413–421 (2005)
23. Klein, P.N., Mozes, S., Sommer, C.: Structural recursive separator decompositions for planar graphs in linear time. In: Proceedings of the 2013 Annual ACM Symposium on the Theory of Computing (STOC2013), pp. 505–514 (2013)
24. Kammer, F., Tholey, T.: Approximate tree decompositions of planar graphs in linear time. In: Proceedings of the 2012 Annual ACM-SIAM Symposium on Discrete Algorithms (SODA2012), pp. 683–698 (2012)
25. Kammer, F., Tholey, T.: Approximate tree decompositions of planar graphs in linear time. arXiv:1104.2275v2, May 2013
26. Mozes, S., Sommer, C.: Exact distance oracles for planar graphs. In: Proceedings of the 2012 Annual ACM-SIAM Symposium on Discrete Algorithms (SODA 2012), pp. 209–222 (2012)
27. Robertson, N., Seymour, P.D.: Graph minors X. Obstructions to tree decomposition. J. Comb. Theory, Ser. B 52, 153–190 (1991)
28. Robertson, N., Seymour, P.D.: Graph minors XIII. The disjoint paths problem. J. Comb. Theory, Ser. B 63, 65–110 (1995)
29. Robertson, N., Seymour, P.D., Thomas, R.: Quickly excluding a planar graph. J. Comb. Theory, Ser. B 62, 323–348 (1994)
30. Seymour, P.D., Thomas, R.: Call routing and the ratcatcher. Combinatorica 14(2), 217–241 (1994)
31. Smith, J.C., Ulusal, E., Hicks, I.V.: A combinatorial optimization algorithm for solving the branchwidth problem. Comput. Optim. Appl. 51(3), 1211–1229 (2012)
32. Tamaki, H.: A linear time heuristic for the branch-decomposition of planar graphs. In: Di Battista, G., Zwick, U. (eds.) ESA 2003. LNCS, vol. 2832, pp. 765–775. Springer, Heidelberg (2003)

Parameterized Directed k-Chinese Postman Problem and k Arc-Disjoint Cycles Problem on Euler Digraphs

Gregory Gutin$^{(\boxtimes)}$, Mark Jones, Bin Sheng, and Magnus Wahlström

Royal Holloway, University of London, Egham, Surrey TW20 0EX, UK
gutin@cs.rhul.ac.uk

Abstract. In the Directed k-Chinese Postman Problem (k-DCPP), we are given a connected weighted digraph G and asked to find k non-empty closed directed walks covering all arcs of G such that the total weight of the walks is minimum. Gutin, Muciaccia and Yeo (Theor. Comput. Sci. **513**, 124–128 (2013)) asked for the parameterized complexity of k-DCPP when k is the parameter. We prove that the k-DCPP is fixed-parameter tractable.

We also consider a related problem of finding k arc-disjoint directed cycles in an Euler digraph, parameterized by k. Slivkins (ESA 2003) showed that this problem is W[1]-hard for general digraphs. Generalizing another result by Slivkins, we prove that the problem is fixed-parameter tractable for Euler digraphs. The corresponding problem on vertex-disjoint cycles in Euler digraphs remains W[1]-hard even for Euler digraphs.

1 Introduction

A digraph H is *connected* if the underlying undirected graph of H is connected. Let $G = (V, A)$ be a connected digraph, where each arc $a \in A$ is assigned a non-negative integer weight $\omega(a)$ (G is a *weighted digraph*). The DIRECTED CHINESE POSTMAN PROBLEM is a well-studied polynomial-time solvable problem in combinatorial optimization [1,6,9].

DIRECTED CHINESE POSTMAN PROBLEM (DCPP)
Input: A connected weighted digraph $G = (V, A)$.
Task: Find a minumum total weight closed directed walk T on G such that every arc of G is contained in T.

In this paper, we will investigate the following generalisation of DCPP.

DIRECTED k-CHINESE POSTMAN PROBLEM (k-DCPP)
Input: A connected weighted digraph $G = (V, A)$ and an integer k.
Task: Find a minimum total weight set of exactly k non-empty closed directed walks such that every arc of G is contained in at least one of them.

A full version of this paper appears at http://arxiv.org/abs/1402.2137.

© Springer International Publishing Switzerland 2014
D. Kratsch and I. Todinca (Eds.): WG 2014, LNCS 8747, pp. 250–262, 2014.
DOI: 10.1007/978-3-319-12340-0_21

Note that the k-DCPP can be extended to directed multigraphs (that may include parallel arcs but no loops), but the extended version could be reduced to the one on digraphs by subdividing parallel arcs and adjusting weights appropriately. Since it is more convenient, we consider the k-DCPP for digraphs only.

In the literature, the undirected version of k-DCPP, abbreviated k-UCPP, has also been studied. If a vertex v of G is part of the input and we require that each of the k walks contains v then the k-DCPP and k-UCPP are polynomial-time solvable [11,20]. However, in general the k-DCCP is NP-complete [8], as is the k-UCPP [8,18].

Lately research in parameterized algorithms and complexity[1] for the CPP and its generalizations was summarized in [2] and reported in [15]. Several recent results described there are of Niedermeier's group who identified a number of practically useful parameters for the CPP and its generalizations, obtained several interesting results and posed some open problems, see, e.g. [5,16,17]. van Bevern *et al.* [2] and Sorge [15] suggested to study the k-UCPP as a parameterized problem with parameter k and asked whether the k-UCPP is fixed-parameter tractable, i.e. can be solved by an algorithm of running time $O(f(k)n^{O(1)})$, where f is a function of k only and $n = |V|$ (we say such an algorithm is *fixed parameter*).

Gutin, Muciaccia and Yeo [8] proved that the k-UCPP is fixed-parameter tractable. Observing that their approach for the k-UCPP is not applicable to the k-DCPP, the authors of [8] asked for the parameterized complexity of k-DCPP parameterized by k. In this paper, we show that the k-DCPP is also fixed-parameter tractable.

Theorem 1. *The k-DCPP is fixed-parameter tractable.*

Our proof is very different from that in [8] for the k-UCPP. While the latter was based on a simple reduction to a polynomial-size kernel, we give a fixed-parameter algorithm directly using significantly more powerful tools. In particular, we use an *approximation* algorithm of Grohe and Grüber [7] for the problem of finding the maximum number $\nu_0(D)$ of vertex-disjoint directed cycles in a digraph D (this algorithm is based on the celebrated paper by Reed *et al.* [12] on bounding $\nu_0(D)$ by a function of $\tau_0(D)$, the minimum size of a feedback vertex set of D). We also use the well-known fixed-parameter algorithm of Chen *et al.* [3] for the feedback vertex set problem on digraphs.

We also consider the following well-known problem related to the k-DCPP.

---

k-ARC-DISJOINT CYCLES PROBLEM (k-ADCP)
Input: A digraph D and an integer k.
Task: Decide whether D has k arc-disjoint directed cycles.

Crucially, we are interested in the k-ADCP because given a set of k arc-disjoint cycles, we can solve the k-DCPP in polynomial time (see Lemma 5). However, this problem is important in its own right.

---

[1] For terminology and results on parameterized algorithms and complexity we refer the reader to, e.g., the monograph [4].

The problem is NP-hard in general but polynomial-time solvable for planar digraphs [10]. In fact, for planar digraphs the maximum number of arc-disjoint directed cycles equals the minimum size of a feedback arc set, see, e.g., [1]. It is natural to consider k as the parameter for the k-ADCP. It follows easily from the results of Slivkins [14] that the k-ADCP is W[1]-hard. It remains W[1]-hard for quite restricted classes of directed multigraphs, e.g., for directed multigraphs which become acyclic after deleting two sets of parallel arcs [14]. Here we show that the k-ADCP-EULER, the k-ADCP on Euler digraphs, is fixed-parameter tractable, generalizing a result in [14] (Theorem 3.1). k-ADCP-EULER was shown to be NP-hard by Vygen [19].

Theorem 2. *The k-ADCP-EULER is fixed-parameter tractable.*

Interestingly, the problem of deciding whether a digraph has k vertex-disjoint directed cycles, which is W[1]-hard (also easily follows from the results of Slivkins [14]), remains W[1]-hard on Euler digraphs. Indeed, consider a non-Euler digraph D and let $\nu_0(D)$ denote the maximum number of vertex-disjoint directed cycles in D. Construct a new digraph H from D by adding two new vertices x and y, arcs xy and yx and the following extra arcs between x and the vertices of D: for each $v \in V(D)$ add $\max\{d^-(v) - d^+(v), 0\}$ parallel arcs vx and $\max\{d^+(v) - d^-(v), 0\}$ parallel arcs xv, where $d^-(v)$ and $d^+(v)$ are the in-degree and out-degree of v, respectively. To eliminate parallel arcs, it remains to subdivide all arcs between x and $V(D)$. Now it is sufficient to observe that H is Euler and $\nu_0(H) = \nu_0(D) + 1$.

To prove Theorems 1 and 2 we study the following problem that generalizes the k-DCPP (in the case when an optimal solution exists in which the number of times each arc is visited by every closed walk is restricted) and k-ADCP. Let $b \leq c$ be non-negative integers.

DIRECTED k-WALK $[b, c]$-COVERING PROBLEM ($k[b, c]$-DWCP)
Input: A connected weighted digraph $G = (V, A)$ and
             an integer k.
Task: Find a minimum total weight set of k non-empty
             closed directed walks in which every arc of G appears
             between b and c times.

Let D be a digraph. For a vertex ordering $\theta = (v_1, v_2, \ldots, v_n)$ of $V(D)$, the *cutwidth* of θ is the maximum number of arcs between $\{v_1, \ldots, v_i\}$ and $\{v_{i+1}, \ldots v_n\}$ over all $i \in [n]$. The *cutwidth* of D is the minimum cutwidth of all vertex orderings of $V(D)$.

In Sect. 3 we will prove the following theorem.

Theorem 3. *Let (G, k) be an instance of $k[b, c]$-DWCP and suppose we are given a vertex ordering $\theta = (v_1, v_2, \ldots, v_n)$ of G with cutwidth at most p. Then, in time $O^*((c2^k)^p 2^k)$, we can solve (G, k) and find an optimal feasible solution if one exists.*

Note that when c and p are upper-bounded by functions of k, the algorithm of this theorem is fixed-parameter.

The paper is organised as follows. In Sect. 2, we prove six lemmas providing structural results for the k-DCPP and k-ADCP-EULER, which will later be used to reduce these problems to $k[b, c]$-DWCP. In Sect. 3, we prove Theorem 3. In Sect. 4, we put the results of the previous two sections together to prove Theorems 1 and 2. We conclude the paper with brief discussions of open problems in Sect. 5.

The key results of Sect. 2 are as follows. Lemma 3 shows that, given an Euler directed graph, we can either find k arc-disjoint cycles or a vertex ordering with cutwidth bounded by a function of k. This allows us to either solve the k-ADCP-EULER directly or reduce it to the $k[0, 1]$-DWCP on a graph of bounded cutwidth, allowing us to apply Theorem 3. Lemmas 5 and 6 concern the Eulerian graph G_T derived from a solution T to the DCPP on G. Lemma 5 shows that given k arc-disjoint cycles in G_T, we can solve the k-DCPP on G in polynomial time. Lemma 6 shows that if no arc appears in G_T more than k times (in particular if there are fewer than k arc-disjoint cycles in G_T), there is an optimal solution for the k-DCPP such that no arc is visited more than k times in total by the k walks of the solution. This allows us to reduce the k-DCPP to the $k[1, k]$-DWCP, and Lemma 3 allows us to bound the cutwidth of the graph. Thus, in this case we can again apply Theorem 3.

In what follows, all walks and cycles in directed multigraphs are directed. For a positive integer p, $[p]$ will denote the set $\{1, 2, \ldots, p\}$. For integers $a \leq b$, $[a, b]$ will denote the set $\{a, a + 1, \ldots, b\}$. Given a directed graph D, a *feedback vertex set* for D is a set S of vertices such that $D - S$ contains no directed cycles. A *feedback arc set* for D is a set F of arcs such that $D - F$ contains no directed cycles. A vertex v of a digraph is *balanced* if the in-degree of v equals its out-degree. A digraph D is *balanced* if every vertex of D is balanced. A directed graph is Euler if and only if it is connected and balanced [1].

2   Structural Results and Fixed-Parameter Algorithms

The next lemma is a simple sufficient condition for an Euler digraph to contain k arc-disjoint cycles.

Lemma 1. *Every balanced digraph D having a vertex of out-degree at least $k \geq 1$, contains k arc-disjoint cycles that can be found in polynomial time.*

Proof. For $k = 1$, it is true as D has a cycle that can be found in polynomial time. Let $k \geq 2$ and let C be a cycle in D. Observe that after deleting the arcs of C, D has a vertex of out-degree at least $k - 1$ and we are done by induction hypothesis.   □

It follows from Reed *et al.* [12] and Propositions 13.3.1 and 15.3.1 in [1] that there is a function $f : \mathbb{N} \to \mathbb{N}$ such that for every k, if a digraph D does not have k arc-disjoint cycles, then it has a feedback arc set with at most $f(k)$ arcs. This result can be easily extended to directed multigraphs by subdividing parallel arcs. Using this result, Grohe and Grüber [7] showed that there is a non-decreasing and unbounded function $h : \mathbb{N} \to \mathbb{N}$ and a fixed-parameter algorithm that for a digraph D returns at least $h(k)$ arc-disjoint cycles if D has at least k arc-disjoint cycles (k is the parameter).

Let $h^{-1} : \mathbb{N} \to \mathbb{N}$ be defined by $h^{-1}(q) = \min\{p : h(p) \geq q\}$. Since h is a non-decreasing and unbounded function, h^{-1} is a non-decreasing and unbounded function. Combining the above results, we find that for every digraph D, either the algorithm of Grohe and Grüber returns at least k arc-disjoint cycles, or D has a feedback arc set of size at most $f(h^{-1}(k))$.

Chen *et al.* [3] designed a fixed-parameter algorithm that decides whether a digraph D contains a feedback vertex set of size k (k is the parameter). As this is an iterative compression algorithm, it can be easily modified to an algorithm for finding a minimum feedback vertex set in D (the running time of the latter algorithm is $q(\tau_0(D))n^{O(1)}$, where $\tau_0(D)$ is the minimum size of a feedback vertex set in D, $n = |V(D)|$ and $q(k) = 4^k k!$). The modified algorithm can be used for finding a minimum feedback arc set in D as D can be transformed, in polynomial time, into another digraph H such that D has a feedback arc set of size k if and only if H has a feedback vertex set of size k, see, e.g., [1] (Proposition 15.3.1).

Lemma 2. *There is a function $g : \mathbb{N} \to \mathbb{N}$ and a fixed-parameter algorithm such that for a directed multigraph D, the algorithm returns either k arc-disjoint cycles or a feedback arc set of size at most $g(k)$ (here k is the parameter).*

Proof. By subdividing arcs, we may assume that D is a digraph, i.e. D has no parallel arcs. Let $\kappa := k - 1$ and perform the following loop: for $\kappa := \kappa + 1$ run both Grohe-Grüber algorithm and Chen *et al.* algorithm on D with parameter κ until we get either at least k arc-disjoint cycles or a feedback arc set of size at most κ. Note that by [12], the loop will be completed for $\kappa \leq f(h^{-1}(k))$. Thus, our procedure is a fixed-parameter algorithm with respect to parameter k and we may set $g(k) = f(h^{-1}(k))$.   □

Lemma 3. *Let $g : \mathbb{N} \to \mathbb{N}$ be the function in Lemma 2. Let D be an Euler directed multigraph. We can obtain either k arc-disjoint cycles of D or a vertex ordering of cutwidth at most $2g(k)$.*

Proof. Let us run the procedure of Lemma 2 for D and k. If we get k arc-disjoint cycles, we are done. Otherwise, we get a feedback arc set F of D such that $|F| \leq g(k)$. Then $D' = D - F$ is an acyclic directed multigraph. We let $\theta = (v_1, \ldots, v_n)$ be an acyclic ordering of D', i.e., D' has no arc of the form $v_i v_j$, $i > j$, (it is well-known that such an ordering exists [1]). Now θ is a vertex ordering for D with at most $|F|$ arcs from $\{v_{i+1}, \ldots, v_n\}$ to $\{v_1, \ldots, v_i\}$ for each $i \in [n-1]$, and because D is Euler there are the same number of arcs from $\{v_1, \ldots, v_i\}$ to $\{v_{i+1}, \ldots, v_n\}$ [1, Corollary 1.7.3]. So θ is a vertex ordering with cutwidth at most $2g(k)$.   □

In the rest of this section, $G = (V, A)$ is a connected weighted directed graph. For a solution $T = \{T_1, \ldots, T_k\}$ to the k-DCPP on G ($k \geq 1$), let $G_T = (V, A_T)$, where A_T is a multiset containing all arcs of A, each as many times as it is traversed in total by $T_1 \cup \cdots \cup T_k$.

Lemmas 4 and 5 are similar to two simple results obtained for the k-UCPP in [8]. Note that given k closed walks which cover all the arcs of a digraph, their union is a closed walk covering all the arcs and, therefore, it is a solution for the DCPP. Hence, the following proposition holds.

Lemma 4. *The weight of an optimal solution for the k-DCPP on G is not smaller than the weight of an optimal solution for the DCPP on G.*

Lemma 5. *Let T be an optimal solution for the DCPP on G. If G_T contains at least k arc-disjoint cycles, then the weight of an optimal solution for the k-DCPP on G is equal to the weight of an optimal solution of the DCPP on G. Furthermore if k arc-disjoint cycles in G_T are given, then an optimal solution for the k-DCPP can be found in polynomial time.*

Proof. Note that G_T is an Euler directed multigraph and so every vertex of G_T is balanced. Let $\mathcal{C}$ be any collection of k arc-disjoint cycles in G_T. Delete all arcs of $\mathcal{C}$ from G_T and observe that every vertex in the remaining directed multigraph G' is balanced. Find an optimal DCPP solution for every connected component of G' and append each such solution F to a cycle in $\mathcal{C}$ which has a common vertex with F. As a result, in polynomial time, we obtain a collection Q of k closed walks for the k-DCPP on G of the same weight as T. So Q is optimal by Lemma 4. □

For a directed multigraph D, let $\mu_D(xy)$ denote the multiplicity of an arc xy of D. The *multiplicity* $\mu(D)$ of D is the maximum of the multiplicities of its arcs. Thus, Lemmas 1 and 5 imply that if $\mu(G_T) \geq k$ for any optimal solution T of the DCPP on G, then there is an optimal solution of the k-DCPP on G with weight equal to the weight of G_T. The next lemma helps us in the case that $\mu(G_T) \leq k - 1$.

Lemma 6. *Let T be an optimal solution of the DCPP on G such that $\mu(G_T) \leq k - 1$. Then there is an optimal solution W for the k-DCPP on G such that $\mu(G_W) \leq k$.*

Proof. Let T be an optimal solution of DCPP on G and let $\mu(G_T) \leq k - 1$. Suppose that there is an optimal solution W of the k-DCPP on G such that $\mu(G_W) > k$.

Let $\delta(xy) = \mu_{G_W}(xy) - \mu_{G_T}(xy)$ for each arc xy of G. Consider a directed multigraph H' with the same vertex set as G and in which xy is an arc of multiplicity $|\delta(xy)|$ if it is an arc in G and $\delta(xy) \neq 0$. We say that an arc xy of H' is *positive* (*negative*) if $\delta(xy) > 0$ ($\delta(xy) < 0$).

For a digraph D and its vertex x, let $N_D^+(x)$ and $N_D^-(x)$ denote the sets of out-neighbors and in-neighbors of x, respectively. As G_W and G_T are both Euler graphs, we have that

$$\sum_{y \in N_{H'}^+(x)} \delta(xy) - \sum_{y \in N_{H'}^-(x)} \delta(yx)$$

$$= \sum_{y \in N_G^+(x)} (\mu_{G_W}(xy) - \mu_{G_T}(xy)) - \sum_{y \in N_G^-(x)} (\mu_{G_W}(yx) - \mu_{G_T}(yx)) = 0$$

for each vertex x in G. Now create the directed multigraph H by reversing every negative arc of H' (i.e., replace every negative arc uv by the negative arc vu, keeping the weight of the arcs the same), and observe that H is balanced.

Thus, the arcs of H can be decomposed into a collection $\mathcal{C} = \{C_1, \ldots, C_t\}$ of cycles. We define the weight $\omega(C_i)$ of a cycle C_i of $\mathcal{C}$ as the sum of the weights of its positive arcs minus the sum of the weights of its negative arcs, and assume that $\omega(C_1) \leq \cdots \leq \omega(C_t)$.

Set $F_0 = G_T$ and for $i \in [t]$, construct F_i from F_{i-1} as follows: for each arc xy of C_i, if xy is a positive arc in H add a copy of xy to F_{i-1} and if xy is a negative arc in H remove a copy of yx from F_{i-1}. Thus $F_0, F_1, \ldots, F_t$ is a sequence of graphs with $F_0 = G_T, F_t = G_W$, and F_i is an Euler graph for each $i \in [t]$. Furthermore, the multiplicity of each arc xy changes by at most 1 between F_{i-1} and F_i for each $i \in [t]$, and no arc will have its multiplicity both increase and decrease over the course of $F_0, F_1, \ldots, F_t$. Therefore, every arc uv has multiplicity between $\mu_{G_T}(uv)$ and $\mu_{G_W}(uv)$ in each F_i, and so each F_i is a feasible solution for DCPP on G.

Since T is optimal, $\omega(F_0) \leq \omega(F_1) = \omega(F_0) + \omega(C_1)$ and so $\omega(C_1) \geq 0$. Due to the ordering of cycles of $\mathcal{C}$ according to their weights, $\omega(C_i) \geq 0$ for $i \in [t]$ and so $\omega(F_i) \geq \omega(F_{i-1})$ for $i \in [t]$.

Since $\mu(F_0) \leq k-1$ and $\mu(F_t) > k$, and as the multiplicity of each arc changes by at most 1 each time, there is an index j such that $\mu(F_j) = k$. Then the out-degree of some vertex of F_j is at least k and so by Lemma 1, F_j has k arc-disjoint cycles. Similarly to Lemma 5, it is not hard to show that there is a solution U of k-DCPP on G of weight $\omega(F_j)$. Since W is optimal and $\omega(F_j) \leq \omega(F_t) = \omega(G_W)$, U is also optimal and we are done.   □

3   Proof of Theorem 3

Theorem 3 is proved by providing a dynamic programming (DP) algorithm of required complexity. We first make an observation to simplify the DP algorithm.

Lemma 7. *Let $G = (V, A)$ define an instance of $k[b, c]$-DWCP. The instance has a solution of weight at most ρ if and only if there exist (not necessarily connected) non-empty directed multigraphs $G_1, \ldots, G_k$ with the following properties:*

– All multigraphs $G_1, \ldots, G_k$ use only arcs of G (each, possibly, multiple times);
– G_1 is a balanced multigraph;

– *For $2 \leq i \leq k$, G_i is a balanced digraph (with no parallel arcs);*
– *Each arc $a \in A$ occurs between b and c times in the multigraph[2] $G_1 \cup \cdots \cup G_k$, and the total weight of this multigraph is at most ρ.*

Proof. On the one hand, let $W_1, \ldots, W_k$ be a solution to the $k[b, c]$-DWCP instance of weight at most ρ, where each W_i is a closed directed walk. For each $i \in [k]$, let Q_i be the directed multigraph whose vertices are the vertices visited by W_i and which contains an arc uv of multiplicity μ if uv is traversed exactly μ times by W_i. For each $i \geq 2$, if Q_i has parallel arcs, let G_i be a cycle in Q_i and let $Q_i' = Q_i \setminus A(G_i)$ and, otherwise (i.e., Q_i has no parallel arcs), let $G_i = Q_i$ and let Q_i' be empty. Now let $G_1 = Q_1 \cup Q_2' \cup \cdots \cup Q_k'$. Observe that all properties of the lemma are satisfied.

On the other hand, consider directed multigraphs $G_1, \ldots, G_k$ satisfying the properties of the lemma. If all multigraphs G_i are connected, then they are all Euler. Therefore we can find an Euler tour W_i for each graph G_i, which forms the solution to the $k[b, c]$-DWCP instance. If $b = 0$, then we may replace each graph G_i with a cycle C_i contained in G_i, and produce a solution to $k[b, c]$-DWCP that consists of k (not necessarily pairwise arc-disjoint) cycles.

Finally, if not all multigraphs are connected and $b > 0$, we proceed as follows. First, select for each multigraph G_i, $i > 1$ an arbitrary connected component H_i, and move all other components of G_i to G_1, increasing arc multiplicity as appropriate. Next, as long as G_1 remains unconnected, let H be an arbitrary connected component of G_1. As $b > 0$ and G is connected, some component H_i, $i > 1$ must intersect a vertex of H; we may move H to the multigraph G_i and maintain that G_i is connected. Repeat this until G_1 (and hence each multigraph G_i) is connected. Note that this does not change the arc multiplicity or the weight of the solution. Now each multigraph G_i is Euler, and again we can find a solution.   □

Our DP algorithm will calculate a function $\Phi : A(G) \times [k] \rightarrow [0, c]$ corresponding to an optimal solution to the $k[b, c]$-DWCP on G. More precisely, $\Phi(a, j)$ will be the number of copies of arc a in walk number j, for each $a \in A(G), j \in [k]$. The following definitions and the next lemma allow us to express the result of Lemma 7 in terms of this function.

Given a set of arcs M and a function $\phi : M \times [k] \rightarrow [0, c]$, we say that ϕ is *valid* if for each arc $a \in M$, we have that $\sum_{j \in [k]} \phi(a, j) \in [b, c]$, and $\phi(a, j) \leq 1$ for $2 \leq j \leq k$.

Given a vertex v, we say ϕ is *balanced* for v if for each $j \in [k]$,

$$\sum_{uv \in M} \phi(uv, j) = \sum_{vu \in M} \phi(vu, j)$$

that is, v is a balanced vertex in the directed mulitgraph containing $\phi(a, j)$ copies of each arc a.

---

[2] Here, as in the proof, the union of multigraphs means that the multiplicity of an arc in the union equals the sum of multiplicities of this arc in the multigraphs of the union.

Lemma 8. *Let* $G = (V, A)$ *define an instance of* $k[b, c]$-*DWCP. The instance has a solution of weight at most* ρ *if and only if there exists a function* Φ : $A \times [k] \to [0, c]$ *such that*

1. Φ *is valid;*
2. Φ *is balanced for each vertex in* V;
3. $\sum_{a \in A} \Phi(a, j) > 0$ *for each* $j \in [k]$; *and*
4. $\sum_{j \in [k]} \sum_{a \in A} \Phi(a, j) \cdot \omega(a) \leq \rho$.

Proof. Suppose first there is a solution of weight at most ρ, and let $G_1, \ldots, G_k$ be the directed multigraphs given by Lemma 7. Let $\phi : A \times [k] \to [0, c]$ be the function such that $\phi(a, j)$ is the number of copies of an arc a in the graph G_j. As each arc appears between b and c times in $G_1 \cup \cdots \cup G_k$ and G_j has no parallel arcs for $j \geq 2$, we have that ϕ is valid. As each multigraph G_j is balanced, we have that ϕ is balanced for each vertex. As each multigraph is non-empty, we have that $\sum_{a \in A} \phi(a, j) > 0$ for each $j \in [k]$. Finally, $\sum_{j \in [k]} \sum_{a \in A} \phi(a, j) \cdot \omega(a)$ is exactly the total weight of $G_1 \cup \cdots \cup G_k$, which is at most ρ. Therefore, ϕ satisfies the conditions of the lemma.

Conversely, let $\phi : A \times [k] \to [0, c]$ be a function satisfying the conditions of the lemma, and for each $j \in [k]$, let G_j be the directed multigraph containing $\phi(a, j)$ copies of each arc a. As $\sum_{a \in A} \phi(a, j) > 0$, each multigraph G_j is non-empty. By construction, each multigraph G_j uses only arcs of G. As ϕ is balanced for each vertex, we have that each multigraph G_j is balanced. As ϕ is valid, we have that G_j has no parallel arcs for $j \geq 2$, and each arc $a \in A$ occurs between b and c times in $G_1 \cup \cdots \cup G_k$. Finally, the total weight of $G_1 \cup \cdots \cup G_k$ is $\sum_{j \in [k]} \sum_{a \in A} \phi(a, j) \cdot \omega(a) \leq \rho$. So by Lemma 7 there is a solution to the $k[b, c]$-DWCP instance of weight at most ρ.   □

Let $\theta = (v_1, v_2, \ldots, v_n)$ be a vertex ordering of a digraph G of cutwidth at most p. For each $i \in [n-1]$, let E_i be the set of arcs of the form $v_j v_h$ or $v_h v_j$, where $j \leq i$ and $h > i$. In addition let $E_0 = \emptyset$ and $E_n = \emptyset$. As θ has cutwidth at most p, $|E_i| \leq p$ for each i. We refer to $E_0, E_1, \ldots, E_n$ as the *arc bags* of θ. For each $i \in \{0, 1, \ldots, n\}$, let $E_{\leq i} = \bigcup_{0 \leq j \leq i} E_j$.

We now give an intuitive description of the DP algorithm before giving technical details. Our DP algorithm will process each arc bag of θ in turn, from E_0 to E_n. For each arc bag E_i, we store the weights of a range of partial solutions. A function ϕ is used to represent how many times each arc in the bag E_i is used by each walk in the solution. Finally, a set S provides a guarantee that certain walks are non-empty. This is to ensure we don't produce a solution which uses less than k non-empty walks.

Given $i \in [n]$, a valid function $\phi : E_i \times [k] \to [0, c]$ and a subset S of $[k]$, we define $\chi(E_i, \phi, S)$ to be the minimum integer ρ for which there exists a function $\Phi : E_{\leq i} \times [k] \to [0, c]$ satisfying the following conditions:

1. Φ *is valid;*
2. Φ *extends* ϕ, i.e. $\Phi(a, j) = \phi(a, j)$ for each $a \in E_i, j \in [k]$;
3. For each $h \leq i$, Φ *is balanced for* v_h;

4. $\sum_{a \in E_{\leq i}} \Phi(a, j) > 0$ for each $j \in S$; and
5. $\sum_{j \in [k]} \sum_{a \in E_{\leq i}} \Phi(a, j) \cdot \omega(a) \leq \rho$.

If no such integer ρ exists, then we let $\chi(E_i, \phi, S) = \infty$.

Observe that if Φ is a function satisfying the above conditions, then $\chi(E_i, \phi, S) \leq \rho$. In such a case we will call Φ a *witness* for $\chi(E_i, \phi, S) \leq \rho$. Thus, $\chi(E_i, \phi, S)$ is the minimum ρ such that there exists a witness for $\chi(E_i, \phi, S) \leq \rho$.

Note that if Φ is a witness for $\chi(E_i, \phi, S) \leq \rho$, it may be the case that $\sum_{a \in E_{\leq i}} \Phi(a, j) > 0$ for some $j \notin S$. In particular, any witness for $\chi(E_i, \phi, S) \leq \rho$ is also a witness for $\chi(E_i, \phi, S') \leq \rho$, for any $S' \subseteq S$. This allows us to simplify the recursion step in Lemma 10.

The next lemma follows from Lemma 8 and the fact that $E_n = \emptyset$ and $E_{\leq n} = A(G)$.

Lemma 9. *Let $\phi : E_n \times [k] \to [b, c]$ be the empty function. Then there is a solution for the $k[b, c]$-DCPP on G of weight at most ρ if and only if $\chi(E_n, \phi, [k]) \leq \rho$.*

We prove the following lemma in the full version of the paper.

Lemma 10. *Consider an arc bag E_i, for $i \geq 1$. Let $E_i^* = E_i \setminus E_{i-1}$. For any valid $\phi : E_i \times [k] \to [0, c]$ and $S \subseteq [k]$, let $Y = \sum_{j \in S} \sum_{a \in E_i^*} \phi(a, j) \cdot \omega(a)$, and let $S' = \{j \in S : \sum_{a \in E_i^*} \phi(a, j) = 0\}$.*
Then the following recursion holds:

$$\chi(E_i, \phi, S) = Y + \min_{\phi'} \chi(E_{i-1}, \phi', S')$$

where the minimum is taken over all valid $\phi' : E_{i-1} \times [k] \to [0, c]$ satisfying the following conditions:

- *For all $a \in E_i \cap E_{i-1}$ and all $j \in [k]$, $\phi'(a, j) = \phi(a, j)$; and*
- *The function $\phi \cup \phi'$ is balanced for v_i.*

If there is no ϕ' satisfying these conditions, then $\chi(E_i, \phi, S) = \infty$.

Furthermore, if there exists ϕ' satisfying the above conditions and we are given a witness Φ' for $\chi(E_{i-1}, \phi', S') \leq \rho'$, then we can construct a witness for $\chi(E_i, \phi, S) \leq Y + \rho'$ in polynomial time.

Note that in the above lemma, we do not need to guess the set S', as any witness Φ for $\chi(E_i, \phi, S) \leq \rho$ must have $\sum_{a \in E_{\leq i-1}} \Phi(a, j) > 0$ for each j in S' as defined in the lemma, and if a function is a witness for $\chi(E_{i-1}, \phi', S'') = \rho'$ for any $S'' \supseteq S'$ then it is also a witness for $\chi(E_{i-1}, \phi', S') = \rho'$.

We are now ready to prove Theorem 3.

Theorem 3. *Let (G, k) be an instance of $k[b, c]$-DWCP and suppose we are given a vertex ordering $\theta = (v_1, v_2, \ldots, v_n)$ of G with cutwidth at most p. Then, in time $O^*((c2^k)^p 2^k)$, we can solve (G, k) and find an optimal feasible solution if one exists.*

Proof. Our DP algorithm calculates all values $\chi(E_i, \phi, S)$ with $\phi(\cdot, j) \leq 1$ for $j > 1$ in a bottom-up manner, that is, we only calculate values $\chi(E_i, \cdot, \cdot)$ after all values $\chi(E_j, \cdot, \cdot)$ have been calculated for $0 \leq j < i$ (we use the recursion of Lemma 10).

Each arc bag E_i of θ contains at most p arcs. For each arc a, there are $c + 1$ options for $\phi(a, 1)$ and 2 options for $\phi(a, j)$ for each $j > 1$, i.e., $(c+1)2^{k-1} \leq c2^k$ options per arc. Thus there are at most $(c2^k)^p$ valid choices for $\phi : E_i \times [k] \to [0, c]$. As there are 2^k choices for a set $S \subseteq [k]$, the total size of each DP table is $O((c2^k)^p 2^k)$.

Since $E_0 = \emptyset$, the only function $\phi : E_0 \times [k] \to [0, c]$ is the empty function. It is easy to see that $\chi(E_0, \phi, S) = 0$ if $S = \emptyset$, and ∞ otherwise. To speed up the application of Lemma 10 for E_i, $1 \leq i \leq n$, we form an intermediate data structure (e.g. a hash table) T from the data for bag E_{i-1}. Call two entries $\chi(E_i, \phi, S)$ and $\chi(E_{i-1}, \phi', S')$ *compatible* when the conditions in Lemma 10 are met (i.e., $\chi(E_{i-1}, \phi', S')$ is one of the entries included in the minimisation for $\chi(E_i, \phi, S)$). Let the *signature* of entry $\chi(E_{i-1}, \phi', S')$ be $(\phi'', d_1, \ldots, d_k, S')$, where ϕ'' is ϕ' restricted to arcs $E_{i-1} \cap E_i$, and where $d_j = \sum_{a \in A^+(v_i) \cap E_{i-1}} \phi'(a, j) - \sum_{a \in A^-(v_i) \cap E_{i-1}} \phi'(a, j)$ (i.e. d_j is the imbalance at v_i in walk number j). Observe that whether an entry $\chi(E_{i-1}, \phi', S')$ is compatible with the entry $\chi(E_i, \phi, S)$ can be determined from the signature alone, and that for each $\chi(E_i, \phi, S)$ there is at most one compatible signature. Thus, for every occurring signature $(\phi'', d_1, \ldots, d_k, S')$ we let $T(\phi'', d_1, \ldots, d_k, S')$ contain the minimum value over all entries $\chi(E_{i-1}, \cdot, \cdot)$ with matching signature; this can be computed in a single loop over the entries $\chi(E_{i-1}, \cdot, \cdot)$. Then, for every entry $\chi(E_i, \phi, S)$ of the new table, we look in T for the value associated with the compatible signature (and add Y to it, by Lemma 10). Note that the size of the intermediate table T is immaterial; the time taken consists of first one loop through $\chi(E_{i-1}, \cdot, \cdot)$, then a single query to T for each entry in $\chi(E_i, \cdot, \cdot)$. Thus, the entries $\chi(E_i, \cdot, \cdot)$ can all be computed in total time $O^*((c2^k)^p 2^k)$. As $E_n = \emptyset$ there is only one function $\phi : E_n \times [k] \to [b, c]$. By Lemma 9, $\chi(E_n, \phi, [k])$ is the minimum total weight of a solution for $k[b, c]$-DCPP, and ∞ if there is no such solution. Thus to solve $k[b, c]$-DCPP it suffices to check the value of $\chi(E_n, \phi, [k])$.

Thus the algorithm finds the value ρ in time $O^*((c2^k)^p 2^k)$.

Using the method of Lemma 10, we can easily find an optimal solution to $k[b, c]$-DCPP For each arc bag $E_i, \phi : E_i \times [k] \to [0, c], S \subseteq [k]$, in addition to calculating the value $\chi(E_i, \phi, S) = \rho$, we also calculate a witness for $\chi(E_i, \phi, S) \leq \rho$, in the cases where $\rho \neq \infty$. Just as we can calculate the values of all $\chi(E_i, \cdot, \cdot)$ given the values of all $\chi(E_{i-1}, \cdot, \cdot)$, we may construct witnesses for all $\chi(E_i, \cdot, \cdot)$ given witnesses for all $\chi(E_{i-1}, \cdot, \cdot)$, using an intermediate table T as before. (Note that when $\phi : E_0 \times [k] \to [0, c]$ is the empty function, ϕ is itself a witness for $\chi(E_0, \phi, \emptyset) \leq 0$. This gives us the base case in our construction of witnesses.) Given a witness Φ for $\chi(E_n, \phi, [k]) \leq \rho$, Φ satisfies the conditions of Lemma 8. Lemma 8 shows how to construct a solution to $k[b, c]$-DCPP on G of weight at most ρ from this witness. $\qquad\square$

4   Proofs of Theorems 1 and 2

Theorem 2. *The k-ADCP-EULER is fixed-parameter tractable.*

Proof. Let D be an Euler digraph. We may assume that D has no vertex of out-degree at least k as otherwise we are done by Lemma 1. By Lemma 3, for D we can either obtain k arc-disjoint cycles or a vertex ordering θ of cutwidth at most $2g(k)$ for some function $g : \mathbb{N} \to \mathbb{N}$. Note that D is a positive instance of the k-ADCP-EULER if and only if (D, k) has a finite solution for $k[0, 1]$-DWCP (as every closed walk contains a cycle). It remains to observe that the algorithm of Theorem 3 for the $k[0, 1]$-DWCP is fixed-parameter when the out-degree of every vertex of D is upper-bounded by k and the cutwidth of θ is bounded by a function of k.   □

Theorem 1. *The k-DCPP admits a fixed-parameter algorithm.*

Proof. Let $G = (V, A)$ be a digraph and let T be an optimal solution of DCPP on G. Using Lemma 3, we can obtain either k arc-disjoint cycles of D or a vertex ordering of cutwidth bounded by a function of k. If we get a collection $\mathcal{C}$ of k arc-disjoint cycles in G_T, then using $\mathcal{C}$, by Lemma 5, we can solve the k-DCPP on G in (additional) polynomial time. So now assume we have a vertex ordering of G_T of bounded cutwidth. We may assume that every vertex of G_T is of out-degree at most $k-1$ (otherwise by Lemma 1, G_T has a collection of k arc-disjoint cycles). Since every vertex of G_T is of out-degree at most $k - 1$, the multiplicity of G_T is at most $k - 1$. Now Lemma 6 implies that there is an optimal solution W for the k-DCPP on G such that the multiplicity of G_W is at most k. Thus, we may treat the k-DCPP on G as an instance (G, k) of $k[1, k]$-DWCP. It remains to observe that the algorithm of Theorem 3 to solve the $k[1, k]$-DWCP on G will be fixed-parameter.   □

5   Discussions

Our algorithms for solving both k-DCPP and k-ADCP on Euler digraphs have very large running time bounds, mainly because the bound $f(h^{-1}(k))$ on the size of feedback arc set is very large. The function $f(k)$ obtained in [12] is a multiply iterated exponential, where the number of iterations is also a multiply iterated exponential and, as a result, $h^{-1}(k)$ grows very quickly. So obtaining a significantly smaller upper bound for $f(k)$ on Euler digraphs would significantly reduce $h^{-1}(k)$ as well and is of certain interest in itself. In particular, is it true that $f(k) = O(k^{O(1)})$ for Euler digraphs? Note that for planar digraphs, $f(k) = k$ [1, Corollary 15.3.10] and Seymour [13] proved the same result for a wide family of Euler digraphs. It would also be interesting to check whether the k-DCPP or k-ADCP admits a polynomial-size kernel.

Acknowledgement. Research of GG was supported by Royal Society Wolfson Research Merit Award.

References

1. Bang-Jensen, J., Gutin, G.: Digraphs: Theory, Algorithms and Applications, 2nd edn. Springer, New York (2009)
2. van Bevern, R., Niedermeier, R., Sorge, M., Weller, M.: Complexity of arc routing problems. In: Corberán, A., Laporte, G. (eds.) Arc Routing: Problems, Methods and Applications, SIAM, Phil. (in press)
3. Chen, J., Liu, Y., Lu, S., O'Sullivan, B., Razgon, I.: A fixed-parameter algorithm for the directed feedback vertex set problem. J. ACM **55**(5), 1–19 (2008)
4. Downey, R.G., Fellows, M.R.: Fundamentals of Parameterized Complexity. Springer, London (2013)
5. Dorn, F., Moser, H., Niedermeier, R., Weller, M.: Efficient algorithms for Eulerian extension. SIAM J. Discrete Math. **27**(1), 75–94 (2013)
6. Edmonds, J., Johnson, E.L.: Matching, Euler tours and the Chinese postman. Math. Program. **5**, 88–124 (1973)
7. Grohe, M., Grüber, M.: Parameterized approximability of the disjoint cycle problem. In: Arge, L., Cachin, C., Jurdziński, T., Tarlecki, A. (eds.) ICALP 2007. LNCS, vol. 4596, pp. 363–374. Springer, Heidelberg (2007)
8. Gutin, G., Muciaccia, G., Yeo, A.: Parameterized complexity of k-Chinese postman problem. Theor. Comput. Sci. **513**, 124–128 (2013)
9. Lin, Y., Zhao, Y.: A new algorithm for the directed Chinese postman problem. Comput. Oper. Res. **15**(6), 577–584 (1988)
10. Lucchesi, C.L.: A minimax equality for directed graphs. Ph.D. thesis, University of Waterloo, Ontario, Canada (1976)
11. Pearn, W.L.: Solvable cases of the k-person Chinese postman problem. Oper. Res. Lett. **16**(4), 241–244 (1994)
12. Reed, B., Robertson, N., Seymour, P.D., Thomas, R.: Packing directed circuits. Combinatorica **16**(4), 535–554 (1996)
13. Seymour, P.D.: Packing circuits in Eulerian digraphs. Combinatorica **16**(2), 223–231 (1996)
14. Slivkins, A.: Parameterized tractability of edge-disjoint paths on directed acyclic graphs. SIAM J. Discrete Math. **24**(1), 146–157 (2010)
15. Sorge, M.: Some Algorithmic Challenges in Arc Routing. Talk at NII Shonan Seminar no. 18, May 2013
16. Sorge, M., van Bevern, R., Niedermeier, R., Weller, M.: From few components to an Eulerian graph by adding arcs. In: Kolman, P., Kratochvíl, J. (eds.) WG 2011. LNCS, vol. 6986, pp. 307–318. Springer, Heidelberg (2011)
17. Sorge, M., van Bevern, R., Niedermeier, R., Weller, M.: A new view on Rural Postman based on Eulerian Extension and Matching. J. Discrete Alg. **16**, 12–33 (2012)
18. Thomassen, C.: On the complexity of finding a minimum cycle cover of a graph. SIAM J. Comput. **26**(3), 675–677 (1997)
19. Vygen, J.: NP-completeness of some edge-disjoint paths problems. Discrete Appl. Math. **61**(1), 83–90 (1995)
20. Zhang, L.: Polynomial algorithms for the k-Chinese postman problem. In: Information Processing '92, vol. 1, pp. 430–435 (1992)

Colored Modular and Split Decompositions of Graphs with Applications to Trigraphs

Michel Habib and Antoine Mamcarz[✉]

LIAFA UMR 7089, CNRS and Université Paris Diderot - Paris 7, 75205 Paris, France
{Habib,Mamcarz}@liafa.univ-paris-diderot.fr

Abstract. We introduce the colored decompositions framework, in which vertices of the graph can be equipped with colors, and in which the goal is to find decompositions of this graph that do not separate the color classes. In this paper, we give two linear time algorithms for the colored modular and split decompositions of graphs, and we apply them to give linear time algorithms for the modular and split decompositions of trigraphs, which improves a result of Thomassé, Trotignon and Vuskovic (2013). As a byproduct, we introduce the non-separating families that allow us to prove that those two decompositions have the same properties on graphs and on trigraphs.

1 Introduction

Modular decomposition has been introduced in [15]. Modules (also known as homogeneous sets) can help in proving structural results on many classes of graphs like comparability graphs, perfect graphs, cographs, P_4-sparse graphs, permutation graphs, interval graphs, ... Modular decomposition is also useful for solving optimization problems. Homogeneous sets (or at least some particular homogeneous sets) also appear in some decomposition theorems of trigraph classes, for example bull-free trigraphs [7] and claw-free trigraphs [8]. The split decomposition, also known as 1-join decomposition, has been introduced in [11]. It is another decomposition that has a large range of applications, from NP-hard optimization to the recognition of certain classes of graphs such as distance hereditary graphs, circle graphs and parity graphs. 1-join, or at least some particular cases of 1-joins also appears in several decomposition theorems, like for example unichord free graphs [21], claw-free trigraphs [8], or bull-free trigraphs [7].

In this paper, we introduce a generalization of these decompositions: the colored modular and split decompositions. Here, each vertex of the graph receives a color, and we want to find decompositions of this graph that do not separate the color classes. Both modular and split decomposition trees of graphs can be computed in linear time [3,10,12,13,18,19]. In the following, we will use these algorithms as black-boxes in order to derive linear time algorithms to compute the colored modular and colored split decomposition trees of graphs.

© Springer International Publishing Switzerland 2014
D. Kratsch and I. Todinca (Eds.): WG 2014, LNCS 8747, pp. 263–274, 2014.
DOI: 10.1007/978-3-319-12340-0_22

Trigraphs have been introduced in [5,6]. They have proven to be a useful tool to handle recursive graph decompositions. Even though trigraphs have received more and more attention in the past few years, their algorithmic aspects have not been completely investigated [9,20]. We will give algorithms for the modular and split decomposition of trigraphs, using the colored decompositions framework. We solve here a problem asked in [22], about the complexity of modular decomposition of trigraphs. These algorithms improve a result of [20] that finds one minimal homogeneous set of a trigraph in $O(n^2)$, while no algorithm explicitly exists to compute 1-joins of trigraphs. To our knowledge, the complexity of trigraphs decomposition problems has only been established in the case of H-joins in [17], where it is shown that the algorithms of [16] can be applied to trigraphs, yelding an $O(mn^{h-1})$ algorithm for any H-join problem, with $h = |V(H)|$.

In the following, some proofs have been omitted. It is possible to find them in [17].

2   Definitions and Preliminaries

2.1   Set Families

It has been shown in [1,4,14] that it is very helpful to first study the set families generated by a given graph decomposition since their structure characterizes these decompositions. Let us first recall some usual definitions:

Given a ground set V, two sets $X, Y \subsetneq V$ are said to *overlap* if $X \cap Y \neq \varnothing$, $X \backslash Y \neq \varnothing$, and $Y \backslash X \neq \varnothing$.

Given a ground set V, $\mathcal{P} \subseteq 2^V$ is said to be a *partitive family* if $\varnothing, V, \{v\} \in \mathcal{P}$ for every $v \in V$, and for every $X, Y \in \mathcal{P}$ such that X overlaps Y, $X \cap Y$, $X \cup Y$, $X \Delta Y$, $X \backslash Y$ and $Y \backslash X \in \mathcal{P}$.

$\{P^0, P^1\}$ is a *bipartition* of a set V if $P^0 \cap P^1 = \varnothing$ and $P^0 \cup P^1 = V$. A bipartition is said to be *elementary* if it is of the form $\{\{v\}, V \backslash v\}$.

Two bipartitions P_1 and P_2 are said to be *crossing* if for every $i \in \{0, 1\}$, for every $j \in \{0, 1\}$ P_1^i overlaps P_2^j.

Let $\mathcal{B}$ be a family of bipartitions of a ground set V. A bipartition $B \in \mathcal{B}$ is said to be *strong* if no other bipartition $B' \in B$ crosses it.

A family $\mathcal{B} = \{P_1 \ldots P_m\}$ of bipartitions of a ground set V, with $P_i = \{P_i^0, P_i^1\}$, $i = 1, \ldots, m$ is *bipartitive* if:

1. every elementary bipartition belongs to $\mathcal{B}$
2. for every crossing bipartitions P_i, P_j $\{P_i^0 \cap P_j^0, P_i^1 \cup P_j^1\} \in \mathcal{B}$, $\{P_i^0 \cap P_j^1, P_i^1 \cup P_j^0\} \in \mathcal{B}$, $\{P_i^1 \cap P_j^0, P_i^0 \cup P_j^1\} \in \mathcal{B}$, $\{P_i^1 \cap P_j^1, P_i^0 \cup P_j^0\} \in \mathcal{B}$, and $\{P_i \Delta P_j^0, P_i^0 \Delta P_j^1\} \in \mathcal{B}$

Theorem 1 *[11]. The strong members of a bipartitive family $\mathcal{B} \subset 2^V$ can be represented by a tree T such that the leaves of T are in bijection with V, and that the nodes of T are labeled* prime *or* complete *in such a way that every member of $\mathcal{B}$ is either strong or corresponds to a bipartition of the leaves of the subtrees rooted at the neighborhood of a complete node.*

Now, let us introduce non-separating families.

Let V be a ground set. Let $\mathcal{P} \subseteq 2^V$ and $\mathcal{C} \subset 2^V$ be any two families. We define $NS(\mathcal{P}, \mathcal{C})$, the *non-separating family of $\mathcal{P}$ with respect to $\mathcal{C}$* to be $\{P \in \mathcal{P}$ such that $|P| = 1 \vee (\forall C \in \mathcal{C}, P \cap C = \varnothing \vee C \subseteq P)\}$.

We will call the members of $NS(\mathcal{P}, \mathcal{C})$ the *non-separating members of $\mathcal{P}$ with respect to $\mathcal{C}$*.

Let us now study the analogous concept for families of bipartitions:

Let V be a ground set. Let $\mathcal{P} = \{P_1 \ldots P_m\}$ be any family of bipartitions of V, and let $\mathcal{C} =\subset 2^V$ be any family. We define $NS(\mathcal{P}, \mathcal{C})$, the *non-separating family of $\mathcal{P}$ with respect to $\mathcal{C}$* to be $\{P = \{X_1, X_2\} \in \mathcal{P}$ such that $|X_1| = 1 \vee |X_2| = 1 \vee (\forall C \in \mathcal{C}, C \subseteq X_1 \vee C \subseteq X_2)\}$.

We will call the members of $NS(\mathcal{P}, \mathcal{C})$ the *non-separating members of $\mathcal{P}$ with respect to $\mathcal{C}$*.

Note that, for non-separating families, and for non-separating families of bipartitions, it is enough to have $\mathcal{C}$ to be partition of V. Indeed, $X \in NS(\mathcal{P}, \mathcal{C})$ if and only if $X \in NS(\mathcal{P}, (\mathcal{C} \backslash \{C_i, C_j\}) \cup \{C_i \cup C_j\})$ for any two $C_i, C_j \in \mathcal{C}$ such that $C_i \cap C_j \neq \varnothing$, and $X \in NS(\mathcal{P}, \mathcal{C})$ if and only if $X \in NS(\mathcal{P}, \mathcal{C} \cup \{\{x\}\})$, for every $x \in V$.

Lemma 1. *Let V be a ground set. If $\mathcal{P} \subseteq 2^V$ is a partitive family, then, for every other family $\mathcal{C} = \{C_1 \ldots C_k\} \subset 2^V$, $NS(\mathcal{P}, \mathcal{C})$ is also a partitive family.*

Proof. Let X and Y be two overlapping members of $NS(\mathcal{P}, \mathcal{C})$. We want to show that $X \cap Y$, $X \cup Y$, $X \Delta Y$, $X \backslash Y$ and $Y \backslash X$ belong to $NS(\mathcal{P}, \mathcal{C})$. By definition, X and Y are also members of $\mathcal{P}$, and since $\mathcal{P}$ is a partitive family, $X \cap Y$, $X \cup Y$, $X \Delta Y$, $X \backslash Y$ and $Y \backslash X$ belong to $\mathcal{P}$. Clearly, if X and Y are members of $NS(\mathcal{P}, \mathcal{C})$, $X \cup Y \in NS(\mathcal{P}, \mathcal{C})$. Now, we only need to show that $X \cap Y$ and $X \backslash Y$ belong to $NS(\mathcal{P}, \mathcal{C})$. Assume, by contradiction, that $X \cap Y \notin NS(\mathcal{P}, \mathcal{C})$ or $X \backslash Y \notin NS(\mathcal{P}, \mathcal{C})$. By definition, there exists $C_i \in \mathcal{C}$ such that $C_i \cap (X \cap Y) \neq \varnothing$ and $C_i \cap (X \backslash Y) \neq \varnothing$, but in this case, C_i overlaps or strictly contains Y, a contradiction.

Similarly, for families of bipartitions, we have:

Lemma 2. *Let V be a ground set. Let $\mathcal{P} = \{P_1 \ldots P_m\}$ be any family of bipartitions of V. If $\mathcal{P}$ is a bipartitive family, then, for every family $\mathcal{C} = \{C_1 \ldots C_k\} \subset 2^V$, $NS(\mathcal{P}, \mathcal{C})$ is also a bipartitive family.*

Proof. Let $P_i = \{P_i^0, P_i^1\}$ and $P_j = \{P_j^0, P_j^1\}$ be two crossing members of $NS(\mathcal{P}, \mathcal{C})$. By definition, P_i and P_j are also members of $\mathcal{P}$, and since $\mathcal{P}$ is a bipartitive family, we have $\{P_i^0 \cap P_j^0, P_i^1 \cup P_j^1\} \in \mathcal{P}$, $\{P_i^0 \cap P_j^1, P_i^1 \cup P_j^0\} \in \mathcal{P}$, $\{P_i^1 \cap P_j^0, P_i^0 \cup P_j^1\} \in \mathcal{P}$, $\{P_i^1 \cap P_j^1, P_i^0 \cup P_j^0\} \in \mathcal{P}$ $\{P_i^0 \Delta P_j^0, P_i^0 \Delta P_j^1\} \in \mathcal{P}$. Now, if no member of $\mathcal{C}$ overlaps or strictly contains $P_i^k \cap P_j^l$ for all $k, l \in \{0, 1\}$, all the above bipartitions belong to $NS(\mathcal{P}, \mathcal{C})$.

Assume, by contradiction, and without loss of generality, that C_h overlaps or contains strictly $P_i^0 \cap P_j^0$. In both cases, there exists $c \in C_h$ such that $c \notin P_i^0 \cap P_j^0$ and $d \in C_h$ such that $d \in P_i^0 \cap P_j^0$. Now, either $c \in P_i^1$, which contradicts the fact that $P_i \in NS(\mathcal{P}, \mathcal{C})$, or $c \in P_j^1$, which contradicts the fact that $P_j \in NS(\mathcal{P}, \mathcal{C})$.

Let us now explain how to compute the colored modular and split decompositions trees.

2.2 Modular Decomposition

Let us now recall some basic definitions of modular decomposition of graphs.

Let $G = (V, E)$ be a graph. A set $M \subseteq V$ is a module of G if and only if for every vertex $x \in V \backslash M$, for every two vertices $u, v \in M$, $xu \in E$ if and only if $xv \in E$.

A module M of a graph $G = (V, E)$ is said to be *non-trivial* or *proper* if $|M| \geq 2$ and $|V \backslash M| \geq 1$. A graph is said to be *prime for modular decomposition* if it contains only trivial modules.

A module M of a graph $G = (V, E)$ is said to be *strong* if no other module of G overlaps it. A module (resp. strong module) is said to be *maximal* if the only module (resp. strong module) in which it is strictly contained is V.

Theorem 2 *[15]. Let $G = (V, E)$ be a graph, exactly one of the following holds:*

1. *G has only one vertex.*
2. *G is disconnected, the maximal strong modules of G are its connected components.*
3. *$\overline{G}$ is disconnected, the maximal strong modules of G are the connected components of $\overline{G}$.*
4. *both G and $\overline{G}$ are connected, the maximal strong modules of G are the maximal modules of G.*

2.3 Split Decomposition

A well-known generalization of the modular decomposition is the split decomposition defined in [11]. Let us recall the main definitions.

A *split* (1-join) of a connected graph $G = (V, E)$ is a partition of V into 4 sets V_1, V_2, V_3, V_4 such that G contains all possible edges between V_2 and V_3 and no other edges between $X_1 = V_1 \cup V_2$ and $X_2 = V_3 \cup V_4$. We say that $\{X_1, X_2\}$ induces a split of G.

A split $S = \{X_1, X_2\}$ is said to be *proper* if S is not elementary.

Theorem 3 *[14]. The family $\mathcal{S}$ of all splits of a graph is a bipartitive family.*

Theorem 4 *[14]. The family of strong splits of a graph G can be represented by a tree T such that the leaves of T are in bijection with V, and that the nodes of T are labeled* prime, clique *or* star *such that every split of G is either strong or corresponds to a bipartition of the leaves of the subtrees rooted at the neighborhood of a clique or a star node.*

3 Colored Modular Decomposition

In the following, $G = (V, E)$ will denote a graph with $|V| = n$ vertices, $|E| = m$ edges.

Definition 1. *Let $G = (V, E)$ be a graph, and $C = \{C_1 \ldots C_k\} \subset 2^V$ be a partition of V. Let $\mathcal{M}$ be the family of all modules of G. A module M of G is said to be a* colored *module of (G, C) if $M \in NS(\mathcal{M}, C)$.*

The problem we want to solve is the following:

Colored modular decomposition:
Input: A graph $G = (V, E)$, and a partition $C = \{C_1 \ldots C_k\}$ of V.
Result: All colored modules of (G, C), represented by a modular decom position tree.

To do so, we will use the following gadget described in Fig. 1:

Given $X = \{x_1 \ldots x_l\} \subset V$, we build the modular-unsplittable gadget associated to X by turning $G[X]$ into a stable set, then, for every x_j, start by adding x'_j, x^p_j, and $x^{p'}_j$, 3 non-adjacent twins of x_j. Then for every such 4 vertices, add the 3 edges $x^p_j x_j, x_j x'_j, x'_j x^{p'}_j$. Finally, add an edge between x'_j and x_{j+1} for every $j \leq l - 1$.

Let us call $\{x'_i, x^p_i, x^{p'}_i\}$ for all x_i the *auxiliary vertices*, and let $AX(x_i) = \{x'_i, x^p_i, x^{p'}_i\}$ if such vertices exists, or $\varnothing$ otherwise.

We define the graph G^{mod} associated to the graph G and the partition C to be the graph G in which every $C_i \in C$ such that $|C_i| \geq 2$ has been replaced by its modular-unsplittable gadget.

For every $C_i = \{x_1 \ldots x_l\}$ of C, $V^{mod}(C_i) = C_i \cup AX(x_1) \cdots \cup AX(x_l)$.

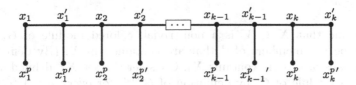

Fig. 1. $G^{mod}[V^{mod}(C_i)]$, with $|C_i| = k$

In order to prove that the above transformation preserves the colored modules of G, we will need the following lemma:

Lemma 3. *Let $(G = (V, E), C = \{C_1 \ldots C_k\})$ be an instance of the colored modular decomposition problem. Let C_i be any member of C. For every non-trivial module M of G^{mod}, either $V^{mod}(C_i) \subseteq M$ or $V^{mod}(C_i) \cap M = \varnothing$.*

Proof. This lemma holds for any C_i such that $|C_i| = 1$. Let us now consider that $|C_i| \geq 2$.

First, we show that no proper module of G^{mod} can contain exactly one vertex of $V^{mod}(C_i)$. Assume, by contradiction, that there exists a non-trivial module M of G^{mod} such that $M \cap V^{mod}(C_i) = x$. By construction, there exists x' and $x^{p'}$, two vertices of $V^{mod}(C_i)$ such that $xx' \in E$, $xx^{p'} \notin E$, and $N(x)\backslash V^{mod}(C_i) = N(x')\backslash V^{mod}(C_i) = N(x^{p'})\backslash V^{mod}(C_i)$. Since M is non-trivial, there exists at least one other vertex $y \in M$. Since M is a module of G^{mod}, and since $xx' \in E$, $x'y \in E$. But now, since $N(x')\backslash V^{mod}(C_i) = N(x^{p'})\backslash V^{mod}(C_i)$, $yx^{p'} \in E$, and since $xx^{p'} \notin E$, M is not a module of G^{mod}, a contradiction.

Now, assume by contradiction that M is a non-trivial module of G^{mod} such that $M \cap V^{mod}(C_i) \neq \varnothing$ and $V^{mod}(C_i) \not\subset M$. Since M contains at least 2 vertices of $V^{mod}(C_i)$, every pendant vertex of $G^{mod}[V^{mod}(C_i)]$ either belongs to M, or is adjacent to no vertex of M. Since every non-pendant vertex of $G^{mod}[V^{mod}(C_i)]$ is adjacent to a pendant vertex of $G^{mod}[V^{mod}(C_i)]$, M contains at least one such pendant vertex p. Let x be the only neighbor of p in $G^{mod}[V^{mod}(C_i)]$. Since $G^{mod}[V^{mod}(C_i)]$ is connected, any cut of $G^{mod}[V^{mod}(C_i)]$ contains at least one edge. As a consequence, there is at least one vertex of $V^{mod}(C_i)\backslash M$ that is adjacent to every vertex of M, and since $p \in M$, this vertex is x. Since M contains at least 2 vertices of $V^{mod}(C_i)$, and since x is adjacent to every vertex of M, M contains at least one other neighbor x' of x. Let p' be the pendant vertex of $G^{mod}[V^{mod}(C_i)]$ that is adjacent to x'. Since $p'x' \in E$ and $p'p \notin E$, p' must belong to M, but now x is no longer adjacent to every vertex of M, a contradiction.

We are now ready to prove the main lemma of this section:

Lemma 4. *Let $(G = (V,E), \mathcal{C} = \{C_1 \ldots C_k\})$ be an instance of the colored modular decomposition problem. Let X be any subset of V, and let $C_1^X \ldots C_l^X$ be the members of $\mathcal{C}$ that are contained in X. X is a non-trivial, colored module of $G, \mathcal{C}$ if and only if $X^+ = (X\backslash(\cup_{i=1}^l C_i^X)) \cup (\cup_{i=1}^l V^{mod}(C_i))$ is a non-trivial module of G^{mod}.*

Proof. Assume that $X \subset V$ is a non-trivial, colored module of G, and let $C_1^X \ldots C_l^X$ be the members of $\mathcal{C}$ that are contained in X. (By definition, no $C_i \in \mathcal{C}$ overlaps or strictly contains X). Consider the set X^+ defined as above. Since X is a module of G, by definition of G^{mod}, for every $v \in V(G)\backslash X$, for every $x, y \in X$, $xv \in E(G^{mod})$ if and only if $yv \in E(G^{mod})$. Moreover, by construction, we have that for every $C_i \in \mathcal{C}$, for every $x \in C_i$, for every $y \in AX(x)$, $N_{G^{mod}}(x)\backslash V^{mod}(C_i) = N_{G^{mod}}(y)\backslash V^{mod}(C_i)$, and so we have that for every $v \in V(G^{mod})\backslash X^+$, for every $x, y \in X^+$, $xv \in E(G^{mod})$ if and only if $yv \in E(G^{mod})$, i.e. X^+ is a module of G^{mod}.

Conversely, assume that some set $X^+ \subset V(G^{mod})$ is a non-trivial module of G^{mod}. Consider the set $X = (X^+\backslash(\cup_{i=1}^l V^{mod}(C_i))\cup(\cup_{i=1}^l C_i^X))$. By Lemma 3, we know that no $V^{mod}(C_i)$ overlaps or strictly contains X^+. Moreover, we have that $X^+ \cap V(G)$ is a module of $G^{mod}[V(G)]$. But by construction, $X = X^+ \cap V(G)$, so X is a module of $G^{mod}[V(G)]$. And since by construction, the only difference between G and the graph $G^{mod}[V(G)]$ are edges and non-edges that are either contained inside of X, or contained in $V(G)\backslash X$, X is a module of G. Since no singleton of $\mathcal{C}$ has been replaced by its modular-unsplittable gadget in G^{mod}, X is indeed non-trivial.

Theorem 5. *Given* $(G = (V, E), \mathcal{C} = \{C_1 \ldots C_k\})$, *an instance of the colored modular decomposition problem, there exists a* $O(n + m)$ *time algorithm that computes the colored modular decomposition tree of* G.

Proof. The graph G^{mod} can be computed in linear time. Indeed, it is possible to remove all the edges that are contained in any C_i in linear time. Adding a twin to a vertex x can be done in $O(d(x))$ time, and there are a constant number of twins to add to each vertex. It is possible to add an edge in constant time, and there are at most $O(n)$ edges to add in order to build the inner structure of the graphs induced by all $V^{mod}(C_i)$.

G^{mod} contains at most $4n$ vertices, and at most $4m + n$ edges, so the modular decomposition tree of G^{mod} can be computed in $O(n + m)$, using for example, algorithms from [10,18,19].

Given the modular decomposition tree of G^{mod}, it is possible to compute the colored modular decomposition tree of G in linear time, by removing the leaves corresponding to the auxiliary vertices.

Correctness of this procedure follows from Lemma 4.

4   Colored Split Decomposition

We will apply the same ideas to the split decomposition. We leave some of the proofs in appendix, as this section is quite similar to the previous one.

Definition 2. *Let* $G = (V, E)$ *be a graph, and* $\mathcal{C} = \{C_1 \ldots C_k\} \subset 2^V$ *be a partition of* V. *Let* $\mathcal{S}$ *be the family of all split of* G. *A split* S *of* G *is said to be a colored split of* $(G, \mathcal{C})$ *if* $S \in NS(\mathcal{S}, \mathcal{C})$.

The problem we want to solve is the following:

Colored split decomposition:
Input: A graph $G = (V, E)$, and a partition $\mathcal{C} = \{C_1 \ldots C_k\}$ of V.
Result: All colored splits of $(G, \mathcal{C})$, represented by a split decomposition tree.

To do so, we will use the following gadget as described in Fig. 2:

Given $C = \{x_1 \ldots x_l\} \subset V$, we build the split-unsplittable gadget associated to X by turning $G[X]$ into a stable set, then, for every x_j, start by adding $x_j^1, \ldots x_j^5$, 5 non-adjacent twins of x_j. Then for every such 6 vertices, add 6 edges such that $x_i, x_i^1, \ldots x_i^5$ induce a C_6. Finally, add an edge between x_i and x_{i+1}^3 for every $i \leq k - 1$, and one more between x_k and x_1^3.

Let us call $\{x_i^1, \ldots x_i^5\}$ for all x_i the *auxiliary vertices*, and let $AX(x_i) = \{x_i^1, \ldots x_i^5\}$ if such vertices exists, or $\emptyset$ otherwise.

We define the graph G^s associated to the graph G and the partition $\mathcal{C}$ to be the graph G in which every $C_i \in \mathcal{C}$ such that $|C_i| \geq 2$ has been replaced by its split-unsplittable gadget.

For every $C_i = \{x_1 \ldots x_l\}$ of $\mathcal{C}$, $V^s(C_i) = C_i \cup AX(x_1) \cdots \cup AX(x_l)$.
We invite the reader to consult the proofs of the 3 following claims in the appendix.

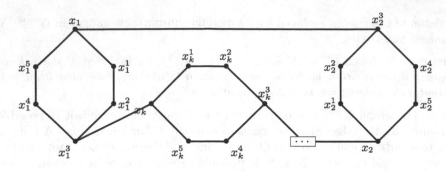

Fig. 2. $G^s[V^s(C_i)]$, with $|C_i| = k$

Lemma 5. *Let $G = (V, E)$ be a graph, and $C = \{C_1 \ldots C_k\} \subset 2^V$ be a partition of V. For every proper split $S = \{X_1, X_2\}$ of G^s, either $V^s(C_i) \subseteq X_1$, or $V^s(C_i) \subseteq X_2$.*

Lemma 6. *Let $G = (V, E)$ be a graph, and $C = \{C_1 \ldots C_k\} \subset 2^V$ be a partition of V. Let $\{X_1, X_2\}$ be any bipartition of V, let $C_1^1 \ldots C_l^1$ be the members of C that are contained in X_1, and let $C_1^2 \ldots C_m^2$ be the members of C that are contained in X_2. The bipartition $\{X_1, X_2\}$ induces a proper colored split of G, C if and only if $\{X_1^+, X_2^+\} = \{(X_1 \backslash (\cup_{i=1}^l C_i^1)) \cup (\cup_{i=1}^l V^s(C_i^1)), (X_2 \backslash (\cup_{i=1}^m C_i^2)) \cup (\cup_{i=1}^m V^s(C_i^2))\}$ induces a proper split of G^s.*

Theorem 6. *Given a graph G, there exists a $O(n + m)$ time algorithm that computes its colored split decomposition tree.*

5   Trigraphs

In this section, we will apply the colored decomposition framework to produce algorithms for the modular and split decomposition of trigraphs. These objects are slightly more general than graphs since, informally, they are graphs in which some edges are left undecided. Those edges are often called *switchable pairs*.

Definition 3. *A trigraph G is an ordered pair (V, Θ), where V is a finite set, called the vertex-set of G, and $\Theta : V \times V \to \{-1, 0, 1\}$ is a map, called the adjacency function of G, such that for all $u, v \in V, \theta(u, v) = \theta(v, u)$.*

We will consider *loopless* trigraphs, i.e. for all $u \in V, \theta(u, u) = -1$.

Definition 4. *Let $G = (V, \Theta)$ be a trigraph. A graph $G_d = (V, E)$ is said to be a realization of G if for every two vertices $u, v \in V$, $\Theta(u, v) = 1 \to uv \in E$, and $\Theta(u, v) = -1 \to uv \notin E$.*

We generalize to trigraphs a few concepts of graph theory:

Definition 5. *The complement of a trigraph $G = (V, \Theta)$ is the trigraph $\overline{G} = (V, \overline{\Theta})$ where $\overline{\Theta}(x, y) = -\Theta(x, y)$.*

Definition 6. *Let $G = (V, \Theta)$ be a trigraph. We say that two vertices x and y are semiadjacent (resp. stronglyadjacent, 0adjacent) if $\Theta(x, y) \neq -1$ ($\Theta(x, y) = 1$, $\Theta(x, y) = 0$). As usual, the* semiconnected components *(stronglyconnected components, 0connected components) of G are the equivalence classes of the transitive closure of the semiadjacency (strongadjacency, 0adjacency) relation in G, and we say that G is* semiconnected *(stronglyconnected, 0connected) if all its vertices belong to the same semiconnected (stronglyconnected, 0connected) component.*

Definition 7. *Let $G = (V, \Theta)$ be a trigraph. We say that xy is a* strong-edge *(resp.strong-non-edge, weak edge or switchable pair), if $\Theta(x, y) = 1$ ($\Theta(x, y) = -1$, $\Theta(x, y) = 0$).*

In the following, $G = (V, \Theta)$ will denote a trigraph with $|V| = n$ vertices, $m_1 = |\{uv|\Theta(u, v) = 1\}|$, and $m_0 = |\{uv|\Theta(u, v) = 0\}|$. We will assume all trigraphs are given by two adjacency lists, one containing the edges between the pairs u, v such that $\Theta(u, v) = 1$, and one containing the edges between the pairs u, v such that $\Theta(u, v) = 0$.

5.1 Modular Decomposition of Trigraphs

First, let us give the related definitions.

Definition 8. *Let $G = (V, \Theta)$ be a trigraph. A set $M \subseteq V$ is a* module *of G if and only if M is a module of every realization of G.*

It should be noticed that if we change the definition to : "there exists at least one realization for which M is a module", then we are back to the homogeneous Set sandwich problem, as studied in [2].

Note that the definitions of proper, maximal, and strong modules, as well as the definition of a prime graph can be applied to trigraphs without modifications.

We can now express this definition in terms of non-separating families:

Lemma 7. *Let $G = (V, \Theta)$ be a trigraph. A set $M \subseteq V$ is a module of G if and only if M is a colored module of (G_d, C_0), where G_d is any realization of G, and C_0 is the set of all 0connected components of G.*

Proof. First, we show that for every non-trivial module M of G, for every $C_i \in C_0$, either $C_i \subseteq M$ or $C_i \cap M = \emptyset$. Assume, by contradiction that M is a module of every realization of G, and that there exists $x \in V \backslash M$ and $y \in M$ such that $\Theta(x, y) = 0$. Let z be any other vertex of M. Without loss of generality, consider $G_d = (V, E)$, a realization of G in which $xy \in E$. Since M is a module of G_d, we have $xz \in E$. Now, consider $G'_d = (V, E' = E \backslash \{xy\})$ G'_d is still a realization of G, but M is no longer a module of G'_d since $xy \notin E'$ and $xz \in E'$.

Conversely, assume that M is a non-trivial and colored module of (G_d, C_0), where G_d is any realization of G, and C_0 is the set of all 0connected components of G. Let $G'_d = (V, E')$ be any other realization of G. By assumption, for every $x \in V \backslash M$, for every $y \in M$, $\Theta(x, y) \neq 0$, but now, by definition of a realization, we have that for every $x \in V \backslash M$, for every $y \in M$, $xy \in E$ if and only if $xy \in E'$, and since M was a module of G_d (by assumption) M is also a module of G'_d.

Corollary 1. *The modular decomposition tree of a trigraph $G = (V, \Theta)$ can be computed in $O(n + m_1 + m_0)$-time.*

Proof. The 0connected components of G can be computed in $O(n + m_0)$. We can build a realization of G in $O(n + m_1)$, and then, using Theorem 5, we can compute the modular decomposition tree of G in $O(n + m_0 + m_1)$.

Now, let us prove that this decomposition has the same properties on graphs and on trigraphs. From Lemmas 1 and 7, we have:

Corollary 2. *The set of all modules of a trigraph forms a partitive family.*

Theorem 7. *Let $G = (V, \Theta)$ be a trigraph. Exactly one of the following is true:*

1. *G has only one vertex.*
2. *G is not semiconnected, the maximal strong modules of G are its semiconnected components.*
3. *$\overline{G}$ is not semiconnected, the maximal strong modules of G are the semiconnected components of $\overline{G}$.*
4. *G contains more than one vertex, and both G and $\overline{G}$ are semiconnected, the maximal strong modules of G are the maximal modules of G.*

The proof of this theorem relies heavily on Corollary 2 and is left in appendix.

5.2   Split Decomposition of Trigraphs

Definition 9. *A bipartition of the vertices of a stronglyconnected trigraph $G = (V, \Theta)$ into X_1 and X_2 induces a split of G if and only if $\{X_1, X_2\}$ induces a split of every realization of G.*

Here again, we can reformulate this statement in term of non-separating families. We leave the proof of the equivalence in appendix.

Lemma 8. *Let $G = (V, \Theta)$ be a strongly connected trigraph. A bipartition $S = \{X_1, X_2\}$ of V induces a split of G if and only if S is a colored split of (G_d, C_0), where G_d is any realization of G, and C_0 is the set of all 0connected components of G.*

Corollary 3. *Given a trigraph G, there exists a $O(n + m_0 + m_1)$ time algorithm that computes its split decomposition tree.*

Proof. The 0connected components of G can be computed in $O(n + m_0)$. We can build a realization of G in $O(n + m_1)$, and then, using Theorem 6, we can compute the split decomposition tree of G in $O(n + m_0 + m_1)$.

Correctness of this procedure follows from Lemma 6.

And finally, we show that the split decomposition has the same behavior on graphs and trigraphs. From Lemmas 2 and 8, we have:

Corollary 4. *The family of all splits of a trigraph is a bipartitive family.*

From Theorem 1, we have:

Corollary 5. *[11] The family of strong splits of a trigraph G can be represented by a tree with nodes labeled* prime, clique *or* star *such that every split of G is either strong or corresponds to a bipartition of the neighborhood of a clique or a star node.*

Proof. We only need to show that the trigraphs G such that any bipartition of $V(G)$ induces a split are stars and cliques. Indeed, any trigraph with this property cannot contain a 0edge, and so by Theorem 4, these are exactly the stars and the cliques.

6   Conclusion and Perspectives

We gave two algorithms for usual decompositions of trigraphs, with the same running time than their graph counterparts. As a byproduct, we now have a better understanding of the modular and split decomposition of trigraphs, as we were able to prove that those decompositions have the same properties on both objects.

There is still much to be done concerning the algorithmic of trigraphs. Some interesting problems could be the skew cutset or the clique cutset problem. It would also be nice to know whether or not there exists a problem that is polynomial on graphs and NP-complete on trigraphs. Given the results of this paper and of [17], let us conjecture that the complexity of an algorithm that solves a problem on a graph and on a trigraph is the same.

References

1. Bui-Xuan, B.-M., Habib, M., Rao, M.: Tree-representation of set families and applications to combinatorial decompositions. Eur. J. Comb. **33**(5), 688–711 (2012)
2. Cerioli, M.R., Everett, H., de Figueiredo, C.M.H., Klein, S.: The homogeneous set sandwich problem. IPL **67**(1), 31–35 (1998)
3. Charbit, P., de Montgolfier, F., Raffinot, M.: Linear time split decomposition revisited. SIAM J. Discrete Math. **26**(2), 499–514 (2012)
4. Chein, M., Habib, M., Maurer, M.C.: Partitive hypergraphs. Discrete Math. **37**(1), 35–50 (1981)
5. Chudnovsky, M.: Berge trigraphs and their applications. Ph.D. thesis, Princeton University (2003)
6. Chudnovsky, M.: Berge trigraphs. J. Graph Theory **48**, 85–111 (2005)
7. Chudnovsky, M.: The structure of bull-free graphs i. J. Comb. Theory Ser. B **102**(1), 233–251 (2012)
8. Chudnovsky, M., Seymour, P.: Claw-free graphs. iv. decomposition theorem. J. Comb. Theory, Ser. B **98**(5), 839–938 (2008)
9. Chudnovsky, M., Trotignon, N., Trunck, T., Vuskovic, K.: Coloring perfect graphs with no balanced skew-partitions. CoRR, abs/1308.6444 (2013)
10. Cournier, A., Habib, M.: A new linear algorithm for modular decomposition. In: Tison, Sophie (ed.) CAAP 1994. LNCS, vol. 787, pp. 68–84. Springer, Heidelberg (1994)

11. Cunningham, W.H., Edmonds, J.: A combinatorial decomposition theory. Canad. J. Math **32**(3), 734–765 (1980)
12. Dahlhaus, E.: Efficient parallel and linear time sequential split decomposition (extended abstract). In: Thiagarajan, P.S. (ed.) FSTTCS 1994. LNCS, vol. 880, pp. 171–180. Springer, Heidelberg (1994)
13. Dahlhaus, E.: Parallel algorithms for hierarchical clustering and applications to split decomposition and parity graph recognition. J. Algorithms **36**(2), 205–240 (2000)
14. de Montgolfier, F.: Décomposition modulaire des graphes: théorie, extensions et algorithmes. Ph.D. thesis, Université Montpellier 2 (2003)
15. Gallai, T.: Transitiv orientierbare graphen. Acta Math. Acad. Scientiarum Hung. **18**(1–2), 25–66 (1967)
16. Habib, M., Mamcarz, A., Montgolfier, F.: Computing h-joins with application to 2-modular decomposition. Algorithmica **70**(2), 245–266 (2014)
17. Mamcarz, A.: Some applications of vertex splitting for graph algorithms. Ph.D. thesis, University of Paris 7 (2014)
18. McConnell, R.M., Spinrad, J.P.: Modular decomposition and transitive orientation. Discrete Math. **201**(1–3), 189–241 (1999)
19. Tedder, M., Corneil, D.G., Habib, M., Paul, C.: Simpler linear-time modular decomposition via recursive factorizing permutations. In: Aceto, L., Damgård, I., Goldberg, L.A., Halldórsson, M.M., Ingólfsdóttir, A., Walukiewicz, I. (eds.) ICALP 2008, Part I. LNCS, vol. 5125, pp. 634–645. Springer, Heidelberg (2008)
20. Thomassé, S., Trotignon, N., Vuskovic, K.: Parameterized algorithm for weighted independent set problem in bull-free graphs. CoRR, abs/1310.6205 (2013)
21. Trotignon, N., Vušković, K.: A structure theorem for graphs with no cycle with a unique chord and its consequences. ArXiv e-prints, September 2013
22. Trotignon, N.: Complexity of some trigraph problems. Private communication, 2013

Edge Elimination in TSP Instances

Stefan Hougardy$^{(\boxtimes)}$ and Rasmus T. Schroeder

Research Institute for Discrete Mathematics, University of Bonn, Bonn, Germany
hougardy@or.uni-bonn.de

Abstract. The Traveling Salesman Problem is one of the best studied NP-hard problems in combinatorial optimization. Powerful methods have been developed over the last 60 years to find optimum solutions to large TSP instances. The largest TSP instance so far that has been solved optimally has 85,900 vertices. Its solution required more than 136 years of total CPU time using the branch-and-cut based Concorde TSP code [1]. In this paper we present graph theoretic results that allow to prove that some edges of a TSP instance cannot occur in any optimum TSP tour. Based on these results we propose a combinatorial algorithm to identify such edges. The runtime of the main part of our algorithm is $O(n^2 \log n)$ for an n-vertex TSP instance. By combining our approach with the Concorde TSP solver we are able to solve a large TSPLIB instance more than 11 times faster than Concorde alone.

Keywords: Traveling salesman problem · Exact algorithm

1 Introduction

An instance of the Traveling Salesman Problem (TSP for short) consists of a complete graph on a vertex set V together with a symmetric length function $l : V \times V \to \mathbb{R}_+$. A *tour* T is a cycle that contains each vertex of the graph exactly once. The length of a tour T with edge set $E(T)$ is defined as $\sum_{e \in E(T)} l(e)$. A tour T for a TSP instance is called *optimum* if no other tour for this instance has smaller length. Finding such an optimum TSP tour is a well known NP-hard problem [5].

The Traveling Salesman Problem is one of the best studied problems in combinatorial optimization. Many exact and approximate algorithms have been developed over the last 60 years. In this paper we present several theoretical results that allow us to eliminate edges from a TSP instance that provably cannot be contained in any optimum TSP tour. Based on these results we present a combinatorial algorithm that identifies such edges. As the runtime of our main algorithm is only $O(n^2 \log n)$ for an n-vertex instance, it can be used as a preprocessing step to other TSP algorithms. On large instances our algorithm can speed up the runtime of existing exact TSP algorithms significantly. It also can improve the performance of heuristic algorithms for the TSP. We present examples for both applications in Sect. 7. For a good description of the state of the art in algorithms for the Traveling Salesman Problem see [1].

© Springer International Publishing Switzerland 2014
D. Kratsch and I. Todinca (Eds.): WG 2014, LNCS 8747, pp. 275–286, 2014.
DOI: 10.1007/978-3-319-12340-0_23

In this paper we present our results for the 2-dimensional Euclidean TSP instances, i.e., instances where the vertices are points in the Euclidean plane and the length of an edge is the (rounded) Euclidean distance between the two corresponding points. However, most of our results hold for arbitrary symmetric TSP instances that even do not need to be metric.

The idea of eliminating edges that cannot occur in any optimum TSP tour already appears in the seminal paper of Dantzig, Fulkerson, and Johnson [4]. Some additional results on eliminating such edges are proved in [10]. However, these results are useful only in combination with branch-and-bound based TSP algorithms. On the complete graph of a TSP instance almost no edge can be eliminated using these results.

Our Contribution. We present several results that allow to prove that certain edges in a TSP instance cannot belong to any optimum TSP tour. The Main Edge Elimination Theorem that we prove in Sect. 3, allows to reduce the $n(n-1)/2$ edges of an n-vertex TSP instance to about $30n$ edges or less for the TSPLIB [11] instances. In Sect. 5 we show how a weaker form of the Theorem can be applied in constant time per edge. This reduces the running time of this step to a few hours for instances containing around 100,000 vertices. Some additional methods for eliminating edges are presented in Sect. 4. We combine these with a backtrack search which we present in Sect. 6. This will allow us to reduce the number of edges in the TSPLIB instances to about $5n$ edges. The total runtime on 100,000 point instances is less than three days on a single processor. Our algorithm can be highly parallelized. For every edge an independent job can be run.

Section 7 contains the results of our algorithm on TSPLIB [11] instances as well as on a 100,000 vertex instance. Here we also show how our approach can speed up finding optimum solutions to large TSP instances significantly. The TSP solver Concorde [1] is the fastest available algorithm to solve large TSP instances optimally. Concorde needs more than 199 CPU days for the TSPLIB instance d2103. After running our edge elimination algorithm for 2 CPU days the runtime of Concorde decreases to slightly more than 16 CPU days. The total speed up we obtain is more than a factor of 11.

We also report two other successful applications of the edge elimination approach in Sect. 7.

2   Notation and Preliminaries

Most of our results do not depend on the type of TSP instance used. Some results can be improved in the Euclidean case. For convenience in this paper we restrict ourselves to Euclidean TSP instances. More precisely we use the discretized Euclidean distance function EUC_2D from the well known TSPLIB [11]. The vertices of the instance, also called points, lie in the 2-dimensional Euclidean plane. The length $l(p, q)$ of an edge between two points p and q, simply denoted by pq, results from their Euclidean distance rounded to the nearest integer. Note that the EUC_2D distance function is not metric and that an optimum TSP tour for such an instance may contain two crossing edges, which is not possible for purely Euclidean instances. To avoid some degenerate cases we assume in this paper that a TSP instance contains at least four vertices. Edges that do not belong to any optimum TSP tour will be called *useless*. An instance (V, E) consists of a set of points V and edges E, where E contains all edges of the complete graph on V except some useless edges. This implies that *all* optimum TSP tours on the complete graph on V are contained in the graph (V, E). Our edge elimination algorithm will start with some TSP instance (V, E) and return an instance (V, E') such that E' is a subset of E and contains all optimum TSP tours.

Currently, the most successful heuristic TSP algorithms [7] are based on the concept of *k-opt moves*. Given a TSP tour a k-opt move makes local changes to the tour by replacing k edges of the tour by k other edges. For a k-opt move we require, that after the replacement of the k edges the new subgraph is 2-regular. If the new subgraph is connected and thus is a tour we call the k-opt move *valid*. If a tour T allows a valid k-opt move resulting in a shorter tour, then T cannot be an optimum tour. This simple observation is the core of our algorithm for proving the existence of useless edges.

Let pq and xy be two edges in a TSP instance. We call pq and xy *compatible*, denoted by $pq \sim xy$, if

$$\max\left(l(px) + l(qy), l(py) + l(qx)\right) \geq l(pq) + l(xy). \tag{1}$$

Otherwise pq and xy are called *incompatible*. Note that two edges that have at least one vertex in common are always compatible.

Lemma 1. *Any two edges in an optimum TSP tour are compatible.*

Proof. Assume pq and xy are two incompatible edges in an optimum TSP tour T. By (1) we have $l(px) + l(qy) < l(pq) + l(xy)$ and $l(py) + l(qx) < l(pq) + l(xy)$. Thus T can be improved by a 2-opt move, that replaces edges pq and xy by either px and qy or by py and qx. One of these two 2-opt moves must be valid. This contradicts the assumption that T is an optimum TSP tour. $\square$

For $k > 2$ we call a set of k edges k-*incompatible*, if they cannot belong to the same optimum TSP tour.

3   The Main Edge Elimination Theorem

To be able to formulate our Main Edge Elimination Theorem, we need to introduce the concept of *potential points* first. Given a TSP instance (V, E), an edge pq, a point r and disjoint sets $R_1, R_2 \subset V$. We call r *potential* with respect to pq and R_1 and R_2, if for every optimum tour containing pq, one neighbor of r lies in R_1 and the other neighbor lies in R_2. We say that the sets R_1 and R_2 *certify* the potentiality of r. For r we define the set of compatible neighbors with respect to pq as

$$R_{pq}^r := \{x \in V \mid rx \in E \wedge pq \sim rx\}. \tag{2}$$

A naive approach to certify potentiality is the following lemma.

Lemma 2. *Let (V, E) be a TSP instance, $pq \in E$ and $r \in V \setminus \{p, q\}$. Let $R_1, R_2 \subseteq V$ with $R_1 \cap R_2 = \emptyset$ and $R_{pq}^r \subseteq R_1 \cup R_2$. If for $i \in \{1, 2\}$:*

$$l(pq) + l(rx) + l(ry) > l(pr) + l(rq) + l(xy) \quad \text{for all } x, y \in R_i, \qquad (3)$$

then R_1 and R_2 certify the potentiality of r.

Proof. Assume that pq is contained in an optimum tour T. Let $rx, ry \in T$. Then $x, y \in R_{pq}^r$ and hence $x, y \in R_1 \cup R_2$. Assume that $x, y \in R_i$ for $i \in \{1, 2\}$. Then replacing the edges pq, rx and ry by the edges pr, rq and xy is a valid 3-opt move. By inequality (3) this 3-opt move yields a shorter tour, contradicting the optimality of T.   □

In Sect. 5 we will develop a method that certifies the potentiality of a point in constant time.

Theorem 3 (Main Edge Elimination). *Let (V, E) be a TSP instance and $pq \in E$. Let r and s be two different potential points with respect to pq with covering R_1 and R_2 respectively S_1 and S_2. Let $r \notin S_1 \cup S_2$ and $s \notin R_1 \cup R_2$. If*

$$l(pq) - l(rs) + \min_{z \in S_1} \{l(sz) - l(pz)\} + \min_{y \in R_2} \{l(ry) - l(qy)\} > 0 \qquad (4)$$

and

$$l(pq) - l(rs) + \min_{x \in R_1} \{l(rx) - l(px)\} + \min_{w \in S_2} \{l(sw) - l(qw)\} > 0, \qquad (5)$$

then the edge pq is useless.

Proof. Assume that the edge pq is contained in an optimum TSP tour T. Let $rx, ry, sz, sw \in T$ be the incident edges of r and s. We may assume that the vertices x, y, z, and w are labeled in such a way that $x \in R_1$, $y \in R_2$, $z \in S_1$, and $w \in S_2$. As r and s are potential, we have $r, s \notin \{p, q\}$. By assumption we have $rs \notin T$, making the four edges rx, ry, sz and sw distinct.

Now there exist two possible 3-opt moves as shown in Fig. 1. The first is to replace pq, rx, and sw with px, rs, and qw. The second is to replace pq, ry, and sz with pz, rs, and qy. It is easy to verify that for every tour containing the edges pq, rx, ry, sz and sw, one of these two 3-opt moves must be valid.

The two 3-opt moves are decreasing the length of the tour T by

$$l(pq) - l(rs) + l(rx) - l(px) + l(sw) - l(qw), \qquad (6)$$

$$l(pq) - l(rs) + l(ry) - l(pz) + l(sz) - l(qy) \text{ respectively.} \qquad (7)$$

By inequalities (4) and (5), both terms are strictly positive. Since one of these 3-opt moves is valid, this yields a tour shorter than T, contradicting the optimality of T.   □

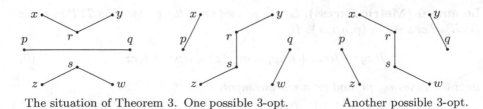

The situation of Theorem 3.   One possible 3-opt.   Another possible 3-opt.

Fig. 1. Two possible 3-opt moves that imply that the edge pq is useless.

4   The Close Point Elimination Theorems

The Main Edge Elimination Theorem will be our primary tool to prove that an edge in a TSP instance is useless. As soon as many edges of a TSP instance are known to be useless other methods can be applied. In this section we present our so called *Close Point Elimination*. When applied to the complete graph of a TSP instance it will eliminate almost no edge. However, in combination with the Main Edge Elimination Theorem it will allow to identify additional useless edges. Starting with an edge pq and a vertex r, the idea is to show that no outgoing edge pair of r can be in the same optimum tour together with the edge pq.

Theorem 4 (Close Point Elimination). *Let (V, E) be a TSP instance and $pq \in E$. Let $r \in V \setminus \{p, q\}$. If for all $x, y \in R_{pq}^r$ with $\{x, y\} \neq \{p, q\}$ the edges pq, rx and ry are 3-incompatible, then the edge pq is useless.*

Proof. Assume that an optimum tour T contains the edge pq. Let rx and ry be the two edges in T that are incident with r. Since T is a tour, we have $\{p, q\} \neq \{x, y\}$. By Lemma 1 we have $pq \sim rx$ and $pq \sim ry$. Hence $x, y \in R_{pq}^r$. The condition of the theorem gives that pq, rx and ry are 3-incompatible, contradicting the optimality of the tour. $\qquad\square$

A straight forward way to show that three edges for which two have a common vertex are 3-incompatible, is to use a simple 3-opt move.

Lemma 5 *Let pq, rx and ry be three edges of a TSP instance. If*

$$l(xy) + l(pr) + l(qr) < l(pq) + l(rx) + l(ry) \tag{8}$$

the three edges are 3-incompatible.

The proof is obvious. For the degenerate case with $x = p$ we obtain a stronger result by using the notion of *metric excess*. The metric excess $m_{pq}(z)$ of a vertex z with respect to an edge pq corresponds to the minimum length difference of a 3-opt move which shortcuts the eulerian walk which is obtained when the edge pq is inserted in a tour by adding the two edges zp and zq. It is defined as

$$m_{pq}(z) = \min_{x, y \in N(z) \setminus \{p, q\}} \max \{l(xz) + l(zp) - l(xp), l(yz) + l(zp) - l(yp),$$

$$l(xz) + l(zq) - l(xq), l(yz) + l(zq) - l(yq) \}.$$

Lemma 6 (Metric Excess). *Let pq, pr and rx be three edges of a TSP instance (V, E). Let $z \in V \setminus \{p, q, r, x\}$. If*

$$l(xq) + l(rz) + l(zp) - m_{pr}(z) < l(pq) + l(rx), \tag{9}$$

then the edges pq, pr and rx are 3-incompatible.

Proof. Assume that an optimum tour T contains the edges pq, pr and rx. We show that there exists a 3-opt move yielding a tour shorter than T. Delete the edges pq and rx and insert the edges qx, pz and rz. Note that this edge set is eulerian but not a TSP tour as vertex z has degree four. But as $l(xq) + l(rz) + l(zp) - m_{pr}(z) < l(pq) + l(rx)$ a short cut is possible that yields a tour shorter than T. This contradicts the optimality of the tour T. $\square$

5   Certifying Potential Points

The aim of this section is to show that one can prove in constant time that a point r is potential with respect to an edge pq. This helps to adopt the Main Edge Elimination Theorem such that most edges can be excluded efficiently. Remember that $l(p, q)$ denotes the rounded Euclidean distance between p and q. By $|pq|$ we denote the truly Euclidean distance of p and q. Note that

$$l(pq) - \frac{1}{2} \le |pq| \le l(pq) + \frac{1}{2}. \tag{10}$$

We now want to find a covering for a point r and an edge pq, i.e. two sets R_1, R_2 containing the set $R := \{x \in V \mid rx \in E \wedge pq \sim rx\}$ which was already defined in Sect. 3. Observe that in the truly Euclidean case the compatibility of the edges pq and rs implies that for every point t on the line segment between r and s, the edges pq and rt are also compatible. Since we use EUC_2D lengths, this only holds after adding some constants. For each vertex r choose δ_r s.t. no vertex apart from r lies in the interior of the circle around r with radius δ_r. One can for example use

$$\delta_r := \frac{1}{2} + \max\{d \in \mathbb{Z}_+ \mid \forall s \in V \setminus \{r\} \; l(rs) > d\}. \tag{11}$$

For an edge pq and a point $r \in V \setminus \{p, q\}$ define the two lengths

$$l_p := \delta_r + l(pq) - l(qr) - 1 \quad \text{and} \quad l_q := \delta_r + l(pq) - l(pr) - 1. \tag{12}$$

For each vertex $s \in V \setminus \{r\}$ define a point s_r in the Euclidean plane lying on the line segment between r and s which satisfies $|rs_r| = \delta_r$.

Lemma 7. *Let (V, E) be a TSP instance, $pq \in E$, $r \in V \setminus \{p, q\}$ and $s \in V \setminus \{r\}$. If $|ps_r| < l_p$ and $|qs_r| < l_q$ then the edges pq and rs are incompatible.*

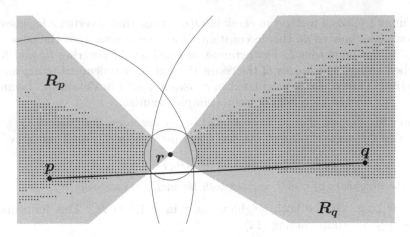

Fig. 2. The small dots indicate the set R^r_{pq}. By Lemma 7 all these vertices must be contained in the gray cones R_p and R_q.

Proof. We show that both 2-opt moves involving the edges pq and rs have shorter length. Using (10), (12), and the triangle inequality we get:

$$l(ps) + l(qr) \leq |ps_r| + |ss_r| + \frac{1}{2} + l(qr) < l_p + |ss_r| + \frac{1}{2} + l(qr)$$

$$= \delta_r + l(pq) - l(qr) - \frac{1}{2} + |ss_r| + l(qr) \leq l(pq) + l(rs)$$

$l(qs) + l(pr) < l(pq) + l(rs)$ is proven analogously. Hence pq and rs are incompatible. $\qquad\square$

Figure 2 illustrates Lemma 7. It shows two cones R_p and R_q for which $R^r_{pq} \subseteq R_p \cup R_q$. The cones are defined as $R_p := \{t \mid |qt_r| \geq l_q\}$ and $R_q := \{t \mid |pt_r| \geq l_p\}$. We do not need that the cones R_p and R_q are disjoint, although this condition is not yet sufficient to show that r is potential. However we need that they actually exist, i.e. the circles around p and q have to intersect the circle around r.

Lemma 8. *Let* (V, E) *be a TSP instance,* $pq \in E$ *and* $r \in V \setminus \{p, q\}$*. If*

$$l_p + l_q \geq l(pq) - \frac{1}{2}, \tag{13}$$

then the circle with center r *and radius* δ_r *intersects both circles with centers* p *and* q *and radii* l_p *respectively* l_q.

Proof. It suffices to show $|ir| - \delta_r \leq l_i \leq |ir| + \delta_r$ for $i \in \{p, q\}$.

$$|pr| - \delta_r \leq l(pr) + \frac{1}{2} - \delta_r \stackrel{(12)}{=} l(pq) - \frac{1}{2} - l_q \stackrel{(13)}{\leq} l_p$$

$$\stackrel{(12)}{\leq} (|pr| + |qr| + \frac{1}{2}) - (|qr| - \frac{1}{2}) + \delta_r - 1 = |pr| + \delta_r$$

Analogously for $i = q$. $\qquad\square$

Lemma 2 yields a method to check in $O(n^2)$ time that a vertex r is potential. We now show how to do this in constant time. We assume that the edge pq is part of an optimum tour T. Furthermore we consider the covering R_p and R_q as described above. The angles of the cones R_p and R_q are denoted by α_p and α_q respectively. They can be calculated in constant time (for details see Lemma 9 in [8]). The certification follows from a simple argument.

Lemma 9. *Assume that pq is contained in an optimum TSP tour T, $r \in V \setminus \{p, q\}$ and the angle γ between the two edges incident with r in T satisfies*

$$\gamma > max\{\alpha_p, \alpha_q\}. \tag{14}$$

Then the neighbors of r in T cannot both lie in R_p respectively R_q.

Proof. W.l.o.g. assume both neighbors of r in T lie in R_p. This immediately implies $\gamma \leq \alpha_p$, contradicting (14). □

It now suffices to show that inequality (14) holds for every optimum tour containing pq. This can be checked using the following statement.

Lemma 10. *Let (V, E) be a TSP instance and T an optimum tour. Let $pq \in T$ and $r \in V \setminus \{p, q\}$. Assume that inequality (13) holds. Define the angle γ_r as*

$$\gamma_r := \arccos \left(1 - \frac{\left(l_p + l_q - l(pq) + \frac{1}{2}\right)^2}{2\delta_r^2} \right). \tag{15}$$

Then the angle γ between the two edges of T incident with vertex r satisfies

$$\gamma \geq \gamma_r. \tag{16}$$

Observe that by the definitions of l_p and l_q the argument of the arccos term in (15) is not less than -1.

Proof. Let $rx, ry \in T$ be the two incident edges of r. Let $\mu := |x_r y_r|$. The cosine formula yields the following equation:

$$\mu^2 = 2\delta_r^2 - 2\delta_r^2 \cos \gamma \tag{17}$$

As T is an optimum tour, there is no valid 3-opt move which yields a shorter tour. Hence we get:

$$l(pq) + l(rx) + l(ry) \leq l(pr) + l(qr) + l(xy)$$
$$\Rightarrow \quad l_p + l_q + |x_r x| + |y_r y| - l(pq) + 1 \leq l(xy)$$
$$\Rightarrow \quad l_p + l_q - l(pq) + \frac{1}{2} \leq \mu$$
$$\overset{(13)}{\Rightarrow} \quad \left(l_p + l_q - l(pq) + \frac{1}{2}\right)^2 \leq \mu^2 = 2\delta_r^2 - 2\delta_r^2 \cos \gamma$$
$$\Rightarrow \quad \cos \gamma \leq 1 - \frac{\left(l_p + l_q - l(pq) + \frac{1}{2}\right)^2}{2\delta_r^2}$$
$$\Rightarrow \quad \gamma \geq \gamma_r. \qquad □$$

From Lemmas 9 and 10 we immediately get the following result.

Lemma 11. *Let pq be an edge contained in some optimum TSP tour T and $r \in V \setminus \{p, q\}$. Assume that inequality (13) holds. If*

$$\gamma_r > max\{\alpha_p, \alpha_q\}, \tag{18}$$

then the sets R_p and R_q certify the potentiality of r.

The results so far provide a way to prove in constant time that a given vertex is potential. Although not all potential points can be detected using this approach, sufficiently many are found as shown in Sect. 7. For simplifying notation we introduce the following concept: Let pq be an edge and $r \in V \setminus \{p, q\}$. The vertex r is called *strongly potential* (with respect to pq), if the conditions (13) and (18) hold.

Thus, checking whether a point r is strongly potential can be done in constant time (assuming that the value δ_r is known, which can be computed in a preprocessing step for all vertices). Verifying the inequalities in the Main Edge Elimination Theorem still needs $O(n)$ time.

The aim now is to show that this can bedone in constant time by computing appropriate lower bounds for (4) and (5).

Lemma 12. *Let (V, E) be a TSP instance and r strongly potential with respect to pq. Let R_p and R_q be the covering certifying r. Then*

$$\min_{x \in R_p}\{l(rx) - l(px)\} \geq \delta_r - 1 - max\{|px_r| : x \in R_p\} \text{ and} \tag{19}$$

$$\min_{y \in R_q}\{l(ry) - l(qy)\} \geq \delta_r - 1 - max\{|qy_r| : y \in R_q\}. \tag{20}$$

Proof. Let $x \in R_p$. Then

$$l(rx) - l(px) \geq |rx| - |px| - 1 \geq \delta_r + |x_r x| - (|px_r| + |x_r x|) - 1$$
$$\geq \delta_r - 1 - max\{|px_r| : x \in R_p\}$$

Similarly one can prove this for the set R_q. □

Let C_r be the circle around r with radius δ_r. Define the two arcs

$$B_p := \{x \in C_r \mid |qx| \geq l_q\} \quad \text{and} \quad B_q := \{y \in C_r \mid |py| \geq l_p\}.$$

Further let $\tilde{p}$ and $\tilde{q}$ be the points on C_r with greatest distance to p respectively q. Since B_p and B_q are connected, the maxima in the inequalities (19) and (20) can only be attained at $\tilde{p}$ respectively $\tilde{q}$, or at the endpoints of B_p and B_q respectively. Hence the equations

$$\max\{|px_r| : x \in R_p\} \leq \max\{|pt| : t \in B_p\} \quad \text{and} \tag{21}$$

$$\max\{|qy_r| : y \in R_q\} \leq \max\{|qt| : t \in B_q\} \tag{22}$$

hold if and only if

$$|p\tilde{q}| \leq l_p \quad \text{and} \quad |q\tilde{p}| \leq l_q. \tag{23}$$

The right hand sides of Eqs. (21) and (22) can easily be calculated in constant time. For details see Sect. 8.2 in [8].

6 The Algorithm

Our algorithm that eliminates useless edges consists of three independent steps. Step 1 uses the results of Sects. 3 and 5 and eliminates the majority of all edges. Step 2 applies the Main Edge Elimination in combination with the Close Point Elimination to eliminate additional edges. Finally in Step 3 we use a backtrack search of bounded depth to eliminate even more edges.

Step 1: Fast Elimination. To prove that an edge pq is useless we need to find two potential points r and s satisfying the conditions of the Main Edge Elimination Theorem. For a point r we use the method described in Sect. 5, to prove that it is potential. In fact we will only use r if it is strongly potential. This can be checked in constant time. The next step is to calculate the minima appearing in the Main Edge Elimination Theorem using Lemma 12. This can be done separately for each potential point in constant time.

Once two potential points r and s with their corresponding minima are calculated, one can check in constant time whether the inequalities of the Main Edge Elimination Theorem are satisfied. Since only two potential points which satisfy the conditions of the theorem are needed, a smart ordering and stopping criterion for checking the potentiality of points can speed up the algorithm drastically. We select the points ordered by their distance from the midpoint of the edge pq and stop after at most 10 points that have been considered.

Using a 2-d tree we compute in a preprocessing step the values δ_r for all vertices of the instance. In most cases it turns out that at most three strongly potential points have to be considered to prove that an edge is useless.

Step 2: Direct Elimination. For an edge pq we consider two vertices r and s with all their incident edge pairs. The methods of Lemmas 5 and 6 are used to eliminate as many edge pairs leaving r and s. In case no edge pairs are left for one of the vertices, the edge pq can be eliminated. Otherwise a direct application of the Main Edge Elimination Theorem can be used. For all combinations of edge pairs one has to verify the 3-opt moves from Eqs. (6) and (7) to eliminate the edge pq.

Step 3: Backtrack Search. In this step starting with an edge pq we extend a set of disjoint paths recursively. We allow two operations for the extension. Either we select one of the existing paths and add an edge incident to one of its endpoints, or we add a vertex not yet contained in any of the disjoint paths and two edges incident with this vertex. After each extension we check, similar as in step 2, whether the Main Edge Elimination or the Close Point Elimination allows to eliminate one of the path edges. In this case we backtrack. Moreover we check that the collection of paths is minimal in the sense that no collection of paths exists that has shorter length and that connects the same pairs of endpoints and uses the same set of interior points. We use an extension of the Held-Karp algorithm [6] for this. We always select the extension of the set of paths that has the smallest number of possibilities. If all extensions have been examined without reaching a predefined extension depth, then we have proven that edge pq is useless.

7   Experimental Results

We applied our algorithm to all TSPLIB [11] instances which use the EUC_2D metric as well as to some larger EUC_2D instances from [2]. Table 1 contains the results on some of these instances.

The TSP solver Concorde [1] is the fastest available algorithm to solve large TSP instances optimally. It can be downloaded at [2]. We applied Concorde to the TSPLIB instance d2103. The total runtime needed by Concorde was $17,219,190\,\text{s}$[1]. This agrees with the runtime reported for this instance on page 503 of [1]. Then we ran the three steps of our algorithm as described in Sect. 6. For Step 3 we used an extension depth of 12. After $168,153\,\text{s}$ all but $16,566$ edges were eliminated. We changed the length of all eliminated edges to some large

Table 1. Results for some TSPLIB instances with at least 1,000 vertices as well as a 100,000 vertex instance from [2]. The first three columns contain the instance name, the number of vertices and the number of edges. Then for each of the three steps as described in Sect. 6 we list the number of edges that remain after this step as well as the runtime. For Step 3 we used an extension depth of 10 which gave a reasonable trade off between the runtime and the number of eliminated edges. The last two columns contain the total runtime of our algorithm and the ratio of the number of edges remaining after Step 3 divided by the number of vertices. All runtimes are given in the format **hh:mm:ss** and are measured on a single core of a 2.9 GHz Intel Xeon.

Instance	n	m	Step 1		Step 2		Step 3		Total	Ratio
			Edges	Time	Edges	Time	Edges	Time	Runtime	
pr1002	1002	501501	42636	1	5810	2:13	4521	2:28:07	2:30:21	4.2
u1060	1060	561270	43887	1	6063	2:24	4619	3:30:48	3:33:13	4.4
vm1084	1084	586986	40958	1	6035	3:28	4610	1:17:35	1:21:05	4.3
pcb1173	1173	687378	32533	1	7662	33	6084	3:10:17	3:10:51	5.2
d1291	1291	832695	122897	4	12552	52:21	11317	13:33:14	14:25:40	8.8
rl1304	1304	849556	124561	3	21689	11:54	14527	12:53:14	13:05:11	11.1
rl1323	1323	874503	106860	2	16743	5:33	12691	9:41:18	9:46:53	9.6
nrw1379	1379	950131	28468	1	7199	1:58	5752	2:12:44	2:14:44	4.2
u1432	1432	1024596	21970	1	7817	2:51	6495	2:13:02	2:15:55	4.5
d1655	1655	1368685	230855	9	14345	37:30	12103	10:05:46	10:43:26	7.3
vm1748	1748	1526878	144681	6	12303	23:10	7691	2:48:14	3:11:30	4.4
u1817	1817	1649836	109056	5	13201	10:58	11736	6:19:22	6:30:25	6.5
rl1889	1889	1783216	206768	9	23410	3:15:40	18673	25:18:52	28:34:41	9.9
d2103	2103	2210253	166866	8	19631	55:01	18105	18:19:34	19:14:44	8.6
u2152	2152	2314476	117030	5	15101	11:17	13170	7:07:45	7:19:08	6.1
u2319	2319	2687721	21698	3	9919	44	9473	1:41:41	1:42:22	4.1
pr2392	2392	2859636	121514	7	15598	13:03	12088	7:41:45	7:44:55	5.1
pcb3038	3038	4613203	95576	8	17940	11:05	14869	5:44:08	5:55:22	4.9
fnl4461	4461	9948030	128527	15	23963	9:30	19082	7:14:21	7:24:07	4.3
brd14051	14051	98708275	2661869	4:39	93497	18:50:15	64486	28:39:22	47:34:16	4.6
d15112	15112	114178716	1703765	5:51	130110	10:25:44	66010	38:37:42	49:09:17	4.4
d18512	18512	171337816	1449877	5:30	112681	1:49:35	84203	32:38:07	34:33:13	4.5
mona-lisa100k	100000	4999950000	2071297	3:45:21	476001	22:51	322716	55:42:51	59:51:04	3.2

---

[1] Runtime on a 2.9 GHz Intel Xeon [2]. We took the average runtime of two independent runs. Log-files are at http://www.or.uni-bonn.de/~hougardy/EdgeElimination.

value and gave this new instance again to Concorde. On this instance Concorde needed $1,392,582$ s. Thus the total runtime was improved by our edge elimination algorithm by more than a factor of 11.

Two other successful applications of our edge elimination approach were reported to us by Cook [3]. First, the edge elimination approach in combination with the LKH algorithm [7] improved the so far best known TSP tour for the DIMACS instance E100k.0 [9]. The shortest tour known so far had length $225,786,982$. It was found using the LKH algorithm. Cook's implementation of our edge elimination approach eliminated all but $274,741$ edges in the TSP instance E100k.0. By applying the LKH algorithm to this edge set a tour of length $225,784,127$ was found [9]. Secondly, Cook applied the edge elimination approach to a truly Euclidean instance (i.e., a Euclidean instance where the point distances are not rounded). Finding optimum TSP tours in such instances is much harder than in instances with rounded Euclidean norm. The largest truly Euclidean instance that Cook was able to solve so far had 500 points. With the help of the edge elimination approach he solved an instance with 1000 points.

Acknowledgement. We are very grateful to Bill Cook for supplying us with some data and several helpful comments. We also thank our reviewers for their careful reading and useful comments.

References

1. Applegate, D.L., Bixby, R.E., Chvátal, V., Cook, W.J.: The Traveling Salesman Problem: A Computational Study. Princeton University Press, Princeton (2006)
2. W.J. Cook's TSP website at http://www.math.uwaterloo.ca/tsp/
3. W.J. Cook. Personal communication, December 2013
4. Dantzig, G., Fulkerson, R., Johnson, S.: Solution of a large-scale traveling-salesman problem. J. Oper. Res. Soc. Am. **2**(4), 393–410 (1954)
5. Garey, M.R., Johnson, D.S.: Computers and Intractability. Freeman, New York (1979)
6. Held, M., Karp, R.M.: A dynamic programming approach to sequencing problems. J. Soc. Ind. Appl. Math. **10**(1), 196–210 (1962)
7. Helsgaun, K.: General k-opt submoves for the Lin-Kernighan TSP heuristic. Math. Program. Comput. **1**(2–3), 119–163 (2009)
8. Hougardy, S., Schroeder, R.: Edge Elimination in TSP Instances (2014). arXiv:1402.7301v1
9. Helsgaun, K.: http://www.akira.ruc.dk/~keld/research/LKH/DIMACS_results. html
10. Jonker, R., Volgenant, T.: Nonoptimal edges for the symmetric traveling salesman problem. Oper. Res. **32**(4), 837–846 (1984)
11. Reinelt, G.: TSPLIB 95. Interdisziplinäres Zentrum für Wissenschaftliches Rechnen (IWR), Heidelberg (1995)

The Parameterized Complexity of the Rainbow Subgraph Problem

Falk Hüffner, Christian Komusiewicz$^{(\boxtimes)}$, Rolf Niedermeier,
and Martin Rötzschke

Institut für Softwaretechnik und Theoretische Informatik,
TU Berlin, Berlin, Germany
{falk.hueffner,christian.komusiewicz,rolf.niedermeier}@tu-berlin.de

Abstract. The NP-hard RAINBOW SUBGRAPH problem, motivated from bioinformatics, is to find in an edge-colored graph a subgraph that contains each edge color exactly once and has at most k vertices. We examine the parameterized complexity of RAINBOW SUBGRAPH for paths, trees, and general graphs. We show, for example, APX-hardness even if the input graph is a properly edge-colored path in which every color occurs at most twice. Moreover, we show that RAINBOW SUBGRAPH is W[1]-hard with respect to the parameter k and also with respect to the dual parameter $\ell := n - k$ where n is the number of vertices. Hence, we examine parameter combinations and show, for example, a polynomial-size problem kernel for the combined parameter ℓ and "maximum number of colors incident with any vertex".

1 Introduction

The RAINBOW SUBGRAPH problem is defined as follows.

RAINBOW SUBGRAPH
Instance: An undirected graph $G = (V, E)$, an edge coloring col $: E \to \{1, \dots, p\}$ for some $p \geq 1$, and an integer $k \geq 0$.
Question: Is there a subgraph G' of G that contains each edge color exactly once and has at most k vertices?

We call a subgraph G' with these properties a *solution* of order at most k. In the problem name, the term *rainbow* refers to the fact that all edges of G' have a different color. For convenience, we define a *rainbow cover* as a subgraph where every color occurs at least once. Note that every rainbow cover G' of order at most k has a subgraph that is a solution: Simply remove any edge whose color appears more than once in G'. Repeating this operation as long as possible yields a solution of the same order as G'.

Falk Hüffner: Supported by DFG project ALEPH (HU 2139/1).
Christian Komusiewicz: Partially supported by a post-doctorial grant funded by the Région Pays de la Loire.

D. Kratsch and I. Todinca (Eds.): WG 2014, LNCS 8747, pp. 287–298, 2014.
DOI: 10.1007/978-3-319-12340-0_24

Table 1. Complexity overview for RAINBOW SUBGRAPH. The $O^*()$-notation suppresses factors polynomial in the input size.

Par.	Paths	Trees	General graphs
p	$O^*(3^p)$ (Theorem 4)	$O^*(3^p)$ (Theorem 4)	W[1]-hard (Theorem 2)
p, Δ	—"—	—"—	$O^*((4\Delta - 4)^p)$ (Theorem 3)
k	$O^*(3^k)$ (Theorem 4+(2))	$O^*(3^k)$ (Theorem 4+(3))	W[1]-hard (Theorem 2+(1))
k, Δ	—"—	—"—	$O^*(2^{k\Delta/2})$ (Theorem 3)
ℓ	$O^*(5^\ell)$ (Theorem 8)	W[1]-hard (Theorem 5)	W[1]-hard (Theorem 5)
ℓ, Δ_C	—"—	$O^*((2\Delta_C + 1)^\ell)$ (Theorem 8)	$O^*((2\Delta_C + 1)^\ell)$ (Theorem 8)
			$O(\Delta_C^3 \ell^4)$-vertex kernel (Theorem 7)
ℓ, q	—"—	W[1]-hard (Theorem 5)	W[1]-hard (Theorem 5)
Δ, q	APX-hard (Theorem 1)	APX-hard (Theorem 1)	APX-hard ([8])

RAINBOW SUBGRAPH arises in bioinformatics: The (POPULATION) PARSIMONY HAPLOTYPING problem can be reduced to RAINBOW SUBGRAPH [11]; note, however, that depending on the input, this reduction might not produce a polynomial-size instance. Another bioinformatics application appears in the context of PCR primer set design for spotted microarray experiments [5].

Previous work. The optimization version of RAINBOW SUBGRAPH has been mostly studied in terms of polynomial-time approximability. Here the optimization goal is to minimize the number of vertices in the solution; we refer to this problem as MINIMUM RAINBOW SUBGRAPH. MINIMUM RAINBOW SUBGRAPH is APX-hard even on graphs with maximum vertex degree $\Delta \geq 2$ in which every color occurs at most twice [8]. Moreover, MINIMUM RAINBOW SUBGRAPH cannot be approximated within a factor of $c \ln \Delta$ for some constant c unless NP has slightly superpolynomial time algorithms [12].

The more general MINIMUM-WEIGHT MULTICOLORED SUBGRAPH problem has a randomized $\sqrt{q \log p}$-approximation algorithm, where q is the maximum number of times any color occurs in the input graph [7]. MINIMUM RAINBOW SUBGRAPH can be approximated within a ratio of $(\delta + \ln\lceil \delta \rceil + 1)/2$, where δ is the average vertex degree in the solution [9], and within a factor of $\max(\sqrt{2n}, \sqrt{\Delta}(1+\sqrt{\ln \Delta/2}))$ [12]. Katrenič and Schiermeyer [8] present an exact algorithm for RAINBOW SUBGRAPH that has a running time of $n^{O(1)} \cdot 2^p \cdot \Delta^{2p}$, where Δ is the maximum vertex degree of the input.

Our contributions. Since RAINBOW SUBGRAPH is NP-hard even on collections of paths and cycles [8], we perform a broad parameterized complexity analysis. Table 1 gives an overview on the complexity of MINIMUM RAINBOW SUBGRAPH on paths, trees, and general graphs, when parameterized by

- p: number of colors;
- k: number of vertices in the solution;
- $\ell := n - k$: number of vertex deletions to obtain a solution;
- Δ: maximum vertex degree;
- $\Delta_C := \max_{v \in V} |\{c \mid \exists \{u, v\} \in E : \mathrm{col}(\{u, v\}) = c\}|$: maximum color degree;
- q: maximum number of times any color occurs in the input graph.

For each parameter and some parameter combinations, we give either a fixed-parameter algorithm, show W[1]-hardness, or show NP-hardness for constant parameter values.

Our main results are as follows: RAINBOW SUBGRAPH is APX-hard even if the input graph is a properly edge-colored path with $q = 2$. RAINBOW SUBGRAPH is W[1]-hard on general graphs for each of the considered parameters. For the number of colors p, solution order k, and number ℓ of vertex deletions, the complexity seems to depend on the density of the graph as the problem is W[1]-hard for each of these parameters but it becomes tractable if any of these parameters is combined with the maximum degree Δ. For the parameter ℓ, W[1]-hardness holds even if the input graph is a tree.

Preliminaries. APX is the class of optimization problems that allow constant-factor approximations. If a problem is APX-hard, then it cannot be approximated in polynomial time to arbitrary constant factors, unless P = NP. A problem is called *fixed-parameter tractable* (FPT) with respect to some problem-specific parameter x if it can be solved in $f(x) \cdot |I|^{O(1)}$ time, where $|I|$ is the instance size and f is an arbitrary computable function. A *kernel* for a parameterized problem is, roughly, a polynomial-time self-reduction that results in an instance whose size is bounded only in the parameter. Analogously to NP, the class W[1] captures parameterized hardness. It is widely assumed that if a problem is W[1]-hard, then it is not fixed-parameter tractable.

We will use the following simple observation several times.

Observation 1. *Let $G' = (V', E')$ be a solution for a RAINBOW SUBGRAPH instance with $G = (V, E)$. If there are two vertices u, v in V' such that $\{u, v\} \in E$ but $\{u, v\} \notin E'$, then there is a solution G'' that does contain the edge $\{u, v\}$ and has the same number of vertices.*

Observation 1 is true since replacing the edge in G' that has the same color as $\{u, v\}$ by $\{u, v\}$ is a solution. Next, we list some easy to see observations regarding parameter bounds:

$$p \le k(k - 1)/2, \tag{1}$$
$$p \le k\Delta/2, \tag{2}$$
$$p \le k - 1 \qquad \text{if } G \text{ is acyclic.} \tag{3}$$

Due to lack of space, some proofs are deferred to a long version of this article.

2   Parameterization by Color Occurrences

We now consider the complexity of RAINBOW SUBGRAPH parameterized by the maximum number of color occurrences q. Indeed, the value q is bounded in some applications: For example in the graph formulation of PARSIMONY HAPLOTYP-ING, q depends on the maximum number of ambiguous positions in a genotype which can be assumed to be small. Unfortunately, RAINBOW SUBGRAPH remains hard under q-parameterization, even for heavily restricted graph classes.

Katrenič and Schiermeyer [8] showed that MINIMUM RAINBOW SUBGRAPH is APX-hard for $\Delta = 2$. The instances produced by their reduction contain precisely two edges of each color, so APX-hardness even holds for $q = 2$. However, the resulting graph contains cycles and is not properly edge-colored, so the complexity on acyclic graphs and on properly edge-colored graphs (like those resulting from PARSIMONY HAPLOTYPING instances) remains to be explored. We show that neither restriction is helpful as RAINBOW SUBGRAPH is APX-hard for properly edge-colored paths with $q = 2$. This strengthens the hardness result of Katrenič and Schiermeyer [8]. For this purpose, we develop an L-reduction from the following special case of MINIMUM VERTEX COVER:

MINIMUM VERTEX COVER IN CUBIC GRAPHS
Instance: An undirected graph $H = (W, F)$ in which every vertex has degree three.
Task: Find a minimum-cardinality vertex cover of G.

MINIMUM VERTEX COVER IN CUBIC GRAPHS is APX-complete [1].

Theorem 1. MINIMUM RAINBOW SUBGRAPH *is APX-hard even when the input is a properly edge-colored path in which every color occurs at most twice.*

Proof. Given an instance $H = (W = \{w_1, \ldots, w_n\}, F)$ of MINIMUM VERTEX COVER IN CUBIC GRAPHS, construct an edge-colored path $G = (V, E)$ as follows. The vertex set is $V := \{v_1, \ldots, v_{16n+2}\}$. The edge set is $E := \{\{v_i, v_{i+1}\} \mid 1 \leq i \leq 16n + 1\}$, that is, vertices with successive indices are adjacent. It remains to specify the edge colors. Herein, we use u^* to denote *unique* colors, that is, if an edge is u^*-colored, then it receives an edge color that does not appear anywhere else in G. In addition to these unique colors, introduce five colors for each vertex of H, that is, for each $w_i \in W$ create edge colors c_i, c_i', c_i'', x_i, and y_i. The colors c_i, c_i', and c_i'' are "filling" colors which are needed because G is connected. Furthermore, for each edge $f_i \in F$ introduce an edge color ϕ_i.

Now, color the first $6n + 1$ edges of G by the sequence

$$u^* \, c_1 \, u^* \, c_1' \, u^* \, c_1'' \, u^* \, c_2 \, u^* \, c_2' \, u^* \, c_2'' \, u^* \, \cdots \, c_2 \, u^* \, c_2' \, u^* \, c_2'' \, u^*.$$

That is, the edge between v_0 and v_1 is u^*-colored, the edge between v_1 and v_2 receives color c_1, and so on. The u^*-colors are unique and thus occur only once in G. Thus, both endpoints of these colors are contained in every solution.

Now for each vertex w_i in H color 10 edges in G according to the edges that are incident with w_i. More precisely, for each w_i color the edges from $v_{6n+2+10(i-1)}$ to $v_{6n+2+10i}$. We call the subpath of G with these vertices the w_i-*part* of G. Let $\{f_r, f_s, f_r\}$ denote the set of edges incident with w_i. Then color the edges between $v_{6n+2+10(i-1)}$ and $v_{6n+2+10i}$ by the sequence

$$c_i \, \phi_r \, x_i \, \phi_s \, c_i' \, y_i \, \phi_t \, c_i'' \, x_i \, y_i.$$

That is, the edge between $v_{6n+2+10(i-1)}$ and $v_{6n+2+10(i-1)+1}$ receives color c_i, the edge between $v_{6n+2+10(i-1)+1}$ and $v_{6n+2+10(i-1)+2}$ receives color ϕ_r, and so on.

The resulting graph is a path with exactly $16 \cdot n + 1$ edges and $p = 8 \cdot n + |F| + 1$ colors.

The idea of the construction is that we may use the vertices of the w_i-part to "cover" the colors corresponding to the edges incident with w_i. If we do so, then the solution has two connected components in the w_i-part. Otherwise, it is sufficient to include one connected component from the w_i-part. Since the solution graph is acyclic and the number of edges in a minimal solution is fixed, the number of connected components in the solution and its order are equal up to an additive constant.

We now show formally that the reduction fulfills the two properties of L-reductions [14]. Let S^* be an optimal vertex cover for the MINIMUM VERTEX COVER IN CUBIC GRAPHS instance and let G^* be an optimal solution to the constructed MINIMUM RAINBOW SUBGRAPH instance.

The first property we need to show is that $|V(G^*)| = O(|S^*|)$. As observed above, the number of colors p in G is $O(n + |F|)$ and thus $|V(G^*)| \leq 2p = O(n + |F|)$. Clearly, S^* contains at least $|F|/3$ vertices, since every vertex in H covers at most three edges. Moreover, since H is cubic we have $n < 2|F|$ and thus $|S^*| = \Theta(n + |F|)$. Consequently, $|V(G^*)| = O(|S^*|)$.

The second property we need to show is the following: given a solution G' to G, we can compute in polynomial time a solution S' to G such that

$$|S'| - |S^*| = O(|V(G')| - |V(G^*)|).$$

Let G' be a solution to G. The proof outline is as follows. We show that G' has order at least $p + n + 1 + x$, $x \geq 0$, and that, given G', we can compute in polynomial time a size-x vertex cover S' of H. Then we show that, conversely, there is a solution of order at most $p + n + 1 + |S^*|$. Thus, the differences between the solution sizes in the MINIMUM VERTEX COVER IN CUBIC GRAPHS instance and in the MINIMUM RAINBOW SUBGRAPH instance are essentially the same. We omit the details. $\qquad\square$

3 Parameterization by Number of Colors

We now consider the parameter number of colors p. We show that RAINBOW SUBGRAPH is generally W[1]-hard with respect to p but becomes fixed-parameter tractable if the input graph is sparse. By Eq. (1), and the fact that we can always construct a solution by arbitrarily selecting one edge of each color, implying $k \leq 2p$, the parameter p is polynomially upper- and lower-bounded by the solution order k. In consequence, while our main focus is on parameter p, every parameterized complexity result for p also implies the corresponding parameterized complexity result for k.

A graph G is called d-degenerate if every subgraph of G has a vertex of degree at most d. We can show that even on 2-degenerate bipartite graphs, the decision problem RAINBOW SUBGRAPH is W[1]-hard for parameter p (and thus also for parameter k) by a parameterized reduction from the MULTICOLORED CLIQUE problem.

Theorem 2. MINIMUM RAINBOW SUBGRAPH *is W[1]-hard with respect to the number of colors* p, *even if the input graph is 2-degenerate and bipartite.*

Replacing degeneracy by the larger parameter maximum degree Δ of G yields fixed-parameter tractability: Katrenič and Schiermeyer [8] proposed an algorithm that solves MINIMUM RAINBOW SUBGRAPH in $(2\Delta^2)^p \cdot n^{O(1)}$ time. We show an improved bound:

Theorem 3. *Let* (G, col) *be an instance of* MINIMUM RAINBOW SUBGRAPH *with* p *colors and maximum vertex degree* Δ. *An optimal solution can be computed in* $O((4\Delta - 4)^p \cdot \Delta n^2)$ *time or in* $O((4\Delta - 4)^k \cdot n^2 + 2^{k\Delta/2} \cdot (k\Delta)^3 \log(k\Delta))$ *time, where* k *is the order of the solution.*

To prove Theorem 3, we follow a two-step approach: First, we enumerate connected candidate subgraphs exploiting the sparseness constraint. Second, we select from these candidate subgraphs a minimum-order set with all colors, exploiting techniques by Björklund et al. [2].

The algorithm by Katrenič and Schiermeyer [8] has a somewhat different structure, but can also be understood in terms of a subgraph enumeration process and a combinatorial part: It employs a method for enumerating all connected rainbow subgraphs in $O(\Delta^{2p} \cdot np)$ time and finds a solution via dynamic programming. In contrast, we consider only connected *induced* subgraphs in the first step, which improves efficiency.

In the second step, we select from the computed set of connected subgraphs a minimum order subset with all colors. Clearly, those subgraphs correspond to the connected components of some optimal solution, which can be retrieved by stripping edges with redundant colors. The second step reduces to MINIMUM-WEIGHT SET COVER when we consider the induced subgraphs as sets (of colors) which are weighted (by the order of the subgraph). We first describe an algorithm for MINIMUM-WEIGHT EXACT COVER using fast subset convolution and then use it to solve MINIMUM-WEIGHT SET COVER. To improve efficiency, we apply techniques by Björklund et al. [2].

Step one: enumerating induced subgraphs. We make use of the following lemma:

Lemma 1 ([10, Lemma 2]). *Let* G *be a graph with maximum degree* Δ *and let* v *be a vertex in* G. *There are at most* $4^k \cdot (\Delta - 1)^k$ *connected (induced) subgraphs of* G *that contain* v *and have order at most* k. *Furthermore, these subgraphs can be enumerated in* $O(4^k \cdot (\Delta - 1)^k \cdot n)$ *time.*

Clearly, we can enumerate all connected induced subgraphs of G of order at most k by applying Lemma 1 for each vertex $v \in V(G)$.

Step two: MINIMUM-WEIGHT SET COVER. We consider MINIMUM-WEIGHT SET COVER instances with input sets $\mathcal{C} = \{C_1, \ldots, C_m\}$ and weight function w, where n denotes the cardinality of the ground set $U := \bigcup_{C_i \in \mathcal{C}} C_i$, and $w(\mathcal{C}')$ for $\mathcal{C}' \subseteq \mathcal{C}$ denotes the sum of weights of the sets in $\mathcal{C}'$.

MINIMUM-WEIGHT SET COVER can be solved in $O(2^m)$ time using polynomial space by exhaustive search and in $O(2^n m)$ time using exponential space by dynamic programming. Cygan et al. [4] presented a polynomial-space algorithm with running time $O^*(\min\{4^n m^{\log n}, 9^n\})$. For our application of MINIMUM-WEIGHT SET COVER these algorithms are somewhat ill-suited since m may be potentially as large as 2^n, resulting in $4^n \cdot n^{O(1)}$ running times. Better algorithms are known for the unweighted MINIMUM SET COVER problem which can be solved in $O(2^{0.299(n+m)})$ time [6] and $2^n n^{O(1)}$ time [3], where the second running time avoids the m factor. In the following, we use *fast subset convolution* [2] to obtain an $2^n (nW)^{O(1)}$-time algorithm for MINIMUM-WEIGHT SET COVER, where W is the maximum weight.

We use the following lemma due to Björklund et al. [2]:

Lemma 2 ([2]). *Consider a set U with $|U| = n$ and a mapping $Q : 2^U \to \{0, \ldots, W\}$. The mapping Q^1 with $Q^1[U'] = \min_{U'' \subseteq U'}(Q[U''] + Q[U' \setminus U''])$ for every $U' \subseteq U$ is called the* convolution *of Q and can be computed in $O(2^n n^3 W \log^2(nW))$ time.*

Björklund et al. [2] did not give precise running time estimates, but Lemma 2 can be derived using their Theorem 1, assuming $O(n \log^2 n)$ time for addition and multiplication of n-bit integers.

As Björklund et al. [2] noted, partitioning problems over the set U can be solved by computing multiple convolutions. We describe in the following the algorithm for MINIMUM-WEIGHT EXACT COVER (the variant of weighted SET COVER where each element needs to be covered by exactly one set) and then how to use the result to solve MINIMUM-WEIGHT SET COVER.

MINIMUM-WEIGHT EXACT COVER
Instance: A family $\mathcal{C}$ of sets with weight function $w : \mathcal{C} \to \{0, \ldots, W\}$.
Task: Find a minimum-weight subfamily $\mathcal{S} \subseteq \mathcal{C}$ such that each element of $\bigcup_{C_i \in \mathcal{C}} C_i$ occurs in exactly one set in $\mathcal{S}$.

Lemma 3. MINIMUM-WEIGHT EXACT COVER *with weight function $w : \mathcal{C} \to \{0, \ldots, W\}$ can be solved in $O(2^n \cdot n^3 W \log(n) \log^2(nW))$ time.*

Proof. We define an x-*cover* of a subset $U' \subseteq U$ to be a minimum-weight subfamily $\mathcal{C}' \subseteq \mathcal{C}$ containing at most x sets such that each element of U' occurs in exactly one set of $\mathcal{C}'$ and $\bigcup_{C_i \in \mathcal{C}'} C_i = U'$. In these terms MINIMUM-WEIGHT EXACT COVER is to find an n-cover for U.

Consider a mapping $Q : 2^U \to \{0, \ldots, W\}$ and let initially $Q[C_i] = w(C_i)$ for $C_i \in \mathcal{C}$ and $Q[U'] = \infty$ for the remaining $U' \subseteq U$. Now let Q^x denote the mapping resulting from x consecutive convolutions of Q, that is, $Q^0 = Q$ and Q^{x+1} is the convolution of Q^x. We prove by induction that (for all $U' \subseteq U$ and all $x \geq 0$) $Q^x[U']$ is the weight of a 2^x-cover for U' if such a cover exists and $Q^x[U'] = \infty$ otherwise. This implies in particular that $Q^{\lceil \log_2 n \rceil}[U]$ is the weight of an optimal solution to $\mathcal{C}$, if a solution exists.

Clearly the mapping $Q^0 = Q$ meets the claim. Assume that $Q^x[U']$ is the weight of a 2^x-cover for $U' \subseteq U$ if such a cover exists, and $Q^x[U'] = \infty$ otherwise. Now let $\mathcal{C}'$ be a 2^{x+1}-cover for some $U' \subseteq U$. Let $\mathcal{C}_\alpha, \mathcal{C}_\beta \subseteq \mathcal{C}'$ be disjoint subfamilies, $\mathcal{C}_\alpha \cup \mathcal{C}_\beta = \mathcal{C}'$, such that $|\mathcal{C}_\alpha| \leq 2^x$ and $|\mathcal{C}_\beta| \leq 2^x$. (If $|\mathcal{C}'| = 1$, then $\mathcal{C}_\alpha = \mathcal{C}'$ and $\mathcal{C}_\beta = \emptyset$). Let $U_\alpha = \bigcup_{C_i \in \mathcal{C}_\alpha} C_i$, $U_\beta = \bigcup_{C_i \in \mathcal{C}_\beta} C_i$. Now $\mathcal{C}_\alpha$ is a 2^x-cover for U_α: it covers each element of U_α exactly once, and if there was an exact cover with lower weight, we could combine it with $\mathcal{C}_\beta$ to get an exact cover for $\bigcup_{C_i \in \mathcal{C}'} C_i$ with lower weight than $\mathcal{C}'$, contradicting that $\mathcal{C}'$ is a 2^{x+1}-cover. The same holds for $\mathcal{C}_\beta$. Hence, $Q^x[U_\alpha] = w(\mathcal{C}_\alpha)$ and $Q^x[U_\beta] = w(\mathcal{C}_\beta)$, therefore $w(\mathcal{C}') = Q[U_\alpha] + Q[U_\beta]$, and due to the minimality of $w(\mathcal{C}')$ we obtain (by convolution) $Q^{x+1}[U'] = \min_{U'' \subseteq U'}(Q[U''] + Q[U' \setminus U'']) = w(\mathcal{C}')$. So $Q^{x+1}[U']$ is the weight of a 2^{x+1}-cover for U'. If no 2^{x+1}-cover for U' exists, then there is no $U'' \subseteq U'$ such that $Q^x[U''] \neq \infty$ and $Q^x[U' \setminus U''] \neq \infty$, hence $Q^{x+1}[U'] = \infty$.

To retrieve the actual solution family, we search for some $U' \subseteq U$ such that $Q^{\lceil \log_2 n \rceil}[U'] + Q^{\lceil \log_2 n \rceil}[U \setminus U'] = Q^{\lceil \log_2 n \rceil}[U]$. We repeat this step for U' and $U \setminus U'$ recursively, until we obtain subsets of U that have a 1-cover. The union of those 1-covers are the sets of the solution family.

The initial mapping Q can be constructed within $O(nm) = O(2^n n)$ time. Next, we compute $\lceil \log_2 n \rceil$ convolutions of Q, each of which takes $O(2^n n^3 W \log^2(nW))$ time, by Lemma 2. Retrieving the solution family takes $O(2^n n)$ time, so we obtain an overall running time of $O(2^n n^3 W \log(n) \log^2(nW))$.   □

To convert a table of minimum exact cover weights to a table of minimum (not necessarily exact) cover weights, we iterate over each set $U' \subseteq U$ in increasing order of size, and for each $u \in U'$ replace $Q[U']$ by $\min(Q[U'], Q[U' \setminus \{u\}])$. Together with Lemma 1, this concludes the proof of Theorem 3.

For acyclic inputs, we can use dynamic programming to speed up the enumeration of connected rainbow subgraphs of G, avoiding the dependency on Δ.

Theorem 4. *For an acyclic instance (G, col) of* MINIMUM RAINBOW SUBGRAPH *an optimal solution can be computed within* $O(3^p pn + 2^p p^3 n \log^2(pn))$ *time.*

4   Parameterization by Number of Vertex Deletions

In this section, we consider the dual parameter $\ell := n - k$ (where k is the solution order and n is the order of the input graph), that is, the number of vertices that are *not* part of a solution and thus are "deleted" from the input graph. In Sect. 3, we showed that RAINBOW SUBGRAPH is W[1]-hard for the parameter k, but that it becomes fixed-parameter tractable for the parameter (Δ, k). We show that both results also hold when replacing k by ℓ. Hence, parameter ℓ is useful when we ask for the existence of relatively large solutions in sparse graphs.

In contrast to the parameter k, for which RAINBOW SUBGRAPH becomes fixed-parameter tractable on trees, we observe W[1]-hardness for parameter ℓ even on very restricted input trees.

Theorem 5. RAINBOW SUBGRAPH *is W[1]-hard with respect to the dual parameter ℓ even when the input is a tree of height three and every color occurs at most twice.*

By Theorem 5, parameterization by ℓ alone does not yield fixed-parameter tractability. Hence, we consider combinations of ℓ with two parameters. One is the maximum degree Δ, and the other one is the *maximum color degree* $\Delta_C :=$ $\max_{v \in V} |\{c \mid \exists \{u, v\} \in E : \mathrm{col}(\{u, v\}) = c\}|$, which is the maximum number of colors incident with any vertex in G. This parameter was also considered by Schiermeyer [13] for obtaining bounds on the size of minimum rainbow subgraphs. Note that the maximum color degree is upper-bounded by both the maximum degree and by the number of colors in G and that it may be much smaller than either parameter.

First, we show that for the combined parameter (Δ, ℓ) the problem has a polynomial-size problem kernel. To our knowledge, this is the first non-trivial kernelization result for RAINBOW SUBGRAPH. As it is common for kernelizations, it is based on a set of data reduction rules. The main idea of the kernelization is as follows. We first remove edges whose colors appear very often compared to Δ and ℓ. Afterwards, deleting any vertex v "influences" only a bounded number of other vertices: at most Δ edges are incident with v and for each of these edges the number of other edges that have the same color depends only on Δ and ℓ. We then consider some vertices that are in every rainbow cover. To this end, we call a vertex v *obligatory* if there is some edge color such that all edges with this color are incident with v. In the data reduction rules, we reduce those obligatory vertices that have only obligatory neighbors. Together with the previous reduction rules, we then obtain the kernel by the following argument: If there are many non-obligatory vertices, then we can find a greedy solution since any vertex deletion has bounded "influence". Otherwise, the overall instance size is bounded as every other vertex is a neighbor of some non-obligatory vertex and each non-obligatory vertex has at most Δ neighbors.

As mentioned above, the first rule removes edges whose color appears very often compared to Δ and ℓ.

Rule 1. *If there is an edge color c such that there are more than $\Delta \ell$ edges with color c, then remove all edges with color c from G.*

We now deal with obligatory vertices. The first simple rule identifies edge colors that are already covered by obligatory vertices.

Rule 2. *If G contains an edge $\{u, v\}$ of color c such that u and v are obligatory, then remove all other edges with color c from G.*

We now work on instances that are reduced with respect to Rule 2. Observe that in such instances every edge between two obligatory vertices has a unique color. This observation is crucial for showing the correctness of the following rules. Their aim is to remove obligatory vertices that have only obligatory neighbors. When removing a vertex in these rules, we decrease k and n by one, thus the value of ℓ remains the same. The correctness of the first rule is obvious.

Rule 3. *Let (G, col) be an instance that is reduced with respect to Rule 2. Then, remove all connected components of G that consist of obligatory vertices only.*

The next two rules remove edges between obligatory vertices.

Rule 4. *Let* (G, col) *be an instance of* RAINBOW SUBGRAPH *that is reduced with respect to Rule 2. If* G *contains three obligatory vertices* u, v, *and* w *such that* $\{u, v\}, \{v, w\} \in E$ *and* u *has only obligatory neighbors, then remove* $\{u, v\}$ *from* G. *If* u *has degree zero now, then remove* u *from* G.

Rule 5. *Let* (G, col) *be an instance of* RAINBOW SUBGRAPH *that is reduced with respect to Rule 2. If* G *contains four obligatory vertices* u, v, w *and* x *such that* $\{u, v\} \in G$ *and* $\{w, x\}$ *in* G *and* u *and* x *have only obligatory neighbors, then do the following.*

Remove $\{w, x\}$ *from* G. *If* v *and* w *are not adjacent, then insert* $\{v, w\}$ *and assign it a unique color. If* x *has now degree zero, then remove* x *from* G.

Note that application of Rule 4 does not increase the maximum degree of the instance and decreases the degree of v and w. Furthermore, note that application of Rule 5 may increase the degree of v by one but directly triggers an application of Rule 4 which reduces the degree of v and u again by one. Hence, both rules can be exhaustively applied without increasing the overall maximum degree.

We now show that the instance either has a rainbow cover or that it has bounded size.

Lemma 4. *Let* (G, col) *be an instance that is reduced with respect to Rule 1–5. Then,* (G, col) *is a yes-instance or it contains at most* $2\Delta \cdot (\Delta + 1) \cdot \Delta_C \cdot \ell^2$ *vertices.*

Proof. We consider a special type of vertex sets that can be safely deleted. To this end, call a vertex set S a *colorful packing* if

1. no vertex in S is obligatory, and
2. for all u and v in S the set of colors incident with u is disjoint from the set of colors incident with v.

Assume that (G, col) has a *colorful packing* of size ℓ. Then, $G - S$ is a rainbow cover of order k: For each color incident with some vertex v in S, there are two other vertices in V that are connected by an edge with this color (as v is not obligatory). By the second condition, these two vertices are not in S. Hence, this edge color is contained in $G - S$. Summarizing, if (G, col) contains a colorful packing of size at least ℓ, then (G, col) is a yes-instance.

Now, assume that a maximum-cardinality colorful packing S in G has size less than ℓ. Each vertex in S is incident with at most Δ_C colors. For each of these colors, the graph induced by the edges of this color has at most $\Delta \ell$ edges and thus at most $2\Delta \ell$ vertices, since the instance is reduced with respect to Rule 1.

Let T denote the set of vertices in $V \setminus S$ that are incident with at least one edge that has the same color as an edge incident with some vertex in S. By the above discussion,

$$|T| \leq 2\Delta \cdot \Delta_C \cdot \ell \cdot (\ell - 1).$$

Note that T includes all neighbors of vertices in S. By the maximality of S, all vertices in $V \setminus (S \cup T)$ are obligatory. Now partition $V \setminus (S \cup T)$ into the set X

that has neighbors in T and the set Y that has only neighbors in $(X \cup Y)$. The set X has size at most $(2\Delta_C \cdot \Delta \cdot \ell \cdot (\ell - 1)) \cdot \Delta$ since the maximum degree in G is Δ. The set Y has size at most 1 since otherwise one of the Rule 3–5 applies (we omit the details).

Hence, since S has size at most $\ell - 1$, G contains at most vertices.

$$\ell - 1 + 2\Delta \cdot \Delta_C \cdot \ell \cdot (\ell - 1) + 2\Delta^2 \cdot \Delta_C \cdot \ell \cdot (\ell - 1) + 1 < 2\Delta \cdot (\Delta + 1) \cdot \Delta_C \cdot \ell^2$$

Thus, if any instance contains more vertices, then a colorful packing of size at least ℓ exists and the instance is a yes-instance.   □

Using Lemma 4, we obtain the following theorem.

Theorem 6. RAINBOW SUBGRAPH *admits a problem kernel with at most* $2\Delta \cdot (\Delta + 1) \cdot \Delta_C \cdot \ell^2$ *vertices that can be computed in* $O(m^2 + mn)$ *time.*

We now consider parameterization by (Δ_C, ℓ) (recall that the color degree Δ_C can be much smaller than Δ). First, by performing the following additional rule, we can use the kernelization result for (Δ, ℓ) to obtain a polynomial problem kernel for (Δ_C, ℓ).

Rule 6. *If G contains a vertex v such that at least $\ell + 2$ edges incident with v have the same color c, then delete an arbitrary one of these edges.*

Rule 6 can be exhaustively performed in linear time. Afterwards, the maximum degree Δ of G is at most $\Delta_C \cdot (\ell + 1)$. In combination with Theorem 6, this immediately implies the following.

Theorem 7. RAINBOW SUBGRAPH *has a problem kernel with at most* $2(\Delta_C + 1)^3 \ell^2 (\ell + 1)^2$ *vertices that can be computed in* $O(m^2 + mn)$ *time.*

Finally, we describe a simple branching for the parameter (Δ_C, ℓ). Herein, *deleting a vertex* means to remove it from G and to decrease ℓ by one; thus, a deleted vertex is *not* part of a rainbow cover of order k of the original instance.

Branching Rule 1. *If G contains a non-obligatory vertex u, then branch into the following cases. First, recursively solve the instance obtained from deleting u from G. Then, for each color c that is incident with u pick an edge $\{v, w\}$ with color c. If v (w) is non-obligatory, then recursively solve the instance obtained from deleting v (w).*

Note that the parameter ℓ decreases by one in each branch. Exhaustively applying Branching Rule 1 until either every vertex is obligatory or $\ell \leq 0$ yields an algorithm with the following running time.

Theorem 8. RAINBOW SUBGRAPH *can be solved in* $O((2\Delta_C + 1)^\ell \cdot (n + m))$ *time.*

5   Outlook

Considering its biological motivation, it would be interesting to gain further, potentially data-driven parameterizations of MINIMUM RAINBOW SUBGRAPH that may help identifying further practically relevant and tractable special cases. From a more graph-theoretic point of view, we left open a deeper study of parameters measuring the degree of acyclicity of the underlying graph, such as treewidth or feedback set numbers. A further question is whether for our fixed-parameter tractability result in Theorem 3 we can avoid exponential memory consumption.

Acknowledgments. We thank the reviewers of WG' 14 for their thorough and valuable feedback.

References

1. Alimonti, P., Kann, V.: Some APX-completeness results for cubic graphs. Theor. Comput. Sci. **237**(1–2), 123–134 (2000)
2. Björklund, A., Husfeldt, T., Kaski, P., Koivisto, M.: Fourier meets Möbius: fast subset convolution. In: Proceedings of 39th STOC, pp. 67–74. ACM (2007)
3. Björklund, A., Husfeldt, T., Koivisto, M.: Set partitioning via inclusion-exclusion. SIAM J. Comput. **39**(2), 546–563 (2009)
4. Cygan, M., Kowalik, Ł., Wykurz, M.: Exponential-time approximation of weighted set cover. Inf. Process. Lett. **109**(16), 957–961 (2009)
5. Fernandes, R., Skiena, S.: Microarray synthesis through multiple-use PCR primer design. Bioinformatics **18**(suppl 1), 128–135 (2002)
6. Fomin, F.V., Grandoni, F., Kratsch, D.: A measure & conquer approach for the analysis of exact algorithms. J. ACM **56**(5), 25:1–25:32 (2009)
7. Hajiaghayi, M.T., Jain, K., Lau, L.C., Măndoiu, I.I., Russell, A., Vazirani, V.V.: Minimum multicolored subgraph problem in multiplex PCR primer set selection and population haplotyping. In: Alexandrov, V.N., van Albada, G.D., Sloot, P.M.A., Dongarra, J. (eds.) ICCS 2006. LNCS, vol. 3992, pp. 758–766. Springer, Heidelberg (2006)
8. Katrenič, J., Schiermeyer, I.: Improved approximation bounds for the minimum rainbow subgraph problem. Inf. Process. Lett. **111**(3), 110–114 (2011)
9. Koch, M., Camacho, S.M., Schiermeyer, I.: Algorithmic approaches for the minimum rainbow subgraph problem. ENDM **38**, 765–770 (2011)
10. Komusiewicz, C., Sorge, M.: Finding dense subgraphs of sparse graphs. In: Thilikos, D.M., Woeginger, G.J. (eds.) IPEC 2012. LNCS, vol. 7535, pp. 242–251. Springer, Heidelberg (2012)
11. Camacho, S.M., Schiermeyer, I., Tuza, Z.: Approximation algorithms for the minimum rainbow subgraph problem. Discrete Math. **310**(20), 2666–2670 (2010)
12. Popa, A.: Better lower and upper bounds for the minimum rainbow subgraph problem. Theor. Comput. Sci. **543**, 1–8 (2014)
13. Schiermeyer, I.: On the minimum rainbow subgraph number of a graph. Ars Mathematica Contemporanea **6**(1), 83–88 (2012)
14. Williamson, D.P., Shmoys, D.B.: The Design of Approximation Algorithms. Cambridge University Press, New York (2011)

Kernelizations for the Hybridization Number Problem on Multiple Nonbinary Trees

Leo van Iersel[1]([✉]) and Steven Kelk[2]

[1] Centrum Wiskunde and Informatica (CWI), P.O. Box 94079,
1090 GB Amsterdam, The Netherlands
l.j.j.v.iersel@gmail.com
[2] Department of Knowledge Engineering (DKE), Maastricht University,
P.O. Box 616, 6200 MD Maastricht, The Netherlands
steven.kelk@maastrichtuniversity.nl

Abstract. A well-studied problem in phylogenetics is to determine the minimum number of hybridization events necessary to explain conflicts among several evolutionary trees, e.g. from different genes. An evolutionary history with hybridization events (or, more generally, reticulations) can be described by a rooted leaf-labelled directed acyclic graph, which is called a phylogenetic network. The reticulation number of such a phylogenetic network can be defined as the sum of all indegrees minus the number of vertices plus one. The considered problem can now formally be stated as follows. Given a finite set X, a collection $\mathcal{T}$ of rooted phylogenetic trees on X and $k \in \mathbb{N}^+$, the HYBRIDIZATION NUMBER problem asks if there exists a rooted phylogenetic network on X that displays all trees from $\mathcal{T}$ and has reticulation number at most k. We show that HYBRIDIZATION NUMBER admits a kernel of size $4k(5k)^t$ if $\mathcal{T}$ contains t (not necessarily binary) rooted phylogenetic trees. In addition, we show a slightly different kernel of size $20k^2(\Delta^+ - 1)$ with Δ^+ the maximum outdegree of the input trees.

1 Introduction

In phylogenetics, the central challenge is to construct a plausible evolutionary history for a set of contemporary species X given incomplete data. This usually concerns biological evolution, but the paradigm is equally applicable to more abstract form s of evolution, e.g. natural languages [16]. Classically an evolutionary history is modelled by a *rooted phylogenetic tree*, essentially a rooted tree in which the leaves are bijectively labelled by X [18]. In recent years, however, there has been growing interest in generalizing this model to directed acyclic graphs, i.e., to *rooted phylogenetic networks* [1,9,15]. In the latter model, *reticulations* are of central importance, which are vertices of indegree 2 (or higher); these are used to represent non-treelike evolutionary phenomena such as hybridization

Leo van Iersel was supported by a Veni grant of the Netherlands Organization for Scientific Research (NWO).

D. Kratsch and I. Todinca (Eds.): WG 2014, LNCS 8747, pp. 299–311, 2014.
DOI: 10.1007/978-3-319-12340-0_25

and lateral gene transfer. This has naturally given rise to the HYBRIDIZATION NUMBER problem: given a set of phylogenetic trees $\mathcal{T}$ on the same set of taxa X, construct a phylogenetic network on X with as few indegree-2 vertices as possible, such that an image of every tree in $\mathcal{T}$ is embedded in the network [2]. We defer formal definitions to the preliminaries.

HYBRIDIZATION NUMBER has attracted considerable interest in a short space of time. Even in the case when $\mathcal{T}$ consists of two binary (i.e., bifurcating) trees the problem is NP-hard, APX-hard [5] and in terms of approximability is a surprisingly close relative of the problem DIRECTED FEEDBACK VERTEX SET [13,19]. On the positive side, this variant of the problem is fixed-parameter tractable (FPT) in parameter k, the minimum number of indegree-2 vertices required. Initially this was established via kernelization [4], but more recently efficient bounded-search algorithms have emerged with $O(3.18^k \cdot poly(n))$ being the current state of the art, where $n = |X|$ [21].

In this article we focus on the general case when $|\mathcal{T}| \geq 2$ and the trees in $\mathcal{T}$ are allowed to be nonbinary (i.e., not necessarily binary). This causes complications for two reasons. First, when $|\mathcal{T}| > 2$ the popular "maximum acyclic agreement forest" abstraction breaks down, a central pillar of algorithms for the $|\mathcal{T}| = 2$ case. Second, in the nonbinary case the images of the trees in the network are allowed to be more "resolved" than the original trees. (More formally, an input tree T is seen as being embedded in a network N if T can be obtained from a subgraph of N by contracting edges.) The reason for this is that vertices with outdegree greater than two are used by biologists to model uncertainty in the order that species diverged. Both factors complicate matters considerably. Consequently, progress has been more gradual.

For the case of multiple binary trees, there exists a polynomial kernel [20], various heuristics [6,7,22] and an exact approach without running-time bound [23].

For the case of two nonbinary trees, there is also a polynomial kernel [14], based on a highly technical kernelization argument, and a simpler FPT algorithm based on bounded search [17].

This leaves the case of an unbounded number of nonbinary trees as the main variant for which it is unclear whether the problem is FPT. There has, however, been some partial progress: for fixed k the problem is polynomial-time solvable and the problem is FPT if the number of trees is bounded *or* the maximum outdegree of the trees is bounded [11]. The main problem with the result from [11] is its theoretical character: it is indirect (based on [12]) and yields a bounded-search algorithm with astronomical running time.

Here we mirror the bounded-search result from [11] by showing that HYBRIDIZATION NUMBER admits a kernel of size $4k(5k)^t$ if $\mathcal{T}$ contains t nonbinary rooted phylogenetic trees. In addition, we show a slightly different kernel of size $20k^2(\Delta^+ - 1)$ with Δ^+ the maximum outdegree. We believe this result is important for several reasons.

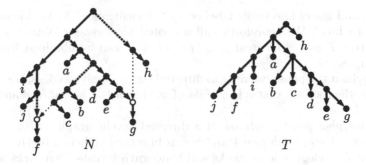

Fig. 1. A (rooted phylogenetic) network N and a (rooted phylogenetic) tree T. Network N is binary, has two reticulations (unfilled) and reticulation number 2. Tree T is displayed by N because it can be obtained from N by deleting the dotted edges and contracting the dashed edges.

First, it is the first polynomial kernel for any fixed number of nonbinary trees, and the first polynomial kernel for an unbounded number of trees with outdegrees bounded by any constant.

Second, it represents a significant step forward in our understanding of the complexities associated with nonbinary trees. In particular, the result of [14] is so technical due to the difficulties of dealing with so-called *common chains*, which in the case of binary trees are much easier to deal with [4,20]. The sister result of [11] avoids this technical analysis by exhaustive guessing which is mathematically unsatisfying and is one of the reasons for its purely theoretical running time. Here, for the first time, we present a simple and unified kernelization strategy for dealing with common chains which avoids technical case analysis (and exhaustive guessing) and can cope with the chains as they unfold across many trees.

Third, the $4k(5k)^t$ kernel introduces an interesting way to deal with multiple parameters simultaneously. It is based on searching, for decreasing q, for certain substructures called "q-star chains", which are chains that are common to all t input trees and form stars in q of the input trees. When we encounter such substructures we truncate them to a size that is a function of q and k. Since we loop through all possible values of q ($0 \leq q \leq t$), we eventually truncate all common substructures. The correctness of each step heavily relies on the fact that substructures for larger values of q have already been truncated. However, when q decreases, the size to which substructures can be reduced increases (as will become clear later). This has the effect that the size of kernelized instances is a function of k and t and not of k only. For the $20k^2(\Delta^+ - 1)$ kernel, we use a similar but simpler technique.

2   Preliminaries

Let X be a finite set. A *rooted phylogenetic X-tree* is a rooted tree with no vertices with indegree 1 and outdegree 1, a root with indegree 0 and outdegree at least 2, and leaves bijectively labelled by the elements of X. We identify each leaf with its label. We henceforth call a rooted phylogenetic X-tree a *tree* for short. A tree T is a *refinement* of a tree T' if T' can be obtained from T by contracting edges.

Throughout the paper, we refer to directed edges simply as edges. If $e = (u, v)$ is an edge, then we say that v is a *child* of u, that u is a *parent* of v and that v is the *head* of e.

A *rooted phylogenetic network* is a directed acyclic graph with no vertices with indegree 1 and outdegree 1 and leaves bijectively labelled by the elements of X. Rooted phylogenetic networks will henceforth be called *networks* for short in this paper. A tree T is *displayed* by a network N if T can be obtained from a subgraph of N by contracting edges. Note that, without loss of generality, we may assume that edges incident to leaves are not contracted. See Fig. 1 for an example. Using $d^-(v)$ to denote the indegree of a vertex v, a *reticulation* is a vertex v with $d^-(v) \geq 2$. The *reticulation number* of a network N with vertex set V is given by

$$r(N) = \sum_{v \in V : d^-(v) \geq 2} (d^-(v) - 1).$$

Given a set of trees $\mathcal{T}$ on X, we use $r(\mathcal{T})$ to denote the minimum value of $r(N)$ over all phylogenetic networks N on X that display $\mathcal{T}$. We are now ready to formally define the problem we consider.

Problem: HYBRIDIZATION NUMBER.

Instance: A finite set X, a collection $\mathcal{T}$ of rooted phylogenetic trees on X and $k \in \mathbb{N}^+$.

Question: Is $r(\mathcal{T}) \leq k$, i.e., does there exist a phylogenetic network N on X that displays $\mathcal{T}$ and has $r(N) \leq k$?

A network is called *binary* if each vertex has indegree and outdegree at most 2 and if each vertex with indegree 2 has outdegree 1. By the following lemma we may restrict to binary networks.

Observation 1 [11]. *If there exists a network N on X that displays $\mathcal{T}$ then there exists a binary network N' on X that displays $\mathcal{T}$ such that $r(N) = r(N')$.*

The observation follows directly from noting that, for each network N, there exists a binary network N' with $r(N') = r(N)$ such that N can be obtained from N' by contracting edges. Hence, any tree displayed by N is also displayed by N'.

A subtree T' of a network N (or of a tree T) is said to be *pendant* if no vertex of T' other than possibly its root has a child that is not in T'. A pendant subtree is called *trivial* if it has only one leaf.

The notion of "generators" is used to describe the underlying structure of a network without nontrivial pendant subtrees [12]. Let $k \in \mathbb{N}^+$. A *binary k-reticulation generator* is defined as an acyclic directed multigraph with a single root with indegree 0 and outdegree 1, exactly k vertices with indegree 2 and

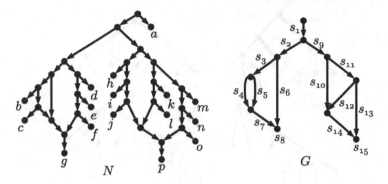

Fig. 2. A network N and the 4-reticulation generator G underlying N. Generator G has two vertex sides s_8 and s_{15} and 13 edge sides. For example, leaves d, e and f are on edge side s_6 and leaf g is on vertex side s_8.

outdegree at most 1, and all other vertices have indegree 1 and outdegree 2. See Fig. 2 for an example. Let N be a binary network with no nontrivial pendant subtrees and with $r(N) = k$. Then, a binary k-reticulation generator is said to be the *generator underlying* N if it can be obtained from N by adding a new root with an edge to the old root, deleting all leaves and suppressing all resulting indegree-1 outdegree-1 vertices. In the other direction, N can be reconstructed from its underlying generator by subdividing edges, adjoining a leaf to each vertex that subdivides an edge, or has indegree 2 and outdegree 0, via a new edge, and deleting the outdegree-1 root. The *sides* of a generator are its edges (the *edge sides*) and its vertices with indegree 2 and outdegree 0 (the *vertex sides*). Thus, each leaf of N is on a certain side of its underlying generator. To formalize this, consider a leaf x of a binary network N without nontrivial pendant subtrees and with underlying generator G. If the parent p of x has indegree 2, then p is a vertex side of G and we say that x *is on side p*. If, on the other hand, the parent p of x has indegree 1 and outdegree 2, then p is used to subdivide an edge side e of G and we say that x *is on side e*. We say that two leaves x and y (with $x \neq y$) are *on the same side* of N if the underlying generator of N has an edge side e such that x and y are both on side e. The following lemma from [20] will be useful.

Lemma 1 [20]. *If N is a binary phylogenetic network with no nontrivial pendant subtrees and $r(N) = k > 0$ and if G is its underlying generator, then G has at most $4k - 1$ edge sides, at most k vertex sides and at most $5k - 1$ sides in total.*

A *kernelization* of a parameterized problem is a polynomial-time algorithm that maps an instance x with parameter k to an instance x' with parameter k' such that (1) (x', k') is a yes-instance if and only if (x, k) is a yes-instance, (2) the size of x' is bounded by a function f of k, and (3) the size of k' is bounded by a function of k [8]. A kernelization is usually referred to as a *kernel* and the function f

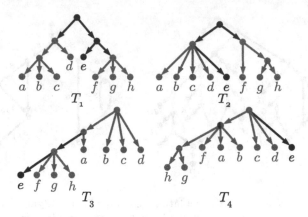

Fig. 3. Example instance of HYBRIDIZATION NUMBER consisting of four trees that have a common pendant subtree on $\{f, g, h\}$ and a common 1-star chain (d, c, b, a). Chain (d, c, b, a) is pendant in T_1 and T_2 but not in T_3 and T_4. It is a 1-star chain because all its leaves have a common parent in only T_2.

as the *size* of the kernel. Thus, a parameterized problem admits a polynomial kernel if there exists a kernelization with f being a polynomial. A parameterized problem is *fixed-parameter tractable* (FPT) if there exists an algorithm that solves the problem in time $O(g(k)|x|^{O(1)})$, with g being some function of k and $|x|$ the size of x. It is well known that a parameterized problem is fixed-parameter tractable if and only if it admits a kernelization and is decidable. However, there exist fixed-parameter tractable problems that do not admit a kernel of *polynomial* size unless the polynomial hierarchy collapses [3]. Kernels are of practical interest because they can be used as polynomial-time preprocessing which can be combined with any algorithm (usually an exponential-time exact algorithm) solving the problem.

3 A Polynomial Kernel for a Bounded Number of Trees

We first introduce the following key definitions. Let $\mathcal{T}$ be a set of trees. A tree T' is said to be a *common pendant subtree* of $\mathcal{T}$ if it is a refinement of a pendant subtree of each $T \in \mathcal{T}$ and T' is said to be *nontrivial* if it has at least two leaves.

Definition 1. *If T is a tree on X, $p \geq 2$ and $x_1, \ldots, x_p \in X$, then $(x_1, \ldots, x_p)$ is a* chain *of T if:*

1. *there exists a directed path $(v_1, ..., v_t)$ in T, for some $t \geq 1$;*
2. *each x_i is a child of some v_j;*
3. *if x_i is a child of v_j and $i < p$, then x_{i+1} is either a child of v_j or of v_{j+1};*
4. *for $i \in \{2, \ldots, t-1\}$, the children of v_i are all in $\{v_{i+1}, x_1, x_2, \ldots, x_p\}$.*

If, in addition, $t = 1$ or the children of v_t are all in $\{x_1, \ldots, x_p\}$, then $(x_1, \ldots, x_p)$ is said to be a pendant chain *of T. The length of the chain is p.*

Algorithm 1. Kernelization algorithm for $t := |\mathcal{T}|$ trees

---

1 **Subtree Reduction: if** *there is a nontrivial maximal common pendant subtree T' of $\mathcal{T}$* **then**
2   Let $x \notin X$. In each $T \in \mathcal{T}$, if T'' is the pendant subtree of T that T' is a refinement of, replace T'' by a single leaf labelled x. Remove the labels labelling leaves of T' from X and add x to X.
3   **go to** Line 1

4 **Chain Reduction: for** $q = t - 1, t - 2, \ldots, 0$ **do**
5   **if** *there exists a maximal common q-star chain $(x_1, \ldots, x_p)$ of $\mathcal{T}$ with $p > (5k)^{t-q}$* **then**
6   Delete leaves $x_{(5k)^{t-q}+1}, \ldots, x_p$ from X and from each tree in $\mathcal{T}$ and repeatedly suppress outdegree-1 vertices and delete unlabelled outdegree-0 vertices until no such vertices remain.
7   **go to** Line 1

A chain is said to be a *common chain* of $\mathcal{T}$ if it is a chain of each tree in $\mathcal{T}$. The following observations follow easily from the definition of a chain.

Observation 2. *If $(x_1, \ldots, x_p)$ is a common chain of $\mathcal{T}$ and $1 \leq i < j \leq p$, then $(x_i, \ldots, x_j)$ is a common chain of $\mathcal{T}$.*

Observation 3. *If $(x_1, \ldots, x_p)$ is a chain of a tree T, $1 \leq i < j \leq p$ and x_i and x_j have a common parent in T, then $x_i, \ldots, x_j$ have a common parent in T.*

Definition 2. *If $\mathcal{T}$ is a set of trees on X and $x_1, \ldots, x_p \in X$, then $(x_1, \ldots, x_p)$ is a common q-star chain of $\mathcal{T}$ if:*

(a) $(x_1, \ldots, x_p)$ is a common chain of $\mathcal{T}$ and
(b) in precisely q trees of $\mathcal{T}$, all of $x_1, \ldots, x_p$ have a common parent.

We say that a common q-star chain $(x_1, \ldots, x_p)$ of $\mathcal{T}$ is *maximal* if there is no common q-star chain $(y_1, \ldots, y_{p'})$ of $\mathcal{T}$ with $\{x_1, \ldots, x_p\} \subsetneq \{y_1, \ldots, y_{p'}\}$. Notice that a common 0-star chain is a common chain that does not form a star in any tree. An illustration of the above definitions is in Fig. 3.

We are now ready to describe the kernelization, which is in Algorithm 1.

It is not too difficult to see that the subtree reduction preserves the reticulation number and can be applied in polynomial time.

To prove correctness of the chain reduction, we use two lemmas which have been omitted due to space constraints and can be found in the full version of this paper. The idea of these lemmas is illustrated in Fig. 4. The two trees T_1 and T_2 in this figure have a common chain (a, b, c, d, e). Both trees are displayed by network N. However, the leaves of the chain are spread out over different sides of the underlying generator G of N. To prove correctness of the chain reduction, we want to argue that there exists a modified network N' in which the leaves of the chain (a, b, c, d, e) all lie on the same side. Moreover, network N' should

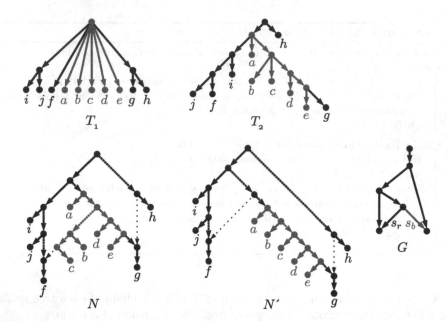

Fig. 4. Two trees T_1 and T_2 with a common chain (a, b, c, d, e) highlighted in blue (grey), a network N that displays these trees, the network N' as constructed in Lemmas 4 and 5, and the underlying generator G of both networks. Dashed and dotted edges are used to indicate that T_2 can be obtained from either of N and N' by deleting the dotted edges and contracting the dashed edges (Color figure online).

display all input trees and its reticulation number should not be higher than the reticulation number of N.

In T_1, all leaves of the chain have a common parent. For this case, Lemma 4 (omitted) argues that all leaves of the chain can be moved to any side that contains at least one of its leaves, and the resulting network still displays T_1.

In T_2, there are two leaves $x_i = d$ and $x_j = e$ that are on the same side of G (the blue side s_b) and that do not have a common parent in T_2. For this case, Lemma 5 (omitted) argues that all the leaves of the chain can be moved to side s_b, and the resulting network will still display T_2. (Note that we cannot move all the leaves of the chain to the red side s_r, even though it contains two leaves b, c of the chain, because b and c have a common parent in T_2).

Hence, the network N' obtained by moving all leaves of the chain to the blue side s_b displays both T_1 and T_2. Furthermore, $r(N') = r(N) = 2$.

The next lemma shows correctness of the chain reduction, and thereby of Algorithm 1. It is based on the idea that, if a chain is long enough, one of Lemmas 4 and 5 applies for each tree.

Lemma 2. *Let $q \in \{0, \ldots, t-1\}$ (with $t = |\mathcal{T}|$) and let $(X, \mathcal{T}, k)$ be an instance of* Hybridization Number *without nontrivial common pendant subtrees or maximal common q'-star chains of more than $(5k)^{t-q'}$ leaves, for $q < q' \leq t-1$.*

Let (X', T', k) be the instance obtained after applying the chain reduction to a maximal common q-star chain $C = (x_1, \ldots, x_p)$ of T with $p > (5k)^{t-q}$. Then $r(T) \leq k$ if and only if $r(T') \leq k$.

Proof. It is clear that if $r(T) \leq k$ then $r(T') \leq k$ because the chain reduction only deletes leaves (and suppresses and deletes vertices).

It remains to prove the other direction. Assume that $r(T') \leq k$, i.e., there exists a network N' that displays T' and has $r(N') \leq k$. Define $m := (5k)^{t-q-1}$. Hence, there are no common chains of T of more than m leaves that have a common parent in more than q of the trees.

Let $C' = (x_1, \ldots, x_{5km})$. First observe that C' is a common chain of T' and, moreover, that C' is a common q'-star chain of T' with $q' \geq q$. Moreover, we claim the following.

Claim (1). Any two leaves in $\{x_1, \ldots, x_{5km-1}\}$ have a common parent in a tree $T \in \mathcal{T}$ if and only if they have a common parent in the corresponding tree $T' \in \mathcal{T}'$.

This claim follows directly from the observation that, in the chain reduction, the parents of $x_1, \ldots, x_{5km-1}$ cannot become outdegree-1 and are therefore not being suppressed. Correctness of the next claim can be verified in a similar way.

Claim (2). If C is not pendant in $T \in \mathcal{T}$, then any two leaves in $\{x_1, \ldots, x_{5km}\}$ have a common parent in T if and only if they have a common parent in the corresponding tree $T' \in \mathcal{T}'$.

Now define
$$C^* := (x_1, x_{1+m}, x_{1+2m}, \ldots, x_{1+(5k-1)m}),$$
i.e., C^* contains $5k$ leaves and the indices of any two subsequent leaves are m apart.

Let G' be the generator underlying N'. Each leaf of C^* is on a certain side of G'. Since G' has at most $5k - 1$ sides (by Lemma 1) and C^* contains $5k$ leaves, there exist two leaves x_i, x_j of C^* that are on the same side of G' by the pigeonhole principle. Assume without loss of generality that $j > i$. Then, by the construction of C^*, $j \geq i + m$.

We modify network N' to a network N'' by moving the whole chain C' to the side of the network containing x_i and x_j. To describe this modification more precisely, let v_{5km} be the parent of x_i in N'. Then, N'' is the network obtained from N' by deleting the leaves $x_1, \ldots, x_{5km}$, subdividing the edge entering v_{5km} by a directed path $v_1, \ldots, v_{5km-1}$, adding the leaves $x_1, \ldots, x_{5km}$ by edges $(v_1, x_1), \ldots, (v_{5km}, x_{5km})$ and cleaning up the resulting directed graph.

For each tree $T' \in \mathcal{T}'$ in which all of $x_1, \ldots, x_{5km}$ have a common parent, Lemma 4 shows that N'' displays T'. There are at least q such trees. In fact, it follows from the following claim that there are precisely q such trees. Moreover, the claim shows that in all other trees x_i and x_j do *not* have a common parent. Therefore, it follows from Lemma 5 that these trees are also displayed by N''.

Claim (3). The number of trees of T' in which x_i and x_j have a common parent is at most q.

To prove the claim, consider $C^{**} := (x_i, \ldots, x_j)$. Since C^{**} is a subchain of C', it is a chain of each tree in T' by Observation 2.

First consider the case $q = t - 1$ and assume that x_i and x_j have a common parent in more than q trees in T' and hence in all trees in T'. Then, $x_i, \ldots, x_j$ all have a common parent in all trees in T', by Observation 3. Since C is a q-star chain of T, there are $q = t - 1$ trees in T in which all leaves of C have a common parent. Let T^* be the only tree in T in which the leaves of C do not all have a common parent. Then C is not pendant in T^* or its leaves would form a non-trivial common pendant subtree of T. Hence, x_i and x_j have a common parent in T^* by Claim (2). However, this is a contradiction because then x_i and x_j form a nontrivial common pendant subtree of T.

To finish the proof of Claim (3), consider the case $q < t - 1$. In that case, $m > 1$ and hence C^{**} contains only leaves in $\{x_1, \ldots, x_{5km-1}\}$. Because C^{**} contains more than m leaves, the number of trees of T in which all the leaves of C^{**} have a common parent is at most q here we use the fact that there are no common q'-star chains for $q' > q$ that have more than m leaves). Hence, it follows from Claim (1) that the number of trees of T' in which the leaves of C^{**} have a common parent is at most q. Claim (3) then follows by Observation 3.

Hence, we have shown that N'' displays T'. We now construct a network N from N'' by replacing the reduced chain by the unreduced chain. More precisely, let e_{5km} be the edge of N'' that leaves v_{5km} but is not the edge (v_{5km}, x_{5km}). Subdivide e_{5km} by a directed path $(v_{5km+1}, \ldots, v_p)$ and add leaves $x_{5km+1}, \ldots, x_p$ by edges $(v_{5km+1}, x_{5km+1}), \ldots, (v_p, x_p)$. This gives N. Then, by a similar argument as in the proof of Lemma 5, N displays T. Moreover, since none of the applied operations increase the reticulation number, we have $r(N) \leq r(N')$. □

The next lemma, whose proof has been omitted, shows that the chain reduction can be performed in polynomial time.

Lemma 3. *There exists a polynomial-time algorithm that, given a set T of trees on X and $q \in \mathbb{N}$, decides if there exists a common q-star chain of T and constructs such a chain of maximum size if one exists.*

To see that Algorithm 1 runs in polynomial time, it remains to observe that at least one leaf is removed in each iteration and hence that the number of iterations is bounded by $|X|$.

Let (X', T', k) be a kernelized instance of HYBRIDIZATION NUMBER. If there exists a network N' displaying T' with $r(N') \leq k$ then N' has at most one leaf per vertex side of the underlying generator (since common pendant subtrees have been reduced) and at most $(5k)^{|T|}$ leaves per edge side (since common chains have been reduced). Hence,

$$|X'| \leq k + (4k - 1)(5k)^{|T|} \leq 4k(5k)^{|T|}.$$

Correctness of the following theorem now follows from Lemmas 2–5.

Theorem 1. *The problem* HYBRIDIZATION NUMBER *on* $|\mathcal{T}| = t$ *trees admits a kernel with at most* $4k(5k)^t$ *leaves.*

4 A Polynomial Kernel for Bounded Outdegrees

Algorithm 2 describes a polynomial kernel for HYBRIDIZATION NUMBER if not the number of input trees but their maximum outdegree is bounded. Let Δ^+ be the maximum outdegree over all vertices of all trees in $\mathcal{T}$.

Algorithm 2. Kernelization algorithm for bounded outdegree

---

1 **Subtree Reduction: if** *there is a maximal common pendant subtree* T' *of* $\mathcal{T}$ **then**

2   Let $x \notin X$. In each $T \in \mathcal{T}$, if T'' is the pendant subtree of T that T' is a refinement of, replace T'' by a single leaf labelled x. Remove the labels labelling leaves of T' from X and add x to X.

3   **go to** Line 1

4 **Chain Reduction: if** *there is a maximal common chain* $(x_1, \ldots, x_p)$ *of* $\mathcal{T}$ *with* $p > 5k(\Delta^+ - 1)$ **then**

5   Delete leaves $x_{5k(\Delta^+-1)+1}, \ldots, x_p$ from X and from each tree in $\mathcal{T}$ and repeatedly suppress outdegree-1 vertices and delete unlabelled outdegree-0 vertices until no such vertices remain.

6   **go to** Line 1

The proof of the following theorem follows the same ideas as the proofs of Lemmas 2–5 and has been omitted.

Theorem 2. *The problem* HYBRIDIZATION NUMBER *on trees with maximum outdegree* Δ^+ *admits a kernel with at most* $20k^2(\Delta^+ - 1)$ *leaves.*

5 Discussion and Open Problems

The main open question remains whether HYBRIDIZATION NUMBER has a polynomial kernel for an unbounded number of nonbinary trees with unbounded outdegrees. A related question is whether this problem is fixed-parameter tractable.

Note that when the input trees are not required to have the same label set X, HYBRIDIZATION NUMBER is not fixed-parameter tractable unless P = NP. The reason for this is that it is NP-hard to decide if $r(\mathcal{T}) = 1$ for sets $\mathcal{T}$ consisting of rooted phylogenetic trees with three leaves each [10, Theorem 7].

Another question is whether the kernel size can be reduced for certain fixed $|\mathcal{T}|$. For $|\mathcal{T}| = 2$, our results give a cubic kernel, while Linz and Semple [14] showed a linear kernel of a modified, weighted problem, by analyzing carefully how common chains can look in two trees. Can something like this be done for more than two trees? In particular, does there exist a quadratic kernel for three trees?

Finally, there is the problem of solving the kernelized instances. For this, a fast exponential-time exact algorithm is needed (or a good heuristic). However, it is not known if there exists an $O(c^n)$-algorithm for HYBRIDIZATION NUMBER for any constant c and $n = |X|$, even for three binary trees. A related, but possibly more ambitious goal would be an $O(c^k n^{O(1)})$-algorithm for the same problem. Note that such algorithms do exist for the case $|\mathcal{T}| = 2$ [21].

References

1. Bapteste, E., van Iersel, L., Janke, A., Kelchner, S., Kelk, S., McInerney, J.O., Morrison, D.A., Nakhleh, L., Steel, M., Stougie, L., Whitfield, J.: Networks: expanding evolutionary thinking. Trends Genet. **29**(8), 439–441 (2013)
2. Baroni, M., Grünewald, S., Moulton, V., Semple, C.: Bounding the number of hybridisation events for a consistent evolutionary history. Math. Biol. **51**, 171–182 (2005)
3. Bodlaender, H.L., Downey, R.G., Fellows, M.R., Hermelin, D.: On problems without polynomial kernels. J. Comput. Syst. Sci. **75**(8), 423–434 (2009)
4. Bordewich, M., Semple, C.: Computing the hybridization number of two phylogenetic trees is fixed-parameter tractable. IEEE/ACM Trans. Comput. Biol. Bioinf. **4**(3), 458–466 (2007)
5. Bordewich, M., Semple, C.: Computing the minimum number of hybridization events for a consistent evolutionary history. Discrete Appl. Math. **155**(8), 914–928 (2007)
6. Chen, Z.-Z., Wang, L.: Algorithms for reticulate networks of multiple phylogenetic trees. IEEE/ACM Trans. Comput. Biol. Bioinf. **9**(2), 372–384 (2012)
7. Chen, Z.-Z., Wang, L.: An ultrafast tool for minimum reticulate networks. J. Comput. Biol. **20**(1), 38–41 (2013)
8. Downey, R.G., Fellows, M.R.: Parameterized Complexity. Springer, New York (1999)
9. Huson, D.H., Rupp, R., Scornavacca, C.: Phylogenetic Networks: Concepts, Algorithms and Applications. Cambridge University Press, Cambridge (2011)
10. Jansson, J., Nguyen, N.B., Sung, W.-K.: Algorithms for combining rooted triplets into a galled phylogenetic network. SIAM J. Comput. **35**(5), 1098–1121 (2006)
11. Kelk, S., Scornavacca, C.: Towards the fixed parameter tractability of constructing minimal phylogenetic networks from arbitrary sets of nonbinary trees (2012). arXiv:1207.7034 [q-bio.PE]
12. Kelk, S., Scornavacca, C.: Constructing minimal phylogenetic networks from softwired clusters is fixed parameter tractable. Algorithmica **68**(4), 886–915 (2014)
13. Kelk, S., van Iersel, L., Lekić, N., Linz, S., Scornavacca, C., Stougie, L.: Cycle killer.. qu'est-ce que c'est? on the comparative approximability of hybridization number and directed feedback vertex set. SIAM J. Discrete Math. **26**(4), 1635–1656 (2012)
14. Linz, S., Semple, C.: Hybridization in non-binary trees. IEEE/ACM Trans. Comput. Biol. Bioinf. **6**(1), 30–45 (2009)
15. Morrison, D.: Introduction to Phylogenetic Networks. RJR Productions, Uppsala (2011)
16. Nakhleh, L., Ringe, D., Warnow, T.: Perfect phylogenetic networks: a new methodology for reconstructing the evolutionary history of natural languages. Language **81**(2), 382–420 (2005)

17. Piovesan, T., Kelk, S.: A simple fixed parameter tractable algorithm for computing the hybridization number of two (not necessarily binary) trees. IEEE/ACM Trans. Comput. Biol. Bioinf. **10**(1), 18–25 (2013)
18. Semple, C., Steel, M.: Phylogenetics. Oxford University Press, Oxford (2003)
19. van Iersel, L., Kelk, S., Lekić, N., Stougie, L.: Approximation algorithms for non-binary agreement forests. SIAM J. Discrete Math. **28**(1), 49–66 (2014)
20. van Iersel, L., Linz, S.: A quadratic kernel for computing the hybridization number of multiple trees. Inf. Process. Lett. **113**(9), 318–323 (2013)
21. Whidden, C., Beiko, R.G., Zeh, N.: Fixed-parameter algorithms for maximum agreement forests. SIAM J. Comput. **42**(4), 1431–1466 (2013)
22. Yufeng, W.: Close lower and upper bounds for the minimum reticulate network of multiple phylogenetic trees. Bioinformatics **26**, i140–i148 (2010)
23. Yufeng, W.: An algorithm for constructing parsimonious hybridization networks with multiple phylogenetic trees. J. Comput. Biol. **20**(10), 792–804 (2013)

Graph-TSP from Steiner Cycles

Satoru Iwata[1], Alantha Newman[2]([✉]), and R. Ravi[3]

[1] Department of Mathematical Informatics, University of Tokyo,
Tokyo 113-8656, Japan
iwata@mist.i.u-tokyo.ac.jp
[2] CNRS-Université Grenoble Alpes and G-SCOP, 38000 Grenoble, France
alantha.newman@grenoble-inp.fr
[3] Tepper School of Business, Carnegie Mellon University, Pittsburgh, USA
ravi@cmu.edu

Abstract. We present an approach for the traveling salesman problem with graph metric based on Steiner cycles. A Steiner cycle is a cycle that is required to contain some specified subset of vertices. For a graph G, if we can find a spanning tree T and a simple cycle that contains the vertices with odd-degree in T, then we show how to combine the classic "double spanning tree" algorithm with Christofides' algorithm to obtain a TSP tour of length at most $\frac{4n}{3}$. We use this approach to show that a graph containing a Hamiltonian path has a TSP tour of length at most $4n/3$.

Since a Hamiltonian path is a spanning tree with two leaves, this motivates the question of whether or not a graph containing a spanning tree with few leaves has a short TSP tour. The recent techniques of Mömke and Svensson imply that a graph containing a depth-first-search tree with k leaves has a TSP tour of length $4n/3 + O(k)$. Using our approach, we can show that a $2(k-1)$-vertex connected graph that contains a spanning tree with at most k leaves has a TSP tour of length $4n/3$. We also explore other conditions under which our approach results in a short tour.

1 Introduction

We consider the well studied Traveling Salesman problem with graph metric, also known as graph-TSP. Throughout this paper, the input graph $G = (V, E)$ is assumed to be an undirected, unweighted, 2-(vertex) connected graph, and all edge lengths in the complete graph can be obtained via the shortest path metric on the given graph. Our goal is to find a tour of minimum length that visits each vertex at least once. In this paper, we focus on a connection between graph-TSP and that of finding *Steiner cycles*.

1.1 Background

Graph-TSP has received much attention recently. Oveis Gharan, Saberi and Singh were the first to improve on the approximation ratio of 3/2 by an infini-

---

R. Ravi: Supported in part by NSF grants CCF1143998 and CCF1218382.

D. Kratsch and I. Todinca (Eds.): WG 2014, LNCS 8747, pp. 312–323, 2014.
DOI: 10.1007/978-3-319-12340-0_26

Fig. 1. In this (non-simple) cycle C, the number of unique vertices $|C| = 8$, but the length of the cycle $\ell(C) = 10$.

tesimal, but constant, factor [12]. This was quickly followed by the breakthrough work of Mömke and Svensson, who introduced a new approach leading to a substantial improvement in the approximation ratio [19]. Subsequently, Mucha gave a refined analysis of their approach, proving an approximation ratio of 13/9 for graph-TSP [20]. More recently, Sebő and Vygen presented an approximation algorithm with ratio 7/5 for the problem [23].

It is widely believed that an approximation ratio of at most 4/3 should be efficiently computable. The approach of Mömke and Svensson is based on setting up a circulation network and showing that a low-cost circulation leads to a low cost TSP tour. They obtained a 4/3-approximation for subcubic graphs, but high-degree graphs appear to be more challenging for their framework. Vishnoi recently gave a randomized algorithm that finds a TSP tour very close to n with high probability for a k-regular graph when k is sufficiently large [26]. Our goal is to consider other techniques that are applicable for graphs that are not low-degree or regular.

2   Steiner Cycles

The Steiner cycle problem has been previously, but not extensively, studied under varying definitions [8,13,25]. For our purposes, a Steiner cycle is defined to be a simple cycle that contains a specified subset $S \subseteq V$ of vertices. It may also contain any subset of vertices from the set $V \setminus S$. We use the following definition:

Definition 1. *Given a graph $G = (V, E)$ and a subset of vertices, $S \subseteq V$, a Steiner cycle, $C \subset E$, is a simple cycle whose vertices contains the set S.*

It is important to observe that in our definition of a Steiner cycle, there are no repeated vertices, since a Steiner cycle is a simple cycle. We define an approximate Steiner cycle as one in which we are allowed to repeat vertices. For a cycle C, we will use $|C|$ to denote the number of unique vertices it contains. We define the cycle *length*, $\ell(C)$, to be total length of a traversal of the cycle. If C is a simple cycle, then $|C| = \ell(C)$. For example, in Fig. 1, the non-simple cycle has eight unique vertices and has length ten. Our definition of cycle length is the same as the standard definition for the length of a TSP tour in the graph metric. Now we can define an approximate Steiner cycle.

Definition 2. *Given a graph $G = (V, E)$ and a subset of vertices, $S \subseteq V$, an approximate Steiner cycle, $C \subset E$, with relative length $\beta \geq 1$ is a cycle whose vertices contains the set S and for which $\ell(C)/|C| \leq \beta$.*

In an approximate Steiner cycle, since we are allowed to repeatedly visit vertices as we traverse the cycle, it may be the case that the number of unique vertices will be smaller than the length, $|C| < \ell(C)$. Throughout this paper, whenever we refer simply to a "cycle", we mean a simple cycle.

Other natural definitions of the Steiner cycle problem are concerned with such aspects as minimizing the number of non-required (Steiner) vertices in the cycle. In our definition of the approximate Steiner cycle problem, the only objective that we wish to minimize is the ratio of the length of a cycle, $\ell(C)$, to the number of unique vertices, $|C|$, it contains. Thus, the measure of an optimal solution is independent of the size of the set of required vertices. The work that appears to be most related to the Steiner cycle problem as we have defined it concerns the concept of *cyclability*: A set of vertices $X \subseteq V$ is called *cyclable* if it is contained in some cycle. The quantity $cyc(G)$ is the maximum number such that all subsets containing at most $cyc(G)$ vertices are cyclable. Note that $cyc(G) = n$ if and only G is Hamiltonian. It seems that most of the work on cyclability has been done with the intention of eventually using it to prove that certain graphs are Hamiltonian or because it can be viewed as a relaxation of Hamiltonicity. An interesting list of theorems on cyclability can be found in [21]. Here, we explore cyclability as a tool to obtain approximate TSP tours.

2.1   Our Approach

Graph-TSP can clearly be cast as a special case of the Steiner cycle problem in which all of the vertices in V are required to belong to the Steiner cycle. In this paper, we show that even if the required set of vertices is possibly much smaller than the entire vertex set V, an (approximation) algorithm for the Steiner cycle problem can still be used to approximate graph-TSP.

Suppose we can find a spanning tree T for the graph G and a simple cycle C_T that contains all of the vertices that have an odd-degree in the tree T. When $|C_T|$ is large, we show that we can use the folklore "double spanning tree" algorithm to find a short tour. When $|C_T|$ is small, then there is a small matching on the odd-degree vertices in T and we can therefore show that Christofides algorithm [5] yields a short tour. Thus, our algorithm, described in Sect. 3, can be viewed as a combination of these two standard algorithms for graph-TSP.

We are not aware of any previous work studying how to combine these two classic algorithms for graph-TSP. However, a similar algorithm that combines these two algorithms was given by Guttman-Beck, Hassin, Khuller and Raghavachari for the s, t-path TSP [15]. In their algorithm, they first find an MST for the input graph. If the path from s to t in this MST is long, they double edges in the MST that do not belong to this path. If the path from s to t is short, they modify the input graph by adding an edge from s to t with length equal to the shortest s, t-path in G and run Christofides on this modified graph as in the algorithm by Hoogeveen [17]. Taking the better of these two algorithms results in a 5/3-approximation for the s, t-path problem, which does not improve on the worst-case approximation ratio of Hoogeveen's algorithm. Nevertheless, this approach was used to design algorithms for special variants of

Fig. 2. A graph G with a spanning tree T (second figure, blue edges) and a simple cycle C_T (third figure, purple edges) containing all of the odd-degree nodes of T (Color figure online).

the path TSP problem [15], and the ideas were also eventually used to obtain improved approximation guarantees for the s,t-path TSP itself [22]. In our algorithm, rather than basing the subcases on the path length from s to t in an MST, we are basing the two subcases on the length of a cycle containing the nodes with odd degree in a particular MST.

2.2 Overview of Our Results

In Sect. 3, we give a complete description of our algorithm. In Sect. 4, we use this algorithm to show that if the input graph contains a Hamiltonian path, then it has a TSP tour of length at most $4n/3$. Moreover, if we are given the Hamiltonian path, then we can efficiently find such a tour. This theorem was first proved by Gupta using a different approach [14].

One can view a Hamiltonian path as a spanning tree with two leaves. A natural question is how well we can approximate a TSP tour in a graph that contains a spanning tree with few leaves. In Sect. 5, we show how our approach can be used to address this question in some special cases. In Sect. 6, we discuss how approximate Steiner cycles can also be used to obtain an approximation guarantee for graph-TSP. Finally, in Sect. 7, we consider some examples (Fig. 2).

3 TSP Tours from Steiner Cycles

Given an undirected, unweighted graph, $G = (V, E)$, with graph metric, our goal is to find a TSP tour of minimum length. A TSP tour must visit each vertex at least once. As stated previously in the introduction, we assume that G is a 2-connected graph and we define $n = |V|$.

Let T be a spanning tree of G and let $S_T \subset V$ be the vertices that have odd degree in T. Suppose there is a simple cycle C_T that contains all the vertices in S_T. Note that the simple cycle C_T can be of arbitrary length, i.e. can contain arbitrarily many vertices in $V \setminus S_T$.

Theorem 1. *For a given graph G, suppose we have a minimum spanning tree T and a simple cycle C_T that contains all vertices with odd degree in T. Then we can construct a TSP tour of G with length at most $4n/3$.*

Proof: Consider the following cases. Recall that $|C_T|$ denotes the number of unique vertices contained in the cycle C_T. Since C_T is a simple cycle, $|C_T|$ also denotes its length.

(i) $|C_T| > 2n/3$. In this case, we can contract the cycle C_T to a single vertex. The resulting graph has at most $n/3$ vertices. We can then find a minimum spanning tree on this graph and double each edge. When we uncontract the vertex corresponding to the cycle C_T, we obtain an Eulerian tour whose total length is at most $4n/3$.

(ii) $|C_T| \leq 2n/3$. In this case, since all of the vertices of S_T are contained in C_T, there is a matching of the vertices in S_T with length at most $n/3$. Using this matching plus T, we obtain an Eulerian tour of G of length at most $4n/3$.

$\square$

We can therefore see that if G has a tree T and a simple cycle C_T that contains all of the vertices with odd degree in T, then G has a TSP tour of length at most $4n/3$. We now show how to apply this theorem to some special classes of graphs.

4   Graphs Containing a Hamiltonian Path

Recall that a Hamiltonian path in G is a path that visits each vertex in V exactly once. Note that the first and last vertices on the path might not be adjacent vertices in G. More generally, G might not be Hamiltonian. In this section, we show that for an unweighted graph $G = (V, E)$ with graph metric, if G contains a Hamiltonian path, then G has a TSP tour of length at most $4n/3$.

Theorem 2. *Suppose G contains a Hamiltonian path. Then G has a TSP tour of length at most $4n/3$.*

Proof: Suppose that the first and last vertices of the Hamiltonian path are adjacent in the graph. Then G is Hamiltonian and, moreover, given the Hamiltonian path, we can find this tour.

If the first and last vertices of the Hamiltonian path are not adjacent in G, then since G is 2-vertex connected, we can use Menger's theorem [10,18], which states that there are two vertex disjoint paths between any two non-adjacent vertices in a 2-vertex connected graph. Thus, we have a simple cycle including the odd-degree nodes on the tree (the first and last nodes in the Hamiltonian path) and the proof of the theorem follows directly from applying Theorem 1. $\square$

Since there are constructive proofs of Menger's Theorem, Theorem 2 results in an efficient algorithm, assuming the Hamiltonian path is given.

5   Graphs Containing a Spanning Tree with k Leaves

A Hamiltonian path can be viewed as a spanning tree with two leaves. A natural extension is to ask what happens when a graph does not contain a Hamiltonian path but rather a spanning tree with few leaves. Does it still have a short TSP tour? Suppose G has a spanning tree with k leaves. If G is well-connected, we can use a well-known theorem of Dirac to obtain an upper bound on the length of a TSP tour of G.

Theorem 3. *Suppose G is $2(k-1)$-connected and contains a spanning tree with k leaves. Then G has a TSP tour of length at most $4n/3$.*

Proof: A spanning tree with k leaves contains at most $2(k-1)$ vertices with odd degree. A theorem of Dirac states that if a graph is c-vertex connected, then any subset $X \subseteq V$ of vertices with $|X| \le c$ is contained in some simple cycle [3,9]. Thus, if $c = 2(k-1)$, then G is c-connected by the assumption of the theorem. Moreover, G has at most c odd-degree vertices if it has k leaves. We can therefore let X be the set of odd-degree vertices and the theorem follows directly from applying Theorem 1. □

Finding a simple cycle containing c vertices in a c-connected graph can be done efficiently (see Chap. 9 in [3]). Thus, Theorem 3 results in an efficient algorithm assuming the spanning with k leaves is given.

More generally, Steiner cycles have been studied by the Graph Theory community and if a set of vertices $X \subseteq V$ is contained in a cycle, then the set X is called *cyclable*. This terminology is attributed to Chvatal [6]. Moreover, cyclability of a graph G, i.e. $cyc(G)$, is the maximum number such that every subset of at most $cyc(G)$ vertices is cyclable. If a graph G has a cyclable number $c = cyc(G)$ and it also contains a spanning tree with at most $c/2+1$ leaves, then this spanning tree contains at most c odd-degree vertices. Thus, it will contain a TSP tour of length $4n/3$ via Theorem 1. Considerable effort has been invested in computing the cyclablity of certain graph classes. For example, we cite the following two theorems:

Theorem 4 *[16].* *For every 3-connected cubic graph G, $cyc(G) \ge 9$. This bound is sharp (the Petersen graph).*

Theorem 5 *[2].* *For every 3-connected cubic planar graph G, $cyc(G) \ge 23$. This bound is sharp.*

Theorem 4 implies that if a 3-connected, cubic graph G contains a spanning tree with at most five leaves, then G has a TSP tour of length at most $4n/3$. Theorem 5 shows that if a 3-connected, planar, cubic graph G contains a spanning tree with at most 12 leaves, then G has a TSP tour of length at most $4n/3$. We remark that showing that a 3-connected cubic graph has a spanning trees with at most five leaves as a means to bounding the length of a TSP tour would only be an alternative approach, as it is already known that a cubic graph has a TSP tour of length at most $4n/3$ [1,4,14,19].

A well-known theorem of Dirac states that every graph with minimum degree at least $n/2$ is Hamiltonian. A analogous theorem can be shown for cyclability. Let $X \subseteq V$ be a subset of vertices and define $\sigma_2(X) := \min\{\sum_{y \in Y} d(y) : Y \subseteq X, |Y| = 2, Y \text{ is an independent set}\}$. In other words, if we choose each pair of non-adjacent vertices in X and add up their degrees, $\sigma_2(X)$ is the minimum of this quantity. This is used in the following theorem due to Shi:

Theorem 6 *[24].* *Let $G = (V, E)$ be a 2-connected graph and $X \subset V$. If $\sigma_2(X) \ge n$, then X is cyclable in G.*

If we find a spanning tree T such that all non-adjacent pairs of vertices with odd-degree in T have total degree at least n (in G), then G has a TSP tour of length at most $4n/3$. The vertices that have an even degree in the tree are allowed to have low degree in G. Another nice theorem on cyclability is due to Fournier:

Theorem 7 *[11]. Let G be a 2-connected graph and $X \subseteq V$. If $\alpha(X) \le \kappa(G)$, then X is cyclable in G.*

Here, $\alpha(X)$ means the largest independent set in X, and $\kappa(G)$ is the connectivity of G. It is known that if $\alpha(G) \le \kappa(G)$, then G is Hamiltonian [7]. Theorem 7 implies that if the set of odd-degree vertices in a spanning tree has a maximum independent set that is smaller than the connectivity of G, then G has a TSP tour of length at most $4n/3$.

In relation to Theorem 3, it is reasonable to ask if, for sufficiently large k, a $2(k-1)$-connected graph has a spanning tree with k leaves. This is not the case as demonstrated by the following example. Consider the complete bipartite graph $G = K_{c,n}$ where $n >> c$. Then G is c-connected, but the minimum length TSP tour is roughly $2n$. So G cannot contain a spanning tree with at most $c/2+1$ leaves.

5.1   Graphs Containing a k-Leaf DFS Spanning Tree

If G has a depth-first-search (DFS) spanning tree with k leaves, then we note that the techniques of Mömke and Svensson [19] can be used to obtain a TSP tour of length at most $4n/3 + 2k/3$. Specifically, in this case, it is not difficult to see that there is a circulation (as defined by Mömke and Svensson) of cost at most k. This implies that one can also use the techniques from Mömke and Svensson to prove Theorem 2. We emphasize that a DFS spanning tree must be used to directly apply the techniques of Mömke and Svensson. In comparison, in Theorem 3, we can use any spanning tree with k leaves. The proof of Lemma 1 is straightforward, but we include it for the sake of completeness.

Lemma 1. *If G has a DFS spanning tree with at most k leaves, then it has a circulation, as defined by Mömke and Svensson [19], of cost at most k.*

Proof: We will demonstrate a 2-connected subgraph of G such that the cost of a circulation on this subgraph is at most k.

Consider a path from the root of the DFS tree to a leaf. Let us call this path p_1. Suppose that the vertices on p_1 are labeled sequentially from the root to the leaf in increasing order, $1, 2, ... \ell(p_1)$, where $\ell(p_1)$ denotes the number of vertices in the path p_1. We find a back-edge from the leaf or the vertex labeled $\ell(p_1)$ to a vertex with the smallest label. Suppose that this edge goes from $\ell(p_1)$ to h. Then at the next step, we find the back-edge (i, j) where $\ell(p_1) > i > h$ and $j < i$ and j is as small as possible. Since G is 2-connected, we will always be able to find such an edge. Otherwise G would contain a cut vertex, which would contradict the 2-connectivity of G.

Now consider a path on the DFS tree from some vertex on p_1 to another leaf. Call the path from the root to this leaf p_2. Perform the same procedure as above: starting at the leaf, find some back-edges, so that the resulting subgraph containing paths p_1 and p_2 and these back-edges is 2-connected. At some point, we will add a back edge that intersects with the path p_1. If this is a branching node, i.e. the last node that belongs to both p_1 and p_2, we will add one more back edge so that the resulting subgraph is 2-connected.

Note that each vertex in p_2 that is below this branching node, i.e. has a higher label, has only one back-dge coming into it. The only vertices that may have more than one back-edge coming into them are the branch node and another node with a lower label. However, since in Lemma 4.1 of [19], each subtree of a branch node is accounted separately in the circulation network, if the branch node now has, say, two back-edges, it also has two subtrees, so its contribution to the circulation is still zero. A node above the branch node with B back-edges coming into it will contribute at most $B - 1$ to the cost of the circulation.

As we add each root-leaf path in the DFS tree, and we add the new path and a set of back-edges to make the subgraph 2-connected, we will add at most one back-edge to a vertex that already has incoming back-edges. Thus, the circulation is upper bounded by k if the DFS tree has k leaves.   □

Theorem 8. *If G has a DFS spanning tree with at most k leaves, then it has a TSP tour of at most $4n/3 + 2k/3$.*

Proof: This follows from Lemma 1 and Lemma 4.1 of Mömke and Svensson [19].   □

6   Cycle Length and Approximation Ratio Tradeoff

We have shown that a simple cycle that contains the odd-degree nodes in some spanning tree yields a TSP tour of length at most $4n/3$. Suppose we can only obtain an approximate Steiner cycle. Then what is the guarantee on the length of the TSP tour? We now show that we can obtain the following tradeoff. For a cycle C in G that is not necessarily simple, recall that $|C|$ is the number of unique vertices in the cycle C and $\ell(C)$ denotes its length.

Theorem 9. *Given G, a minimum spanning tree T and an approximate Steiner cycle C_T that contains all the odd-degree vertices in T such that $\ell(C_T) \leq (1 + \gamma)|C_T|$, we can construct a TSP tour of G of length at most $\frac{4n}{3-\gamma}$.*

Proof: We consider two cases based on the number of unique vertices in the cycle C_T:

(i) $|C_T| > \frac{2n}{3-\gamma}$. Then we contract the cycle C_T to a single vertex, find a minimum spanning tree on the resulting graph and double each edge in this spanning tree. Since the length of cycle $\ell(C_T) \leq (1+\gamma)|C_T|$, the total length of the resulting Eulerian tour is at most:

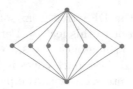

Fig. 3. Any spanning tree of this graph has too many leaves to be spanned by a simple cycle. However, note that the solution to the Held-Karp LP relaxation will be $|E| = 2n$ for this graph, certifying that the lower bound is much greater than $4n/3$ in this case.

$$\ell(TSP) \le (1+\gamma)|C_T| + 2(n - |C_T|) \tag{1}$$
$$= 2n + (1 + \gamma - 2)|C_T| \tag{2}$$
$$= 2n - (1 - \gamma)|C_T| \tag{3}$$
$$< 2n - \frac{2n}{(3 - \gamma)}(1 - \gamma) \tag{4}$$
$$= \frac{4n}{3 - \gamma}. \tag{5}$$

(ii) $|C_T| \le \frac{2n}{3-\gamma}$. In this case, we find a matching of the odd-degree vertices in T with length at most $(1+\gamma)|C_T|/2$. The total length of the resulting Eulerian tour S is at most:

$$\ell(TSP) \le n + (1+\gamma)\frac{|C_T|}{2} \tag{6}$$
$$\le n + \frac{(1+\gamma)}{2}\frac{2n}{(3 - \gamma)} \tag{7}$$
$$= \frac{4n}{3 - \gamma}. \tag{8}$$

$\square$

6.1 Approximation Guarantees from LP Bounds

In general, it could be the case that there does not exist a spanning tree whose odd-degree vertices can be contained in a simple cycle. An example of such a graph can be found in Fig. 3. However, suppose we can compute, via an LP relaxation or some other means, a lower bound on the length of a TSP tour, e.g. $OPT \ge (1 + \alpha)n$ for $0 \le \alpha \le 1$. Then the following Corollary of Theorem 9 states a sufficient condition for a $\frac{4}{3}$-approximation to the optimal TSP tour.

Corollary 1. *If an optimal tour is lowerbounded by $OPT \ge (1 + \alpha)n$ and G contains a spanning tree T and a cycle C_T containing the odd-degree nodes of T such that $\ell(C_T) \le (1 + 4\alpha)|C_T|/(1 + \alpha)$, then G has a TSP tour of length at most $\frac{4}{3} \cdot OPT$.*

Note that Theorem 9 says that if we can find a tree T and a cycle C_T such that $\ell(C_T)/|C_T| < 4/3$, then we can find a TSP tour less than $3n/2$. To find a tour shorter than $7n/5$ (which is currently the best known bound when the solution to the standard LP relaxation equals n [23]), we require that $\ell(C_T)/|C_T| < 8/7$.

7   Discussion

We have reduced the problem of finding a short TSP tour to the problem of finding an (approximate) Steiner cycle where the required vertices are the odd-degree nodes in some spanning tree, and we have flexibility as to whether or not we include the non-required vertices in the cycle. But is this problem any easier than graph-TSP itself? For example, in Fig. 4, we give an example of a graph and a spanning tree such that the odd-degree vertices of the spanning tree is the entire vertex set! Thus, finding a Steiner cycle for these vertices is no easier than finding a TSP tour. However, in this example, we can see that there are many other possible spanning trees. Figure 5 shows two other possible spanning trees and corresponding Steiner cycles. We note that given a spanning tree, the Steiner cycle including the odd-degree nodes may not be unique. Another example of a graph G and a spanning tree in which every vertex can have odd degree is shown in Fig. 6. But, again, there are many other spanning trees in which only a subset of the vertices have odd degree (Fig. 7).

Each of the examples we have considered so far actually contains a Hamiltonian path. Thus, by applying Theorem 2, we can see that they have a TSP tour of length at most $4n/3$. There are actually interesting examples of cubic, 3-edge connected graphs that do not contain a Hamiltonian path. The graph shown in Fig. 8 is such a graph due to Zamfirescu [27]. We see that we can construct a spanning tree and Steiner cycle containing all of the vertices that have odd degree in the spanning tree.In conclusion, let us consider the following question:

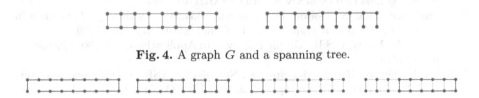

Fig. 4. A graph G and a spanning tree.

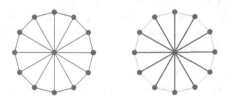

Fig. 5. Alternative spanning trees for G and corresponding Steiner cycles.

Fig. 6. The wheel graph has a spanning tree in which all vertices have odd degree.

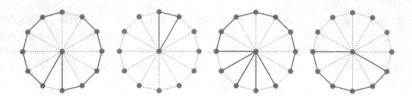

Fig. 7. Alternative spanning trees with fewer odd-degree vertices for the wheel graph.

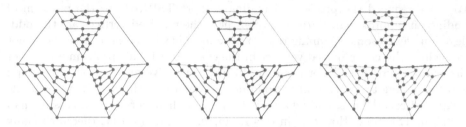

Fig. 8. A cubic, 3-edge connected graph with no Hamiltonian paths. We show a spanning tree and a corresponding Steiner cycle containing all the nodes with odd degree in the spanning tree.

Suppose the standard linear programming relaxation for Graph TSP has value n on a fixed graph. Then is there a spanning tree T and a simple cycle C_T that contains all of the vertices that are odd-degree in T? If a graph is Hamiltonian, then this is (trivially) true for any spanning tree.

References

1. Aggarwal, N., Garg, N., Gupta, S.: A 4/3-approximation for TSP on cubic 3-edge-connected graphs (2011). arXiv preprint arXiv:1101.5586
2. Aldred, R.E., Bau, S., Holton, D.A., McKay, B.D.: Cycles through 23 vertices in 3-connected cubic planar graphs. Graphs Comb. **15**(4), 373–376 (1999)
3. Bondy, J.A., Murty, U.S.R.: Graph Theory with Applications, vol. 290. Macmillan, London (1976)
4. Boyd, S., Sitters, R., van der Ster, S., Stougie, L.: TSP on cubic and subcubic graphs. In: Günlük, O., Woeginger, G.J. (eds.) IPCO 2011. LNCS, vol. 6655, pp. 65–77. Springer, Heidelberg (2011)
5. Christofides, N.: Worst-case analysis of a new heuristic for the travelling salesman problem. Technical report, DTIC Document (1976)
6. Chvatal, V.: New directions in Hamiltonian graph theory. In: Harary, F. (ed.) New Directions in Graph Theory. Academic press, London (1973)
7. Chvátal, V., Erdös, P.: A note on Hamiltonian circuits. Discrete Math. **2**(2), 111–113 (1972)
8. Cornuéjols, G., Fonlupt, J., Naddef, D.: The traveling salesman problem on a graph and some related integer polyhedra. Math. program. **33**(1), 1–27 (1985)
9. Dirac, G.: Some theorems on abstract graphs. Proce. Lond. Math. Soc. **3**(1), 69–81 (1952)

10. Dirac, G.: Short proof of Menger's graph theorem. Mathematika **13**(1), 42–44 (1966)
11. Fournier, I.: Cycles et numérotations de graphes. These d'Etat, LRI, Université de Paris-Sud (1985)
12. Gharan, S.O., Saberi, A., Singh, M.: A randomized rounding approach to the traveling salesman problem. In: 2011 IEEE 52nd Annual Symposium on Foundations of Computer Science (FOCS), pp. 550–559. IEEE (2011)
13. González, J.J.S.: The Steiner cycle polytope. Eur. J. Oper. Res. **147**(3), 671–679 (2003)
14. Gupta, S.: Towards a $\frac{4}{3}$-approximation for the metric traveling salesman problem. Master's thesis, Indian Institute of Technology, Delhi, May 2011
15. Guttmann-Beck, N., Hassin, R., Khuller, S., Raghavachari, B.: Approximation algorithms with bounded performance guarantees for the clustered traveling salesman problem. Algorithmica **28**(4), 422–437 (2000)
16. Holton, D.A., McKay, B.D., Plummer, M.D., Thomassen, C.: A nine point theorem for 3-connected graphs. Combinatorica **2**(1), 53–62 (1982)
17. Hoogeveen, J.: Analysis of Christofides' heuristic: some paths are more difficult than cycles. Oper. Res. Lett. **10**(5), 291–295 (1991)
18. Menger, K.: Zur allgemeinen Kurventheorie. Fundam. Math. **10**(1), 96–115 (1927)
19. Mömke, T., Svensson, O.: Approximating graphic TSP by matchings. In: IEEE 52nd Annual Symposium on Foundations of Computer Science, pp. 560–569 (2011)
20. Mucha, M.: $\frac{13}{9}$-approximation for graphic TSP. Theory Comput. Syst. 1–18 (2012)
21. Ozeki, K., Yamashita, T.: A degree sum condition concerning the connectivity and the independence number of a graph. Graphs Comb. **24**(5), 469–483 (2008)
22. Sebő, A.: Eight-fifth approximation for the path TSP. In: Goemans, M., Correa, J. (eds.) IPCO 2013. LNCS, vol. 7801, pp. 362–374. Springer, Heidelberg (2013)
23. Sebő, A., Vygen, J.: Shorter tours by nicer ears: 7/5-approximation for graphic TSP, 3/2 for the path version, and 4/3 for two-edge-connected subgraphs (2012). arXiv preprint arXiv:1201.1870
24. Shi, R.: 2-neighborhoods and Hamiltonian conditions. J. Graph Theory **16**(3), 267–271 (1992)
25. Steinová, M.: Approximability of the minimum Steiner cycle problem. Comput. Inf. **29**(6+), 1349–1357 (2012)
26. Vishnoi, N.K.: A permanent approach to the traveling salesman problem. In: 2012 IEEE 53rd Annual Symposium on Foundations of Computer Science (FOCS), pp. 76–80. IEEE (2012)
27. Zamfirescu, T.: Three small cubic graphs with interesting Hamiltonian properties. J. Graph Theory **4**(3), 287–292 (1980)

A Characterization of Mixed Unit Interval Graphs

Felix Joos[✉]

Institut für Optimierung und Operations Research, Universität Ulm, Ulm, Germany
felix.joos@uni-ulm.de

Abstract. We give a complete characterization of mixed unit interval graphs, the intersection graphs of closed, open, and half-open unit intervals of the real line. This is a proper superclass of the well known unit interval graphs. Our result solves a problem posed by Dourado, Le, Protti, Rautenbach and Szwarcfiter (Mixed unit interval graphs. Discrete Math. **312**, 3357–3363 (2012)). Our characterization also leads to a polynomial-time recognition algorithm for mixed unit interval graphs.

Keywords: Unit interval graph · Proper interval graph · Intersection graph

1 Introduction

A graph G is an *interval graph*, if there is a function I from the vertex set of G to the set of intervals of the real line such that two vertices are adjacent if and only if their assigned intervals intersect. The function I is an *interval representation* of G. Interval graphs are well known and investigated – algorithmically as well as structurally [4,6,9]. There are several efficient algorithms that decide, if a given graph is an interval graph. See for example [2].

An important subclass of interval graphs are unit interval graphs. An interval graph G is a *unit interval graph*, if there is an interval representation I of G such that I assigns to every vertex a closed interval of unit length. This subclass is well understood and also easy to characterize structurally [11] as well as algorithmically [1].

Frankl and Maehara [5] showed that it does not matter, if we assign the vertices of G only to closed intervals or only to open intervals of unit length. Rautenbach and Szwarcfiter [10] characterized, by a finite list of forbidden induced subgraphs, all interval graphs G such that there is an interval representation of G that uses only open and closed unit intervals.

Dourado et al. [3] gave a characterization of all diamond-free interval graphs that have an interval representation such that all vertices are assigned to unit intervals, where all kinds of unit intervals are allowed and a diamond is a complete graph on four vertices minus an edge. Furthermore, they made a conjecture concerning the general case.

© Springer International Publishing Switzerland 2014
D. Kratsch and I. Todinca (Eds.): WG 2014, LNCS 8747, pp. 324–335, 2014.
DOI: 10.1007/978-3-319-12340-0_27

We prove that their conjecture is not completely correct and give a complete characterization of this class. Since the conjecture is rather technical and not given by a list of forbidden subgraphs, we refer the reader to [3] for a detailed formulation of the conjecture, but roughly speaking, they missed the class of forbidden subgraphs shown in Fig. 6. Moreover, we provide a polynomial-time recognition algorithm for this graph class.

In Sect. 2 we introduce all definitions and relate our results to other work. In Sect. 3 we state and prove our results.

2 Preliminary Remarks

We only consider finite, undirected, and simple graphs. Let G be a graph. We denote by $V(G)$ and $E(G)$ the vertex and edge set of G, respectively. If C is a set of vertices, then we denote by $G[C]$ the subgraph of G induced by C. Let $\mathcal{M}$ be a set of graphs. We say G is $\mathcal{M}$-free, if for every $H \in \mathcal{M}$, the graph H is not an induced subgraph of G. For a vertex $v \in V(G)$, let the *neighborhood* $N_G(v)$ of v be the set of all vertices that are adjacent to v and let the *closed neighborhood* $N_G[v]$ be defined by $N_G(v) \cup \{v\}$. Two distinct vertices u and v are *twins* (in G) if $N_G[u] = N_G[v]$. If G contains no twins, then G is *twin-free*.

Let $\mathcal{N}$ be a family of sets. We say a graph G has an $\mathcal{N}$-*intersection representation*, if there is a function $f : V(G) \to \mathcal{N}$ such that for any two distinct vertices u and v, there is an edge joining u and v if and only if $f(u) \cap f(v) \neq \emptyset$. If there is an $\mathcal{N}$-intersection representation for G, then G is an $\mathcal{N}$-*graph*. Let $x, y \in \mathbb{R}$. We denote by

$$[x, y] = \{z \in \mathbb{R} : x \leq z \leq y\}$$

the *closed interval*, by

$$(x, y) = \{z \in \mathbb{R} : x < z < y\}$$

the *open interval*, by

$$(x, y] = \{z \in \mathbb{R} : x < z \leq y\}$$

the *open-closed interval*, and by

$$[x, y) = \{z \in \mathbb{R} : x \leq z < y\}$$

the *closed-open interval* of x and y. For an interval A, let $\ell(A) = \inf\{x \in \mathbb{R} : x \in A\}$ and $r(A) = \sup\{x \in \mathbb{R} : x \in A\}$. If I is an interval representation of G and $v \in V(G)$, then we write $\ell(v)$ and $r(v)$ instead of $\ell(I(v))$ and $r(I(v))$, respectively, if there are no ambiguities. Let $\mathcal{I}^{++}$ be the set of all closed intervals, $\mathcal{I}^{--}$ be the set of all open intervals, $\mathcal{I}^{-+}$ be the set of all open-closed intervals, $\mathcal{I}^{+-}$ be the set of all closed-open intervals, and $\mathcal{I}$ be the set of all intervals. In addition, let $\mathcal{U}^{++}$ be the set of all closed unit intervals, $\mathcal{U}^{--}$ be the set of all open unit intervals, $\mathcal{U}^{-+}$ be the set of all open-closed unit intervals, $\mathcal{U}^{+-}$ be the set of all closed-open unit intervals, and $\mathcal{U}$ be the set of all unit intervals. We call a $\mathcal{U}$-graph a *mixed unit interval graph*.

By a result of [3,10], every interval graph is an $\mathcal{I}^{++}$-graph. With our notation unit interval graphs equals $\mathcal{U}^{++}$-graphs. An interval graph G is a *proper interval graph* if there is an interval representation of G such that $I(u) \not\subseteq I(v)$ for every distinct $u, v \in V(G)$.

The next result due to Roberts characterizes unit interval graphs.

Theorem 1 (Roberts [11]). *The classes of unit interval graphs, proper interval graphs, and $K_{1,3}$-free interval graphs are the same.*

The second result shows that several natural subclasses of mixed unit interval graphs actually coincide with the class of unit interval graphs.

Theorem 2 (Dourado et al., Frankl and Maehara [3,5]). *The classes of $\mathcal{U}^{++}$-graphs, $\mathcal{U}^{--}$-graphs, $\mathcal{U}^{+-}$-graphs, $\mathcal{U}^{-+}$-graphs, and $\mathcal{U}^{+-} \cup \mathcal{U}^{-+}$-graphs are the same.*

A graph G is a *mixed proper interval graph* (respectively an *almost proper interval graph*) if G has an interval representation $I : V(G) \to \mathcal{I}$ (respectively $I : V(G) \to \mathcal{I}^{++} \cup \mathcal{I}^{--}$) such that

- there are no two distinct vertices u and v of G with $I(u), I(v) \in \mathcal{I}^{++}$, $I(u) \subseteq I(v)$, and $I(u) \neq I(v)$, and
- for every vertex u of G with $I(u) \notin \mathcal{I}^{++}$, there is a vertex v of G with $I(v) \in \mathcal{I}^{++}$, $\ell(u) = \ell(v)$, and $r(u) = r(v)$.

A natural class extending the class of unit interval graphs are $\mathcal{U}^{++} \cup \mathcal{U}^{--}$-graphs. These were characterized by Rautenbach and Szwarcfiter.

Theorem 3 (Rautenbach and Szwarcfiter [10]). *For a twin-free interval graph G, the following statements are equivalent.*

- *G is a $\{K_{1,4}, K_{1,4}^*, K_{2,3}^*, K_{2,4}^*\}$-free graph. (See Fig. 1 for an illustration.)*
- *G is an almost proper interval graph.*
- *G is a $\mathcal{U}^{++} \cup \mathcal{U}^{--}$-graph.*

Note that an interval representation can assign the same interval to twins and hence the restriction to twin-free graphs does not weaken the statement but simplifies the description.

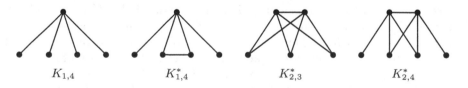

$K_{1,4}$ $K_{1,4}^*$ $K_{2,3}^*$ $K_{2,4}^*$

Fig. 1. Forbidden induced subgraphs for twin-free $\mathcal{U}^{++} \cup \mathcal{U}^{--}$-graphs.

Fig. 2. A graph, which is a $\mathcal{U}$-graph, but not a $\mathcal{U}^{++} \cup \mathcal{U}^{--}$-graph.

The next step is to allow all different types of unit intervals. The class of $\mathcal{U}$-graphs is a proper superclass of the $\mathcal{U}^{++} \cup \mathcal{U}^{--}$-graphs, because the graph illustrated in Fig. 2 is a $\mathcal{U}$-graph, but not a $\mathcal{U}^{++} \cup \mathcal{U}^{--}$-graph (it contains a $K_{1,4}^{*}$). Dourado et al. already made some progress in characterizing this class.

Theorem 4 (Dourado et al. [3]). *For a graph G, the following two statements are equivalent.*

- *G is a mixed proper interval graph.*
- *G is a mixed unit interval graph.*

They also characterized diamond-free mixed unit interval graphs. There is another approach by Le and Rautenbach [8] to understand the class of $\mathcal{U}$-graphs by restricting the ends of the unit intervals to integers. They found a infinite list of forbidden induced subgraphs, which characterize these so-called *integral $\mathcal{U}$-graphs*.

3   Results

In this section we state and prove our main results. We start by introducing a list of forbidden induced subgraphs. See Figs. 3, 4, 5, and 6 for illustration. Let $\mathcal{R} = \bigcup_{i=0}^{\infty}\{R_i\}$, $\mathcal{S} = \bigcup_{i=1}^{\infty}\{S_i\}$, $\mathcal{S}' = \bigcup_{i=1}^{\infty}\{S_i'\}$, and $\mathcal{T} = \bigcup_{i \geq j \geq 0}\{T_{i,j}\}$. For $k \in \mathbb{N}$ let the graph Q_k arise from the graph R_k by deleting two vertices of degree one, which have a common neighbor. We call the common neighbor of the two deleted vertices and its neighbor of degree two *special vertices* of Q_k. Note that if a graph G is twin-free, then the interval representation of G is injective.

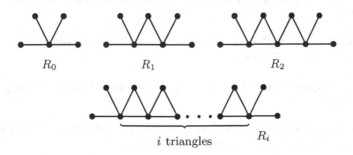

Fig. 3. The class $\mathcal{R}$.

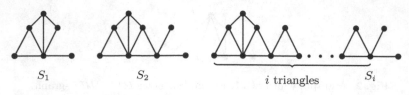

Fig. 4. The class $\mathcal{S}$.

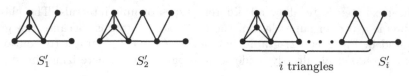

Fig. 5. The class $\mathcal{S}'$.

Lemma 5. (Dourado et al. [3]). *Let* $k \in \mathbb{N}$.

(a) *Every* $\mathcal{U}$*-representation of the claw* $K_{1,3}$ *arises by translation of the following* $\mathcal{U}$*-representation* $I : V(K_{1,3}) \to \mathcal{U}$ *of* $K_{1,3}$, *where* $I(V(K_{1,3}))$ *consists of the following intervals*
  - *either* $[0,1]$ *or* $(0,1]$,
  - $[1,2]$ *and* $(1,2)$, *and*
  - *either* $[2,3]$ *or* $[2,3)$.

(b) *Every injective* $\mathcal{U}$*-representation of* Q_k *arises by translation and inversion of one of the two injective* $\mathcal{U}$*-representations* $I : V(Q_k) \to \mathcal{U}$ *of* Q_k, *where* $I(V(Q_k))$ *consists of the following intervals*
  - *either* $[0,1]$ *or* $(0,1]$,
  - $[1,2]$ *and* $(1,2)$, *and*
  - $[i, i+1]$ *and* $[i, i+1)$ *for* $2 \le i \le k+1$.

(c) *The graphs in* $\{T_{0,0}\} \cup \mathcal{R}$ *are minimal forbidden subgraphs for the class of* $\mathcal{U}$*-graphs with respect to induced subgraphs.*

(d) *If* G *is a* $\mathcal{U}$*-graph, then every induced subgraph* H *in* G *that is isomorphic to* Q_k *and every vertex* $u^* \in V(G) \setminus V(H)$ *such that* u^* *is adjacent to exactly one of the two special vertices of* H, *the vertex* u^* *has exactly one neighbor in* $V(H)$.

Lemma 6. *If a graph* G *is a twin-free mixed unit interval graph, then* G *is* $\{K_{2,3}^*\} \cup \mathcal{R} \cup \mathcal{S} \cup \mathcal{S}' \cup \mathcal{T}$*-free.*

For the sake of space restrictions, we omit the proof of Lemma 6 and proceed to our main result.

Theorem 7. *A twin-free graph* G *is a mixed unit interval graph if and only if* G *is a* $\{K_{2,3}^*\} \cup \mathcal{R} \cup \mathcal{S} \cup \mathcal{S}' \cup \mathcal{T}$*-free interval graph.*

Proof of Theorem 7: By Lemma 6, we know if G is a twin-free mixed unit interval graph, then G is a $\{K_{2,3}^*\} \cup \mathcal{R} \cup \mathcal{S} \cup \mathcal{S}' \cup \mathcal{T}$-free interval graph. Let now G be a twin-free $\{K_{2,3}^*\} \cup \mathcal{R} \cup \mathcal{S} \cup \mathcal{S}' \cup \mathcal{T}$-free interval graph. We show that G is a

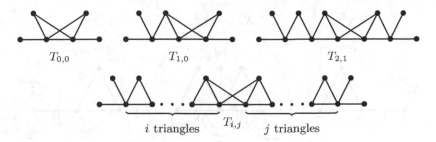

Fig. 6. The class $\mathcal{T}$.

mixed proper interval graph. By Theorem 4, this proves Theorem 7. Since G is an interval graph, G has an $\mathcal{I}^{++}$-representation I. As in [10] we call a pair (u, v) of distinct vertices a *bad pair* if $I(u) \subseteq I(v)$. Let I be such that the number of bad pairs is as small as possible. If I has no bad pair, then we are done by Theorem 1. Hence we assume that there is at least one bad pair. The strategy of the proof is as follows. Claims 1 to 6 collect properties of G and I, before we modify our interval representation of G to show that G is a mixed proper interval graph. In Claims 7 to 10 we prove that our modification of the interval representation preserves all intersections and non-intersections. Claims 1 to 3 are similar to Claims 1 to 3 in [10], respectively. For the sake of space restrictions we omit the proofs.

Claim 1. *If (u, v) is a bad pair, then there are vertices x and y such that $\ell(v) \leq r(x) < \ell(u)$ and $r(u) < \ell(y) \leq r(v)$.*

Let a_1 and a_2 be two distinct vertices. Claim 1 implies that $\ell(a_1) \neq \ell(a_2)$ and $r(a_1) \neq r(a_2)$. Suppose $\ell(a_1) < \ell(a_2)$. Let ϵ be the smallest distance between two distinct endpoints of intervals of I. If $r(a_1) = \ell(a_2)$, then $I' : V(G) \to \mathcal{I}^{++}$ be such that $I'(a_1) = [\ell(a_1), r(a_1) + \epsilon/2]$, and $I'(z) = I(z)$ for $z \in V(G) \setminus \{a_1\}$. By the choice of ϵ, we conclude that I' is an interval representation of G with as many bad pairs as I. Therefore, we assume without loss of generality that we chose I such that all endpoints of the intervals of I are distinct. Hence the inequalities in Claim 1 are strict inequalities.

Claim 2. *If (u, w) and (v, w) are bad pairs, then $u = v$, that is, no interval contains two distinct intervals.*

Claim 3. *If (u, v) and (u, w) are bad pairs, then $v = w$, that is, no interval is contained in two distinct intervals.*

A vertex x is to the *left* (respectively *right*) of a vertex y (in I), if $r(x) < \ell(y)$ (respectively $r(y) < \ell(x)$). Two adjacent vertices x and y are *distinguishable* by vertices to the left (respectively right) of them, if there is a vertex z, which is adjacent to exactly one of them and to the left (respectively right) of one of them. The vertex z *distinguishes* x and y. Next, we show that for a bad pair (u, v) there is the structure as shown in Fig. 7 in G. We introduce a positive integer $\ell_{u,v}^{\max}$ that, roughly speaking, indicates how large this structure is.

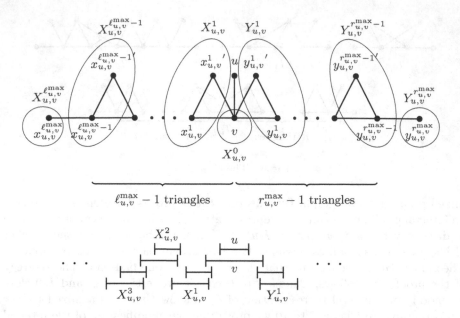

Fig. 7. The structure in G forced by a bad pair (u, v).

For a bad pair (u, v) let $v = X^0_{u,v}$ and let $X^1_{u,v}$ be the set of vertices that are adjacent to v and to the left of u. Let $y_{u,v}$ be a vertex to the right of u and adjacent to v. Claim 1 guarantees $|X^1_{u,v}| \geq 1$ and the existence of $y_{u,v}$. If $|X^1_{u,v}| = 1$, then let $\ell^{\max}_{u,v} = 1$ and we stop here. Suppose $|X^1_{u,v}| \geq 2$. Since G is R_0-free, $X^1_{u,v}$ is a clique and since G is S'_1-free, we conclude $|X^1_{u,v}| = 2$. Let $\{x, x'\} = X^1_{u,v}$ such that $r(x) < r(x')$. For contradiction, we assume that there is a vertex z to the right of x that distinguishes x and x'. We conclude $\ell(v) < \ell(z)$. By Claim 2, $r(v) < r(z)$. This implies that (u, z) is a bad pair, which contradicts Claim 3. Thus z does not exist. In addition (x, x') is not a bad pair, otherwise Claim 1 guarantees a vertex z such that $r(x) < \ell(z) < r(x')$, which is a contradiction. Thus $\ell(x) < \ell(x') < r(x) < r(x')$. Let $x^1_{u,v} = x$ and $x^{1}_{u,v}{}' = x'$. Note that $N_G(x^{1}_{u,v}{}') \subset N_G(x^1_{u,v})$.

Let $X^2_{u,v} = N_G(x^1_{u,v}) \setminus N_G(x^{1}_{u,v}{}')$. Note that all vertices in $X^2_{u,v}$ are to the left of $x^{1}_{u,v}{}'$. Since G is twin-free, $|X^2_{u,v}| \geq 1$. If $|X^2_{u,v}| = 1$, then let $\ell^{\max}_{u,v} = 2$ and we stop here. Suppose $|X^2_{u,v}| \geq 2$. Since G is R_1-free, $X^2_{u,v}$ is a clique and since G is S'_2-free, we conclude $|X^2_{u,v}| = 2$. Let $\{x, x'\} = X^2_{u,v}$ such that $r(x) < r(x')$. For contradiction, we assume that there is a vertex z to the right of x that distinguishes x and x'. Since $z \notin X^2_{u,v}$, we conclude $\ell(x^{1}_{u,v}{}') < r(z)$. If $r(z) < \ell(v)$, then $G[\{z, x, x', x^1_{u,v}, x^{1}_{u,v}{}', v, u, y_{u,v}\}]$ is isomorphic to S_2, which is a contradiction. Thus $\ell(v) < r(z)$. If $r(z) < \ell(u)$, then $|X^1_{u,v}| = 3$, which is a contradiction. Thus $\ell(u) < r(z)$. If $r(u) < r(z)$, then (u, v) and (u, z) are bad pairs, which is a contradiction to Claim 3. Thus $\ell(u) < r(z) < r(u)$. Now $G[\{z, x', x^{1}_{u,v}{}', v, u, y_{u,v}\}]$ is isomorphic to $T_{0,0}$, which is the final contradiction.

Note that (x, x') is not a bad pair, otherwise Claim 1 guarantees a vertex z such that $r(x) < \ell(z) < r(x')$, which is a contradiction. Thus $\ell(x) < \ell(x') < r(x) < r(x')$. Let $x_{u,v}^2 = x$ and $x_{u,v}^2{}' = x'$. Note that $N_G(x_{u,v}^2{}') \subset N_G(x_{u,v}^2)$. Let $X_{u,v}^3 = N_G(x_{u,v}^2) \setminus N_G(x_{u,v}^2{}')$. Note that all vertices in $X_{u,v}^3$ are to the left of $x_{u,v}^2{}'$.

We assume that for $k \geq 3$, $i \in [k-1]$ and $j \in [k]$

- we defined $X_{u,v}^j$,
- $|X_{u,v}^i| = 2$ holds,
- we defined $x_{u,v}^i$ and $x_{u,v}^i{}'$,
- $\ell(x_{u,v}^i) < \ell(x_{u,v}^i{}') < r(x_{u,v}^i) < r(x_{u,v}^i{}')$ holds,
- the vertices in $X_{u,v}^{i+1}$ are to the left of $x_{u,v}^i{}'$, and
- the vertices in $X_{u,v}^i$ are not distinguishable to the right.

If $|X_{u,v}^k| = 1$, then let $\ell_{u,v}^{\max} = k$ and we stop here. Suppose $|X_{u,v}^k| \geq 2$. Since G is R_{k-1}-free, $X_{u,v}^k$ is a clique and since G is S_k'-free, we obtain $|X_{u,v}^k| = 2$. Let $\{x, x'\} = X_{u,v}^k$ such that $r(x) < r(x')$. For contradiction, we assume that there is a vertex z to the right of x that distinguishes x and x'. Since $z \notin X_{u,v}^k$, we conclude $\ell(x_{u,v}^{k-1}{}') < r(z)$. If $r(z) < \ell(x_{u,v}^{k-2})$, then $G[\{z, x, x', v, u, y_{u,v}\} \cup \bigcup_{i=1}^{k-1} X_{u,v}^i]$ is isomorphic to S_k, which is a contradiction. Thus $\ell(x_{u,v}^{k-2}) < r(z)$. If $r(z) < \ell(x_{u,v}^{k-2}{}')$, then $|X_{u,v}^{k-1}| = 3$, which is a contradiction. Thus $\ell(x_{u,v}^{k-2}{}') < r(z)$. If $r(z) < \ell(x_{u,v}^{k-3})$, then $G[\{z, x', x_{u,v}^{k-1}{}', v, u, y_{u,v}\} \cup \bigcup_{i=1}^{k-2} X_{u,v}^i]$ is isomorphic to $T_{k-3,0}$, which is a contradiction. Thus $\ell(x_{u,v}^{k-3}) < r(z)$. If $r(z) < r(x_{u,v}^{k-2})$, then $|X_{u,v}^{k-2}| = 3$, which is a contradiction. Thus $r(x_{u,v}^{k-2}) < r(z)$ and hence $(x_{u,v}^{k-1}{}', z)$ and $(x_{u,v}^{k-2}, z)$ are bad pairs, which is a contradiction to Claim 2. Thus x, x' are not distinguishable to the right. We obtain that (x, x') is not a bad pair, otherwise Claim 1 guarantees a vertex z such that $r(x) < \ell(z) < r(x')$, which is a contradiction. Thus $\ell(x) < \ell(x') < r(x) < r(x')$. Let $x_{u,v}^k = x$ and $x_{u,v}^k{}' = x'$. Note that $N_G(x_{u,v}^k{}') \subset N_G(x_{u,v}^k)$. Let $X_{u,v}^{k+1} = N_G(x_{u,v}^k) \setminus N_G(x_{u,v}^k{}')$. Note that all vertices in $X_{u,v}^{k+1}$ are to the left of $x_{u,v}^k{}'$. By induction, this leads to the following properties.

Claim 4. *If (u, v) is a bad pair, $k \in [\ell_{u,v}^{\max} - 1]$, then the following holds:*

(a) $|X_{u,v}^k| = 2$.
(b) The vertices in $X_{u,v}^k$ are not distinguishable by vertices to the right of them.
(c) We have $\ell(x_{u,v}^i) < \ell(x_{u,v}^i{}') < r(x_{u,v}^i) < r(x_{u,v}^i{}')$, that is $(x_{u,v}^k, x_{u,v}^k{}')$ and $(x_{u,v}^k{}', x_{u,v}^k)$ are not bad pairs.

Note that $\ell_{u,v}^{\max}$ is the smallest integer k such that $|X_{u,v}^{k-1}| \geq 2$ and $|X_{u,v}^k| = 1$. Due to space restrictions, we omit the proofs of Claims 5 and 6.

Claim 5. *If (u,v) is a bad pair and $k \in [\ell_{u,v}^{\max} - 1]$, then the following holds.*

(a) $x_{u,v}^{k}{}'$ is not contained in a bad pair.
(b) There is no vertex $z \in V(G)$ such that $(x_{u,v}^{k}, z)$ is a bad pair.

For a bad pair (u,v) define $Y_{u,v}^{k}$ as $X_{u,v}^{k}$ by interchanging in the definition right by left. Let $r_{u,v}^{\max}$ be the smallest integer k such that $|Y_{u,v}^{k-1}| = 2$ and $|Y_{u,v}^{k}| = 1$. By symmetry, one can prove a "y"-version of Claims 4, 5 and 6(a) and (b). Let $\{y_{u,v}^{k}, y_{u,v}^{k}{}'\} = Y_{u,v}^{k}$ such that $N_G(y_{u,v}^{k}{}') \subset N_G(y_{u,v}^{k})$ for $k \le r_{u,v}^{\max} - 1$.

Claim 6. *Let (u,v) and (w,z) be bad pairs and $k \in [\ell_{u,v}^{\max}]$.*

(a) If $X_{u,v}^{k} \cap X_{w,z}^{\tilde{k}} \neq \emptyset$, then $x_{u,v}^{k-1} = x_{w,z}^{\tilde{k}-1}$ for $\tilde{k} \in [\ell_{w,z}^{\max}]$.
(b) If $X_{u,v}^{k} \cap X_{w,z}^{\tilde{k}} \neq \emptyset$, then $X_{u,v}^{k} = X_{w,z}^{\tilde{k}}$ for $\tilde{k} \in [\ell_{w,z}^{\max}]$.
(c) If $X_{u,v}^{k} \cap Y_{w,z}^{\tilde{k}} \neq \emptyset$, then $X_{u,v}^{k} \cap Y_{w,z}^{\tilde{k}} = x_{u,v}^{k} = y_{w,z}^{\tilde{k}}$ for $\tilde{k} \in [r_{w,z}^{\max}]$

Next, we define step by step new interval representations of G as follows. First we shorten the intervals of $X_{u,v}^{k}$ for every bad pair (u,v) and $k \in [\ell_{u,v}^{\max}]$. Let $I' : V(G) \to \mathcal{I}^{++}$ be such that $I'(x) = [\ell(x), \ell(x_{u,v}^{k-1})]$ if $x \in X_{u,v}^{k}$ for some bad pair (u,v) and $I'(x) = I(x)$ otherwise. By Claim 6(a), I' is well-defined; that is, if $x \in X_{u,v}^{k} \cap X_{w,z}^{\tilde{k}}$, then $\ell(x_{u,v}^{k-1}) = \ell(x_{w,z}^{\tilde{k}-1})$. Let $\ell'(x)$ and $r'(x)$ be the left and right endpoint of the interval $I'(x)$ for $x \in V(G)$, respectively.

Claim 7. *I' is an interval representation of G.*

Proof of Claim 7: Trivially, if two intervals do not intersect in I, then they do not intersect in I'. For contradiction, we assume that there are two vertices $a, b \in V(G)$ such that $I(a) \cap I(b) \neq \emptyset$ and $I'(a) \cap I'(b) = \emptyset$. At least one interval is shorten by changing the interval representation. Say $a \in X_{u,v}^{k}$ for some bad pair (u,v) and $k \in [\ell_{u,v}^{\max}]$. Hence $b \neq x_{u,v}^{k-1}$ and $\ell(x_{u,v}^{k-1}) < \ell(b)$ and by Claim 4(b), $\ell(b) < r(x_{u,v}^{k})$. We conclude that $(b, x_{u,v}^{k-1})$ is not a bad pair, otherwise Claim 1 implies the existence of a vertex $z \in X_{u,v}^{k}$ to the left of b, but $z \notin \{x_{u,v}^{k}, x_{u,v}^{k}{}'\}$, which is a contradiction to Claim 4(a). Thus $r(x_{u,v}^{k-1}) < r(b)$. If $k = 1$, then (u, b) is also a bad pair, which is a contradiction to Claim 3. Thus $k \geq 2$. Since $\ell(b) < r(x_{u,v}^{k})$, we obtain $\ell(b) < \ell(x_{u,v}^{k-1}{}')$. Since $(x_{u,v}^{k-1}{}', b)$ is not a bad pair by Claim 5(a), $r(b) < r(x_{u,v}^{k-1}{}')$. Thus $b \in X_{u,v}^{k-1}$, which is a contradiction to $|X_{u,v}^{k-1}| = 2$. $\square$

Claim 8. *The change of the interval representation of G from I to I' creates no new bad pair (a,b) such that $\{a,b\} \neq X_{u,v}^{k}$ for some $k \in [\ell_{u,v}^{\max}]$ and some bad pair (u,v).*

Proof of Claim 8: For contradiction, we assume that (a,b) is a new bad pair and $\{a,b\} \neq X_{u,v}^{k}$. Since (a,b) is a new bad pair, $I'(a)$ is a proper subset of $I(a)$. Thus let $a \in X_{u,v}^{k}$ and $b \notin X_{u,v}^{k}$. If $a \in X_{u,v}^{k}$ and $|X_{u,v}^{k}| = 2$, then $\ell(b) < \ell(x_{u,v}^{k}{}')$ and $r'(a) = \ell(x_{u,v}^{k-1}) < r(b) < r(x_{u,v}^{k}{}')$, because of Claim 5(a). Thus

$b \in X_{u,v}^k$, which is a contradiction. If $a \in X_{u,v}^k$ and $|X_{u,v}^k| = 1$, then $\ell(b) < \ell(x_{u,v}^k)$ and $r'(a) = \ell(x_{u,v}^{k-1}) < r(b) < r(x_{u,v}^k)$. Thus $b \in X_{u,v}^k$, which is the final contradiction. □

In a second step, we shorten the intervals of $Y_{u,v}^i$ for every bad pair (u,v) and $i \in [r_{u,v}^{\max}]$. Let $I'' : V(G) \to \mathcal{I}^{++}$ be such that $I''(y) = [r'(y_{u,v}^{k-1}), r'(y)]$ if $y \in Y_{u,v}^k$ for some bad pair (u,v) and $I''(y) = I'(y)$ else. Note that bad pairs are only referred to the interval representation I. Let $\ell''(x)$ and $r''(x)$ be the left and right endpoints of the interval $I''(x)$ for $x \in V(G)$, respectively.

Claim 9. *I'' is an interval representation of G.*

Due to space restrictions, we omit the proof of Claim 9.

Claim 10. *The change of the interval representation of G from I to I'' creates no new bad pair (a, b) such that $\{a, b\} \neq X_{u,v}^k$ for some $k \in [\ell_{u,v}^{\max}]$ or $\{a, b\} \neq Y_{u,v}^i$ for some $i \in [r_{u,v}^{\max}]$ and some bad pair (u, v).*

Proof of Claim 10: For contradiction, we assume that (a, b) is a new bad pair and $Y_{u,v}^i \neq \{a, b\} \neq X_{u,v}^k$. Thus $a \in X_{u,v}^k$ or $a \in Y_{u,v}^k$ and $b \notin X_{u,v}^k$ or $b \notin Y_{u,v}^i$, respectively. If $a \in X_{u,v}^k$ and $|X_{u,v}^k| = 2$, then $\ell(b) < \ell(x_{u,v}^k{}')$ and $\ell(x_{u,v}^k{}') < r(b) < r(x_{u,v}^k{}')$. Thus $b \in X_{u,v}^k$, which is a contradiction. If $a \in X_{u,v}^k$ and $|X_{u,v}^k| = 1$, then $\ell(b) < \ell(x_{u,v}^k)$ and $\ell(x_{u,v}^{k-1}) < r(b) < r(x_{u,v}^k)$. Thus $b \in X_{u,v}^k$, which is a contradiction. If $a \in Y_{u,v}^i$ the proof is almost exactly the same. □

Now we are in a position to blow up some intervals to open or half-open intervals to get a mixed proper interval graph. Let $I^* : V(G) \to \mathcal{I}$ be such that

$$
I^*(x) = \begin{cases}
(\ell(v), r(v)), & \text{if } (x, v) \text{ is a bad pair,} \\
(\ell''(x_{u,v}^k), r''(x_{u,v}^k)], & \text{if } x = x_{u,v}^k{}' \text{ for some bad pair } (u, v) \text{ and} \\
& \quad k \in [\ell_{u,v}^{\max} - 1], \\
[\ell''(y_{u,v}^i), r''(y_{u,v}^i)), & \text{if } x = y_{u,v}^i{}' \text{ for some bad pair } (u, v) \text{ and} \\
& \quad i \in [r_{u,v}^{\max} - 1], \\
[\ell''(x), r''(x)], & \text{else.}
\end{cases}
$$

Note that I^* is well-defined by Claims 5 and 6; that is, the four cases in the definition of I^* induces a partition of the vertex set of G. Moreover, the interval representation I^* defines a mixed proper interval graph. As a final step, we prove that I'' and I^* define the same graph. Since we make every interval bigger, we show that for every two vertices a, b such that $I''(a) \cap I''(b) = \emptyset$, we still have $I^*(a) \cap I^*(b) = \emptyset$. For contradiction, we assume the opposite. Let a, b be two vertices such that $I''(a) \cap I''(b) = \emptyset$ and $I^*(a) \cap I^*(b) \neq \emptyset$. It follows by our approach and definition of our interval representation I'', that both a and b are blown up intervals.

First we suppose a and b are intervals that are blown up to open intervals, that is, there are distinct vertices $\tilde{a}$ and $\tilde{b}$ such that $(a, \tilde{a})$ and $(b, \tilde{b})$ are bad pairs. Furthermore, the intervals of $\tilde{a}$ and $\tilde{b}$ intersect not only in one point. By Claims 2 and 3, we assume without loss of generality, that $\ell''(\tilde{a}) < \ell''(\tilde{b}) < r''(\tilde{a}) < r''(\tilde{b})$.

Therefore, by the construction of I'', we obtain a is adjacent to $\tilde{b}$ and $\tilde{a}$ is adjacent to b, and in addition they intersect in one point, respectively. Now, $G[\{x^1_{a,\tilde{a}}, a, \tilde{a}, b, \tilde{b}, y^1_{b,\tilde{b}}\}]$ is isomorphic to $T_{0,0}$, which is a contradiction.

Now we suppose a is blown up to an open interval and b is blown up to an open-closed interval (the case closed-open is exactly symmetric). Let $\tilde{a}$ be the vertex such that $(a, \tilde{a})$ is a bad pair. Let $\tilde{b}, u, v \in V(G)$ and $k \in \mathbb{N}$ such that $\{b, \tilde{b}\} = X^k_{u,v}$. We suppose $\tilde{a} \neq \tilde{b}$. We conclude $\ell''(\tilde{a}) < \ell''(\tilde{b}) < r''(\tilde{a}) < r''(\tilde{b})$. As above, we conclude a is adjacent to $\tilde{b}$ and $\tilde{a}$ is adjacent to b, and in addition they intersect in one point, respectively. Thus $G[\{x^1_{a,\tilde{a}}, a, \tilde{a}, v, u, y^1_{u,v}\} \cup \bigcup_{i=1}^{k} X^i_{u,v}]$ induces a $T_{k,0}$, which is a contradiction. Now we suppose $\tilde{a} = \tilde{b}$. We conclude that $G[\{x^1_{a,\tilde{a}}, a, v, u, y^1_{u,v}\} \cup \bigcup_{i=1}^{k} X^i_{u,v}]$ is isomorphic to R_k, which is a contradiction.

It is easy to see that a and b cannot be both blown up to closed-open or both open-closed intervals, because G is R_k-free for $k \geq 0$ and the definition of I''.

Therefore, we consider finally the case that a is blown up to a closed-open and b to an open-closed interval. Let $\tilde{a}, \tilde{b}, u, v, w, z \in V(G)$ and $k, \tilde{k} \in \mathbb{N}$ such that $\{a, \tilde{a}\} = Y^k_{u,v}$ and $\{b, \tilde{b}\} = X^{\tilde{k}}_{w,z}$. First we suppose $\tilde{a} \neq \tilde{b}$. Again, we obtain $\ell''(\tilde{a}) < \ell''(\tilde{b}) < r''(\tilde{a}) < r''(\tilde{b})$ and a is adjacent to $\tilde{b}$ and $\tilde{a}$ is adjacent to b, and furthermore they intersect in one point, respectively. Thus $G[\{x^1_{u,v}, u, v, w, z, y^1_{w,z}\} \cup \bigcup_{i=1}^{k} Y^i_{u,v} \cup \bigcup_{i=1}^{\tilde{k}} X^i_{w,z}]$ is isomorphic to $T_{k,\tilde{k}}$. Next we suppose $\tilde{a} = \tilde{b}$ and hence $G[\{x^1_{u,v}, u, v, w, z, y^1_{w,z}\} \cup \bigcup_{i=1}^{k} Y^i_{u,v} \cup \bigcup_{i=1}^{\tilde{k}} X^i_{w,z}]$ is isomorphic to $R_{k+\tilde{k}}$. This is the final contradiction and completes the proof of Theorem 7.   □

In Theorem 7 we only consider twin-free $\mathcal{U}$-graphs to reduce the number of case distinctions in the proof. In Corollary 8 we resolve this technical condition. See Figs. 8 and 9 for illustration. Let $\mathcal{S}'' = \bigcup_{i=2}^{\infty} \{S''_i\}$. For the sake of space restrictions, we omit the proof.

Corollary 8. *A graph G is a mixed unit interval graph if and only if G is a $\{G_1\} \cup \mathcal{R} \cup \mathcal{S} \cup \mathcal{S}'' \cup \mathcal{T}$-free interval graph.*

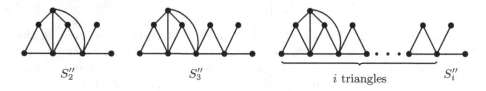

$$S''_2 \qquad\qquad S''_3 \qquad\qquad \underbrace{\qquad\qquad}_{i \text{ triangles}} \qquad S''_i$$

Fig. 8. The class $\mathcal{S}''_i$.

$$G_1$$

Fig. 9. The graph G_1.

It is possible to extract a polynomial-time algorithm from the proof of Theorem 7. Given a graph G, then first start with a polynomial-time algorithm [2] which decides whether G is an interval graph and if yes computes an interval representation I of G. Second, go along the claims of Theorem 7. By suitable modifications of I either I becomes a mixed proper interval representation or the algorithm finds a forbidden induced subgraph. Note that by Theorem 4, the class of mixed proper interval graphs coincides with the class of mixed unit interval graphs.

Theorem 9. *There is a polynomial-time algorithm which decides whether a graph has an interval representation using unit intervals only.*

Remark 1: I was informed by Alan Shuchat, Randy Shull, Ann Trenk and Lee West that they independently found a proof for a characterization of mixed unit interval graphs by forbidden induced subgraphs.

Remark 2: A full version of this paper appeared in [7].

References

1. Corneil, D.G.: A simple 3-sweep LBFS algorithm for the recognition of unit interval graphs. Discrete Appl. Math. **138**, 371–379 (2004)
2. Corneil, D.G., Olariu, S., Stewart, L.: The LBFS structure and recognition of interval graphs. SIAM J. Discrete Math. **23**, 1905–1953 (2009)
3. Dourado, M.C., Le, V.B., Protti, F., Rautenbach, D., Szwarcfiter, J.L.: Mixed unit interval graphs. Discrete Math. **312**, 3357–3363 (2012)
4. Fishburn, P.C.: Interval Orders and Interval Graphs. Wiley, New York (1985)
5. Frankl, P., Maehara, H.: Open interval-graphs versus closed interval-graphs. Discrete Math. **63**, 97–100 (1987)
6. Golumbic, M.C.: Algorithmic Graph Theory and Perfect Graphs. Annals of Discrete Mathematics, vol. 57. Elsevier, Amsterdam (2004)
7. Joos, F.: A characterization of mixed unit interval graphs. J. Graph Theory (2014). doi:10.1002/jgt.21831
8. Le, V.B., Rautenbach, D.: Integral mixed unit interval graphs. In: Gudmundsson, J., Mestre, J., Viglas, T. (eds.) COCOON 2012. LNCS, vol. 7434, pp. 495–506. Springer, Heidelberg (2012)
9. Lekkerkerker, C.G., Boland, J.C.: Representation of a finite graph by a set of intervals on the real line. Fund. Math. **51**, 45–64 (1962)
10. Rautenbach, D., Szwarcfiter, J.L.: Unit interval graphs of open and closed intervals. J. Graph Theor. **72**(4), 418–429 (2013)
11. Roberts, F.S.: Indifference graphs. In: Harary, F. (ed.) Proof Techniques in Graph Theory, pp. 139–146. Academic Press, New York (1969)

On the Number of Connected Sets
in Bounded Degree Graphs

Kustaa Kangas[1]([✉]), Petteri Kaski[2], Mikko Koivisto[1], and Janne H. Korhonen[1]

[1] Helsinki Institute for Information Technology HIIT,
Department of Computer Science, University of Helsinki,
P.O. Box 68, 00014 Helsinki, Finland
{mikko.koivisto,juho-kustaa.kangas,janne.h.korhonen}@cs.helsinki.fi
[2] Helsinki Institute for Information Technology HIIT, Department of Information
and Computer Science, Aalto University, P.O. Box 15400, 00076 Aalto, Finland
petteri.kaski@aalto.fi

Abstract. A set of vertices in a graph is *connected* if the set induces a
connected subgraph. Using Shearer's entropy lemma, we show that the
number of connected sets in an n-vertex graph with maximum vertex
degree d is $O(1.9351^n)$ for $d = 3$, $O(1.9812^n)$ for $d = 4$, and $O(1.9940^n)$
for $d = 5$. Dually, we construct infinite families of generalized ladder graphs whose number of connected sets is bounded from below by
$\Omega(1.5537^n)$ for $d = 3$, $\Omega(1.6180^n)$ for $d = 4$, and $\Omega(1.7320^n)$ for $d = 5$.

1 Introduction

A *connected set* in an undirected graph G is a subset of vertices that induces
a connected subgraph. Besides being fundamental combinatorial objects, connected sets play a key role in various exponential-time graph algorithms. For
instance, for an n-vertex graph one can solve the traveling salesman problem [5],
solve the maximum internal spanning tree problem [2], and evaluate the Tutte
polynomial [3] in time that is within an $n^{O(1)}$ factor of the number of connected
sets of the graph. Within the same time bound an algorithm also finds an optimal
Bayesian network having the input graph as its super-structure [17].

Surprisingly little is known about extremal combinatorics of connected sets
in different graph classes. What is immediate, however, is that an n-vertex graph
can have at most 2^n connected sets and that this bound is achieved by complete
graphs. It is also easy to see that sparsity alone does not imply a much smaller
number of connected sets: an n-star has an average degree less than 2, but the
number of connected sets is $2^{n-1}+n$. In this light, graphs of bounded degree form
a natural graph class to study; we define the *degree* of a graph as the maximum
degree of a vertex. Parameterizing by the size of the connected set, Bollobás [6,
pp. 129–130] provides two ways to prove that any graph of degree $d \geq 3$ has at
most $(e(d-1))^k$ connected sets with $k+1$ vertices, one of which is a given vertex.

K.K., M.K., and J.K. supported by the Academy of Finland, grants 125637, 218153,
and 255675. P.K. supported by the Academy of Finland, grants 252083 and 256287.

D. Kratsch and I. Todinca (Eds.): WG 2014, LNCS 8747, pp. 336–347, 2014.
DOI: 10.1007/978-3-319-12340-0_28

For large k, this bound is, however, loose and of no use for bounding the total number of connected sets of an n-vertex graph. The first nontrivial upper bound, namely $\beta_d^n + n$, where $\beta_d = (2^{d+1} - 1)^{1/(d+1)}$, was given by Björklund et al. [5]. In particular, we have $\beta_d = 1.9680, 1.9874, 1.9948$ for $d = 3, 4, 5$, respectively. We are not aware of better bounds, prior to this work.

There is no reason to believe that the Björklund et al. bound is tight. First, its proof applies Shearer's entropy lemma, in essence, by taking the product of the number of possible projections of connected sets to the closed neighborhood of each vertex. Specifically, the proof provides no means to construct a graph that would attain the upper bound. Also, connectivity is an inherently global property that cannot be captured by looking at individual local neighborhoods. Second, while it is easy to construct arbitrarily large graphs that have an exponential number of connected sets, getting near the upper bounds appears to be challenging.

In this paper, we seek improved upper bounds for the number of connected sets by applying Shearer's entropy lemma in an expanded context. Namely, we are interested in projecting the connected sets not only to the immediate closed neighborhood of each vertex but rather to the ball of radius $r \geq 2$ (the r-neighborhood) around each vertex. By carrying out a computer search over the possible projections of connected sets to r-neighborhoods, we obtain improved upper bounds for $d \leq 5$.

Theorem 1. *Every n-vertex graph with maximum degree $d \leq 5$ has at most $b_d^n + n(2^{d^2+1} - 1)$ connected sets, where $b_3 = 1.9351$, $b_4 = 1.9812$, and $b_5 = 1.9940$.*

Dually, we show the following lower bounds for $d \leq 5$.

Theorem 2. *For each $d \leq 5$ there exists an infinite family of graphs with maximum degree d such that each graph of n vertices has at least a_d^n connected sets, where $a_3 = 1.5537$, $a_4 = 1.6180$, and $a_5 = 1.7320$.*

Related work. The maximum number of subsets of vertices satisfying a given property has been studied for many different types of properties. In the case of maximal independent sets (or dually, maximal cliques), the classical Moon–Moser [16] upper bound is known to be tight. However, in the case of minimal dominating sets [9], minimal feedback vertex sets (in general graphs [8] or in tournaments [14]), maximal bicliques [13], potential maximal cliques [10], and minimal separators [11], the gap between the known lower and upper bounds remains relatively large. In each of these cases the upper bounds are obtained by a careful analysis of an appropriate branching algorithm.

Entropy methods, applied in the present work, have previously yielded tight bounds for certain properties in bounded degree graphs. For independent sets in d-regular graphs with n vertices the bound $(2^{d+1} - 1)^{n/2d}$ was conjectured to be tight by Alon [1]. Kahn [15] showed that the conjecture holds for bipartite graphs. Recently, Zhao [18] confirmed that the conjecture holds in general by presenting a surprisingly simple reduction to the bipartite case; Galvin [12] reviews earlier developments. Björklund *et al.* [4] used Shearer's entropy lemma to show that the bound $(2^{d+1}-1)^{n/(d+1)}$ is tight for the number of dominating sets in n-vertex graphs of degree at most d.

In the case of connected sets, Björklund *et al.* [5] give further upper bounds for the number of connected sets when the connected sets are also required to be dominating or "transient", or when the graph is assumed to be triangle-free. In particular, the traveling salesman problem can also be solved within an $n^{O(1)}$ factor of the number of transient connected sets, improving upon the bound based on connected sets alone. Motivated by an application to structure learning in Bayesian networks, Perrier, Imoto, and Miyano [17] present an empirical study on the number of connected sets in random bounded-degree graphs.

2   Upper Bounds on the Number of Connected Sets

Our upper bounds are derived by extending the Shearer's entropy lemma -based projection approach of Björklund *et al.* [5] to consider neighborhoods whose radius r is greater than one. Here the essential difficulty and our contribution is to develop computer-assisted analytical tools to study projections of connected sets to neighborhoods of vertices.

We begin by reviewing a basic template suitable for any maximum degree d and any radius r for vertex neighborhoods. We then proceed to characterize in more detail the worst-case graphs induced by the neighborhoods. Making use of the characterization, we give an algorithm that suffices to carry out a complete analysis of the cases $d \leq 5$ and $r \leq 2$, leading to Theorem 1. While a computer search would be feasible beyond these parameters, we conclude this section by showing that our method of studying the "boundary-connected" projections appears to be restricted to the case $d \leq 5$ and $r \leq 2$. That is, beyond these parameters, an analysis based on boundary-connectivity appears not to yield improved upper bounds over those obtained by simply taking $r = 1$.

2.1   The Projection Method

Our main tool for deriving upper bounds for the number of connected sets is Shearer's entropy lemma, which is most conveniently deployed in our context in the following combinatorial form:

Lemma 1 (Chung *et al.* [7]). *Let V be an n-element set and let $A_1, A_2, \ldots, A_k$ be subsets of V such that every $v \in V$ occurs in at least δ of these subsets. Let $\mathcal{F}$ be a set of subsets of V. For each $1 \leq i \leq k$, define the projections $\mathcal{F}_i := \{F \cap A_i : F \in \mathcal{F}\}$. Then,*

$$|\mathcal{F}|^\delta \leq \prod_{i=1}^{k} |\mathcal{F}_i| .$$

Lemma 1 enables us to obtain control over the number of connected sets in a graph by taking the sets A_i to be (augmented) neighborhoods of vertices.

In more precise terms, let G be an undirected graph with vertex set V and let $S \subseteq V$ be a subset of vertices. Let $r = 0, 1, \ldots$ be a radius parameter. Let us write $N_G^r[S]$ for the set of all vertices $u \in V$ such that there exists a vertex $v \in S$ for which the shortest-path distance between u and v is at most r. In particular, when $S = \{v\}$ is a singleton set consisting of the vertex $v \in V$ only, we write $N_G^r[v]$ for $N_G^r[S]$ and say that $N_G^r[v]$ is the *(closed) neighborhood* of the vertex v of *radius* r. When $r = 1$ we may omit the parameter r from the notation. We observe that $N_G^0[S] = S$ and that $N_G^r[S] = N_G[N_G^{r-1}[S]]$ for $r \geq 1$.

The following immediate lemma recalls the Moore bound δ_r for the size of $N_G^r[v]$.

Lemma 2. *Suppose that the graph G has maximum vertex degree d. Then for all $r = 0, 1, \ldots$ and all vertices $v \in V$ it holds that $|N_G^r[v]| \leq \delta_r$, where*

$$\delta_r := 1 + d \sum_{i=0}^{r-1} (d-1)^i = \frac{d(d-1)^r - 2}{d - 2}.$$

Now let $\mathcal{F}$ be a set of subsets of V. For $r = 0, 1, \ldots$ and $v \in V$, let us write $\mathcal{F}_{v,r} = \{F \cap N_G^r[v] : F \in \mathcal{F}\}$ for the *projection* of $\mathcal{F}$ into the neighborhood of v of radius r. We are now ready to prove our main template for upper bounds. In essence, this lemma replaces the application of Jensen's inequality in the Björklund *et al.* [5] analysis with a uniform bound (the parameter ρ) that is easier to deploy over larger neighborhoods.

Lemma 3. *Let $0 \leq \rho \leq 1$ be a number such that $|\mathcal{F}_{v,r}| \leq 2^{|N_G^r[v]|}\rho$ holds for all $v \in V$. Then, $|\mathcal{F}| \leq (2\rho^{1/\delta_r})^n$.*

Proof. Our intent is to apply Lemma 1. Towards this end, start by setting $A_v := N_G^r[v]$ for each $v \in V$. Next, for each $u \in V$, if u is contained in $k \leq \delta_r - 1$ subsets A_v, then add u to $\delta_r - k$ subsets not already containing u (it does not matter which). As a result, each u is contained in exactly δ_r subsets A_v.

Now define for each $v \in V$ the set $\mathcal{F}_v := \{F \cap A_v : F \in \mathcal{F}\}$. Because $N_G^r[v] \subseteq A_v$, we have

$$|\mathcal{F}_v| \leq |\mathcal{F}_{v,r}| \cdot 2^{|A_v| - |N_G^r[v]|} \leq 2^{|N_G^r[v]|}\rho \cdot 2^{|A_v| - |N_G^r[v]|} \leq 2^{|A_v|}\rho.$$

Taking the product over all $v \in V$ and observing that $\sum_{v \in V} |A_v| = \delta_r n$, the claim follows by Lemma 1. □

To illustrate the use of Lemma 3 in a simple setting, let us reprove the Björklund *et al.* [5] upper bound for the number of connected sets of G:

Corollary 1. *Let G be an n-vertex graph with maximum vertex degree at most d. Then G has at most $(2^{d+1} - 1)^{n/(d+1)} + n$ connected sets.*

Proof. Let $\mathcal{F}$ be the family of connected sets of G, with the n singleton sets consisting of each individual vertex removed from $\mathcal{F}$. Take $r = 1$ and observe that then $\delta_r = d+1$. Furthermore, since the singleton sets $\{v\}$ have been removed from $\mathcal{F}$, we must have $F \cap N_G^r[v] \neq \{v\}$ for each $v \in V$ and $F \in \mathcal{F}$. It follows that we can take $\rho = 1 - 1/2^{d+1}$ and the claim follows.   □

2.2 Neighborhoods with Radius $r \geq 2$

Let us now proceed to consider the case $r \geq 2$ and in particular the feasible projections of connected sets to a vertex neighborhood $N_G^r[v]$. Accordingly, assume that $r \geq 2$ is fixed.

Since our focus is on exponential growth rates as a function of the number of vertices, n, we can simplify the analysis by omitting all nonempty connected sets that are completely contained in at least one of the neighborhoods $N_G^r[v]$. Let us call such sets *local* connected sets. The following lemma is immediate.

Lemma 4. *There are at most $(2^{\delta_r} - 1)n$ local connected sets.*

Our interest in what follows is thus to carry out a worst-case analysis of the number of connected sets that are *not* local. Let $\mathcal{F}$ be the family of non-local connected sets of G. Our intent is now to apply the projection method and Lemma 3 to $\mathcal{F}$.

Intuitively, a connected set that is not local must "exit" any neighborhood that it intersects because otherwise the set would be localized in that neighborhood. In particular, such "exit" requires us to have vertices at the "boundary" of the neighborhood.

Let us say that a subset $S \subseteq N_G^r[v]$ is *boundary-connected* relative to v if each connected component of $G[S]$ contains at least one vertex $u \in N_G^r[v] \setminus N_G^{r-1}[v]$ such that u is adjacent to less than d vertices in $N_G^r[v]$. (In particular, degree less than d is necessary so that we can potentially "exit" from u to outside $N_G^r[v]$. Note, however, that the definition does not require that such an exit actually exists in G. In particular we want this to be the case since we want to be able to check for boundary-connectivity without looking beyond the subgraph induced by $N_G^r[v]$). Figure 1 shows examples of sets that are and are not boundary-connected for $d = 3$ and $r = 2$.

Lemma 5. *Let C be a non-local connected set of G. Then it holds for each vertex $v \in V$ that the projection $C \cap N_G^r[v]$ is boundary-connected relative to v.*

Proof. When C is empty the claim is trivial, so suppose that C is nonempty. Because C is non-local, we must have $C \not\subseteq N_G^r[v]$. It suffices to show that $S := C \cap N_G^r[v]$ is boundary-connected relative to v. Let $t \in C \setminus N_G^r[v]$. Let $G[S']$ be a connected component of $G[S]$ and $s \in S'$. Because C is a connected set, there is a path $(v_0, v_1, \ldots, v_k)$ in G such that $v_0 = s$, $v_k = t$ and, for some $i \leq k$, $s' := v_{i-1} \in S'$ and $t' := v_i \in C \setminus S'$. Now, because t' cannot belong to $N_G^r[v]$ (indeed, otherwise we would get a contradiction to the assumption that $G[S']$ is a connected component of $G[S]$), it holds that $s' \in N_G^r[v] \setminus N_G^{r-1}[v]$. It remains to

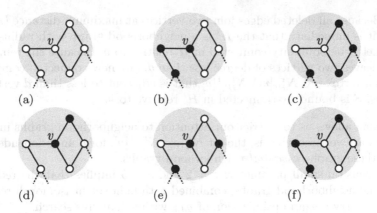

Fig. 1. Vertex subsets that can (a–c) or cannot (d–f) belong to projections of non-local connected sets, with $d = 3$ and $r = 2$

observe that s' is adjacent to at most $d-1$ vertices in $N_G^r[v]$, since s' is adjacent to t' and G is of degree at most d. □

Now observe that Lemmas 3 and 5 together imply that the number of non-local connected sets of G is bounded from above by $(2\rho_{d,r}^{1/\delta_r})^n$, where $\rho_{d,r}$ is a constant such that every neighborhood $N_G^r[v]$ has at most $2^{|N_G^r[v]|}\rho_{d,r}$ boundary-connected sets S.

Our strategy for completing the proof of Theorem 1 is now to optimize the values $\rho_{d,r}$ for $r = 2$ and $d \leq 5$ with computer search. Because boundary-connectivity is intrinsic to each neighborhood $N_G^r[v]$, we can carry out the optimization *without paying attention how this neighborhood is connected to the rest of the graph.*

2.3 Extremal Neighborhood Graphs for $r = 2$

Let us say that a graph H with maximum degree d is a *neighborhood graph* with *radius* r and *root* v if the vertex set of H is $N_H^r[v]$. Clearly, a neighborhood graph has at most δ_r vertices. Thus, for any fixed d and r we can optimize the constant $\rho_{d,r}$ by finding the maximum number of boundary-connected sets (relative to v) admitted by any neighborhood graph (with root v) for the parameters d and r. This is what we proceed to do, using computer search.

The following small observation is useful to reduce the number of neighborhood graphs that need to be considered in the search.

Lemma 6. *Let H be a neighborhood graph with radius r and root v. Let H' be the neighborhood graph with radius r and root v obtained from H by removing each edge of H that joins two vertices in $N_H^r[v] \setminus N_H^{r-1}[v]$. Then H' has at least as many boundary-connected sets relative to v as H.*

Proof. Because all deleted edges join two vertices at maximum distance (r) from v in H, it is immediate that the H' is a neighborhood graph with radius r and root v. Let S be boundary-connected in H relative to v. Because the removal of an edge leaves two vertices of degree less than d, any new connected component contains a vertex $u \in N_{H'}^r[v] \setminus N_{H'}^{r-1}[v]$ that is adjacent to less than d vertices of H'. Thus, S is boundary-connected in H' relative to v.   □

This lemma allows us to restrict our attention to neighborhood graphs in which the boundary vertices, that is, the set $N_G^r[v] \setminus N_G^{r-1}[v]$, form an independent set. We call these graphs *essential* neighborhood graphs.

Our focus on small parameters $r = 2$ and $d \leq 5$ implies that the reduction to essential neighborhood graphs, combined with lightweight isomorph rejection suffices to carry out an optimization of $\rho_{d,r}$ with exhaustive search.

Let us now turn to the details of the algorithm that we use to enumerate the essential neighborhood graphs. Recall that we have fixed $r = 2$ and the maximum degree to be at most $d \leq 5$. Suppose the graph H has n vertices. Since $r = 2$ we can partition the set of vertices V of H into three sets V_0, V_1, V_2 based on distance from the root vertex $v \in V$. Let us write $|V_0| = n_0$, $|V_1| = n_1$, and $|V_2| = n_2$. It is immediate that $V_0 = \{v\}$ and hence $n_0 = 1$. Furthermore, $n = n_0 + n_1 + n_2$. Since the maximum degree is at most d, we have $1 \leq n_1 \leq d$ and $1 \leq n_2 \leq (d-1)n_1$. Thus in particular we observe that $3 \leq n \leq 1 + d^2$. Finally, we observe that we can characterize the edges of H as follows. First, each vertex in V_1 is adjacent to v. Second, the vertices in V_1 may or may not be adjacent to each other, we have to search through all possibilities within the degree bound. Third, each vertex in V_2 must be adjacent to at least one vertex in V_1 and must not be adjacent to v; again we have to search through all possibilities. Finally, because H is an essential neighborhood graph, there are no edges joining the vertices in V_2.

To reduce the number of isomorphic (and hence redundant) graphs encountered in the search, we implement the following lightweight isomorph rejection. Suppose that there is a total order on V_1 and on V_2. In the second stage of the algorithm, when we are searching through all possible ways of joining vertices in V_1 with edges, we require that the degrees of the vertices in V_1 form a non-increasing sequence if listed in the total order of V_1. Furthermore, in the third stage of the algorithm, when we are joining vertices in V_2 by edges to vertices in V_1, we require that with respect to the lexicographic order of subsets of V_1 it holds that $N_H[u] \leq N_H[u']$ whenever $u < u'$ holds for $u, u' \in V_2$. It is immediate that even with this isomorph rejection in place, the algorithm traverses at least one representative from every isomorphism class of neighborhood graphs (with the root individualized).

For each essential neighborhood graph H that survives our isomorph rejection, we test whether the graph is not *maximal*, that is, whether it would be possible to add an edge with both ends in V_1 or an edge joining a vertex in V_1 with a vertex V_2 so that the affected vertices have degree at most d in V_1 and at most $d - 1$ in V_2 after the addition. If H is not maximal, we reject it from further consideration. (Indeed, for a fixed value of d, a maximal graph maximizes the

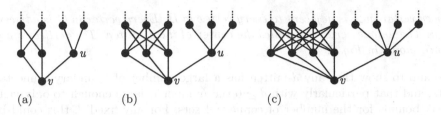

Fig. 2. Worst-case neighborhood graphs of radius 2, for degree 3, 4, and 5

number of boundary-connected sets, and at least one such maximal graph from each isomorphism class of maximal graphs will be encountered in the search.)

Each H that survives the maximality test is passed to a final enumeration of the boundary-connected sets, which proceeds as follows. First we add a new vertex z to H and join it by an edge to every vertex $u \in V_2$, provided that u has degree less than d. Then we count the connected sets that contain z using the folklore algorithm; see, for example, the description given by Björklund *et al.* [3].

The total time required to carry out the search[1] was a few hours on a standard desktop computer with an Intel Core i7–4770K CPU. Figure 2 shows the worst-case neighborhood graphs found for $d = 3, 4, 5$. The corresponding numbers of boundary-connected sets are 184, 1744, and 15136, respectively, yielding the optimal ratios

$$\rho_{3,2} = 184/2^8, \quad \rho_{4,2} = 1744/2^{11}, \quad \rho_{5,2} = 15136/2^{14}.$$

To complete the proof of Theorem 1 it remains to apply Lemma 3 and calculate

$$2\rho_{3,2}^{1/10} = 1.9350\ldots, \quad 2\rho_{4,2}^{1/17} = 1.9811\ldots, \quad 2\rho_{5,2}^{1/26} = 1.9939\ldots.$$

2.4   Limitations of the Method

A fundamental limitation of neighborhood graphs and boundary-connectivity is that we have little control over what happens at the boundary vertices since these vertices may be connected beyond the boundary. With increasing d or r this limitation becomes more severe because the size of the boundary increases implying that we can exclude comparatively less and less projections when applying the projection method.

In fact, we can witness this limitation already for $r = 2$ and $d \geq 6$ as we now proceed to demonstrate. Indeed, we observe that the worst-case neighborhood graphs shown in Fig. 2 follow a pattern which we can generalize as follows:

Definition 1. *An undirected graph G is a d-mitten if its vertices can be partitioned into singletons $\{v\}$ and $\{u\}$ and sets A, B, C, each of size $d - 1$ with exactly the following adjacencies: v is adjacent to u and every vertex in A.*

---

[1] Our implementation is available at http://www.cs.helsinki.fi/u/jwkangas/consets/.

Every vertex in A is adjacent to every vertex in B. Every vertex in C is adjacent to u. The unique vertex v is called the center *of the d-mitten. The vertices v and u are shown in Fig. 2.*

We aim to show that any d-mitten has a large number of boundary-connected sets, and that particularly with $d \geq 6$ the number is large enough to only yield weak bounds for the number of connected sets. For any fixed d, this could be verified by direct calculation, possibly again aided by a computer. However, the simple structure of d-mittens allows us to find a closed-form expression that, not only enables the analysis for an arbitrary d, but also gives a way to check the correctness of the numbers computed for $d = 3, 4, 5$ by the general algorithm.

Lemma 7. *Let G be a d-mitten with center v. Then the number of boundary-connected sets of G relative to v is given by* $2^{3d-1} - 5 \cdot 2^{2d-2} + 2^d$.

Proof. Let $\{u\}$, A, B, and C be the vertex subsets of G guaranteed in the definition of d-mitten. We will count the number of vertex subsets S of G that are *not* boundary-connected relative to v, or *n.b.c.* for short.

Assume first that $v \in S$. Observe that now S is n.b.c. if and only if there is no path in $G[S]$ from v to a vertex in B or C. We consider separately the cases $u \in S$ and $u \notin S$. Suppose $u \in S$. Then S cannot intersect C, since otherwise there would be a path in $G[S]$ from v to a vertex in C. Likewise, S can intersect only A or B but not both. Now, if S does not intersect A, then S may contain any of the 2^{d-1} subsets of B, and vice versa, yielding $2^d - 1$ (i) possibilities for S in total, where the -1 is due to double counting the case where S is disjoint from both A and B. Suppose then that $u \notin S$. Again, we have $2^d - 1$ possible intersections with $A \cup B$, but now, in addition, any subset of C can be contained in S, yielding $(2^d - 1)2^{d-1}$ (ii) possibilities for S in total.

Assume then that $v \notin S$. Now S is n.b.c. if and only if $X := S \cap (\{u\} \cup C)$ is n.b.c. or $Y := S \cap (A \cup B)$ is n.b.c. We count first the cases where X is n.b.c. This holds exactly when $u \in S$ and S does not intersect C. From A and B any subset can be contained in S, yielding $2^{2(d-1)}$ (iii) possibilities. Finally, we count the cases where Y is n.b.c. but X is not. The set Y is n.b.c. exactly when S intersects A but not B, yielding $2^{d-1} - 1$ possibilities. The set X has 2^d possible configurations of which exactly one is n.b.c. Thus, we have $(2^{d-1} - 1)(2^d - 1)$ (iv) possibilities in total.

Summing up (i)–(iv) yields $5 \cdot 2^{2d-2} - 2^d$. It remains to note that G has $3d - 1$ vertices and thus 2^{3d-1} vertex subsets in total.   □

The following lemma shows that d-mittens, indeed, result in weak bounds compared to the simple bound:

Lemma 8. *Let $d \geq 6$. Then*

$$2\left(\frac{2^{3d-1} - 5 \cdot 2^{2d-2} + 2^d}{2^{3d-1}}\right)^{1/(d^2+1)} \geq (2^{d+1} - 1)^{1/(d+1)}.$$

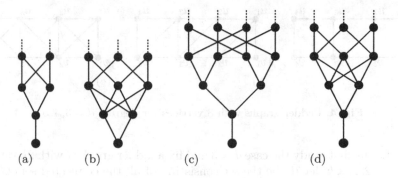

Fig. 3. Neighborhood graphs with a large number of boundary-connected sets

Proof. For $d = 6$, the inequality is verified by direct calculation (details omitted). Suppose $d \geq 7$. We observe that the inequality holds if and only if

$$\left(1 - 2^{-d-1}\right)^{(d^2+1)/(d+1)} \leq 1 - 5 \cdot 2^{-d-1} + 2^{-2d+1}.$$

Denote $p := 2^{-d-1}$ and $k := \lfloor (d^2+1)/(d+1) \rfloor$. Observe that $\binom{k}{2} \leq \binom{d+1}{2} \leq 2^{d+1}$ and $k \geq \lfloor (7^2 + 1)/(7 + 1) \rfloor = 6$. Thus, by a Bonferroni inequality, we have $(1-p)^k \leq 1 - kp + \binom{k}{2}p^2 \leq 1 - (k-1)p \leq 1 - 5 \cdot 2^{-d-1}$, completing the proof. □

Finally, we turn to the case where the radius r is larger than 2. Here we only investigate the cases where r equals 3 or 4 and the maximum degree d equals 3 and 4. For these cases, the graphs shown in Fig. 3 imply

$$\rho_{3,3} \geq 31/2^6, \quad \rho_{4,3} \geq 321/2^9, \quad \rho_{3,4} \geq 1480/2^{12}, \quad \rho_{4,4} \geq 459/2^{10}.$$

Consequently,

$$2\rho_{3,3}^{1/22} \geq 1.9351, \quad 2\rho_{4,3}^{1/53} \geq 1.9824, \quad 2\rho_{3,4}^{1/46} \geq 1.9562, \quad 2\rho_{4,4}^{1/161} \geq 1.9900,$$

exceeding the respective values $b_3 = 1.9351$ and $b_4 = 1.9812$ given in Theorem 1.

3   Lower Bounds on the Number of Connected Sets

We prove Theorem 2 by analyzing generalized ladder graphs. An undirected graph with $2k$ vertices is a *ladder graph* of *degree* $d \geq 3$ if its vertices can be labeled as $u_1, u_2, \ldots, u_k$ and $v_1, v_2, \ldots, v_k$ so that the graph has exactly the following adjacencies between the vertices. First, u_i is adjacent to u_{i+1} and v_i is adjacent to v_{i+1} for $i = 1, 2, \ldots, k - 1$. Second, u_i is adjacent to v_j if and only if $0 \leq i - j \leq \lfloor (d-3)/2 \rfloor$ or $0 \leq j - i \leq \lceil (d-3)/2 \rceil$. Figure 4 shows examples of ladder graphs of small degree d.

Theorem 3. *Every ladder graph with $2k \geq d$ vertices and degree $d \leq 5$ has at least α_d^k connected sets, where $\alpha_3 = 1 + \sqrt{2}$, $\alpha_4 = (3 + \sqrt{5})/2$, and $\alpha_5 = 3$.*

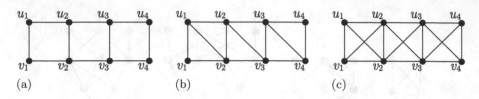

Fig. 4. Ladder graphs with 8 vertices for degrees $d = 3, 4, 5$

Proof. Let us first study the case $d = 3$ and fix a ladder graph G with $2k$ vertices. For $p = 1, 2, \ldots, k$, let $\mathcal{U}_p$ be the set consisting of all the connected sets C in G such that the set $C \cap \{u_i, v_i\}$ has (i) size at least 1 for each $1 \le i \le p$ and (ii) size 0 for each $p + 1 \le i \le k$. Partition $\mathcal{U}_p$ into $\mathcal{S}_p$ and $\mathcal{T}_p$ such that C is in $\mathcal{S}_p$ if and only if $C \cap \{u_p, v_p\}$ has size 1; otherwise C is in $\mathcal{T}_p$. Let us write s_p for the size of $\mathcal{S}_p$ and t_p for the size of $\mathcal{T}_p$. We have $s_1 = 2$ and $t_1 = 1$. For $p \ge 2$ we have

$$s_p = s_{p-1} + 2t_{p-1}, \quad t_p = s_{p-1} + t_{p-1}.$$

In particular, $s_p = 2s_{p-1} + s_{p-2}$, and thus

$$s_p = \frac{1}{\sqrt{2}}(1 + \sqrt{2})^p - \frac{1}{\sqrt{2}}(1 - \sqrt{2})^p.$$

For $p \ge 2$ we observe that $t_p \ge 2s_p/3$ and $(1 - \sqrt{2})^p/\sqrt{2} < 1/2$. Thus,

$$|\mathcal{U}_k| = s_k + t_k \ge \frac{5}{3}\left(\frac{1}{\sqrt{2}}(1 + \sqrt{2})^k - \frac{1}{2}\right) > (1 + \sqrt{2})^k - 1.$$

Next let us consider the case $d = 4$. We proceed as in the previous case, but let x_p, y_p, and z_p be the number of members of $\mathcal{U}_p$ that contain, respectively, u_p but not v_p, or v_p but not u_p, or both u_p and v_p. We have $x_1 = 1$, $y_1 = 1$, and $z_1 = 1$. For $p \ge 2$ we have

$$x_p = x_{p-1} + z_{p-1}, \quad y_p = x_{p-1} + y_{p-1} + z_{p-1}, \quad z_p = x_{p-1} + y_{p-1} + z_{p-1},$$

which implies $z_p = 3z_{p-1} - z_{p-2}$, and thus

$$z_p = \frac{1}{\sqrt{5}}\left(\frac{3 + \sqrt{5}}{2}\right)^p - \frac{1}{\sqrt{5}}\left(\frac{3 - \sqrt{5}}{2}\right)^p.$$

For $p \ge 2$ we observe that $x_p \ge z_p/2$ and $((3 - \sqrt{5})/2)^p/\sqrt{5} < 1/4$. Thus,

$$|\mathcal{U}_k| = x_k + y_k + z_k \ge \frac{5}{2}\left(\frac{1}{\sqrt{5}}\left(\frac{3 + \sqrt{5}}{2}\right)^k - \frac{1}{4}\right) > \left(\frac{3 + \sqrt{5}}{2}\right)^k - 1.$$

Finally, let us consider the case $d = 5$. Define $\mathcal{U}_p$ as above and observe that $C \in \mathcal{U}_p$ if and only if for every $i = 1, 2, \ldots, p$ it holds that at least one of u_i and v_i is in C. Thus, $|\mathcal{U}_p| = 3^p$ and this holds in particular when $p = k$.   □

Theorem 2 now follows as an immediate corollary of Theorem 3.

4   Concluding Remarks

This paper has explored the possibility of extending the projection method to neighborhoods with radius $r \geq 2$ to obtain improved upper bounds for the number of connected sets in bounded-degree graphs. Our improved bounds for $d \leq 5$ present a rather modest improvement, and the upper and lower bounds in Theorems 1 and 2 remain far apart. To strengthen the projection method it would appear that one needs control on how the projections change as one moves from one neighborhood to the neighborhood of an adjacent vertex.

References

1. Alon, N.: Independent sets in regular graphs and sum-free subsets of finite groups. Isr. J. Math. **73**, 247–256 (1991)
2. Binkele-Raible, D., Fernau, H., Gaspers, S., Liedloff, M.: Exact and parameterized algorithms for max internal spanning tree. Algorithmica **65**(1), 95–128 (2013)
3. Björklund, A., Husfeldt, T., Kaski, P., Koivisto, M.: Computing the Tutte polynomial in vertex-exponential time. In: FOCS, pp. 677–686. IEEE Computer Society (2008)
4. Björklund, A., Husfeldt, T., Kaski, P., Koivisto, M.: Trimmed Moebius inversion and graphs of bounded degree. Theor. Comput. Syst. **47**(3), 637–654 (2010)
5. Björklund, A., Husfeldt, T., Kaski, P., Koivisto, M.: The traveling salesman problem in bounded degree graphs. ACM Trans. Algorithms **8**(2), 18:1–18:13 (2012)
6. Bollobás, B.: The Art of Mathematics: Coffee Time in Memphis. Cambridge University Press (2006)
7. Chung, F., Graham, R., Frankl, P., Shearer, J.: Some intersection theorems for ordered sets and graphs. J. Comb. Theor. Ser. A **43**(1), 23–37 (1986)
8. Fomin, F.V., Gaspers, S., Pyatkin, A.V., Razgon, I.: On the minimum feedback vertex set problem: exact and enumeration algorithms. Algorithmica **52**(2), 293–307 (2008)
9. Fomin, F.V., Grandoni, F., Pyatkin, A.V., Stepanov, A.A.: Combinatorial bounds via measure and conquer: Bounding minimal dominating sets and applications. ACM Trans. Algorithms **5**(1), 9:1–9:17 (2008)
10. Fomin, F.V., Villanger, Y.: Finding induced subgraphs via minimal triangulations. In: Marion, J.Y., Schwentick, T. (eds.) STACS. Volume 5 of LIPIcs., Schloss Dagstuhl - Leibniz-Zentrum für Informatik, pp. 383–394 (2010)
11. Fomin, F.V., Villanger, Y.: Treewidth computation and extremal combinatorics. Combinatorica **32**(3), 289–308 (2012)
12. Galvin, D.: An upper bound for the number of independent sets in regular graphs. Discrete Math. **309**(23–24), 6635–6640 (2009)
13. Gaspers, S., Kratsch, D., Liedloff, M.: On independent sets and bicliques in graphs. Algorithmica **62**(3–4), 637–658 (2012)
14. Gaspers, S., Mnich, M.: Feedback vertex sets in tournaments. J. Graph Theory **72**(1), 72–89 (2013)
15. Kahn, J.: An entropy approach to the hard-core model on bipartite graphs. Combin. Probab. Comput. **10**, 219–237 (2001)
16. Moon, J.W., Moser, L.: On cliques in graphs. Isr. J. Math. **3**, 23–28 (1965)
17. Perrier, E., Imoto, S., Miyano, S.: Finding optimal Bayesian network given a superstructure. J. Mach. Learn. Res. **9**, 2251–2286 (2008)
18. Zhao, Y.: The number of independent sets in a regular graph. Combin. Probab. Comput. **19**, 315–320 (2010)

Parameterized Edge Hamiltonicity

Michael Lampis[1], Kazuhisa Makino[1], Valia Mitsou[2(✉)], and Yushi Uno[3]

[1] Research Institute for Mathematical Sciences, Kyoto University, Kyoto, Japan
{mlampis,makino}@kurims.kyoto-u.ac.jp
[2] JST Erato Minato Discrete Structure Manipulation System Project,
Sapporo, Japan
vmitsou@erato.ist.hokudai.ac.jp
[3] Department of Mathematics and Information Sciences, Graduate School of Science,
Osaka Prefecture University, Osaka, Japan
uno@mi.s.osakafu-u.ac.jp

Abstract. We study the parameterized complexity of the classical EDGE HAMILTONIAN PATH problem and give several fixed-parameter tractability results. First, we settle an open question of Demaine et al. by showing that EDGE HAMILTONIAN PATH is FPT parameterized by vertex cover, and that it also admits a cubic kernel. We then show fixed-parameter tractability even for a generalization of the problem to arbitrary hypergraphs, parameterized by the size of a (supplied) hitting set. We also consider the problem parameterized by treewidth or clique-width. Surprisingly, we show that the problem is FPT for both of these standard parameters, in contrast to its vertex version, which is W[1]-hard for clique-width. Our technique, which may be of independent interest, relies on a structural characterization of clique-width in terms of treewidth and complete bipartite subgraphs due to Gurski and Wanke.

1 Introduction

The focus of this paper is the EDGE HAMILTONIAN PATH problem, which can be defined as follows: given an undirected graph $G(V, E)$, does there exist a permutation of E such that every two consecutive edges in the permutation share an endpoint? This is a very well-known graph-theoretic problem, which corresponds to the restriction of (vertex) HAMILTONIAN PATH to line graphs. Despite some superficial similarity to the problem of finding an Eulerian path, this problem has long been known to be NP-complete, even for graphs which are bipartite or have maximum degree 3 [1,25,29].

The EDGE HAMILTONIAN PATH problem is a very natural graph-theoretic problem with a long history (see e.g. [4–8,24]). In this paper, we investigate the

Valia Mitsou: This work was done while the author was at the Graduate Center, CUNY and visiting Kyoto University. Research is (partly) supported by JST ERATO Minato project.

Yushi Uno: Research partially supported by KAKENHI Grand numbers 23500022 and 25106508.

© Springer International Publishing Switzerland 2014
D. Kratsch and I. Todinca (Eds.): WG 2014, LNCS 8747, pp. 348–359, 2014.
DOI: 10.1007/978-3-319-12340-0_29

complexity of this problem from the parameterized complexity perspective. More specifically, we consider the case where some structural parameter of the input graph G, such as its treewidth, has a moderate value. Despite the problem's prominence, to the best of our knowledge, EDGE HAMILTONIAN PATH has never before been studied in this setting. Such an investigation is of inherent interest from the point of view of graph theory and parameterized complexity. Beyond this, we are partially motivated by a specific question recently asked explicitly by Demaine et al. [14]. In their investigation of the card game UNO, the authors of [14] present an XP (i.e. running in $n^{f(k)}$) dynamic programming algorithm for EDGE HAMILTONIAN PATH on bipartite graphs, where k is the size of the smaller part. They then, quite naturally, ask if this can be improved to an FPT algorithm. In this paper, we present a number of results that positively settle not only this, but several other more general such questions (the question from [14] was also independently settled by Dey et al. [15]).

Overview of results. We give fixed-parameter tractability results for EDGE HAMILTONIAN PATH and its variant EDGE HAMILTONIAN CYCLE, which we show to be essentially equivalent. Our first task is to consider the problem parameterized by the size of the vertex cover of the input graph. We establish that, not only is the problem FPT, but it also admits a cubic kernel through an algorithm that locates and deletes irrelevant edges. This result settles the question from [14] as for a bipartite graph, one part being small implies a small vertex cover. We then go on to give a much more general direct FPT algorithm for the problem, which can still be applied even if we consider the problem on arbitrary hypergraphs with the parameter being the size of a hitting set which is supplied with the input. As a corollary, we note that this result implies that (vertex) HAMILTONIAN PATH is FPT when parameterized by the chromatic number of the complement of the input graph.

Our next direction is to consider the problem on graphs parameterized by treewidth and clique-width. The complexity of EDGE HAMILTONIAN PATH for these parameters was previously unknown, since this is also a more general question than the one posed in [14]. Our first observation is that fixed-parameter tractability for EDGE HAMILTONIAN CYCLE parameterized by treewidth can be obtained from standard meta-theorems, if one relies on an alternative characterization of the problem first given by Harary and Nash-Williams almost 50 years ago [22]. This alternative characterization allows one to recast the ordering problem as the problem of finding a connected Eulerian subgraph whose vertices form a vertex cover of the original graph. The alternative problem with a little work, can be expressed in a variant of Monadic Second Order logic. For the sake of completeness, we also sketch a direct treewidth-based dynamic programming algorithm using this formulation.

Having settled the problem for treewidth, the natural next step is to consider EDGE HAMILTONIAN CYCLE parameterized by clique-width, a prominent structural graph parameter that generalizes treewidth. It is important to note here that the (more common) vertex version of the problem exhibits a sharp complexity jump between these two parameters: HAMILTONIAN CYCLE is FPT for treewidth but for clique-width the problem is W[1]-hard and therefore does not admit an FPT algorithm under standard complexity assumptions [19]. In what is perhaps the most surprising result of this paper, we show that EDGE HAMIL-TONIAN CYCLE remains FPT even for clique-width, despite this parameter's additional generality. On a high level, our strategy is to rely on a characterization of bounded clique-width graphs given by Gurski and Wanke [20] which states roughly that if a graph has small clique-width and no large complete bipartite subgraphs, then it has small treewidth. We devise an algorithm that locates and "reduces" large complete bipartite subgraphs in the input graph, without affecting the answer or increasing the clique-width. By repeatedly applying this step, we end up with a graph of small treewidth for which the problem is FPT. This idea, which was also used in [28], is a rare algorithmic application of the characterization of [20], and may be of independent interest.

2   Preliminaries

We assume that the reader is familiar with the basics of parameterized complexity. In particular, we use the definitions of the classes FPT, XP as well as the notion of a kernelization algorithm and of polynomial kernels (see [16,18,26]).

We will use the definition of treewidth, and in particular the notion of "nice" tree decompositions (see the survey [3]). We also use the notion of clique-width (see [13,17,23]). Let us briefly review the definition. The class of graphs of clique-width k contains all single-vertex graphs where the only vertex has a label from $\{1, \ldots, k\}$. Furthermore, the class is closed under the following operations: disjoint union of two graphs; renaming of all vertices with some label i to some label j; and joining by new edges of all vertices with some label i to all vertices with some label j. All graph classes with bounded treewidth also have bounded clique-width, but the reverse is not true [10].

We will also rely on the following theorem of Gurski and Wanke which intuitively states that large complete bipartite graphs are what separates treewidth from clique-width:

Theorem 1 [20]. *Let G be a graph of clique-width k. If G does not contain the complete bipartite graph $K_{t,t}$ as a subgraph, then $tw(G) \leq 3kt$.*

We will consider the EDGE HAMILTONIAN PATH and EDGE HAMILTONIAN CYCLE problems. As mentioned, in these problems we are looking for a permutation of the edges of the input graph so that any two consecutive edges share an endpoint (in the latter problem, also the first and last edge must share an endpoint). We call such a permutation an edge-Hamiltonian path (respectively an edge-Hamiltonian cycle). We will mostly view these as graph problems, but this problem definition applies equally well to hypergraphs, if we require that two consecutive *hyperedges* share a common vertex. Hypergraphs are the subject of Sect. 4. Recall that for a graph or hypergraph $G(V, E)$, its line graph is the graph $G'(E, H)$ where $(e_1, e_2) \in H$ if and only if e_1, e_2 share a vertex in G. The

EDGE HAMILTONIAN PATH problem on G is equivalent to the HAMILTONIAN PATH problem on G'.

For the graph case, it will be useful to recast these ordering problems as subgraph problems. First, recall that a graph is Eulerian if it is connected and all its vertices have even degree. A DOMINATING EULERIAN SUBGRAPH of a graph $G(V, E)$ is a subgraph $G'(V', E')$ of G such that all edges of E have an endpoint in V', that is, V' is a vertex cover of G, and G' is Eulerian. We will use the following classical observation of Harary and Nash-Williams:

Theorem 2 [22]. *A graph has an edge-Hamiltonian cycle if and only if it contains a dominating Eulerian subgraph.*

Finally, let us mention that we will deal with EDGE HAMILTONIAN PATH and EDGE HAMILTONIAN CYCLE interchangeably, depending on which problem makes the description of our algorithms easier. The reader can easily verify that all our arguments apply to both problems with very minor modifications. It is also not hard to show the following:

Lemma 1. *For the following parameters and for sufficiently large graphs,* EDGE HAMILTONIAN PATH *is FPT if and only if* EDGE HAMILTONIAN CYCLE *is FPT: vertex cover, treewidth, clique-width and hypergraph hitting set.*

Proof of Lemma 1 as well as all other missing proofs appears in the full version of the paper.

3 Vertex Cover

In this section we consider the EDGE HAMILTONIAN PATH problem parameterized by the size of the vertex cover k. We show that the problem has a cubic in k kernel. As in the following sections, we assume that together with the input graph $G(V, E)$ we are given a vertex cover S of G with $|S| = k$. Note though, that this assumption is not important, since a 2-approximate vertex cover can be found in polynomial time [9].

Below follow some definitions which will make the presentation of the results smoother. We assume that the vertices of G are labeled in some lexicographically ordered fashion, and in particular that $S = \{u_1, \ldots, u_k\}$.

Definition 1. *An edge $e \in E$ is defined to be of type i if it is incident to $u_i \in S$ but not incident to any other $u_j \in S$ for $j < i$.*

Definition 2. *Let P be an edge-Hamiltonian path of G. For $i \in \{1, \ldots, k\}$, a group of type i is a maximal set of edges of type i which are consecutive in P. We say that an edge is special if it is the first or the last edge of a group.*

The special edges essentially form the backbone of the edge-Hamiltonian path P. A piece of intuition that will become useful later is that, if one fixes these edges in a proper edge-path, the remaining edges will be easy to deal with, because they are allowed to move freely in and out of groups.

Our next goal then is to show that if a graph has an edge-Hamiltonian path P, then it has one where *few* edges are special. This is summarized in Lemma 2 and Corollary 1. Intuitively, the core idea is a flipping argument: if the same group types appear too many times in a solution, we can reverse a sub-path to obtain a solution with fewer groups.

Lemma 2. *Let G be an edge-Hamiltonian graph. Then, there exists an edge-Hamiltonian path P of G with the following property: for any $i, j \in \{1, \ldots, k\}$, an edge of type j appears directly after an edge of type i at most once.*

Proof. (sketch)

Suppose that P' is an edge-Hamiltonian path of G in which some group of type i immediately precedes some group of type i. Then, we can create a valid path P by reversing the middle part of this path and merging the two groups of type i and those of type j.

The new path has strictly fewer groups. Repeating this process at most a linear (in $|E|$) number of times results in an edge-Hamiltonian path P with the stated property. $\square$

Corollary 1. *Let G be an edge-Hamiltonian graph. Then, there exists an edge-Hamiltonian path P of G such that for all $i \in \{1, \ldots, k\}$, P contains at most k groups of type i. Therefore, P contains at most k^2 groups in total, and for each $i \in \{1, \ldots, k\}$ there exist at most $2k$ special edges of type i.*

We have now proved that if a solution exists, it must have a certain nice form. Let us make one more easy observation.

Lemma 3. *Let $G(V, E)$ be an edge-Hamiltonian graph. Then, there exists an edge-Hamiltonian path P such that, for all $i \in \{1, \ldots, k\}$ for which there exist at least $k + 1$ edges of type i, P has a group of type i with size at least 2.*

Let us note that Lemma 2, Corollary 1 and Lemma 3 still hold even if G is a hypergraph. We will make use of this in the next section.

We are now ready to state the main reduction rule and sketch its correctness.

Lemma 4. *Let $G(V, E)$ be a graph, and $S = \{u_1, \ldots, u_k\}$ a vertex cover of G of size k. Suppose that there exists an edge (u_i, w) satisfying the following:*

1. $w \notin S$
2. *There are at least $k + 2$ edges of type i in G*
3. *For all $u_j \in S$ such that $(u_j, w) \in E$ we have $|(N(u_i) \cap N(u_j)) \setminus S| > 4k$*

Then $G(V, E)$ has an edge-Hamiltonian path if and only if $G'(V, E \setminus \{(u_i, w)\})$ does.

Proof. (sketch)

For the one direction, suppose that G has an edge-Hamiltonian path P. We construct a path P' where (u_i, w) is removed. Let e_1, e_2 be the edges appearing immediately before and after (u_i, w) in P. Suppose they do not share an endpoint (if they do, we can construct P' by deleting (u_i, w) from P). Since they both share an endpoint with (u_i, w) we assume without loss of generality that e_1 is incident on u_i and $e_2 = (u_j, w)$. (Observe that here we have used the fact that G is a graph, so the rest of our argument does not generalize to hypergraphs).

We know now by the last condition that $N(u_j) \cap N(u_i)$ contains at least $4k + 1$ vertices of $V \backslash S$. By Corollary 1 and pigeonhole principle, there exists a vertex of $(N(u_i) \cap N(u_j)) \backslash S$, call it z, such that (u_i, z) and (u_j, z) are not special.

Because (u_i, z) is not special, the two edges appearing immediately before and after it are both incident on u_i. Therefore, deleting (u_i, z) still leaves us with a valid edge-path. Similar reasoning can be used for (u_j, z). We construct a path P' as follows: delete $(u_i, w), (u_i, z)$ and (u_j, z) from P and then insert $(u_i, z), (u_j, z)$ between e_1 and e_2. This is a valid solution for G'. □

The proof of the other direction is easy and appears in the full version.

Lemma 4 now leads to the following theorem.

Theorem 3. EDGE HAMILTONIAN PATH *has a kernel with* $O(k^3)$ *edges, where* k *is the size of the input graph's vertex cover.*

4 Hypergraphs

In this section we present an FPT algorithm for EDGE HAMILTONIAN PATH on hypergraphs parameterized by the size of a (supplied) hitting set. As an interesting consequence, our algorithm also establishes fixed-parameter tractability for a novel parameterization of HAMILTONIAN PATH, namely when the parameter is the chromatic number of the input graph's complement.

In this section, $G(V, E)$ will be a hypergraph (that is, E is a collection of arbitrary subsets of V). We assume that the input also contains a hitting set $S \subset V$ of size k, that is, a set of vertices that intersects all hyperedges. Unlike the previous section, this is not an inconsequential assumption, since finding even an approximate hitting set is generally a hard problem. However, observe that for hypergraphs of bounded rank (i.e. hyperedge size), a hitting set can be computed in FPT time and hence this requirement is nullified on such instances.

We will rely on the fact that much of the material of the previous section carries through unchanged. In particular, Definitions 1, 2, also apply to hypergraphs. Then, Lemma 2, Corollary 1, and Lemma 3 hold for the case of hypergraphs as well. Unfortunately, Lemma 4 does not seem to generalize naturally in this case.

Let us thus describe a different algorithm for this problem. As mentioned, one way to proceed is to try to identify the special hyperedges which form the backbone of a path. Once these have been found, the problem becomes much easier. We will use a color-coding scheme to assist us in selecting these special hyperedges. The high-level idea is the following: for every $i \in \{1, \ldots, k\}$ such that there are at least $2k$ hyperedges of type i, color these hyperedges with $2k$ colors uniformly at random. Then, *merge* (that is, take the union) of all hyperedges of type i that took the same color to a single hyperedge. This process results in a hypergraph G' with $O(k^2)$ hyperedges. We want to show that if this hypergraph has an edge-Hamiltonian path then G does as well, while if G has an edge-Hamiltonian path then G' has one with non-negligible probability. The "good colorings" that give us this non-negligible probability are those that assign a different color to each special edge.

We are now ready to state the main result of this section.

Theorem 4. *Given a hypergraph $G(V, E)$ and a hitting set $S = \{u_1, \ldots, u_k\}$ of G, there is an FPT algorithm that decides if G has an* Edge Hamiltonian Path *in time $2^{O(k^2)} n^{O(1)}$.*

An interesting consequence of Theorem 4 is that it implies fixed-parameter tractability for a non-standard parameterization of Hamiltonian Path. The parameterization we are considering is by the *complement chromatic number*, that is, the chromatic number of the input graph's complement. We are naturally led to this observation, because the line graph of a hypergraph with a hitting set of size k has a vertex set that can be partitioned into at most k cliques. To the best of our knowledge, this parameterization of Hamiltonian Path has not been considered before.

Corollary 2. *Given a graph $G(V, E)$ and a proper k-coloring of its complement graph, there exists an FPT algorithm that decides if G has a Hamiltonian Path in time $2^{O(k^2)} n^{O(1)}$.*

5   Treewidth and Clique-Width

In this section we consider the Edge Hamiltonian Cycle problem parameterized by treewidth or clique-width. As is customary for these parameters, we will assume that a decomposition of width k (or a clique-width expression with k labels) is given to us with the input. This assumption is not necessary though, as both parameters can be approximated in FPT time (see [2,27]).

Let us first consider treewidth. One obvious approach we could try to follow is to use the fact that if G has treewidth k, its *line* graph has clique-width $O(k)$ [21]. Since deciding Edge Hamiltonian Cycle on G is equivalent to deciding Hamiltonian Cycle on its line graph, this would give an XP algorithm using known results for the latter problem (this is similar to the approach of [14]). Unfortunately, since Hamiltonian Cycle is W[1]-hard for clique-width, this approach could not lead to an FPT algorithm for Edge Hamiltonian Cycle on treewidth. We thus have to recast the problem.

We will rely on Theorem 2, which states that the existence of an edge-Hamiltonian cycle is equivalent to the existence of a dominating Eulerian subgraph. Thus, we can view Edge Hamiltonian Cycle as a subgraph problem rather than an ordering problem. This formulation allows us to express the problem in a variant of MSO logic, without reference to orderings. We can then invoke standard meta-theorems to obtain fixed-parameter tractability for treewidth.

Let us sketch the basic idea. Recall that MSO_2 logic allows one to express properties involving sets of vertices *or* edges (see [12]). DOMINATING EULERIAN SUBGRAPH is the problem of looking for a set of vertices V' and a set of edges E' such that: all edges of E have an endpoint in V'; the graph $G'(V', E')$ is connected; all vertices of $G'(V', E')$ have even degree. The first two properties are well-known to be expressible in MSO logic. Interestingly, the third property is expressible in Counting MSO_2 ($CMSO_2$) logic, an extension of MSO_2 which is still FPT for treewidth [11,23]. Thus, EDGE HAMILTONIAN CYCLE is expressible in $CMSO_2$ and is therefore FPT for treewidth.

We can use standard techniques to obtain the following:

Theorem 5. *Given a graph G and a tree decomposition of width k, there exists an algorithm deciding if G has an edge-Hamiltonian cycle in time $k^{O(k)}n^{O(1)}$.*

Let us now move to the main result of this section, which is the tractability of EDGE HAMILTONIAN CYCLE parameterized by clique-width. Our high-level strategy will be to eliminate complete bipartite subgraphs from the input graph, without increasing the graph's clique-width and without affecting the answer of the problem. If we can repeat this process, we will in the end have a graph with small clique-width and no large complete bipartite subgraphs. By Theorem 1, the graph will have small treewidth and we can use Theorem 5.

Our main tool will be a reduction lemma (Lemma 6). Roughly speaking, the lemma states that if we find a sufficiently large complete bipartite graph in G with bipartition A, B, we can reduce it as follows: first we remove all its edges and then we add three new vertices which are connected to all vertices of both A and B. This transformation should not affect the answer.

To prove Lemma 6, it will be useful to first prove the following statement. Roughly speaking, it says that if a graph contains a $K_{3,3}$ (or larger) complete bipartite subgraph, then any DOMINATING EULERIAN SUBGRAPH can be edited to produce a solution using all its vertices.

Lemma 5. *Let $G(V, E)$ be a graph and $A, B \subseteq V$, with A, B disjoint sets, $|A|, |B| \geq 3$ and $A \times B \subseteq E$. If G has a dominating Eulerian subgraph then it also has a dominating Eulerian subgraph $G_0(V_0, E_0)$ such that $(A \cup B) \subseteq V_0$ and $E_0 \cap (A \times B) \neq \emptyset$.*

Proof. Suppose that G has a dominating Eulerian subgraph $G_0(V_0, E_0)$. We will edit this solution by adding vertices and adding or removing edges until the stated properties are achieved. In the remainder, when we say that we *flip* an edge e we mean that, if $e \in E_0$ then we remove it from E_0, otherwise we add it to E_0 and add its endpoints to V_0.

Let us first establish that $|(A \cup B) \backslash V_0| \leq 1$ as follows: if V_0 does not fully contain one of the two sets A, B, it must fully contain the other (because V_0 is a vertex cover). Suppose without loss of generality that $B \subseteq V_0$. If there exist $v_1, v_2 \in A \backslash V_0$, then pick two vertices $u_1, u_2 \in B$. We can flip all the edges of $\{u_1, u_2\} \times \{v_1, v_2\}$ and produce a valid solution with more vertices.

Now, if there is a single vertex $v_1 \in A \backslash V_0$ then we have two cases: if there exist $u_1 \in B, v_2 \in A$ such that $(u_1, v_2) \notin E_0$, we pick an arbitrary $u_2 \in B$ and flip the edges $\{u_1, u_2\} \times \{v_1, v_2\}$. This produces a valid dominating Eulerian subgraph that contains v_1. In the final case, all edges of $A \times B$ not incident on v_1 are used in E_0. Then, picking two arbitrary $u_1, u_2 \in B$ and a vertex $v_2 \in A$ and flipping the edges $\{u_1, u_2\} \times \{v_1, v_2\}$ produces a valid solution that includes v_1. We can conclude that $A \subseteq V_0$.

For the second property, observe that if E_0 does not use any edges of $A \times B$ then we can add an arbitrary cycle to E_0 using edges of $A \times B$ producing a valid solution. □

Lemma 6. *Let $G(V, E)$ be a graph and $A, B \subseteq V$ with A, B disjoint sets, $|A|, |B| \geq 5$ and $A \times B \subseteq E$. Let $C = \{c_1, c_2, c_3\}$ be a set of three new vertices. Consider the graph $G'(V', E')$ where $V' = V \cup C$ and $E' = (E \backslash A \times B) \cup (A \times C) \cup (B \times C)$. Then G' has an edge-Hamiltonian cycle if and only if G does.*

Proof. For the first direction, suppose that G has a dominating Eulerian subgraph $G_0(V_0, E_0)$. We will now describe a dominating Eulerian subgraph $G_0'(V_0', E_0')$ of G'. We set $V_0' = V_0 \cup C$, which is clearly a vertex cover of G'. To construct E_0', we begin with the set of edges $E_0 \backslash (A \times B)$. Now, we need to consider the bipartite subgraph $G_0^{A \cup B}$ of G_0 induced by $A \cup B$. In this subgraph, there will be an even number of vertices of odd degree. For each such vertex u, we add an edge in G_0' from u to each of the vertices of C. This ensures that u will still have an odd degree in the subgraph $G_0'^{A \cup B \cup C}$ of G_0' induced by $A \cup B \cup C$. Furthermore, all vertices of C in G_0' should currently have even degree. Let D be the set of remaining vertices of $G_0^{A \cup B}$, with even degree. If $|D|$ is a multiple of 3, we connect a third of these vertices with c_1 and c_2, a third with c_1 and c_3 and a third with c_2 and c_3. If $|D| = 2 \bmod 3$, then we connect two vertices of D with c_1, c_2 and for the rest we act as in the previous case. If $D = 1 \bmod 3$, and $|D| \geq 4$, we connect four vertices of D with c_1, c_2 and act as before for the rest. Last, for the case that there is only one vertex of even degree, we connect it to c_1 and c_2 while at the same time we remove the edges (v, c_1) and (v, c_2) for some other vertex v of odd degree. Observe that this process ensures that in the end all vertices of A, B have degree in $G_0'^{A \cup B \cup C}$ with the same parity as in $G_0^{A \cup B}$ and all vertices of C have even degree in G_0'. Furthermore, the constructed graph is always connected because the bipartite subgraph is sufficiently large.

For the converse direction, suppose we have a dominating Eulerian subgraph $G_0'(V_0', E_0')$ of G'. By Lemma 5, because $C, (A \cup B)$ form two parts of a sufficiently large complete bipartite subgraph we can assume that $(A \cup B \cup C) \subseteq V_0'$.

We build a dominating Eulerian subgraph $G_0(V_0, E_0)$ of G as follows. First, $V_0 = V_0' \backslash C$, which is a vertex cover of G. Let E_C be the set of edges of E_0' incident on C. It must be the case that $|E_C|$ is even, since all vertices of C have even degree in G_0' and C is an independent set. We start building E_0 by including all the edges of $E_0' \backslash E_C$. We will now go through two phases of "fixing" E_0 by adding to it edges of $A \times B$.

Initially, we concentrate on making all degree parities even. We will say that we *flip* an edge e to mean that, if $e \in E_0$ then we remove it from E_0, otherwise we add it to E_0. Observe that, for our current selection of E_0, the number of vertices of $A \cup B$ with odd degree in G_0 is even. This is a consequence of the fact that $|E_C|$ is even and that all vertices have even degree in G_0'. As long as there exist two vertices u, v of $A \cup B$ with odd degree in G_0, select a shortest path connecting u and v in G and flip its edges. This will only change the parity of the degree of u and v in G_0. Repeating this process will eventually produce a set E_0 that makes the degree of all vertices even.

We now need to augment E_0 to make sure that G_0 is connected. It is not hard to see that if G_0 is not connected, there must be two vertices of $A \cup B$ in different components (otherwise, we could find a disconnected component in G_0'). Our intermediate goal is to create a solution where each part (excluding possibly at most one vertex) belongs as a whole in one connected component. Starting from part A, let's assume that it doesn't belong as a whole in one component, in other words assume that there exist two vertices v_1, v_2 such that v_1, v_2 are in different components.

One of v_1, v_2 should have at least two neighbors in B, otherwise we can find two common non-neighbors u_1, u_2 and add the edges of $\{u_1, u_2\} \times \{v_1, v_2\}$ to E_0 to obtain a valid solution with fewer components. So assume that v_2 has at least two neighbors in B, u', u''.

Now, for each additional vertex v_3 of A, if v_3 is not at the same connected component as v_2, we can add all edges between $\{u', u''\}$ and $\{v_1, v_3\}$ and obtain a solution with fewer connected components. Therefore, every vertex of A except for v_1 belongs to the same connected component as v_2.

With similar reasoning, we can conclude that every vertex of B but (possibly) one vertex (call this u_1 if it exists) also belongs in one connected component. Additionally, we can easily conclude that, in the case the big components from each part are disconnected, we connect them by joining two pairs of vertices from each of them.

We are now almost done. We describe the process to attach v_1 to the big connected component (u_1, if it exists, can be handled in a similar way). If there exists at least one vertex of the big component in A with two non-neighbors in B, then we completely join these three vertices together with v_1 in a $K_{2,2}$. In the other case, all vertices of A from the big component have at most one non-neighbor in B. This means that the big component is very well-connected, so we can take an arbitrary vertex of A together with two arbitrary vertices of B and flip all edges between them while adding all edges from v_1 to these two vertices of B. After performing this step, the connectivity of the graph is increased. □

We are now almost ready to proceed with our algorithm. To simplify presentation, we will only apply Lemma 6 to subgraphs which are at least as large as $K_{7,7}$. Observe that in such a case, G' has strictly fewer edges than G. It is then clear that the reduction is making progress and after a bounded number of applications we get a graph with no large complete bipartite subgraphs.

There is, however, one problem that remains. We must also show that we can apply Lemma 6 repeatedly without increasing the graph's clique-width. If we cannot guarantee this, then, even though we will have eliminated large $K_{t,t}$ subgraphs, we will not be able to invoke Theorem 1 in the end. We therefore have to take care to only apply the reduction rule in some specific situations. For this, we will have to work with the given clique-width expression of G.

Our first step is to handle an obvious part of the given clique-width expression where large bipartite subgraphs are constructed, namely, the join operation.

Lemma 7. *Given a graph G and a clique-width expression with k labels, it is possible to produce in polynomial time a graph G' and a clique-width expression with $k + 2$ labels such that:*

1. *G has an edge-Hamiltonian cycle if and only if G' does*
2. *For every join operation in the expression of G', one of the two involved sets of vertices contains at most 6 vertices.*

Unfortunately, Lemma 7 is not enough to guarantee the elimination of large complete bipartite subgraphs, since these may also be constructed gradually. However, eliminating big joins gives our clique-width expression a certain structure which we can leverage to deal with the remaining bi-cliques efficiently.

Lemma 8. *Given a graph $G(V, E)$ and a clique-width expression with k labels and the property that for all join operations one involved set has size at most 6, we can in polynomial time produce a graph G' with clique-width $k + 2$ such that G' does not contain $K_{21k,21k}$ as a subgraph.*

We can now describe our algorithm. Given a graph G and a clique-width expression with k labels, we first invoke the algorithms of Lemmata 7,8. We are thus left with a graph with clique-width at most $k + 4$ and no complete bipartite subgraph larger than $K_{t,t}$ for $t = O(k)$. By Theorem 1, this graph has treewidth $O(k^2)$. We can now apply an FPT algorithm to obtain a reasonable tree decomposition (see e.g. [2]) and then invoke Theorem 5.

Theorem 6. *Given a graph G and a clique-width expression with k labels, there exists an algorithm that decides if G has an edge-Hamiltonian cycle in time $k^{O(k^2)} n^{O(1)}$.*

References

1. Bertossi, A.A.: The edge Hamiltonian path problem is NP-complete. Inf. Process. Lett. **13**(4), 157–159 (1981)
2. Bodlaender, H.L., Drange, P.G., Dregi, M.S., Fomin, F.V., Lokshtanov, D., Pilipczuk, M.: An $O(c^k n)$ 5-Approximation algorithm for treewidth. In: FOCS, pp. 499–508. IEEE Computer Society (2013)
3. Bodlaender, H.L., Koster, A.M.: Combinatorial optimization on graphs of bounded treewidth. Comput. J. **51**(3), 255–269 (2008)
4. Brualdi, R.A., Shanny, R.F.: Hamiltonian line graphs. J. Graph Theory **5**(3), 307–314 (1981)
5. Catlin, P.A.: Supereulerian graphs: a survey. J. Graph Theory **16**(2), 177–196 (1992)
6. Chartrand, G.: On Hamiltonian line-graphs. Trans. Am. Math. Soc. **134**, 559–566 (1968)
7. Chen, Z.-H., Lai, H.-J., Li, X., Li, D., Mao, J.: Eulerian subgraphs in 3-edge-connected graphs and Hamiltonian line graphs. J. Graph Theory **42**(4), 308–319 (2003)

8. Clark, L.: On Hamiltonian line graphs. J. Graph Theory **8**(2), 303–307 (1984)
9. Cormen, T.H., Leiserson, C.E., Rivest, R.L., Stein, C., et al.: Introduction to Algorithms, vol. 2. MIT press Cambridge, Cambridge (2001)
10. Corneil, D.G., Rotics, U.: On the relationship between clique-width and treewidth. SIAM J. Comput. **34**(4), 825–847 (2005)
11. Courcelle, B.: The monadic second-order logic of graphs. I. recognizable sets of finite graphs. Inf. Comput. **85**(1), 12–75 (1990)
12. Courcelle, B., Engelfriet, J.: Graph Structure and Monadic Second-Order Logic: A Language-Theoretic Approach. Cambridge University Press, New York (2012)
13. Courcelle, B., Makowsky, J.A., Rotics, U.: Linear time solvable optimization problems on graphs of bounded clique-width. Theory Comput. Syst. **33**(2), 125–150 (2000)
14. Demaine, E.D., Demaine, M.L., Harvey, N.J.A., Uehara, R., Uno, T., Uno, Y.: UNO is hard, even for a single player. Theor. Comput. Sci. **521**, 51–61 (2014)
15. Dey, P., Goyal, P., Misra, N.: UNO gets easier for a single player. In: Ferro, A., Luccio, F., Widmayer, P. (eds.) FUN 2014. LNCS, vol. 8496, pp. 147–157. Springer, Heidelberg (2014)
16. Downey, R.G., Fellows, M.R.: Parameterized Complexity. Springer, New York (1999)
17. Espelage, W., Gurski, F., Wanke, E.: How to solve NP-hard graph problems on clique-width bounded graphs in polynomial time. In: Brandstädt, A., Le, V.B. (eds.) WG 2001. LNCS, vol. 2204, pp. 117–128. Springer, Heidelberg (2001)
18. Flum, J., Grohe, M.: Parameterized Complexity Theory, vol. 3. Springer, Berlin (2006)
19. Fomin, F.V., Golovach, P.A., Lokshtanov, D., Saurabh, S.: Clique-width: on the price of generality. In: Mathieu, C. (ed.) SODA, pp. 825–834. SIAM, Philadelphia (2009)
20. Gurski, F., Wanke, E.: The tree-width of clique-width bounded graphs without $K_{n,n}$. In: Brandes, U., Wagner, D. (eds.) WG 2000. LNCS, vol. 1928, pp. 196–205. Springer, Heidelberg (2000)
21. Gurski, F., Wanke, E.: Line graphs of bounded clique-width. Discrete Math. **307**(22), 2734–2754 (2007)
22. Harary, F., Nash-Williams, C.S.J.: On Eulerian and Hamiltonian graphs and line graphs. Can. Math. Bull. **8**, 701–709 (1965)
23. Hliněný, P., Oum, S.-I., Seese, D., Gottlob, G.: Width parameters beyond tree-width and their applications. Comput. J. **51**(3), 326–362 (2008)
24. Lai, H.-J.: Eulerian subgraphs containing given vertices and hamiltonian line graphs. Discrete Math. **178**(1), 93–107 (1998)
25. Lai, T.-H., Wei, S.-S.: The edge Hamiltonian path problem is NP-complete for bipartite graphs. Inf. Process. Lett. **46**(1), 21–26 (1993)
26. Niedermeier, R.: Invitation to Fixed-Parameter Algorithms: Oxford Lecture Series in Mathematics and Its Applications. OUP, Oxford (2006)
27. Oum, S.-I., Seymour, P.: Approximating clique-width and branch-width. J. Comb. Theory Ser. B **96**(4), 514–528 (2006)
28. Razgon, I., Petke, J.: Cliquewidth and knowledge compilation. In: Järvisalo, M., Van Gelder, A. (eds.) SAT 2013. LNCS, vol. 7962, pp. 335–350. Springer, Heidelberg (2013)
29. Ryjáček, Z., Woeginger, G.J., Xiong, L.: Hamiltonian index is NP-complete. Discrete Appl. Math. **159**(4), 246–250 (2011)

Polynomial Time Recognition of Squares of Ptolemaic Graphs and 3-sun-free Split Graphs

Van Bang Le[1](✉), Andrea Oversberg[2], and Oliver Schaudt[2]

[1] Institut Für Informatik, Universität Rostock, Rostock, Germany
le@informatik.uni-rostock.de
[2] Institut Für Informatik, Universität zu Köln, Köln, Germany
{oversberg,schaudt}@zpr.uni-koeln.de

Abstract. The square of a graph G, denoted G^2, is obtained from G by putting an edge between two distinct vertices whenever their distance is two. Then G is called a square root of G^2. Deciding whether a given graph has a square root is known to be NP-complete, even if the root is required to be a chordal graph or even a split graph.

We present a polynomial time algorithm that decides whether a given graph has a ptolemaic square root. If such a root exists, our algorithm computes one with a minimum number of edges.

In the second part of our paper, we give a characterization of the graphs that admit a 3-sun-free split square root. This characterization yields a polynomial time algorithm to decide whether a given graph has such a root, and if so, to compute one.

Keywords: Square of graph · Square of ptolemaic graph · Square of split graph · Recognition algorithm

1 Introduction

The *square* of a graph G is the graph G^2 obtained from G by putting an edge between any two distinct vertices of distance 2. Then G is called the *square root* of G^2. While every graph has a square, not every graph admits a square root. In fact, it is NP-complete to decide whether a given graph has a square root, as was shown by Motwani and Sudan [21]. Since then, squares of graphs and square roots have been intensively studied, in both graph theoretic and algorithmic aspects. See, for example, [1,2,7,10,11,14,16,20] for recent results on this topic.

One successful approach to deal with this hardness is to ask for square roots that belong to a particular graph class. This might be useful if one is interested in structural properties of the root graph, such as chordality, bipartiteness or girth conditions. The negative results in this direction tell us that it is NP-complete to determine whether a graph has a square root that is either chordal [15], split [15], or of girth four [11], and, recently announced in [10], of girth five. On the upside, there are polynomial time algorithms for computing a square root that is either a tree (see for example [6,14,17]), a bipartite graph [14], a proper

© Springer International Publishing Switzerland 2014
D. Kratsch and I. Todinca (Eds.): WG 2014, LNCS 8747, pp. 360–371, 2014.
DOI: 10.1007/978-3-319-12340-0_30

interval graph [15], a block graph [17], a strongly chordal split graph [18], or a graph of girth at least six [11].

Ptolemaic square roots. Note the contrast between the linear time algorithm for finding a block square root and the NP-hardness of finding a chordal square root. It seems enticing to investigate what happens *in between* these two classes. Indeed, two reasonable intermediate graph classes are ptolemaic graphs and strongly chordal graphs. In this paper, we solve the square root problem for ptolemaic graphs by proving the following main result.

Theorem 1. *It can be decided in $\mathcal{O}(n^4)$ time whether a given n-vertex graph has a ptolemaic square root. If such a root exists, a ptolemaic square root with a minimum number of edges can be constructed in the same time.*

A long-standing problem in the research on graph powers is characterizing and recognizing powers of distance-hereditary graphs. This problem stems from a paper by Bandelt, Henkmann and Nicolai [3], who where the first to study powers of distance-hereditary graphs. They where, however, not able to give a full characterization of squares of distance-hereditary graphs and this problem remains unsolved until today.

We see our result as an important step towards the solution of the above mentioned problem. The class of ptolemaic graphs by far the largest subclass of distance-hereditary graphs for which the square root problem is solved. Moreover, previous results on subclasses of distance-hereditary graphs, namely the polynomial time algorithms for the recognition of squares of trees and squares of block graphs [17] can also be subsumed under our result in the following sense. An implicit feature of our algorithm is that if the input graph admits a square root that is a block graph or a tree, such a root is indeed computed. (However, the best known algorithm to compute square roots in these two graph classes runs in linear time and is considerably simpler [17].)

The optimization aspect of our result is motivated by recent work of Cochefert, Couturier, Golovach, Kratsch and Paulusma [7]. They introduce the problem of minimizing or maximizing the number of edges in a square root. Among other results, they give a polynomial time algorithm to compute a square root with a minimum number of edges in the class of graphs of maximum degree 6.

3-sun-free split square roots. As mentioned above, it is NP-complete to decide whether a given graph is the square of some split graph [15].

A polynomially solvable case in computing split square roots reads as follows. A *strongly chordal* graph is a chordal graph that does not contain any ℓ-sun as an induced subgraph; here an ℓ-*sun*, $\ell \geq 3$, consists of a stable set $\{u_1, u_2, \ldots, u_\ell\}$ and a clique $\{v_1, v_2, \ldots, v_\ell\}$ such that for $i \in \{1, \ldots, \ell\}$, u_i is adjacent to exactly v_i and v_{i+1} (index arithmetic modulo ℓ). There is a structural characterization of squares of strongly chordal split graphs, which leads to a quadratic time recognition algorithm [16,18].

Our second result, Theorem 2 below, extends the polynomially solvable case of strongly chordal split square roots to 3-sun-free split square roots. This leaves a larger degree of freedom for the square root, pushing our knowledge on poly-nomially solvable cases further towards the NP-complete case of general split square roots.

Theorem 2. *It can be decided in $\mathcal{O}(n^2m)$ time whether a given n-vertex m-edge graph has a 3-sun-free split square root, and if so, such a square root can be constructed in the same time.*

Our paper is structured as follows. In Sect. 2 we collect relevant notations, defin-itions, and basic facts. We also give some more background on ptolemaic graphs.

We prove our main result, Theorem 1, in Sect. 3. The first of the two ingredi-ents of our algorithm is discussed in Sect. 3.1. We show how the structure of the maximal cliques of the input graph already determines an essential part of any ptolemaic square root. The second ingredient we present in Sect. 3.2. We show that for every maximal clique in the square graph, there is some vertex in any ptolemaic square root whose neighborhood spans this clique. The results of the last section enable us to determine these vertices. The whole algorithm is put together in Sect. 3.3, where we also prove its correctness.

In Sect. 4 we first give a structural characterization of squares of 3-sun-free split graphs in terms of four forbidden induced subgraphs and of the maximal clique structure. Then, Theorem 2 will be derived from this characterization.

We close the paper with a short discussion of our results and propose some questions for further research in Sect. 5.

2   Basic Facts and Definitions

All considered graphs are finite and simple. Let G be a graph and $v \in V(G)$. By $N_G(v)$ we denote the set of neighbors of v in G. The *closed neighborhood* of v in G, that is $N_G(v) \cup \{v\}$, we denote by $N_G[v]$. A *clique*, respectively, an *independent set*, in G is a set of pairwise adjacent, respectively, non-adjacent vertices, in G. For a subset $X \subseteq V(G)$, we denote by $G[X]$ the subgraph induced by X. If two graphs G and H are isomorphic, we may simply write $G \cong H$.

Let $u, v \in V(G)$. The distance of u and v in G we denote by $\text{dist}_G(u, v)$. For any $k \geq 1$, G^k denotes the *k-th power* of G. That is the graph on $V(G)$ where any two distinct vertices are adjacent if and only if their distance in G is at most k. G^2 is called the *square* of G and G is called the *square root* of G^2.

Let u, v be two non-adjacent vertices of G. A subset $S \subseteq V(G)$ is a (u, v)-*separator* if u and v belong to different connected components of $G - S$. A *sepa-rator* is a (u, v)-separator for some non-adjacent vertices $u, v \in V(G)$. We speak of a *minimal separator* if it is not properly contained in another (u, v)-separator. A *minimal clique separator* is a minimal separator that is a clique.

A graph G is called *distance-hereditary* if for all vertices $u, v \in V(G)$ any induced path between u and v is a shortest path. Distance-hereditary graphs were introduced by Bandelt and Mulder [4]. It is well-known that this class can be recognized in linear time [12].

Powers of distance-hereditary graphs have been studied by Bandelt, Henkmann and Nicolai [3]. An important subclass of distance-hereditary graphs are the so-called ptolemaic graphs. A connected graph G is called *ptolemaic* if for every four vertices u, v, w, x the *ptolemaic inequality* holds:

$$\text{dist}_G(u, v)\text{dist}_G(w, x) \leq \text{dist}_G(u, w)\text{dist}_G(v, x) + \text{dist}_G(u, x)\text{dist}_G(v, w).$$

We need the following characterization of ptolemaic graphs. For any graph H we say that G is H-*free* if G does not contain an induced subgraph that is isomorphic to H. For a positive integer ℓ, let P_ℓ denote the path on ℓ vertices and $\ell - 1$ edges, and C_ℓ the cycle on ℓ vertices and ℓ edges. A *gem* is the graph displayed in Fig. 1. A graph is *chordal* if it is C_ℓ-free for all $\ell \geq 4$.

Theorem 3 (Howorka [13]). *For every graph G, the following statements are equivalent.*

(i) *G is ptolemaic;*
(ii) *G is gem-free chordal;*
(iii) *G is C_4-free distance-hereditary;*
(iv) *for all vertices $u, v \in V(G)$ of distance two, $N_G(u) \cap N_G(v)$ is a minimal clique (u, v)-separator.*

It follows that a ptolemaic graph is always gem-free chordal. In our proofs we make extensive use of this particular fact without explicitly refering to the above theorem.

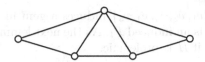

Fig. 1. The gem

A *split graph* is a graph whose vertex set can be partitioned into a clique and an independent set. It is well known that split graphs are exactly the chordal graphs without induced $2K_2$ (the complement of the 4-cycle C_4). For more information on graph classes, their definitions and properties we refer to the book by Brandstädt, Le and Spinrad [5].

3 Ptolemaic Square Roots

We make use of the following property of squares of ptolemaic graphs later.

Theorem 4 ([3,8,19,23]). *Squares of ptolemaic graphs are chordal.*

It is known that ptolemaic graphs are strongly chordal, and squares of strongly chordal graphs are strongly chordal as well (see [8,19,23]).

3.1 Forced Edges in a Ptolemaic Square Root

In this section we show how the structure of the maximal cliques of the square of a ptolemaic graph determines an essential part of any ptolemaic square root. For this, we need the following concept.

Let us say that a *pseudo-P_5* in a graph H is an ordered 5-tuple of distinct vertices $(v_1, v_2, v_3, v_4, v_5)$ such that

(i) $v_2 v_3, v_3 v_4 \in E(H)$,
(ii) $\text{dist}_H(v_1, v_2), \text{dist}_H(v_4, v_5) \leq 2$,
(iii) $\text{dist}_H(v_1, v_3) = \text{dist}_H(v_2, v_4) = \text{dist}_H(v_3, v_5) = 2$,
(iv) and $\text{dist}_H(v_1, v_4), \text{dist}_H(v_1, v_5), \text{dist}_H(v_2, v_5) \geq 3$.

In particular, an induced P_5 is a pseudo-P_5. Figure 2 shows another possible way of how a pseudo-P_5 may appear.

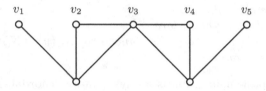

Fig. 2. $(v_1, v_2, v_3, v_4, v_5)$ form a pseudo-P_5.

Note that the set $\{v_1, v_2, v_3, v_4, v_5\}$ induces a gem in H^2. In this gem, the sequence (v_1, v_2, v_4, v_5) is an induced P_4. As the next lemma shows, the converse of this statement holds if H is ptolemaic.

Lemma 1. *Let H be a ptolemaic graph and let $G = H^2$. If a vertex subset $\{v_1, v_2, v_3, v_4, v_5\}$ induces a gem in G where (v_1, v_2, v_4, v_5) is the induced P_4 of this gem, then $(v_1, v_2, v_3, v_4, v_5)$ is a pseudo-P_5 in H.*

A more general version of Lemma 1 reads as follows. Let us say that a *gem-triple* is an ordered triple (A, B, C) of distinct maximal cliques such that

(a) $A \cap C \neq \emptyset$,
(b) $A \cap C \subseteq B$,
(c) $A \cap B \nsubseteq C$, and
(d) $B \cap C \nsubseteq A$.

See Fig. 3 for an illustration.

Lemma 2. *Let H be a ptolemaic graph, let $G = H^2$, and let (A, B, C) be a gem-triple in G. Then for all $u \in A \cap C$ and $v \in (A \cup C) \cap B$ with $u \neq v$, $uv \in E(H)$.*

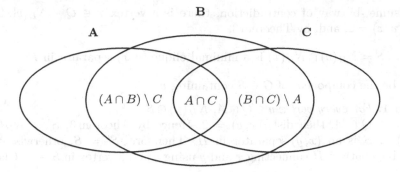

Fig. 3. A gem-triple (A, B, C).

Proof. Let $v_2 \in (A \cap B) \setminus C$, $v_3 \in A \cap C$, $v_4 \in (B \cap C) \setminus A$. Due to the maximality of A and C, there are vertices $v_1 \in A \setminus (B \cup C)$ and $v_5 \in C \setminus (A \cup B)$ with $v_1 v_4, v_2 v_5 \notin E(G)$. Otherwise, $A \cup \{v_4\}$ resp. $C \cup \{v_2\}$ would be a clique, a contradiction.

By Theorem 4, G is chordal. Thus, v_1 and v_5 must be non-adjacent in G, as otherwise $G[\{v_1, v_2, v_4, v_5\}] \cong C_4$. Hence, $G[\{v_1, v_2, v_3, v_4, v_5\}] \cong$ gem. Lemma 1 implies that $(v_1, v_2, v_3, v_4, v_5)$ is a pseudo-P_5 in H. As v_2, v_3, v_4 were arbitrary, every edge between $(A \cap B) \setminus C$, resp. $(B \cap C) \setminus A$, and $A \cap C$ is present in H.

Moreover, if for some $v_3' \neq v_3$ it holds that both (v_2, v_3, v_4) and (v_2, v_3', v_4) are induced P_3 in H, the chordality of H implies $v_3 v_3' \in E(H)$. This means that $A \cap C$ is a clique in H, completing the proof.   □

3.2   Centers of Maximal Cliques

The last section shows that several edges of any ptolemaic square root of a graph are forced. However, a non-trivial degree of freedom remains for the choice of the other edges of the square root. This issue is dealt with in the present section.

Let H be any ptolemaic graph and let $G = H^2$. Let C be a maximal clique of G and $x \in V(H)$ with $N_H[x] = C$. We call x a *center* of C (with respect to H). As Lemma 3 below shows, a center exists for every maximal clique G. Note that for every vertex $v \in V(G)$ being in a maximal clique C of G is equivalent to being identical or adjacent to a center of C in H. In the remainder of this paper, we make use of this fact without explicitly repeating it.

Next we prove a sequence of lemmas in order to prepare our algorithm.

Lemma 3. *Let H be a ptolemaic graph and let $G = H^2$. Every maximal clique of G has a center in H.*

Proof. Let Q be a maximal clique in G, and let $v \in Q$ such that $N_H[v] \cap Q$ is inclusion-maximal. We are going to show that $N_H[v] = Q$. Note that $N_H[v]$ is a clique in G. So, by the maximality of the clique Q, it suffices to show that $Q \subseteq N_H[v]$.

Assume, by way of contradiction, there is a vertex $x \in Q - N_H[v]$. Then $\text{dist}_H(v, x) = 2$, and, by Theorem 3,

$$S = N_H(v) \cap N_H(x) \text{ is a minimal clique } (v, x)\text{-separator in } H.$$

Let A be the component of $G - S$ containing v.

Claim 1. *For every vertex* $y \in Q \cap A$, $N_H(y) \cap S = S$.

If $y \in Q \cap A$, then $\text{dist}_H(x, y) = 2$. Hence by Theorem 3, $S' = N_H(x) \cap N_H(y) \subseteq S$ is an (x, y)-separator in H. Therefore, $S' = S$, otherwise there would be a path in H connecting x and y using v and a vertex in $S - S'$. Claim 1 follows.

Claim 2. *For every vertex* $s \in S$ *and every vertex* $q \in Q$, $\text{dist}_H(s, q) \leq 2$.

If $q \in S$, the claim follows from the fact that S is a clique in H. If $q \in A$, the claim follows from Claim 1. Let $q \in B$, where B is another component of $G - S$. Since $\text{dist}_H(q, v) = 2$, q must have some neighbor in S. Since S is a clique, $\text{dist}_H(s, q) \leq 2$. Claim 2 follows.

Consider now a vertex $s \in S$. The maximality of the clique Q in G and Claim 2 imply that s must belong to Q. On the other hand, by Claim 1, $N_H[v] \cap Q \subseteq N_H[s] \cap Q$, and this inclusion is proper because $sx \in E(H)$ but $vx \notin E(H)$. This contradicts the choice of v. Hence $N_H[v] = Q$ as claimed.   □

Our next lemma enables a key step of our algorithm. It allows to determine the centers of the maximal cliques of G, up to being adjacent twins in G. Here, two vertices are *twins* if they have the same neighbors.

Lemma 4. *Let H be a ptolemaic graph and let $G = H^2$. Let A, C be two maximal cliques of G with $A \cap C \neq \emptyset$, and let v be a center of A. Then $v \in A \setminus C$ if and only if there is a maximal clique B such that (A, B, C) is a gem-triple.*

3.3   The Algorithm

We now state our algorithm and then discuss its logic. Let G be the input graph.

1. Check whether G is chordal. If not, return that G does not have a ptolemaic square root.
2. Compute the maximal cliques of G. Let the set of maximal cliques be denoted $\mathcal{C}$.
3. Initialize the empty graph H on the vertex set $V(G)$.
4. Determine the gem-triples among $\mathcal{C}$.
5. For every gem-triple (A, B, C) in G, add the edge uv to $E(H)$, for all $u \in A \cap C$ and $v \in (A \cup C) \cap B$ with $u \neq v$.
6. Peform the following steps for every $A \in \mathcal{C}$.
   (i) Compute the set $\mathcal{C}_A$ of maximal cliques $C \in \mathcal{C}$ with $A \cap C \neq \emptyset$.
   (ii) Compute the set $\mathcal{C}'_A$ of maximal cliques $C \in \mathcal{C}$ for which there is a $B \in \mathcal{C}$ such that (A, B, C) is a gem-triple.

(iii) Compute the set $\mathcal{C}_A'' = \mathcal{C}_A \setminus \mathcal{C}_A'$.

(iv) Compute the vertex set $X_A = \bigcap \mathcal{C}_A'' \setminus \bigcup \mathcal{C}_A'$.

7. Assign a vertex $x_C \in X_C$ to every $C \in \mathcal{C}$ in an injective way. If this is not possible, return that G does not have a ptolemaic square root.

8. For every $C \in \mathcal{C}$ and $v \in C$: if $v \neq x_C$ and $vx_C \notin E(H)$, add the edge vx_C to $E(H)$.

9. Check whether the graph H is a ptolemaic square root of G. If yes, return H. If not, return that G does not have a ptolemaic square root.

The full discussion of the complexity and the correctness of the above algorithm we omit due to space limitations. In Step 5 the forced egdes are included into the potential square root H according to Lemma 2. Potential centers for H are determined in Steps 6 and 7 according to Lemma 4. Step 8 implements the neighborhoods of these centers.

Lemma 5. *The algorithm can be implemented such that it terminates in $\mathcal{O}(n^4)$ time when applied to an n-vertex graph.*

Thus, the above algorithm terminates in polynomial time.

Lemma 6. *If the input graph G has a ptolemaic square root, the algorithm puts out a ptolemaic square root of G that has a minimum number of edges.*

Finally, Theorem 1 is a direct consequence of the Lemmas 5 and 6.

4   Squares of 3-sun-free Split Graphs

In this section we prove Theorem 2. Recall that deciding if a graph is the square of a strongly chordal split graph can be done in polynomial time [16,18]. This result is based on the following characterization of squares of strongly chordal split graphs; the set of all maximal cliques in a graph G is denoted by $\mathcal{C}(G)$.

Theorem 5 ([16,18])**.** *G is square of a strongly chordal split graph if and only if G is strongly chordal and $\left| \bigcap_{Q \in \mathcal{C}(G)} Q \right| \geq |\mathcal{C}(G)|$.*

We now are going to extend Theorem 5 to 3-sun-free split square roots. Our approach is based on the following fact about maximal cliques in squares of 3-sun-free split graphs. A vertex with inclusion-maximal closed neighborhood is called a *maximal* vertex. For split graphs $H = (V(H), E(H))$ we write $H = (C \cup I, E(H))$, meaning $V(H) = C \cup I$ is a partition of the vertex set of H into a clique C and an independent set I.

Lemma 7 ([16,18])**.** *Let $H = (C \cup I, E(H))$ be a connected split graph without induced 3-sun. Then Q is a maximal clique in H^2 if and only if $Q = N_H[v]$ for some maximal vertex $v \in C$ of H.*

Squares of 3-sun-free split graphs can be characterized as follows (see Fig. 4 for the graphs G_1–G_4).

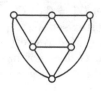 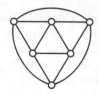

Fig. 4. G_1, G_2, G_3, and G_4.

Theorem 6. *G is the square of a connected 3-sun-free split graph if and only if G is (G_1, G_2, G_3, G_4)-free and satisfies $\left| \bigcap_{Q \in \mathcal{C}(G)} Q \right| \geq |\mathcal{C}(G)|$.*

Proof. A *universal vertex* of G is one that is adjacent to every other vertex of G. Note that, for any connected split-graph $H = (C \cup I, E(H))$, any vertex in C is a universal vertex in H^2.

Assume that $G = H^2$ for some connected 3-sun-free split graph $H = (C \cup I, E(H))$. First, by Lemma 7,

$$|\mathcal{C}(G)| \leq |C|.$$

Furthermore, as C is contained in all maximal cliques in G,

$$|C| \leq \left| \bigcap_{Q \in \mathcal{C}(G)} Q \right|.$$

Therefore,

$$|\mathcal{C}(G)| \leq \left| \bigcap_{Q \in \mathcal{C}(G)} Q \right|.$$

Next, let by way of contradiction, a, b, c, a', b', c' be six vertices such that

$$ab, ac, bc, a'b, a'c, b'a, b'c, c'a, c'b \in E(G), aa', bb', cc' \notin E(G),$$

that is, $G[a, b, c, a', b', c']$ is a G_i for some $i = 1, 2, 3, 4$.

Let Q_1, Q_2, Q_3 be the maximal cliques of G containing $\{a, b, c'\}$, $\{b, c, a'\}$, $\{a, c, b'\}$, respectively. By Lemma 7, $Q_i = N_H[v_i]$ for some (maximal) vertex $v_i \in C$, $i = 1, 2, 3$. In particular,

$$a, b \in N_H[v_1], \ b, c \in N_H[v_2], \ a, c \in N_H[v_3], \ a \notin N_H[v_2], \ b \notin N_H[v_3], \ c \notin N_H[v_1].$$

By noting that $a, b, c, a', b', c' \in I$ (as none of these vertices is universal in G), we conclude that a, b, c, v_1, v_2, v_3 induce a 3-sun in H, a contradiction.

Now, let G be (G_1, G_2, G_3, G_4)-free and satisfy $\left| \bigcap_{Q \in \mathcal{C}(G)} Q \right| \geq |\mathcal{C}(G)|$.

Write $C = \bigcap_{Q \in \mathcal{C}(G)} Q$ and let $\mathcal{C}(G) = \{Q_1, \ldots, Q_q\}$. As $|C| \geq q$, we are able to choose q distinct vertices $c_1, \ldots, c_q$ in C. Let H be the split graph with clique C, independent set $I = V(G) \setminus C$, and edges vc_i for all $v \in I$ and $1 \leq i \leq q$ with $v \in Q_i$.

We claim that $G = H^2$. Indeed, let $xy \in E(G)$. Then there $xy \in Q_i$ for some i. If $x \in C$ or $y \in C$, then clearly $xy \in E(H^2)$. If $x, y \in Q_i \setminus C$, then $xc_i, yc_i \in E(H)$, hence $xy \in E(H^2)$.

Let $xy \in E(H^2)$. If $x \in C$ or $y \in C$, then $xy \in E(G)$ because C is contained in all maximal cliques of G. So, let $x, y \in I$. Hence there is a vertex $c_i \in C$ with $xc_i, yc_i \in E(H)$. By construction of H, $x, y \in Q_i$, showing $xy \in E(G)$.

We have shown that $G = H^2$, as claimed. It remains to prove that H is 3-sun-free. Assume the contrary, and let $v_1 = c_i, v_2 = c_j, v_3 = c_k, u_1, u_2, u_3$ induce a 3-sun in H. Then, by construction of H,

$$u_1 \in (Q_i \cap Q_j) \setminus Q_k, \ u_2 \in (Q_j \cap Q_k) \setminus Q_i, \ u_3 \in (Q_i \cap Q_k) \setminus Q_j.$$

Now, by the maximality of the cliques, u_1 is non-adjacent to some $x \in Q_k \setminus (Q_i \cup Q_j)$, u_2 is non-adjacent to some $y \in Q_i \setminus (Q_j \cup Q_k)$, and u_3 is non-adjacent to some $z \in Q_j \setminus (Q_i \cup Q_k)$. But then $G[u_1, u_2, u_3, x, y, z]$ is one of the G_1, G_2, G_3, G_4, a contradiction. This completes the proof. $\qquad\square$

We now are going to give an interesting reformulation of Theorem 6. A graph G is said to be *clique-Helly* if $\mathcal{C}(G)$ has the Helly property. G is *hereditary clique-Helly* if every induced subgraph of G is clique-Helly. (See [9] for more information on clique-Helly graphs.) Prisner [22] characterized hereditary clique-Helly graphs as follows.

Theorem 7 (Prisner [22]). *G is hereditary clique-Helly if and only if G is (G_1, G_2, G_3, G_4)-free.*

It follows that a split graph is hereditary clique-Helly if and only if it is 3-sun-free. With Theorem 7, Theorem 6 can be reformulated as follows.

Theorem 8. *A graph G is the square of a connected hereditary clique-Helly split graph if and only if G is a hereditary clique-Helly graph satisfying $\left| \bigcap_{Q \in \mathcal{C}(G)} Q \right| \geq |\mathcal{C}(G)|$.*

We can now give the proof of Theorem 2.

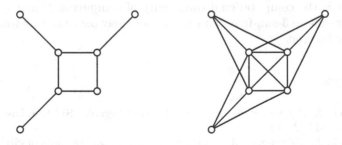

Fig. 5. A distance-hereditary graph (left) and its square (right).

Proof (of Theorem 2). By Lemma 7, G has at most n maximal cliques. By [24], all maximal cliques in G then can be listed in time $\mathcal{O}(n \cdot m \cdot n) = \mathcal{O}(n^2 m)$, and the condition $\left| \bigcap_{Q \in \mathcal{C}(G)} Q \right| \geq |\mathcal{C}(G)|$ can be verified within the same time. Also, testing if G is hereditary clique-Helly can be done in time $\mathcal{O}(n^2 m)$ (see, for instance, [9]). Thus, by Theorem 8, we can decide in time $\mathcal{O}(n^2 m)$ if G is the square of some 3-sun-free split graph, and if so, the proof of Theorem 6 gives a construction for such a square root H within the same time.   □

5   Discussion

In this paper we have presented a polynomial time algorithm to decide whether a given graph has a ptolemaic square root. If such a root exists, our algorithm computes a ptolemaic square root with a minimum number of edges. Let us mention, without a proof, another feature of our algorithm: if the input graph admits a square root that is a block graph or an acyclic graph, such a root is computed. However, the best known algorithm to compute square roots in these two graph classes runs in linear time and is considerably simpler [17].

Several questions arise now that we can compute ptolemaic square roots. It is immediate to ask whether ptolemaic k-th roots can be efficiently computed. We did not tackle this question yet, since a more basic question is apparently unanswered: whether one can compute k-th roots that are block graphs [17].

A question that seems more urgent to us is whether distance-hereditary square roots can be computed efficiently. Distance-hereditary roots have been considered before in the literature [3], yet not from an algorithmic perspective.

Although distance-hereditary graphs share a number of properties with ptolemaic graphs, the two classes behave differently when it comes to graph powers. To give an example, Fig. 5 displays the square of a distance-hereditary graph that does not admit a ptolemaic square root. Indeed, both of our main tools, Lemmas 1 and 3, fail to hold for distance-hereditary graphs (see again Fig. 5).

We also have characterized squares of 3-sun-free split graphs. Our characterization yields a polynomial time recognition algorithm for such squares. Given the hardness of computing split square roots [15] and our polynomial time result (Theorem 2), it is interesting to ask the following question: let F be a fixed split graph, what is the computational complexity of computing F-free split square roots? Our result on 3-sun-free split square roots may pave the way towards such a dichotomy theorem.

References

1. Adamszek, A., Adamszek, M.: Large-girth roots of graphs. SIAM J. Discrete Math. **24**, 1501–1514 (2010)
2. Adamszek, A., Adamszek, M.: Uniqueness of graph square roots of girth six. Electron. J. Combin. **18**, 139 (2011)
3. Bandelt, H.-J., Henkmann, A., Nicolai, F.: Powers of distance-hereditary graphs. Disc. Math. **145**, 37–60 (1995)

4. Bandelt, H.-J., Mulder, H.M.: Distance-hereditary graphs. J. Combin. Theory (Ser. B) **41**, 182–208 (1986)
5. Brandstädt, A., Le, V.B., Spinrad, J.P.: Graph Classes: A Survey. SIAM, Philadelphia (1999)
6. Chang, M.-S., Ko, M.-T., Lu, H.-I.: Linear-time algorithms for tree root problems. In: Arge, L., Freivalds, R. (eds.) SWAT 2006. LNCS, vol. 4059, pp. 411–422. Springer, Heidelberg (2006)
7. Cochefert, M., Couturier, J.-F., Golovach, P.A., Kratsch, D., Paulusma, D.: Sparse square roots. In: Brandstädt, A., Jansen, K., Reischuk, R. (eds.) WG 2013. LNCS, vol. 8165, pp. 177–188. Springer, Heidelberg (2013)
8. Dalhaus, E., Duchet, P.: On strongly chordal graphs. Ars Combin. **24B**, 23–30 (1987)
9. Dourado, M.C., Protti, F., Szwarcfiter, J.L.: Complexity aspects of the helly property: graphs and hypergraphs. Electron. J. Combin. **17**, 1–53 (2009)
10. Farzad, B., Karimi, M.: Square-root finding problem in graphs, a complete dichotomy theorem, arXiv:1210.7684 (2012)
11. Farzad, B., Lau, L.C., Le, V.B., Tuy, N.N.: Complexity of finding graph roots with girth conditions. Algorithmica **62**, 38–53 (2012)
12. Hammer, P.L., Maffray, F.: Completely separable graphs. Disc. App. Math. **27**, 85–99 (1990)
13. Howorka, E.: A characterization of ptolemaic graphs. J. Graph Theory **5**, 323–331 (1981)
14. Lau, L.C.: Bipartite roots of graphs. ACM Trans. Algorithms **2**, 178–208 (2006)
15. Lau, L.C., Corneil, D.G.: Recognizing powers of proper interval, split, and chordal graphs. SIAM J. Discrete Math. **18**, 83–102 (2004)
16. Le, V.B., Nguyen, N.T.: Hardness results and efficient algorithms for graph powers. In: Paul, C., Habib, M. (eds.) WG 2009. LNCS, vol. 5911, pp. 238–249. Springer, Heidelberg (2010)
17. Le, V.B., Tuy, N.N.: The square of a block graph. Disc. Math. **310**, 734–741 (2010)
18. Le, V.B., Tuy, N.N.: A good characterization of squares of strongly chordal split graphs. Inf. Process. Lett. **310**, 120–123 (2011)
19. Lubiw, A.: Doubly lexical orderings of matrices. SIAM J. Comput. **16**, 854–879 (1987)
20. Milanič, M., Schaudt, O.: Computing square roots of trivially perfect and threshold graphs. Disc. App. Math. **161**, 1538–1545 (2013)
21. Motwani, R., Sudan, M.: Computing roots of graphs is hard. Disc. App. Math. **54**, 81–88 (1994)
22. Prisner, E.: Hereditary clique-Helly graphs. J. Comb. Math. Comb. Comput. **14**, 216–220 (1993)
23. Raychaudhuri, A.: On powers of strongly chordal and circular arc graphs. Ars Combin. **34**, 147–160 (1992)
24. Tsukiyama, S., Ide, M., Ariyoshi, H., Shirakawa, I.: A new algorithm for generating all the maximal independent sets. SIAM J. Comput. **6**, 505–517 (1977)

The Maximum Time of 2-Neighbour Bootstrap Percolation: Complexity Results

Thiago Marcilon$^{(\boxtimes)}$, Samuel Nascimento, and Rudini Sampaio

Dept. Computação, Universidade Federal Do Ceará, Fortaleza, Brazil
{thiagomarcilon,sammueln,rudini}@lia.ufc.br

Abstract. In 2-neighbourhood bootstrap percolation on a graph G, an infection spreads according to the following deterministic rule: infected vertices of G remain infected forever and in consecutive rounds healthy vertices with at least 2 already infected neighbours become infected. Percolation occurs if eventually every vertex is infected. The maximum time $t(G)$ is the maximum number of rounds needed to eventually infect the entire vertex set. In 2013, it was proved [7] that deciding if $t(G) \geq k$ is polynomial time solvable for $k = 2$, but is NP-Complete for $k = 4$ and is NP-Complete if the graph is bipartite and $k = 7$. In this paper, we solve the open questions. Let $n = |V(G)|$ and $m = |E(G)|$. We obtain an $\Theta(mn^5)$-time algorithm to decide if $t(G) \geq 3$ in general graphs. In bipartite graphs, we obtain an $\Theta(mn^3)$-time algorithm to decide if $t(G) \geq 3$ and an $O(mn^{13})$-time algorithm to decide if $t(G) \geq 4$. We also prove that deciding if $t(G) \geq 5$ is NP-Complete in bipartite graphs.

Keywords: 2-Neighbour bootstrap percolation · P_3-convexity · Maximum time · Infection on graphs

1 Introduction

We consider a problem in which an infection spreads over the vertices of a connected simple graph G following a deterministic spreading rule in such a way that an infected vertex will remain infected forever. Given a set $S \subseteq V(G)$ of initially infected vertices, we build a sequence $S_0, S_1, S_2, \ldots$ in which $S_0 = S$ and S_{i+1} is obtained from S_i using such spreading rule.

Under r-neighbour bootstrap percolation on a graph G, the spreading rule is a threshold rule in which S_{i+1} is obtained from S_i by adding to it the vertices of G which have at least r neighbours in S_i. We say that a set S_0 percolates G (or that S_0 is a percolating set of G) if eventually every vertex of G becomes infected, that is, there exists a t such that $S_t = V(G)$. In that case, we define $t_r(S)$ as the minimum t such that $S_t = V(G)$. And define, the *percolation time of G* as $t_r(G) = \max\{t_r(S) : S$ percolates $G\}$. In this paper, we shall focus on the case where $r = 2$ and in such case we omit the subscript of the functions $t_r(S)$ and $t_r(G)$.

© Springer International Publishing Switzerland 2014
D. Kratsch and I. Todinca (Eds.): WG 2014, LNCS 8747, pp. 372–383, 2014.
DOI: 10.1007/978-3-319-12340-0_31

Bootstrap percolation was introduced by Chalupa, Leath and Reich [13] as a model for certain interacting particle systems in physics. Since then it has found applications in clustering phenomena, sandpiles [19], and many other areas of statistical physics, as well as in neural networks [1] and computer science [15].

There are two broad classes of questions one can ask about bootstrap percolation. The first, and the most extensively studied, is what happens when the initial configuration S_0 is chosen randomly under some probability distribution? For example, vertices are included in S_0 independently with some fixed probability p. One would like to know how likely percolation is to occur, and if it does occur, how long it takes.

The answer to the first of these questions is now well understood for various graphs. An interesting case is the one of the lattice graph $[n]^d$, in which d is fixed and n tends to infinity, since the probability of percolation under the r-neighbour model displays a sharp threshold between no percolation with high probability and percolation with high probability. The existence of thresholds in the strong sense just described first appeared in papers by Holroyd, Balogh, Bollobás, Duminil-Copin and Morris [3,5,20]. Sharp thresholds have also been proved for the hypercube (Balogh and Bollobás [2], and Balogh, Bollobás and Morris [6]). There are also very recent results due to Bollobás, Holmgren, Smith and Uzzell [10], about the time percolation take on the discrete torus $\mathbb{T}_n^d = (\mathbb{Z}/n\mathbb{Z})^d$ for a randomly chosen set S_0.

The second broad class of questions is the one of extremal questions. For example, what is the smallest or largest size of a percolating set with a given property? The size of the smallest percolating set in the d-dimensional grid, $[n]^d$, was studied by Pete and a summary can be found in [4]. Morris [22] and Riedl [24], studied the maximum size of minimal percolating sets on the square grid and the hypercube $\{0,1\}^d$, respectively, answering a question posed by Bollobás. However, it was proved in [12,14] that finding the smallest percolating set is NP-complete for general graphs. Another type of question is: what is the minimum or maximum time that percolation can take, given that S_0 satisfies certain properties? Recently, Przykucki [23] determined the precise value of the maximum percolation time on the hypercube $2^{[n]}$ as a function of n, and Benevides and Przykucki [8,9] have similar results for the square grid, $[n]^2$, also answering a question posed by Bollobás. In particular, they have a polynomial time algorithm to compute the maximal percolation time on square grids.

Here, we consider the decision version of the maximum time percolation problem, as stated below.

PERCOLATION TIME
Input: A graph G and an integer k.
Question: Is $t(G) \geq k$?

In 2013, Benevides et al. [7] proved that deciding if $t(G) \geq k$ is polynomial time solvable for $k = 2$, but is NP-Complete for $k = 4$ and is NP-Complete if the graph is bipartite and $k = 7$. In this paper, we solve the open questions. Let $n = |V(G)|$ and $m = |E(G)|$. We obtain a $\Theta(mn^5)$-time algorithm to decide if $t(G) \geq 3$ in general graphs. In bipartite graphs, we obtain $O(mn^{13})$-time algorithm to decide if $t(G) \geq 4$ and prove that deciding if $t(G) \geq 5$ is NP-Complete.

1.1   Related Works and Some Notation

It is interesting to notice that infection problems appear in the literature under many different names and were studied by researches of various fields. The particular case in which $r = 2$ in r-neighbourhood bootstrap percolation is also a particular case of a infection problem related to convexities in graph.

A finite *convexity space* [21,25] is a pair $(V, \mathcal{C})$ consisting of a finite ground set V and a set $\mathcal{C}$ of subsets of V satisfying $\emptyset, V \in \mathcal{C}$ and if $C_1, C_2 \in \mathcal{C}$, then $C_1 \cap C_2 \in \mathcal{C}$. The members of $\mathcal{C}$ are called $\mathcal{C}$-*convex sets* and the *convex hull* of a set S is the minimum convex set $H(S) \in \mathcal{C}$ containing S.

A convexity space $(V, \mathcal{C})$ is an *interval convexity* [11] if there is a so-called *interval function* $I : \binom{V}{2} \rightarrow 2^V$ such that a subset C of V belongs to $\mathcal{C}$ if and only if $I(\{x, y\}) \subseteq C$ for every two distinct elements x and y of C. With no risk of confusion, for any $S \subseteq V$, we also denote by $I(S)$ the union of S with $\bigcup_{x,y \in S} I(\{x, y\})$. In interval convexities, the convex hull of a set S can be computed by exhaustively applying the corresponding interval function until obtaining a convex set.

The most studied graph convexities defined by interval functions are those in which $I(\{x, y\})$ is the union of paths between x and y with some particular property. Some common examples are the P_3-convexity [17], geodetic convexity [18] and monophonic convexity [16]. We observe that the spreading rule in 2-neighbours bootstrap percolation is equivalent to $S_{i+1} = I(S_i)$ where I is the interval function which defines the P_3-convexity: $I(S)$ contains S and every vertex belonging to some path of 3 vertices whose extreme vertices are in S. For these reasons, sometimes we call a percolating set by *hull set*.

2 $t(G) \geq 5$ Is NP-Complete in Bipartite Graphs

In [7], it was proved that deciding if $t(G) \geq 7$ is NP-Complete in bipartite graphs. The following theorem improves this result.

Theorem 1. *Deciding if $t(G) \geq k$ is NP-Complete in bipartite graphs for any $k \geq 5$.*

Proof (Sketch of the proof). Given m clauses $\mathcal{C} = \{C_1, \ldots, C_m\}$ on variables $X = \{x_1, \ldots, x_n\}$ as an instance of 3-SAT, we denote the three literals of C_i by $\ell_{i,1}$, $\ell_{i,2}$ and $\ell_{i,3}$. We construct a graph G as follows. For each clause C_i of $\mathcal{C}$, add to G a gadget as the one of Fig. 1. Then, for each pair of literals $\ell_{i,a}, \ell_{j,b}$ such that one is the negation of the other, add a vertex $y_{(i,a),(j,b)}$ adjacent to $w_{i,a}$ and $w_{j,b}$. Let Y be the set of all vertices created this way. Finally, add a vertex z adjacent to all vertices in Y and a pendant vertex z' adjacent to only z.

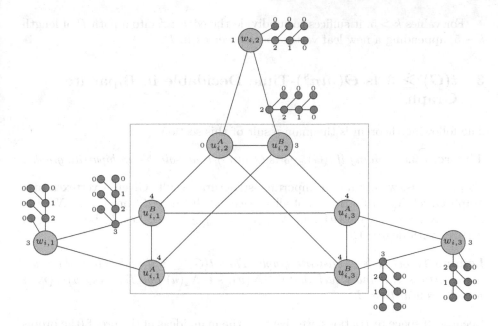

Fig. 1. Bipartite gadget for each clause C_i.

Denote the sets $\{u_{i,1}^A, u_{i,2}^A, u_{i,3}^A, u_{i,1}^B, u_{i,2}^B, u_{i,3}^B\}$ and $\{w_{i,1}, w_{i,2}, w_{i,3}\}$ by U_i and W_i, respectively. Let $U = \cup_{1 \leq i \leq m} U_i$, $W = \cup_{1 \leq i \leq m} W_i$ and L be the set of vertices of degree one in G.

We first consider the case $k = 5$. We show that $\mathcal{C}$ is satisfiable if and only if G contains a hull set with percolation time at least 5.

Suppose that $\mathcal{C}$ has a truth assignment. For each clause C_i, let k_i denote an integer in $\{1, 2, 3\}$ such that ℓ_{i,k_i} is true. Let $S' = \{u_{i,k_i}^A : 1 \leq i \leq m\}$ and $S = S' \cup L$. It is easy to see (from Fig. 1) that all vertices in the clause gadgets are infected in time at most 4. It is also easy to see that $\{w_{i,k_i} : 1 \leq i \leq m\} \subset I^1(S)$ (that is, S infects w_{i,k_i} in time 1 for every $1 \leq i \leq m$). Moreover, $\{w_{i,k_i'} : k_i' \neq k_i, 1 \leq i \leq m\} \subset I^3(S)$, but $\{w_{i,k_i'} : k_i' \neq k_i, 1 \leq i \leq m\} \cap (I^1(S) \cup I^2(S)) = \emptyset$ (that is, S infects all the other $w_{i,k_i'}$ in time exactly 3). Since we used a truth assignment, we have that all vertices of Y are infected in time exactly 4 and consequently the vertex z is infected in time 5. Therefore, G has percolation time at least 5.

Now, suppose that $t(G) \geq 5$ and let S be any hull set of G with $t(S) \geq 5$. Note that $L \subseteq S$; also for any clause C_i, we have $U_i \cap S \neq \emptyset$ because $|N(u_{i,j}^A) - U_i| \leq 1$ and $|N(u_{i,j}^B) - U_i| \leq 1$, for all i, j. This implies that $W \subseteq I^3(S)$, $U \cup Y \subseteq I^4(S)$ and $z \in I^5(S)$. Furthermore, if $Y \cap I^3(S) \neq \emptyset$ then $z \in I^4$ and $t(S) \leq 4$, a contradiction. Then $Y \cap I^3(S) = \emptyset$, which means that no pair $\{u_{i,a}^C, u_{j,b}^D\}$, $C, D \in \{A, B\}$, where $\ell_{i,a}$ is the negation of $\ell_{j,b}$, is in S. This means that assigning true to each $\ell_{i,j}$ for which $u_{i,j}^C \in S$, $C \in \{A, B\}$, gives us an assignment that satisfies $\mathcal{C}$.

For values $k > 5$, it suffices to subdivide the edge zz' into a path P of length $k - 5$, appending a new leaf vertex to each vertex in P. ∎

3 $t(G) \geq 3$ Is $\Theta(mn^3)$-Time Decidable in Bipartite Graphs

The following theorem is the main result of this section.

Theorem 2. *Deciding if $t(G) \geq 3$ is $\Theta(mn^3)$-time solvable in bipartite graphs.*

To prove this, we obtain an important structural result. Given a vertex u of a graph G, let $N_d(u)$ be the set of all vertices at distance d from u. Let $N(u) = N_1(u)$, $N[u] = N(u) \cup \{u\}$ and $N_{\geq d'}(u) = \cup_{d \geq d'} N_d(u)$. Let T_0 be the set of vertices with degree 1.

Lemma 1. *Let G be a bipartite graph. Then $t(G) \geq 3$ if and only if there are three vertices u, v and s such that $v \in N(u)$, $s \in N_2(u)$ and $T_0 \cup N_{\geq 3}(u) \cup \{v, s\}$ percolates u at time 3.*

Because of space restrictions, we give only the main ideas of the proof (the proofs are in the appendix).

Proof (Sketch of the proof). Firstly, suppose that $t(G) \geq 3$. Then there exists a hull set S' and a vertex u such that S' percolates u at time 3. It is not difficult to see that $T_0 \subseteq S'$ and that $S = S' \cup N_{\geq 3}(u)$ is also a hull set which percolates u at time 3. If S contains a vertex in $N(u)$, let v be such a vertex. Otherwise, let v be a neighbour of u with smaller percolating time with respect to the hull set S. Since the graph is bipartite, the distance from v to any other vertex of $N(u)$ is at least two. Then it is not difficult to see that all vertices in $N(u)$ percolated at time ≥ 2 by S are also percolated at time ≥ 2 by $S \cup \{v\}$. By analysing two possibilities about the vertices in $N(u) - \{v\}$, we can conclude (using the fact that the graph is bipartite) that there exists a vertex $s \in N_2(u) \setminus N(v)$ such that $S \cup \{v, s\}$ also percolates u at time 3. Moreover we can prove that $(S \setminus (N(u) \cup N_2(u))) \cup \{v, s\}$ percolates u at time 3 and we are done.

Secondly, suppose that there are three vertices u, v and s such that $v \in N(u)$, $s \in N_2(u)$ and $S_0 = T_0 \cup N_{\geq 3}(u) \cup \{v, s\}$ percolates u at time 3. We then show how to construct a hull set S such that $t(S) \geq 3$. We begin with $S = S_0$. Each step adds one vertex to S and, at the end of each step, it is guaranteed that S_i percolates u at time ≥ 2 and percolates at least one vertex in $\{u\} \cup N(u)$ at time ≥ 3. Let S_i be the constructed set at the end of step i. If S_i is not a hull set, we can prove that there are two adjacent vertices $q \in N_2(u)$ and $w \in N(u)$ which are not percolated by S_i. Let $S_{i+1} = S_i \cup \{q\}$. It is not difficult to see that S_{i+1} also percolates u at time ≥ 2. We then prove that S_{i+1} percolates w at time ≥ 3. If S_{i+1} is a hull set, we are done. Otherwise, repeat the construction until obtaining a hull set. ∎

The idea of the algorithm is as follows. Considering that the graph is connected, the algorithm selects in each step a vertex u and obtains the sets $N(u)$, $N_2(u)$, $N_{\geq 3}(u)$ and T_0 in time $O(m)$. After, the algorithm selects a vertex v in $N(u)$ and a vertex s in $N_2(u)$ and, then, computes the percolation process of $T_0 \cup N_{\geq 3}(u) \cup \{v, s\}$ in time $O(m)$ for, at most, three steps. If, for some triple (u, v, s), u is percolated in time 3, return that $t(G) \geq 3$. Otherwise, return that $t(G) < 3$.

4 $t(G) \geq 3$ Is $\Theta(mn^5)$-Time Decidable in General Graphs

The following theorem is the main result of this section.

Theorem 3. *Deciding if $t(G) \geq 3$ is $\Theta(mn^5)$-time solvable in general graphs.*

To prove this, we obtain an important structural result. Let u and v be vertices of G. Let k be such that $v \in N_k(u)$. The following definitions are technical, but represent a simple fact: if v is a separator (that is, its removal disconnects the graph) and some connected component of $G - v$ contains only vertices of $N_{k+1}(u)$, then any hull set must contain at least one vertex of this component.

Let $\mathcal{T}_0^u$ be the family of subsets of $V(G)$ such that $T_0 \in \mathcal{T}_0^u$ if and only if, for every separator v and every connected component $H_{v,i}$ of $G - v$ such that $u \notin V(H_{v,i})$ and $V(H_{v,i} \subseteq N(v))$, T_0 contains exactly one vertex of $H_{v,i}$, and every vertex of T_0 satisfies this property.

Lemma 2. *Let G be a simple graph. Then $t(G) \geq 3$ if and only if there is a vertex u, a subset $T_0 \in \mathcal{T}_0^u$ and a subset F with $|F| \leq 4$ such that $T_0 \cup N_{\geq 3}(u) \cup F$ percolates u at time 3.*

Moreover, we prove that any set of the family $\mathcal{T}_0^u$ can be chosen. That is, if $T_0 \cup N_{\geq 3}(u) \cup F$ percolates u at time 3 for some $T_0 \in \mathcal{T}_0^u$, then $T_0' \cup N_{\geq 3}(u) \cup F$ also percolates u at time 3 for any $T_0' \in \mathcal{T}_0^u$.

Because of space restrictions, we give only the main ideas of the proof (the proofs are in the appendix).

Proof (Sketch of the proof). Firstly, suppose that $t(G) \geq 3$. Then there exists a hull set S' and a vertex u such that S' percolates u at time 3. Since S' is a hull set, we can prove that there is a subset $T_0 \subseteq S'$ such that $T_0 \in \mathcal{T}_0^u$. It is not difficult to see that $S = S' \cup N_{\geq 3}(u)$ is also a hull set which percolates u at time 3. Let $F' = S \setminus (T_0 \cup N_{\geq 3}(u)) = (S \cap N_{\leq 2}(u)) \setminus T_0$. If $|F'| \leq 4$, then let $F = F'$ and we are done. Otherwise, we can prove with some effort that there exists a subset $F \subseteq N_{\leq 2}(u)$ with $|F| \leq 4$ such that $(S \setminus F') \cup F$ percolates u at time 3, and we are done.

Now suppose that there is a vertex u, a subset $T_0 \in \mathcal{T}_0^u$ and a subset F with $|F| \leq 4$ such that $S_0 = T_0 \cup N_{\geq 3}(u) \cup F$ percolates u at time 3. We then show how to construct a hull set S such that $t(S) \geq 3$. We begin with $S = S_0$. Each step adds one vertex to S and, at the end of each step, it is guaranteed that S_i percolates some vertex u_i at time ≥ 3 ($u_0 = u$) and percolates u at time ≥ 2.

Let S_i be the constructed set at the end of step i. If S_i is a hull set, we are done. So, assume that S_i is not a hull set. Let Y_i be the set of vertices not percolated by S_i.

At first, assume that there exists a vertex $y_i \in Y_i \cap N_2(u_i)$ with no neighbour percolated by S_i at time ≥ 2. Let $S'_{i+1} = S_i \cup \{y_i\}$. Clearly, u_i has at most one neighbour percolated by S_i at time ≤ 1 and, by the choice of y_i, u_i is not adjacent to y_i. It is not difficult to prove that every neighbour of u_i percolated by S_i at time ≥ 2 is also percolated by S'_{i+1} at time ≥ 2. Finally, if some neighbour z of u_i is not percolated by S_i, but is percolated by S'_{i+1}, it is not difficult to prove that its percolating time is ≥ 2, since, otherwise, z should have a neighbour in S_i, a contradiction because z would have two neighbours percolated by S_i. Then S'_{i+1} also percolates u_i at time ≥ 3 (and we let $u_{i+1} = u_i$) and it is not difficult to see that S'_{i+1} also percolates u at time ≥ 2. Let $S_{i+1} = S'_{i+1} \cup N_{\geq 3}(u_{i+1})$. Since the set S'_{i+1} percolates u_{i+1} at time ≥ 3, it is easy to see that the set S_{i+1} also percolates u_{i+1} at time ≥ 3.

Secondly, assume that every vertex $y_i \in Y_i \cap N_2(u_i)$ has exactly one neighbour percolated by S_i and its percolating time is ≥ 2. Let $y_i \in Y_i \cap N_2(u_i)$, let C_i be the connected component of $G[Y_i]$ which contains y_i and let z_i be the neighbour of y_i with percolating time ≥ 2. If every vertex of C_i is adjacent to z_i, then C_i has only vertices in $N(u)$ or only vertices in $N_2(u)$ (otherwise, there would be one vertex in $N_2(u)$ adjacent to u, a contradiction), and every vertex of C_i has no neighbour in $N_3(u)$ (and consequently z_i is a separator). Therefore, T_0 has a vertex ℓ in C_i, a contradiction since there are no vertices percolated by S_i in C_i.

We then conclude that there exist a vertex y'_i in C_i whose neighbour z'_i with percolating time ≥ 2 is distinct from z_i (that is, $z'_i \neq z_i$). Let $S_{i+1} = S_i \cup \{y_i\} \cup N_{\geq 3}(y'_i)$. It is not difficult to see that all vertices in C_i are percolated by S_{i+1}. We can prove that S_{i+1} percolates y'_i at time ≥ 3 and, letting $u_i = y'_i$, we are done.

After some time steps, say t time steps, S_t percolates all vertices of $N_2(u_i)$, since we are only including vertices from $N_2(u_i)$. It is not difficult to see that

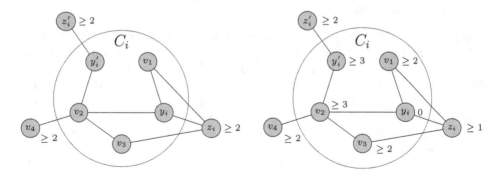

Fig. 2. Vertices of the component C_i before and after the addition of y_i to S.

this fact implies that S_t is a hull set and, since S_t percolates u_t at time ≥ 3, we have that $t(G) \geq 3$.   ■

The idea of the algorithm is as follows. Considering that the graph is connected, the algorithm selects in each step a vertex u and obtains a set $T_0 \in \mathcal{T}_0^u$ in time $O(m)$ (applying breadth-first search, for example). After, the algorithm selects a subset F with at most 4 vertices and computes the percolation process of $T_0 \cup N_{\geq 3}(u) \cup F$ in time $O(m)$ for, at most, three steps. If, for some pair (u, F), u is percolated in time 3, return that $t(G) \geq 3$. Otherwise, return that $t(G) < 3$ (Fig. 2).

5 $t(G) \geq 4$ Is $\Theta(mn^{13})$-Time Decidable in Bipartite Graphs

The following theorem is the main result of this section.

Theorem 4. *Deciding if $t(G) \geq 4$ is $\Theta(mn^{13})$-time solvable in bipartite graphs.*

To prove this, we obtain an important structural result. Let $\mathcal{T}_0$ be the family of subsets of $V(G)$ such that $T_0 \in \mathcal{T}_0$ if and only if T_0 contains all vertices with degree one and, for every pair of adjacent vertices u and v, both with degree two, T_0 has either u or v. It is easy to see that every hull set must contain a set $T_0 \in \mathcal{T}_0$, since each edge uv with that property induces a co-convex set (that is, $V(G) - \{u, v\}$ is convex). Clearly, the size of $\mathcal{T}_0$ can be exponential in the number of vertices. However, we can prove that we need to check only a polynomial number of subsets $T_0 \in \mathcal{T}_0$.

Let V^u be the subset of vertices $v \in N[u]$ such that there is an induced P_3 vxy, where x and y have degree two. Given a vertex u and a vertex $v \in V^u$, we also define the sets C_v^u and D_v^u: for every induced P_3 vxy, where x and y have degree two, $x \in C_v^u$ and $y \in D_v^u$. It is worth noting that, for every $x \in C_v^u$ and y in $D_v^u \cap N(x)$, T_0 must contain either x or y.

Let $\mathcal{T}_0^u$ be a family such that $T_0 \in \mathcal{T}_0^u$ if and only if $N_{\geq 4}(u) \subseteq T_0$, T_0 contains all vertices that have degree 0, and exactly one of the cases below occurs:

- there is a vertex $v \in V^u$ such that T_0 has at least one vertex in D_v^u, at most one vertex in C_v^u and, for every $v' \in V^u$, $v' \neq v$, and every $x \in C_{v'}^u$ and $y \in D_{v'}^u \cap N(x)$ where $x, y \notin C_v^u \cup D_v^u$, T_0 contains $\{x, y\} \cap N_k(u)$, where $k = 2$, if $v \in N(u)$, and $k = 3$, if $v = u$.
- for each vertex $v \in V^u$, T_0 contains all vertices in C_v^u, except at most one vertex $v' \in V^u$, in which case T_0 contains at most one vertex in $C_{v'}^u$.

It is worth noting that the set $\mathcal{T}_0^u$, for any vertex u, is a subset of the set $\{T_0 \cup N_{\geq 4}(u) : \forall T_0 \in \mathcal{T}_0\}$. It is also important to observe that the set $\mathcal{T}_0^u$ can be obtained in $O(n^2)$ time.

Lemma 3. *Let G be a bipartite graph. Then $t(G) \geq 4$ if and only if there is a vertex u, a subset $T_0 \in \mathcal{T}_0^u$ and a subset F with $|F| \leq 10$ such that $T_0 \cup F$ percolates some vertex at time 4.*

Proof (Sketch of the proof). Firstly, suppose that $t(G) \geq 4$. Then there is a hull set S'' and a vertex u such that S'' percolates u at time 4. It is easy to see that the set $S = S'' \cup N_{\geq 4}(u)$ is a also a hull set that percolates u at time 4. With this, there is a set $T \in \mathcal{T}_0$ such that $T_0 = T \cup N_{\geq 4}(u) \subseteq S$.

Assume that there is a vertex $v \in V^u$ percolated at time ≥ 3 by S and a vertex $x \in C_v^u \setminus S$. It is not difficult to see that x is percolated at time ≥ 4 by S. Let $k = 2$, if $v \in N(u)$, or $k = 3$, if $v = u$. Let S' be the union of S with all sets $\{y_1, y_2\} \cap N_k(u)$ such that $y_1 \in C_{v'}^u$ and $y_2 \in D_{v'}^u \cap N(y_1)$ for some $v' \in V^u$, $v' \neq v$, and $y_1, y_2 \notin C_v^u \cup D_v^u$. Since the graph is bipartite, each vertex added to S is either at distance 4 from x or, if it is at distance 2 from x, they share only one common neighbor, which is the only vertex z in the set $\{N(x) \cap D_V^u\}$ (in this case, we have that $z \in S$). Then S' percolates x at time ≥ 4. Therefore, we have that there is a set $T_0' \in \mathcal{T}_0^u$ such that $T_0 \subseteq S'$. Since S' percolates x at time ≥ 4, it percolates some vertex at time 4. Thus, it is possible to prove that there is a set F, with $|F| \leq 10$, such that, $F \cup T_0'$ percolates some vertex at time 4.

Now assume that all vertices in V^u percolated at time ≥ 3 by S are such that all vertices in C_v^u are in S. Since u is percolated at time 4 by S, then either (a) there is a vertex $v \in V^u$ percolated at time ≥ 3 by S and some vertex $x \in C_v^u$ such that $S' = (S - \{x\}) \cup (N(x) \cap D_v^u)$ percolates v at time ≥ 3, or, since there is at most one vertex in V^u that is percolated at time ≤ 2, (b) there is at most one vertex $v \in V^u$ such that there is a vertex $x \in C_v^u \setminus S$. If (a), then x is percolated at time ≥ 4, and we are in the same case of the previous paragraph. If (b), then, if v is percolated at time ≤ 1 by S, then it is easy to see that $S' = S \cup C_v^u$ percolates u at time 4 and, if v is percolated at time 2 by S, then $S' = S$ has at most one vertex in C_v^u. If we have that v is percolated at time ≤ 1 or 2 by S, then there is a set $T_0' \in \mathcal{T}_0^u$ such that $T_0' \subseteq S'$. Since S' percolates some vertex x' at time 4, it is possible to prove that there is a set F, with $|F| \leq 10$, such that, $F \cup T_0'$ percolates x' at time 4.

Now, suppose that there is a vertex u, a set F, with $|F| \leq 10$, and a set $T_0 \in \mathcal{T}_0^u$ such that the set $F \cup T_0$ percolates some vertex x at time 4. Then, we have that the set $S_0 = F \cup T_0 \cup N_{\geq 4}(x)$ percolates x at time 4.

We then show how to construct a hull set S such that $t(S) \geq 4$. We begin with $S = S_0$, and, at each step, we add one vertex to S and, at the end of each step, it is guaranteed that S_i percolates some vertex at time 4. Let S_i be the constructed set at the end of step i. If S_i is a hull set, we are done. So, assume that S_i is not a hull set. Let Y_i be the set of vertices not percolated by S_i.

Suppose that there exists a vertex $y_i \in Y_i \cap N_2(x)$ with no neighbour percolated by S_i at time ≥ 2. Let $S_{i+1} = S_i \cup \{y_i\}$. Clearly, x has at most one neighbour percolated by S_i at time ≤ 1 and, by the choice of y_i, u_i is not adjacent to y_i. It is possible to prove, basing ourselves heavily on the fact that the graph is bipartite, that every neighbour of x that either is percolated by S_i at time ≥ 3 or it is not percolated by S_i, if it is percolated by S_{i+1}, it is percolated by S_{i+1} at time ≥ 3.

When all vertices in the set $Y_i \cap N_2(x)$ have a neighbour percolated by S_i at time ≥ 2, suppose that there exists a vertex $y_i \in Y_i \cap N_3(x)$ with no neighbour percolated by S_i at time ≥ 2. Let $S_{i+1} = S_i \cup \{y_i\}$. Since all vertices in $Y_i \cap N_2(x)$ have a neighbour percolated by S_i at time ≥ 2, it is not difficult to prove that, if a vertex in $N_2(x) \cap Y_i$ is percolated by S_{i+1}, it is percolated by S_{i+1} at time ≥ 2. Thus, all vertices in $N(x)$ that either are percolated at time ≥ 3 by S_i or are not percolated by S_{i+1} are percolated by S_{i+1} at time ≥ 3. Therefore, x is percolated by S_{i+1} at time 4. It is worth noting the fact that y_i is at distance at least two of every vertex that is percolated at time ≥ 2 by S_i and is adjacent to some vertex in $N_2(x) \cap Y_i$, which implies that it is not possible to go back to the previous state, i.e., it is not possible that there is a vertex in $Y_i \cap N_2(x)$ with no neighbour percolated by S_{i+1} at time ≥ 2.

When all vertices in the set $Y_i \cap N_2(x)$ and in the set $Y_i \cap N_3(x)$ have a neighbour percolated by S_i at time ≥ 2, let C_i be any connected component of $G[Y_i]$. We have that every vertex of C_i has exactly one neighbour outside C_i, which is percolated at time ≥ 2 by S_i. We have that C_i has at least 3 vertices because, otherwise, one vertex of C_i would also be in T_0 and, consequently, in X_i. Thus, since the graph is bipartite, there are two vertices y_i and y_i' that are at distance 2 of each other. It is possible to prove that the set $S_i \cup \{y_i\}$ percolates y_i' at time ≥ 4 because it percolates all vertices adjacent to y_i at time ≥ 3. Also, in every connected component of $G[Y_i]$, there is at least one vertex in $N_2(x)$ and one vertex in either $N(x)$ or $N_3(x)$. If y_i' is in $N_2(x)$ (resp. $N(x)$ or $N_3(x)$), let $S_{i+1} = S_i \cup \{y_i\} \cup ((Y_i - V(C_i)) \cap N_2(x))$ (resp. $S_{i+1} = S_i \cup \{y_i\} \cup ((Y_i - V(C_i)) \cap (N(x) \cup N_3(x)))$). It is possible to prove that S_{i+1} percolates all the remaining connected component of $G[Y_i]$ and, thus, it is be a hull set. Also, it is possible to prove that S_{i+1} percolates y_i' at time ≥ 4. Therefore, S_{i+1} is a hull set that percolates y_i' at time ≥ 4.   ∎

The idea of the algorithm is as follows. Considering that the graph is bipartite and connected, the algorithm selects in each step a vertex u, a set $T_0 \in \mathcal{T}_0^u$ and a subset F with at most 10 vertices, and computes the percolation process of $T_0 \cup F$ for at most 4 steps (recall that $T_0 \supseteq N_{\geq 4}(u)$). If, for some triple (u, T_0, F), some vertex x is percolated in time 4, return that $t(G) \geq 4$. Otherwise, return that $t(G) < 4$. Since there are $O(n^2)$ sets in $\mathcal{T}_0^u$ that can also be computed in $O(n^2)$-time, then the algorithm decides if $t(G) \geq 4$ in $O(mn^{13})$ time (Fig. 3).

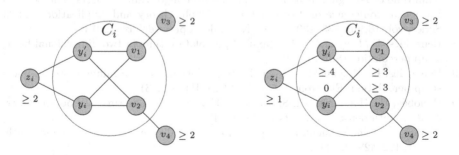

Fig. 3. Vertices of the component C_i before and after the addition of y_i to S.

6 Conclusion

In this paper, we showed the NP-Completeness of the maximum time percolation problem for a fixed $k = 5$ for bipartite graphs, and showed polynomial computable characterizations of general and bipartite graphs for a fixed $k = 3$ and of bipartite graphs for a fixed $k = 4$. Using these results, since the NP-Completeness was proved in [7] for a fixed $k \geq 4$ for general graphs, we were able to solve the remaining open questions regarding the maximum time percolation problem for a fixed k and showed the threshold for polynomiality in general graphs ($k = 3$) and in bipartite graphs ($k = 4$).

We conclude with some interesting directions for future investigation. Can the maximum time percolation problem in induced subgraphs and subgraphs of d-dimensional grids be solved in polynomial time? Can the complexity of the algorithms, which is directly related to the size of the sets that initially percolate some vertex at time k, be improved? Is there a relation between the P_3-Caratheódory number [26] and the size of the sets that initially percolate some vertex at time k?

References

1. Amini, H.: Bootstrap percolation in living neural networks. J. Stat. Phys. **141**(3), 459–475 (2010)
2. Balogh, J., Bollobás, B.: Bootstrap percolation on the hypercube. Probab. Theor. Relat. Fields **134**(4), 624–648 (2006)
3. Balogh, J., Bollobás, B., Duminil-Copin, H., Morris, R.: The sharp threshold for bootstrap percolation in all dimensions. Trans. Amer. Math. Soc. **364**(5), 2667–2701 (2012)
4. Balogh, J., Pete, G.: Random disease on the square grid. Random Struct. Algorithms **13**, 409–422 (1998)
5. Balogh, J., Bollobás, B., Morris, R.: Bootstrap percolation in three dimensions. Ann. Probab. **37**(4), 1329–1380 (2009)
6. Balogh, J., Bollobás, B., Morris, R.: Bootstrap percolation in high dimensions. Combin. Probab. Comput. **19**(5–6), 643–692 (2010)
7. Benevides, F., Campos, V., Dourado, M.C., Sampaio, R.M., Silva, A.: The maximum time of 2-neighbour bootstrap percolation: algorithmic aspects. The Seventh European Conference on Combinatorics, Graph Theory and Applications. CRM Series, vol. 16, pp. 135–139. Scuola Normale Superiore, Pisa (2013)
8. Benevides, F., Przykucki, M.: Maximal percolation time in two-dimensional bootstrap percolation (Submitted)
9. Benevides, F., Przykucki, M.: On slowly percolating sets of minimal size in bootstrap percolation. Electron. J. Comb. **20**(2), P46 (2013)
10. Bollobás, B., Holmgren, C., Smith, P.J., Uzzell, A.J.: The time of bootstrap percolation with dense initial sets (Submitted)
11. Calder, J.: Some elementary properties of interval convexities. J. London Math. Soc. **3**, 422–428 (1971)
12. Centeno, C., Dourado, M.C., Penso, L., Rautenbach, D., Szwarcfiter, J.L.: Irreversible conversion of graphs. Theor. Comput. Sci. **412**, 3693–3700 (2011)

13. Chalupa, J., Leath, P.L., Reich, G.R.: Bootstrap percolation on a bethe lattice. J. Phys. C **12**(1), 31–35 (1979)
14. Chen, N.: On the approximability of influence in social networks. SIAM J. Discrete Math. **23**(3), 1400–1415 (2009)
15. Dreyer, P.A., Roberts, F.S.: Irreversible k-threshold processes: graph-theoretical threshold models of the spread of disease and of opinion. Discrete Appl. Math. **157**(7), 1615–1627 (2009)
16. Duchet, P.: Convex sets in graphs, II. minimal path convexity. J. Comb. Theor. B **44**, 307–316 (1988)
17. Erdős, P., Fried, E., Hajnal, A., Milner, E.C.: Some remarks on simple tournaments. Algebra Univers. **2**, 238–245 (1972)
18. Farber, M., Jamison, R.E.: Convexity in graphs and hypergraphs. SIAM J. Algebraic Discrete Methods **7**, 433–444 (1986)
19. Fey, A., Levine, L., Peres, Y.: Growth rates and explosions in sandpiles. J. Stat. Phys. **138**, 143–159 (2010)
20. Holroyd, A.E.: Sharp metastability threshold for two-dimensional bootstrap percolation. Probab. Theor. Relat. Fields **125**(2), 195–224 (2003)
21. Levi, F.W.: On Helly's theorem and the axioms of convexity. J. Indian Math. Soc. **15**, 65–76 (1951)
22. Morris, R.: Minimal percolating sets in bootstrap percolation. Electron. J. Comb. **16**(1), 20 (2009)
23. Przykucki, M.: Maximal percolation time in hypercubes under 2-bootstrap percolation. Electron. J. Comb. **19**(2), 41 (2012)
24. Riedl, E.: Largest minimal percolating sets in hypercubes under 2-bootstrap percolation. Electron. J. Comb. **17**(1), 13 (2010)
25. Van de Vel, M.L.J.: Theory of Convex Structures. North-Holland, Amsterdam (1993)
26. Barbosa, R.M., Coelho, E.M.M., Dourado, M.C., Rautenbach, D., Szwarcfiter, J.L.: On the carathodory number for the convexity of paths of order three. SIAM J. Discrete Math. **26**, 929–939 (2012)

Parameterized Algorithms for Graph Partitioning Problems

Hadas Shachnai and Meirav Zehavi[✉]

Department of Computer Science, Technion, 32000 Haifa, Israel
{hadas,meizeh}@cs.technion.ac.il

Abstract. We study a broad class of graph partitioning problems, where each problem is specified by a graph $G = (V, E)$, and parameters k and p. We seek a subset $U \subseteq V$ of size k, such that $\alpha_1 m_1 + \alpha_2 m_2$ is at most (or at least) p, where $\alpha_1, \alpha_2 \in \mathbb{R}$ are constants defining the problem, and m_1, m_2 are the cardinalities of the edge sets having both endpoints, and exactly one endpoint, in U, respectively. This class of *fixed-cardinality graph partitioning problems (FGPPs)* encompasses MAX $(k, n - k)$-CUT, MIN k-VERTEX COVER, k-DENSEST SUBGRAPH, and k-SPARSEST SUBGRAPH.

Our main result is an $O^*(4^{k+o(k)} \Delta^k)$ algorithm for any problem in this class, where $\Delta \geq 1$ is the maximum degree in the input graph. This resolves an open question posed by Bonnet et al. [IPEC 2013]. We obtain faster algorithms for certain subclasses of FGPPs, parameterized by p, or by $(k + p)$. In particular, we give an $O^*(4^{p+o(p)})$ time algorithm for MAX $(k, n - k)$-CUT, thus improving significantly the best known $O^*(p^p)$ time algorithm.

1 Introduction

Graph partitioning problems arise in many areas including VLSI design, data mining, parallel computing, and sparse matrix factorization (see, e.g., [1,7,12]). We study the broad class of *fixed-cardinality graph partitioning problems (FGPPs)*, where each problem is specified by a graph $G = (V, E)$, and parameters k and p. We seek a subset $U \subseteq V$ of size k, such that $\alpha_1 m_1 + \alpha_2 m_2$ is at most (or at least) p, where $\alpha_1, \alpha_2 \in \mathbb{R}$ are constants defining the problem, and m_1, m_2 are the cardinalities of the edge sets having both endpoints, and exactly one endpoint, in U, respectively. This class encompasses such fundamental problems as MAX and MIN $(k, n - k)$-CUT, MAX and MIN k-VERTEX COVER, k-DENSEST SUBGRAPH, and k-SPARSEST SUBGRAPH. For example, MAX $(k, n - k)$-CUT is a max-FGPP (i.e., maximization FGPP) satisfying $\alpha_1 = 0$ and $\alpha_2 = 1$, MIN k-VERTEX COVER is a min-FGPP (i.e., minimization FGPP) satisfying $\alpha_1 = \alpha_2 = 1$, k-DENSEST SUBGRAPH is a max-FGPP satisfying $\alpha_1 = 1$ and $\alpha_2 = 0$, and k-SPARSEST SUBGRAPH is a min-FGPP satisfying $\alpha_1 = 1$ and $\alpha_2 = 0$.

A parameterized algorithm with parameter k has running time $O^*(f(k))$ for some function f, where O^* hides factors polynomial in the input size. In this

© Springer International Publishing Switzerland 2014
D. Kratsch and I. Todinca (Eds.): WG 2014, LNCS 8747, pp. 384–395, 2014.
DOI: 10.1007/978-3-319-12340-0_32

paper, we develop a parameterized algorithm with parameter $(k + \Delta)$ for the class of all FGPPs, where $\Delta \geq 1$ is the maximum degree in the graph G. For certain subclasses of FGPPs, we develop algorithms parameterized by p, or by $(k + p)$.

Related Work: Parameterized by k, MAX and MIN $(k, n - k)$-CUT, and MAX and MIN k-VERTEX COVER are W[1]-hard [4,8,11]. Moreover, k-CLIQUE and k-INDEPENDENT SET, two well-known W[1]-hard problems [9], are special cases of k-DENSEST SUBGRAPH (where $p = k(k - 1)/2$), and k-SPARSEST SUBGRAPH (where $p = 0$), respectively. Therefore, parameterized by $(k + p)$, k-DENSEST SUBGRAPH and k-SPARSEST SUBGRAPH are W[1]-hard. Cai et al. [5] and Bonnet et al. [2] studied the parameterized complexity of FGPPs with respect to $(k + \Delta)$. The paper [5] gives $O^*(2^{(k+1)\Delta})$ time algorithms for k-DENSEST SUBGRAPH and k-SPARSEST SUBGRAPH. This result was recently improved in [2] to $O^*(\Delta^k)$ for *degrading* FGPPs. This subclass contains max-FGPPs in which $\alpha_1/2 \leq \alpha_2$, and min-FGPPs in which $\alpha_1/2 \geq \alpha_2$.[1] The authors of [2] also proposed an $O^*(k^{2k}\Delta^{2k})$ time algorithm for all FGPPs, and posed as an open question the existence of constants a and b such that any FGPP can be solved in time $O^*(a^k \Delta^{bk})$. In this paper we answer this question affirmatively, by developing an $O^*(4^{k+o(k)} \Delta^k)$ time algorithm for any FGPP.

Parameterized by p, MAX k-VERTEX COVER can be solved in time $O^*(1.396^p)$, and in randomized time $O^*(1.2993^p)$ [14]. Kneis et al. [14] also show (implicitly) that MIN k-VERTEX COVER can be solved in time $O^*(4^p)$, and in randomized time $O^*(3^p)$. Moreover, by solving any degrading FGPP in time $O^*(\Delta^k)$, Bonnet et al. [2] prove that MAX $(k, n - k)$-CUT can be solved in time $O^*(p^p)$. Recently, Cygan et al. [6] showed that MIN $(k, n - k)$-CUT is also fixed-parameter tractable with respect to p. Parameterized by $(k + p)$, MIN $(k, n - k)$-CUT can be solved in time $O^*(k^{2k}(k + p)^{2k})$ [2].

We note that the parameterized complexity of FGPPs has also been studied with respect to other parameters, such as the treewidth and the vertex cover number of G (see, e.g., [2,3,13]).

Contribution: Our main result is an $O^*(4^{k+o(k)} \Delta^k)$ time algorithm for the class of all FGPPs, answering affirmatively the question posed by Bonnet et al. [2] (see Sect. 2). In Sect. 3, we develop an $O^*(4^{p+o(p)})$ time algorithm for MAX $(k, n - k)$-CUT, which significantly improves the $O^*(p^p)$ running time obtained in [2]. We also present (in Sect. 4) an $O^*(2^{k+\frac{p}{\alpha_2}+o(k+p)})$ time algorithm for the subclass of *positive* min-FGPPs, in which $\alpha_1 \geq 0$ and $\alpha_2 > 0$. Finally, we develop (in Sect. 4) a faster algorithm for non-degrading positive min-FGPPs (i.e., min-FGPPs satisfying $\alpha_2 \geq \alpha_1/2 > 0$). This yields an $O^*(2^{p+o(p)})$ time algorithm for MIN k-VERTEX COVER, improving the previous *randomized* $O^*(3^p)$ time algorithm. Note that all of our algorithms are deterministic.

Due to space constraints, proofs of the results given in Sect. 4 are omitted. We give the full details in [17].

---

[1] A max-FGPP (min-FGPP) is non-degrading if $\alpha_1/2 \geq \alpha_2$ ($\alpha_1/2 \leq \alpha_2$).

Techniques: We obtain our main result by establishing an interesting reduction from non-degrading FGPPs to the WEIGHTED k-EXACT COVER (k-WEC) problem (see Sect. 2). Building on this reduction, combined with an algorithm for degrading FGPPs of [2], and an algorithm given in [19] for k-WEC, we develop an algorithm for any FGPP. To improve the running time of our algorithm, we use a fast construction of representative families [10, 18].

In designing algorithms for FGPPs, parameterized by p or $(k+p)$, we use as a key tool *randomized separation* [5]. Roughly speaking, randomized separation finds a 'good' partition of the nodes in the input graph G via randomized coloring of the nodes in *red* or *blue*. If a solution exists, then, with some positive probability, there is a red colored node-set X that is a solution, such that *all* of the neighbors of nodes in X that are outside X are colored blue. Our algorithm for MAX $(k, n - k)$-CUT makes non-standard use of randomized separation, in requiring that only *some* of the neighbors outside X of nodes in X are blue. This yields the desired improvement in the running time of the algorithm.

Our algorithm for non-degrading positive FGPPs is based on a somewhat different application of randomized separation, in which we randomly color *edges* rather than nodes. If a solution exists, then with some positive probability, there is a node-set X that is a solution, such that *some* edges between nodes in X are red, and *all* of the edges connecting nodes in X and nodes outside X are blue. In particular, we require that the subgraph induced by X, and the subgraph induced by X from which we delete all blue edges, contain the same connected components. We derandomize our algorithms using universal sets [16].

Notation: Given a graph $G = (V, E)$ and a subset $X \subseteq V$, we denote by $E(X)$ the set of edges in E having both endpoints in X, and by $E(X, V \setminus X)$ the set of edges having exactly one endpoint in X. Also, let $\mathrm{val}(X) = \alpha_1|E(X)| + \alpha_2|E(X, V \setminus X)|$.

2 Solving FGPPs in Time $O^*(4^{k+o(k)} \Delta^k)$

In this section we develop an $O^*(4^{k+o(k)} \Delta^k)$ time algorithm for the class of all FGPPs. We proceed in the following steps. In Sect. 2.1 we show that any non-degrading FGPP can be reduced to the WEIGHTED k-EXACT COVER (k-WEC) problem. Applying this reduction, we then show (in Sect. 2.2) how to decrease the size of instances of k-WEC, by using representative families. Finally, we show (in Sect. 2.3) how to solve any FGPP by using the results in Sects. 2.1 and 2.2, an algorithm given in [19] for k-WEC, and an algorithm of [2] for degrading FGPPs.

2.1 From Non-degrading FGPPs to k-WEC

We show below that any non-degrading max-FGPP can be reduced to the maximization version of k-WEC. Given a universe U, a family $\mathcal{S}$ of nonempty subsets of U, a function $w : \mathcal{S} \to \mathbb{R}$, and parameters $k \in \mathbb{N}$ and $p \in \mathbb{R}$, we seek a subfamily $\mathcal{S}'$ of disjoint sets from $\mathcal{S}$ satisfying $|\bigcup \mathcal{S}'| = k$ whose value, given

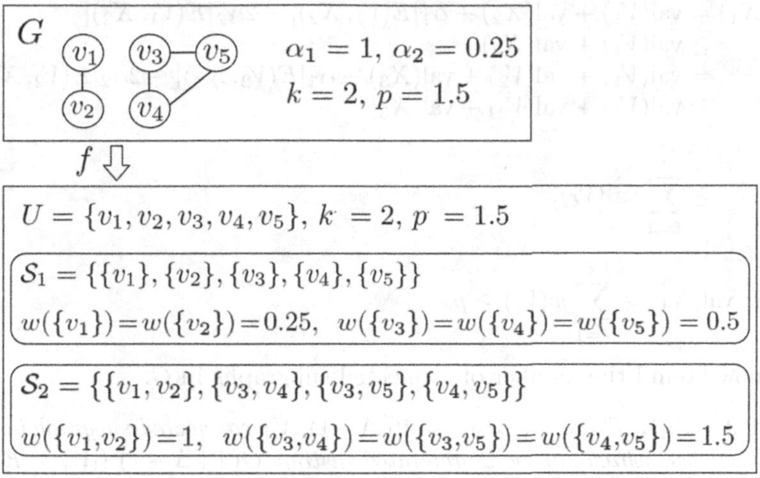

G

(v_1) (v_3)—(v_5) $\alpha_1 = 1,\ \alpha_2 = 0.25$

(v_2) (v_4) $k = 2,\ p = 1.5$

$f \Downarrow$

$U = \{v_1, v_2, v_3, v_4, v_5\},\ k\ = 2,\ p\ = 1.5$

$S_1 = \{\{v_1\}, \{v_2\}, \{v_3\}, \{v_4\}, \{v_5\}\}$

$w(\{v_1\}) = w(\{v_2\}) = 0.25,\ \ w(\{v_3\}) = w(\{v_4\}) = w(\{v_5\}) = 0.5$

$S_2 = \{\{v_1, v_2\}, \{v_3, v_4\}, \{v_3, v_5\}, \{v_4, v_5\}\}$

$w(\{v_1,v_2\}) = 1,\ \ w(\{v_3,v_4\}) = w(\{v_3,v_5\}) = w(\{v_4,v_5\}) = 1.5$

Fig. 1. An illustration of the reduction f, given in Sect. 2.1.

by $\sum_{S \in \mathcal{S}'} w(S)$, is at least p. Any non-degrading min-FGPP can be similarly reduced to the minimization version of k-WEC.

Let Π be a max-FGPP satisfying $\alpha_1/2 \geq \alpha_2$. Given an instance $\mathcal{I} = (G = (V, E), k, p)$ of Π, we define an instance $f(\mathcal{I}) = (U, \mathcal{S}, w, k, p)$ of the maximization version of k-WEC as follows.

- $U = V$
- $\mathcal{S} = \bigcup_{i=1}^{k} \mathcal{S}_i$, where $\mathcal{S}_i$ contains the node-set of any connected subgraph of G on exactly i nodes
- $\forall S \in \mathcal{S} : w(S) = \mathrm{val}(S)$

Note that k and p have the same values in both instances. We illustrate the reduction f in Fig. 1. First, we prove that our reduction is valid.

Lemma 1. $\mathcal{I}$ *is a yes-instance iff* $f(\mathcal{I})$ *is a yes-instance.*

Proof. Assume first that there is a subset $X \subseteq V$ of size k satisfying $\mathrm{val}(X) \geq p$. Let $G_1 = (V_1, E_1), \ldots, G_t = (V_t, E_t)$, for some $1 \leq t \leq k$, be the *maximal* connected components in the subgraph of G induced by X. Then, for all $1 \leq \ell \leq t$, $V_\ell \in \mathcal{S}$. Moreover, $\sum_{\ell=1}^{t} |V_\ell| = |X| = k$, and $\sum_{\ell=1}^{t} w(V_\ell) = \mathrm{val}(X) \geq p$.

Now, assume there is a subfamily of disjoint sets $\{S_1, \ldots, S_t\} \subseteq \mathcal{S}$, for some $1 \leq t \leq k$, such that $\sum_{\ell=1}^{t} |S_\ell| = k$ and $\sum_{\ell=1}^{t} w(S_\ell) \geq p$. Thus, there are connected subgraphs $G_1 = (V_1, E_1), \ldots, G_t = (V_t, E_t)$ of G, such that $V_\ell = S_\ell$, for all $1 \leq \ell \leq t$. Let $X_\ell = \bigcup_{j=\ell}^{t} V_j$, for all $1 \leq \ell \leq t$. Clearly, $|X_1| = k$. Since $\alpha_1/2 \geq \alpha_2$, we get that

$$\begin{aligned}
\mathrm{val}(X_1) &= \mathrm{val}(V_1) + \mathrm{val}(X_2) + \alpha_1|E(V_1, X_2)| - 2\alpha_2|E(V_1, X_2)| \\
&\geq \mathrm{val}(V_1) + \mathrm{val}(X_2) \\
&= \mathrm{val}(V_1) + \mathrm{val}(V_2) + \mathrm{val}(X_3) + \alpha_1|E(V_2, X_3)| - 2\alpha_2|E(V_2, X_3)| \\
&\geq \mathrm{val}(V_1) + \mathrm{val}(V_2) + \mathrm{val}(X_3) \\
&\cdots \\
&\geq \sum_{\ell=1}^{t} \mathrm{val}(V_\ell).
\end{aligned}$$

Thus, $\mathrm{val}(X_1) \geq \sum_{\ell=1}^{t} w(V_\ell) \geq p$. $\square$

We now bound the number of connected subgraphs in G.

Lemma 2 ([15]). *There are at most $4^i(\Delta - 1)^i|V|$ connected subgraphs of G on at most i nodes, which can be enumerated in time $O(4^i(\Delta - 1)^i(|V| + |E|)|V|)$.*

Hence, we have the next result.

Lemma 3. *The instance $f(\mathcal{I})$ can be constructed in time $O(4^k(\Delta - 1)^k(|V| + |E|)|V|)$. Moreover, for any $1 \leq i \leq k$, $|\mathcal{S}_i| \leq 4^i(\Delta - 1)^i|V|$.*

2.2   Decreasing the Size of Inputs for k-WEC

In this section we develop a procedure, called **Decrease**, which compacts the size of an instance $(U, \mathcal{S}, w, k, p)$ of k-WEC. Note that we do not need this procedure to resolve the question posed by Bonnet et al. [2]. Indeed, we use it to improve the running time of our algorithm, from $O^*(11.404^k \Delta^k)$ to the desired $O^*(4^{k+o(k)} \Delta^k)$ steps. To this end, we find a subfamily $\widehat{\mathcal{S}} \subseteq \mathcal{S}$ that contains "enough" sets from $\mathcal{S}$, and thus enables to replace $\mathcal{S}$ by $\widehat{\mathcal{S}}$ without turning a yes-instance into a no-instance. The following definition captures such a subfamily $\widehat{\mathcal{S}}$.

Definition 1. *Given a universe E, nonnegative integers k and r, a family $\mathcal{S}$ of subsets of size r of E, and a function $w : \mathcal{S} \to \mathbb{R}$, we say that a subfamily $\widehat{\mathcal{S}} \subseteq \mathcal{S}$ max (min) represents $\mathcal{S}$ if for any pair of sets $X \in \mathcal{S}$, and $Y \subseteq E \setminus X$ such that $|Y| \leq k - r$, there is a set $\widehat{X} \in \widehat{\mathcal{S}}$ disjoint from Y such that $w(\widehat{X}) \geq w(X)$ $(w(\widehat{X}) \leq w(X))$.*

The next result implies that small representative families can be computed efficiently.[2]

Theorem 1 ([18]). *Given a constant $c \geq 1$, a universe E, nonnegative integers k and r, a family $\mathcal{S}$ of subsets of size r of E, and a function $w : \mathcal{S} \to \mathbb{R}$, a subfamily $\widehat{\mathcal{S}} \subseteq \mathcal{S}$ of size at most $\dfrac{(ck)^k}{r^r(ck-r)^{k-r}} 2^{o(k)} \log|E|$ that max (min) represents $\mathcal{S}$ can be computed in time $O(|\mathcal{S}|(ck/(ck-r))^{k-r} 2^{o(k)} \log|E| + |\mathcal{S}| \log|\mathcal{S}|)$.*

---

[2] This result builds on a powerful construction technique for representative families presented in [10].

Now, consider the maximization version of k-WEC and max representative families. (The minimization version of k-WEC can be similarly handled by using min representative families.) Let $\mathsf{RepAlg}(E, k, r, \mathcal{S}, w)$ denote the algorithm in Theorem 1 with $c = 2$, and let $\mathcal{S}_i = \{S \in \mathcal{S} : |S| = i\}$, for all $1 \leq i \leq k$.

We present below procedure Decrease, which replaces each family $\mathcal{S}_i$ by a family $\widehat{\mathcal{S}}_i \subseteq \mathcal{S}_i$ that represents $\mathcal{S}_i$.

Procedure. Decrease$(U, \mathcal{S}, w, k, p)$

1: **for** $i = 1, 2, \ldots, k$ **do** $\widehat{\mathcal{S}}_i \Leftarrow \mathsf{RepAlg}(U, k, i, \mathcal{S}_i, w)$. **end for**
2: $\widehat{\mathcal{S}} \Leftarrow \bigcup_{i=1}^{k} \widehat{\mathcal{S}}_i$.
3: **return** $(U, \widehat{\mathcal{S}}, w, k, p)$.

In the following, we prove that procedure Decrease is correct.

Lemma 4. $(U, \mathcal{S}, w, k, p)$ *is a yes-instance iff* $(U, \widehat{\mathcal{S}}, w, k, p)$ *is a yes-instance.*

Proof. First, assume that $(U, \mathcal{S}, w, k, p)$ is a yes-instance. Let $\mathcal{S}'$ be a subfamily of disjoint sets from $\mathcal{S}$, such that $|\bigcup \mathcal{S}'| = k$, $\sum_{S \in \mathcal{S}'} w(S) \geq p$, and there is no subfamily $\mathcal{S}''$ satisfying these conditions, and $|\mathcal{S}' \cap \widehat{\mathcal{S}}| < |\mathcal{S}'' \cap \widehat{\mathcal{S}}|$. Suppose, by way of contradiction, that there is a set $S \in (\mathcal{S}_i \cap \mathcal{S}') \setminus \widehat{\mathcal{S}}$, for some $1 \leq i \leq k$. By Theorem 1, there is a set $\widehat{S} \in \widehat{\mathcal{S}}_i$ such that $w(\widehat{S}) \geq w(S)$, and $\widehat{S} \cap S' = \emptyset$, for all $S' \in \mathcal{S}' \setminus \{S\}$. Thus, $\mathcal{S}'' = (\mathcal{S}' \setminus \{S\}) \cup \{\widehat{S}\}$ is a solution to $(U, \mathcal{S}, w, k, p)$. Since $|\mathcal{S}' \cap \widehat{\mathcal{S}}| < |\mathcal{S}'' \cap \widehat{\mathcal{S}}|$, this is a contradiction.

Now, assume that $(U, \widehat{\mathcal{S}}, w, k, p)$ is a yes-instance. Since $\widehat{\mathcal{S}} \subseteq \mathcal{S}$, we immediately get that $(U, \mathcal{S}, w, k, p)$ is also a yes-instance. $\qquad\square$

Next, we show that Theorem 1 implies the following.

Lemma 5. *Procedure* Decrease *runs in time* $O(\sum_{i=1}^{k}(|\mathcal{S}_i|(\frac{2k}{2k-i})^{k-i}2^{o(k)}\log|U|$

$+|\mathcal{S}_i|\log|\mathcal{S}_i|))$. *Moreover,* $|\widehat{\mathcal{S}}| \leq \sum_{i=1}^{k}\frac{(2k)^k}{i^i(2k-i)^{k-i}}2^{o(k)}\log|U| \leq 2.4^{k+o(k)}\log|U|$.

Proof. For any $1 \leq i \leq k$, Theorem 1 implies that Step 1 of procedure Decrease can be executed in time $O(|\mathcal{S}_i|(\frac{2k}{2k-i})^{k-i}2^{o(k)}\log|U| + |\mathcal{S}_i|\log|\mathcal{S}_i|)$.

Moreover, by Theorem 1, $|\widehat{\mathcal{S}}_i| \leq \frac{(2k)^k}{i^i(2k-i)^{k-i}}2^{o(k)}\log|U|$. Denoting $i = \alpha k$, we have that $|\widehat{\mathcal{S}}_i| \leq (\frac{2}{\alpha^\alpha(2-\alpha)^{1-\alpha}})^k 2^{o(k)}\log|U|$. The maximum is obtained at $\alpha \approx 0.6465$, and therefore $|\widehat{\mathcal{S}}_i| \leq 2.4^{k+o(k)}\log|U|$.

Thus, we get the desired upper bounds for $|\widehat{\mathcal{S}}|$ and the running time of procedure Decrease. $\qquad\square$

2.3   An Algorithm for any FGPP

We now present FGPPAlg, an algorithm that solves any FGPP in $O^*(4^{k+o(k)} \cdot \Delta^k)$ steps. Let $\mathsf{DegAlg}(G,k,p)$ denote the algorithm that solves any degrading FGPP in time $O((\Delta+1)^{k+1}|V|)$, given in [2]. Assuming that all the sets in $\mathcal{S}$ have the same size r, the algorithm in Sect. 5 of [19] solves k-WEC in time $O(2.851^{k(r-1)/r} \cdot |\mathcal{S}| \cdot |U| \log^2 |U|)$. This algorithm can be easily modified to solve k-WEC in time $O(2.851^k \cdot |\mathcal{S}| \cdot |U| \log^2 |U|)$, which is good enough for our purpose.

Let Π be an FGPP with parameters α_1 and α_2. Assume w.l.o.g that $\Delta \geq 2$, otherwise Π is clearly solvable in polynomial time, using a simple dynamic programming-based procedure. We now describe algorithm FGPPAlg (see the pseudocode below). First, if Π is a degrading FGPP, then FGPPAlg solves Π by calling DegAlg. Otherwise, by using the reduction f, FGPPAlg transforms the input into an instance of k-WEC. Then, FGPPAlg compacts the size of the resulting instance by calling the procedure Decrease. Finally, FGPPAlg solves Π by calling WECAlg.

Algorithm 1. FGPPAlg$(G = (V,E), k, p)$

1: **if** (Π is a max-FGPP and $\frac{\alpha_1}{2} \leq \alpha_2$) or ($\Pi$ is a min-FGPP and $\frac{\alpha_1}{2} \geq \alpha_2$) **then**
2:     **accept** iff $\mathsf{DegAlg}(G,k,p)$ accepts.
3: **end if**
4: $(U,\mathcal{S},w,k,p) \Leftarrow f(G,k,p)$.
5: $(U,\widehat{\mathcal{S}},w,k,p) \Leftarrow \mathsf{Decrease}(U,\mathcal{S},w,k,p)$.
6: **accept** iff $\mathsf{WECAlg}(U,\widehat{\mathcal{S}},w,k,p)$ accepts.

Theorem 2. *Algorithm* FGPPAlg *solves* Π *in time* $O(4^{k+o(k)}\Delta^k(|V|+|E|)|V|)$.

Proof. The correctness of the algorithm follows immediately from Lemmas 1 and 4, and the correctness of DegAlg and WECAlg.

Note that $2.851^k 2.4^{k+o(k)} = 6.8424^{k+o(k)} \leq 4^{k+o(k)}\Delta^k$. Thus, by Lemmas 3 and 5, and the running times of DegAlg and WECAlg, algorithm FGPPAlg runs in time

$$O(4^k(\Delta-1)^k(|V|+|E|)|V| + \sum_{i=1}^{k}(4^i(\Delta-1)^i|V|(\frac{2k}{2k-i})^{k-i}2^{o(k)}\log|V|)$$
$$+ 2.851^k 2.4^{k+o(k)}|V|\log^3|V|)$$
$$= O(4^{k+o(k)}\Delta^k(|V|+|E|)|V| + 2^{o(k)}|V|\log|V|[\max_{0\leq\alpha\leq1}\{4^\alpha\Delta^\alpha(\frac{2}{2-\alpha})^{1-\alpha}\}]^k)$$
$$= O(4^{k+o(k)}\Delta^k(|V|+|E|)|V| + 4^{k+o(k)}\Delta^k|V|\log|V|)$$
$$= O(4^{k+o(k)}\Delta^k(|V|+|E|)|V|).$$

$\square$

3   Solving MAX $(k, n - k)$-CUT in Time $O^*(4^{p+o(p)})$

We give below an $O^*(4^{p+o(p)})$ time algorithm for MAX $(k, n-k)$-CUT. In Sect. 3.1 we show that it suffices to consider an easier variant of MAX $(k, n - k)$-CUT, that we call NC-MAX $(k, n - k)$-CUT. We solve this variant in Sect. 3.2. Finally, our algorithm for MAX $(k, n - k)$-CUT is given in Sect. 3.3.

3.1   Simplifying MAX $(k, n - k)$-CUT

We first define an easier variant of MAX $(k, n - k)$-CUT. Given a graph $G = (V, E)$, where each node is either red or blue, and positive integers k and p, NC-MAX $(k, n - k)$-CUT asks if there is a subset $X \subseteq V$ of exactly k red nodes and no blue nodes, such that at least p edges in $E(X, V \setminus X)$ have a blue endpoint.

Given an instance (G, k, p) of MAX $(k, n - k)$-CUT, we perform several iterations of coloring the nodes in G; thus, if (G, k, p) is a yes-instance, we generate at least one yes-instance of NC-MAX $(k, n - k)$-CUT. To determine how to color the nodes in G, we need the following definition of universal sets.

Definition 2. *Let $\mathcal{F}$ be a set of functions $f : \{1, 2, \ldots, n\} \to \{0, 1\}$. We say that $\mathcal{F}$ is an (n, t)-universal set if, for every subset $I \subseteq \{1, 2, \ldots, n\}$ of size t and a function $f' : I \to \{0, 1\}$, there is a function $f \in \mathcal{F}$ such that, for all $i \in I$, $f(i) = f'(i)$.*

The following result asserts that small universal sets can be computed efficiently.

Lemma 6 ([16]). *There is an algorithm, UniSetAlg, that given a pair of integers (n, t), computes an (n, t)-universal set $\mathcal{F}$ of size $2^{t+o(t)} \log n$ in time $O(2^{t+o(t)} n \log n)$.*

We now present ColorNodes (see the pseudocode below), a procedure that given an input (G, k, p, q), where (G, k, p) is an instance of MAX $(k, n - k)$-CUT and $q = k + p$, returns a set of instances of NC-MAX $(k, n - k)$-CUT. Procedure ColorNodes first constructs a $(|V|, k + p)$-universal set $\mathcal{F}$. For each $f \in \mathcal{F}$, ColorNodes generates a colored copy V^f of V. Then, ColorNodes returns a set $\mathcal{I}$, including the resulting instances of NC-MAX $(k, n - k)$-CUT.

The next lemma implies the correctness of procedure ColorNodes.

Lemma 7. *An instance (G, k, p) of MAX $(k, n - k)$-CUT is a yes-instance iff ColorNodes$(G, k, p, k + p)$ returns a set $\mathcal{I}$ containing at least one yes-instance of NC-MAX $(k, n - k)$-CUT.*

Proof. If (G, k, p) is a no-instance of MAX $(k, n - k)$-CUT, then clearly, for any coloring of the nodes in V, we get a no-instance of NC-MAX $(k, n - k)$-CUT.

Next suppose that (G, k, p) is a yes-instance, and let X be a set of k nodes in V such that $|E(X, V \setminus X)| \geq p$. Note that there is a set Y of at most p nodes in $V \setminus X$ such that $|E(X, Y)| \geq p$. Let X' and Y' denote the indices of the nodes in X and Y, respectively. Since $\mathcal{F}$ is a $(|V|, k + p)$-universal set, there is

---

Procedure. ColorNodes$(G = (V, E), k, p, q)$

1: let $V = \{v_1, v_2, \ldots, v_{|V|}\}$.
2: $\mathcal{F} \Leftarrow \mathsf{UniSetAlg}(|V|, q)$.
3: **for all** $f \in \mathcal{F}$ **do**
4:     let $V^f = \{v_1^f, v_2^f, \ldots, v_{|V|}^f\}$, where v_i^f is a copy of v_i.
5:     **for** $i = 1, 2, \ldots, |V|$ **do**
6:         **if** $f(i) = 0$ **then** color v_i^f red. **else** color v_i^f blue. **end if**
7:     **end for**
8: **end for**
9: **return** $\mathcal{I} = \{(G_f = (V_f, E), k, p) : f \in \mathcal{F}\}$.

---

a function $f \in \mathcal{F}$ such that: (1) for all $i \in X'$, $f(i) = 0$, and (2) for all $i \in Y'$, $f(i) = 1$. Thus, in G_f, the copies of the nodes in X are red, and the copies of the nodes in Y are blue. We get that (G_f, k, p) is a yes-instance of NC-MAX $(k, n - k)$-CUT. $\qquad\square$

Furthermore, Lemma 6 immediately implies the following result.

Lemma 8. *Procedure* ColorNodes *runs in time* $O(2^{q+o(q)}|V| \log |V|)$, *and returns a set* $\mathcal{I}$ *of size* $O(2^{q+o(q)} \log |V|)$.

3.2   A Procedure for NC-MAX $(k, n - k)$-CUT

We now present SolveNCMaxCut, a procedure for solving NC-MAX $(k, n - k)$-CUT (see below). Procedure SolveNCMaxCut orders the red nodes in V by the number of their blue neighbors in a non-increasing manner. If there are at least k red nodes, and the number of edges between the first k red nodes and blue nodes is at least p, procedure SolveNCMaxCut accepts, and otherwise rejects.

Procedure. SolveNCMaxCut$(G = (V, E), k, p)$

1: **for all** red $v \in V$ **do**
2:     compute the number $n_b(v)$ of blue neighbors of v in G.
3: **end for**
4: let $v_1, v_2, \ldots, v_r$, for some $0 \leq r \leq |V|$, denote the red nodes in V, such that $n_b(v_i) \geq n_b(v_{i+1})$ for all $1 \leq i \leq r - 1$.
5: **accept** iff $(r \geq k$ and $\displaystyle\sum_{i=1}^{k} n_b(v_i) \geq p)$.

Clearly, the following result holds.

Lemma 9. *Procedure* SolveNCMaxCut *solves* NC-MAX $(k, n - k)$-CUT *in time* $O(|V| \log |V| + |E|)$.

3.3   An Algorithm for MAX $(k, n - k)$-CUT

Assume w.l.o.g that G has no isolated nodes. Our algorithm, MaxCutAlg, for MAX $(k, n-k)$-CUT, proceeds as follows (see below). First, if $p < \min\{k, |V|-k\}$, then MaxCutAlg accepts, and if $|V|-k < k$, then MaxCutAlg performs a recursive call with k replaced by $|V| - k$. Then, MaxCutAlg calls ColorNodes to compute a set of instances of NC-MAX $(k, n - k)$-CUT, and accepts iff SolveNCMaxCut accepts at least one of them.

Algorithm 2. MaxCutAlg$(G = (V, E), k, p)$

1: **if** $p < \min\{k, |V| - k\}$ **then accept. end if**
2: **if** $|V| - k < k$ **then accept** iff MaxCutAlg$(G, |V| - k, p)$ accepts. **end if**
3: $\mathcal{I} \Leftarrow$ ColorNodes$(G, k, p, k + p)$.
4: **for all** $(G', k', p') \in \mathcal{I}$ **do**
5:     **if** SolveNCMaxCut(G', k', p') accepts **then accept. end if**
6: **end for**
7: **reject.**

The next lemma implies the correctness of Step 1 in MaxCutAlg.

Lemma 10 ([2]). *In a graph $G = (V, E)$ having no isolated nodes, there is a subset $X \subseteq V$ of size k such that $|E(X, V \setminus X)| \geq \min\{k, |V| - k\}$.*

Our main result is the following.

Theorem 3. *MaxCutAlg solves* MAX $(k, n - k)$-CUT *in time* $O(4^{p+o(p)}(|V| + |E|) \log^2 |V|)$.

Proof. Clearly, (G, k, p) is a yes-instance iff $(G, |V| - k, p)$ is a yes-instance. Thus, Lemmas 7, 9 and 10 immediately imply the correctness of MaxCutAlg.

Denote $m = \min\{k, |V| - k\}$. If $p < m$, then MaxCutAlg runs in time $O(1)$. Next suppose that $p \geq m$. Then, by Lemmas 8 and 9, MaxCutAlg runs in time $O(2^{m+p+o(m+p)}(|V| + |E|) \log^2 |V|) = O(4^{p+o(p)}(|V| + |E|) \log^2 |V|)$. $\square$

4   Algorithms for Positive Min-FGPPs

In this section we summarize our results for positive min-FGPPs.

First, we use a standard application of randomized separation to prove the following.

Theorem 4. *Any positive min-FGPP can be solved in time* $O(2^{k + \frac{p}{\alpha_2} + o(k+p)} \cdot (|V| + |E|) \log |V|)$.

Now, let Π be a non-degrading positive min-FGPP. To solve Π, we use a somewhat different application of randomized separation, in which we randomly color edges rather than nodes. To this end, we define an easier variant of the problem Π, called EC-Π.

In EC-Π, we are given a graph $G = (V, E)$ where each edge is either red or blue, and parameters $k \in \mathbb{N}$ and $p \in \mathbb{R}$. For any subset $X \subseteq V$, let $C(X)$ denote the family containing the node-sets of the maximal connected components in the graph $G_r = (X, E_r)$, where E_r is the set of red edges in E having both endpoints in X. Also, let $\text{val}^*(X) = \sum_{C \in C(X)} \text{val}(C)$. The problem EC-$\Pi$ asks if there is a subset $X \subseteq V$ of exactly k nodes, such that all of the edges in $E(X, V \setminus X)$ are blue, and $\text{val}^*(X) \leq p$.

To solve Π, we first construct a set $\mathcal{I}$ of instances of EC-Π. Then, using a dynamic programming-based procedure, we solve each of the instances in $\mathcal{I}$. We accept iff at least one of the instances in $\mathcal{I}$ is a yes-instance.

This approach leads to the following result, where $x = \max\{\frac{p}{\alpha_2}, \min\{\frac{p}{\alpha_1}, \frac{p}{\alpha_2} + (1 - \frac{\alpha_1}{\alpha_2})k\}\}$.

Theorem 5. *Any non-degrading positive min-FGPP can be solved in time* $O(2^{x+o(x)}(|V|k + |E|) \log |E|)$.

In case $\alpha_1 = \alpha_2 = 1$, we have that $x = p$. Thus, since MIN k-VERTEX COVER is a non-degrading positive min-FGPP which satisfies $\alpha_1 = \alpha_2 = 1$, we have the following.

Corollary 1. MIN k-VERTEX COVER *can be solved in time* $O(2^{p+o(p)}(|V|k + |E|) \log |E|)$.

Acknowledgment. We thank the anonymous referees for valuable comments and suggestions.

References

1. Berkhin, P.: A survey of clustering data mining techniques. In: Kogan, J., Nicholas, C., Teboulle, M. (eds.) Grouping Multidimensional Data: Recent Advances in Clustering, pp. 25–71. Springer, Heidelberg (2006)
2. Bonnet, É., Escoffier, B., Paschos, V.T., Tourniaire, É.: Multi-parameter complexity analysis for constrained size graph problems: using greediness for parameterization. In: Gutin, G., Szeider, S. (eds.) IPEC 2013. LNCS, vol. 8246, pp. 66–77. Springer, Heidelberg (2013)
3. Bourgeois, N., Giannakos, A., Lucarelli, G., Milis, I., Paschos, V.T.: Exact and approximation algorithms for densest k-subgraph. In: Ghosh, S.K., Tokuyama, T. (eds.) WALCOM 2013. LNCS, vol. 7748, pp. 114–125. Springer, Heidelberg (2013)
4. Cai, L.: Parameterized complexity of cardinality constrained optimization problems. Comput. J. **51**(1), 102–121 (2008)
5. Cai, L., Chan, S.M., Chan, S.O.: Random separation: a new method for solving fixed-cardinality optimization problems. In: Bodlaender, H.L., Langston, M.A. (eds.) IWPEC 2006. LNCS, vol. 4169, pp. 239–250. Springer, Heidelberg (2006)

6. Cygan, M., Lokshtanov, D., Pilipczuk, M., Pilipczuk, M., Saurabh, S.: Minimum bisection is fixed parameter tractable. In: STOC, pp. 323–332 (2014)
7. Donavalli, A., Rege, M., Liu, X., Jafari-Khouzani, K.: Low-rank matrix factorization and co-clustering algorithms for analyzing large data sets. In: Kannan, R., Andres, F. (eds.) ICDEM 2010. LNCS, vol. 6411, pp. 272–279. Springer, Heidelberg (2012)
8. Downey, R.G., Estivill-Castro, V., Fellows, M.R., Prieto, E., Rosamond, F.A.: Cutting up is hard to do: the parameterized complexity of k-cut and related problems. Electron. Notes Theor. Comput. Sci. **78**, 209–222 (2003)
9. Downey, R.G., Fellows, M.R.: Fixed-parameter tractability and completeness II: on completeness for W[1]. Theor. Comput. Sci. **141**(1&2), 109–131 (1995)
10. Fomin, F.V., Lokshtanov, D., Saurabh, S.: Efficient computation of representative sets with applications in parameterized and exact agorithms. In: SODA, pp. 142–151 (2014)
11. Guo, J., Niedermeier, R., Wernicke, S.: Parameterized complexity of vertex cover variants. Theory Comput. Syst. **41**(3), 501–520 (2007)
12. Kahng, A.B., Lienig, J., Markov, I.L., Hu, J.: VLSI Physical Design - From Graph Partitioning to Timing Closure. Springer, Netherlands (2011)
13. Kloks, T. (ed.): Treewidth, Computations and Approximations. LNCS, vol. 842. Springer, Heidelberg (1994)
14. Kneis, J., Langer, A., Rossmanîth, P.: Improved upper bounds for partial vertex cover. In: Broersma, H., Erlebach, T., Friedetzky, T., Paulusma, D. (eds.) WG 2008. LNCS, vol. 5344, pp. 240–251. Springer, Heidelberg (2008)
15. Komusiewicz, C., Sorge, M.: Finding dense subgraphs of sparse graphs. In: Thilikos, D.M., Woeginger, G.J. (eds.) IPEC 2012. LNCS, vol. 7535, pp. 242–251. Springer, Heidelberg (2012)
16. Naor, M., Schulman, L.J., Srinivasan, A.: Splitters and near-optimal derandomization. In: FOCS, pp. 182–191 (1995)
17. Shachnai, H., Zehavi, M.: Parameterized algorithms for graph partitioning problems. CoRR abs/1403.0099 (2014)
18. Shachnai, H., Zehavi, M.: Representative families: a unified tradeoff-based approach. In: Schulz, A.S., Wagner, D. (eds.) ESA 2014. LNCS, vol. 8737, pp. 786–797. Springer, Heidelberg (2014)
19. Zehavi, M.: Deterministic parameterized algorithms for matching and packing problems. CoRR abs/1311.0484 (2013)

Between Treewidth and Clique-Width

Sigve Hortemo Sæther[⊠] and Jan Arne Telle

Department of Informatics, University of Bergen, Bergen, Norway
{sigve.sether,telle}@ii.uib.no

Abstract. Many hard graph problems can be solved efficiently when restricted to graphs of bounded treewidth, and more generally to graphs of bounded clique-width. But there is a price to be paid for this generality, exemplified by the four problems MAXCUT, GRAPH COLORING, HAMILTONIAN CYCLE and EDGE DOMINATING SET that are all FPT parameterized by treewidth but none of which can be FPT parameterized by clique-width unless the Exponential Time Hypothesis fails, as shown by Fomin et al. [7]. We therefore seek a structural graph parameter that shares some of the generality of clique-width without paying this price.

Based on splits, branch decompositions and the work of Vatshelle [16] on Maximum Matching-width, we consider the graph parameter sm-width which lies between treewidth and clique-width. Some graph classes of unbounded tree-width, like distance-hereditary graphs, have bounded sm-width. We show that MAXCUT, GRAPH COLORING, HAMILTONIAN CYCLE and EDGE DOMINATING SET are all FPT parameterized by sm-width.

1 Introduction

Many hard problems can be solved efficiently when restricted to graphs of bounded treewidth or even graphs of bounded clique-width. A celebrated algorithmic metatheorem of Courcelle [5] states that any problem expressible in monadic second-order logic (MSO_2) is fixed parameter tractable (FPT) when parameterized by the treewidth of the input graph. This includes many problems like DOMINATING SET, GRAPH COLORING, and HAMILTONIAN CYCLE. Likewise, Courcelle et al. [4] show that the subset of MSO_2 problems expressible in MSO_1-logic, which does not allow quantification over edge sets, is FPT parameterized by clique-width. Originally this required a clique-width expression as part of the input, but this restriction was removed when Oum and Seymour [12] gave an algorithm that, in time FPT parameterized by the clique-width k of the input graph, finds a $2^{O(k)}$-approximation of an optimal clique-width expression.

Clique-width is stronger than treewidth, in the sense that bounded treewidth implies bounded clique-width [3] but not vice-versa, as exemplified by the cliques. Can we hope to find a graph width parameter lying between treewidth and clique-width for which all MSO_2 problems are FPT? Alas no, under the minimal requirement that cliques should have bounded width, Courcelle et al. [4] showed

© Springer International Publishing Switzerland 2014
D. Kratsch and I. Todinca (Eds.): WG 2014, LNCS 8747, pp. 396–407, 2014.
DOI: 10.1007/978-3-319-12340-0_33

that this would imply P=NP for unary languages. There are some basic problems belonging to MSO_2 but not MSO_1, like MAXCUT, GRAPH COLORING, HAMILTONIAN CYCLE and EDGE DOMINATING SET. Fomin et al. [7] showed that none of these four problems can be FPT parameterized by clique-width, unless the Exponential Time Hypothesis collapses. Can we find a graph width parameter lying between treewidth and clique-width for which at least these four problems are FPT? Note that one can define trivial parameters having these properties (e.g. value equal to clique-width if this is at most 3, and otherwise equal to treewidth) but can we find one yielding new FPT algorithms for certain natural graph classes? This is the question motivating the present paper, and the answer is yes. We give a parameter which is low when the graph has low treewidth in local parts, and where each of these parts are connected together in a dense manner.

Before explaining our results, let us mention some related work. A class of graphs can have bounded treewidth only if it is sparse. Indeed, the introduction of clique-width was motivated by the desire to extend algorithmic results for bounded treewidth also to some dense graph classes. Let us say that a parameter x is weaker than parameter y, and y stronger than x, if for any graph class, a bound on x implies a bound on y. Alternatively, x and y are of the same strength, or incomparable. Thus, clique-width is stronger than treewidth. As we discussed above there are limitations inherent in clique-width and there have been several suggestions for width parameters weaker than clique-width but still bounded on some dense graph classes. In particular, let us mention four parameters: neighborhood diversity introduced by Lampis in 2010 [11], twin-cover introduced by Ganian in 2011 [9], shrub-depth introduced by Ganian et al. in 2012 [10], and modular-width proposed by Gajarský et al. in 2013 [8]. All these parameters are bounded on some dense classes of graphs, all of them are weaker than clique-width, but none of them are stronger than treewidth. Modular-width is stronger than both neighborhood diversity and twin-cover, but incomparable to shrub-depth [8]. GRAPH COLORING and HAMILTONIAN CYCLE are W-hard parameterized by shrub-depth but FPT parameterized by modular-width, as recently shown by Gajarský et al. [8] which also leaves as an open problem the complexity of MAXCUT and EDGE DOMINATING SET parameterized by modular-width.

In our quest for a parameter stronger than treewidth and weaker than clique-width, for which the four basic problems MAXCUT, GRAPH COLORING, HAMILTONIAN CYCLE and EDGE DOMINATING SET become FPT, we are faced with two tasks when given a graph G with parameter-value k: we need an FPT algorithm returning a decomposition of width $f(k)$, and we need a dynamic programming algorithm solving each of the four basic problems in FPT time when parameterized by the width of this decomposition. The requirement that the parameter be stronger than treewidth is a guarantee that it shares this property with clique-width and will capture large tree-like classes of graphs, also when some building blocks are dense. Arguably the most natural way to hierarchically decompose a graph are the so-called branch decompositions, originating in work of Robertson and Seymour [14] and used in the definition of both rank-width [12] and boolean-width [1], two parameters of the same strength as clique-width. Branch decompositions can be viewed as a recursive partition of the vertices into two parts, giving a rooted binary tree where each edge of the tree defines the cut given by the vertices in the subtree below the edge. Using any symmetric cut function defined on subsets of vertices we can define a graph width parameter as the minimum, over all branch decompositions, of the maximum cut-value over all edges of the branch decomposition tree. Recently, Vatshelle [16] gave a cut-function based on the size of a maximum matching, whose associated graph width parameter, called MM-width, has the same strength as treewidth.

In Sect. 2, based on the work of Vatshelle, we define the parameter split-matching-width, denoted sm-width, by a cut function based on maximum matching unless the cut is a split, i.e. a complete bipartite graph plus some isolated vertices. The sm-width parameter is stronger than treewidth and weaker than clique-width. It is also stronger than twin-cover but incomparable with neighborhood diversity, shrub-depth and modular-width. We finish Sect. 2 by showing that maximum matching is a submodular cut function. In Sect. 3 this is used together with an algorithm for split decompositions by Cunningham [6] and an algorithm for branch decompositions based on submodular cut functions by Oum and Seymour [12] to design an algorithm that given a graph G with sm-width k computes a branch decomposition of sm-width $O(k^2)$, in time $O^*(8^k)$. To our knowledge the use of split decompositions to compute a width parameter is novel.

In Sect. 4, using a slightly non-standard framework for dynamic programming, we are then able to solve the four basic problems MAXCUT, GRAPH COLORING, HAMILTONIAN CYCLE and EDGE DOMINATING SET, by runtimes $\mathcal{O}^*(8^k), \mathcal{O}^*(k^{5k}), \mathcal{O}^*(2^{24k^2})$, and $\mathcal{O}^*(3^{5k})$ respectively, when given a branch decomposition of sm-width k. In Sect. 5 we show that some well-known graph classes of bounded clique-width also have bounded sm-width, e.g. distance-hereditary graphs have clique-width at most three and sm-width one. We also show that a graph whose twin-cover value is k will have sm-width at most k, and discuss classes of graphs where our results imply new FPT algorithms.

2 Preliminaries

We deal with finite, simple, undirected graphs $G = (V, E)$ and denote also the vertex set by $V(G)$ and the edge set by $E(G)$. For the subgraph of G induced by $S \subseteq V(G)$ we write $G[S]$, and for disjoint sets $A, B \subseteq V(G)$ we denote the induced bipartite subgraph having vertex set $A \cup B$ and edge set $\{uv : u \in A, v \in B\}$ as $G[A, B]$. For $v \in V(G)$ we write $N(v)$ or $N_G(v)$ for the neighbors of v and for $S \subseteq V(G)$ we denote the neighborhood of S by $N(S) = \bigcup_{a \in S} N(a) \setminus S$ or $N_G(S)$; note that $N(S) \cap S = \emptyset$. A matching is a set of edges having no endpoints in common.

A *split* of a connected graph G is a partition of $V(G)$ into two sets V_1, V_2 such that $|V_1| \geq 2, |V_2| \geq 2$ and every vertex in V_1 with a neighbor in V_2 has the same neighborhood in V_2 (this also means every vertex in V_2 with a neighbour in V_1 has the same neighbourhood in V_1). A graph G with a split (V_1, V_2) can be *decomposed* into a graph G_1 and a graph G_2 so that G_1 and G_2 is the induced subgraph of G on V_1 and V_2, respectively, except that an extra vertex v, called a *marker*, is added, and also some extra edges are added to G_1 and G_2, so that $N_{G_1}(v) = N_G(V_2)$ and $N_{G_2}(v) = N_G(V_1)$. If a graph G can be decomposed to the two graphs G_1 and G_2, then G_1 and G_2 *compose* G. We denote this by $G = G_1 * G_2$. A graph that cannot be decomposed (i.e., a graph without a split) is called a *prime*. As all graphs of at most three vertices trivially is a prime, when a prime graph has more than three vertices, it is called a *non-trivial prime graph*. A *split decomposition* of a graph G is a recursive decomposition of G so that all of the obtained graphs are prime. For a split decomposition of G into $G_1, G_2, \ldots, G_k$, a *split decomposition tree* is a tree T where each vertex corresponds to a prime graph and we have an edge between two vertices if and only if the prime graphs they correspond to share a marker. That is, the edge set of the tree is $E(T) = \{v_i v_j : v_i, v_j \in V(T) \text{ and } V(G_i) \cap V(G_j) \neq \emptyset\}$. To see that this is in fact a tree, we notice that T is connected and that we have an edge for each marker introduced. As there are exactly one less marker than there are prime graphs, T must be a tree. See Fig. 1 for an example.

Given a split decomposition of graph G with prime graphs $G_1, G_2, \ldots, G_k$, we define $\text{tot}(v : G_i)$ recursively to be $\{v\}$ if $v \in V(G)$, and otherwise to be $\bigcup_{u \in V(G_j) \setminus \{v\}} \text{tot}(u : G_j)$ for the graph $G_j \neq G_i$ containing the marker v in the split decomposition. Another way of saying this latter part by the use of the split decomposition tree T is: if v is not in $V(G)$, then $\text{tot}(v : G_i)$ is defined to be the vertices of $V(G)$ residing in the prime graphs of the connected component in $T[V(T) - G_i]$ where v is also located. From this last definition, we observe that for a prime graph G_i in a split decomposition of G, the function tot on the vertices of G_i partitions the vertices of $V(G)$. For a set $V' \subseteq V(G_i)$, we define $\text{tot}(V' : G_i)$ to be the union of $\text{tot}(v : G_i)$ for all $v \in V'$. For a set

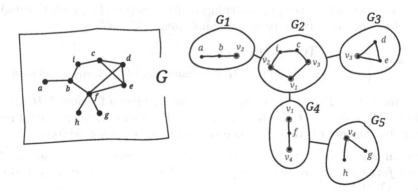

Fig. 1. Split decomposition tree of a graph G. The markers of each prime graph are circled in red. An example of a split decomposition resulting in this tree is: $((G_1 * G_2) * G_3) * (G_4 * G_5)$. Note that $\text{tot}(\{v_1, v_3\} : G_2) = \{d, e, f, g, h\}$, $\text{act}(v_1 : G_4) = \{b, d, e\}$ and $\text{tot}^{-1}(a, f, g : G_2) = \{v_1, v_2\}$ (*Color figure online*).

$S \subseteq V(G)$, the inverse function $\text{tot}^{-1}(S : G_i)$, is defined as the minimal set of vertices $V' \subseteq V(G_i)$ so that $S \subseteq \text{tot}(V' : G_i)$. We define the *active set* of a vertex $v \in G_i$, denoted $\text{act}(v : G_i)$ to be the vertices of $\text{tot}(v : G_i)$ that are contributing to the neighborhood of v in G_i. That is, $\text{act}(v : G_i)$ is defined as $N(V(G) \setminus \text{tot}(v : G_i))$. See Fig. 1 for an example of tot() and act(). Note that if G has a split decomposition into prime graphs $G_1, \ldots, G_k$, then for any marker v there are exactly two prime graphs G_i and G_j containing v, and we have $\text{tot}(v : G_i) \cup \text{tot}(v : G_j) = V(G)$.

A *branch decomposition* (T, δ) of a graph G consists of a subcubic tree T (a tree of maximum degree 3) and a bijective function δ from the leaves of T to the vertices of G. For a graph G a *cut* $(A, \overline{A})$ for $A \subseteq V(G)$ is a bipartition of verices of G. For a cut (A, B) of G, we say the edges in G with one enpoint in A and the other in B *cross* the cut (A, B). In a branch decomposition $(T = (V_T, E_T), \delta)$ of a graph G, each edge $e \in E_T$ partitions $V(G)$ into two parts: the vertices mapped by δ from the leaves of one component of $T - e$, and the vertices mapped by δ from the leaves of the other component. Thus each edge of T induces a cut in G, namely the cut corresponding to that edge's bipartition of $V(G)$. For a graph G, a *cut function* $f : 2^{V(G)} \to \mathbb{N}$ is a symmetric $(f(A) = f(\overline{A}))$ function on subsets of $V(G)$. For a branch decomposition (T, δ) of G its f-width, for a cut function f, is the maximum of $f(A)$ over all cuts $(A, \overline{A})$ of G induced by the edges of T. For a graph G, its f-width, for a cut function f, is the minimum f-width over all branch decompositions of G.

Vatshelle [16] defined the Maximum-Matching-width (MM-width) $\text{mmw}(G)$ of a graph G based on the cut function mm defined for any graph G and $A \subseteq V(G)$ by letting $\text{mm}(A)$ be the cardinality of a maximum matching of the bipartite graph $G[A, \overline{A}]$. In his work, Vatshelle shows that there is a linear dependency between the treewidth of a graph and the Maximum-Matching-width of the graph.

Theorem 1 [16]. *Let G be a graph, then $\frac{1}{3}(\text{tw}(G) + 1) \leq \text{mmw}(G) \leq \text{tw}(G) + 1$*

In this paper we define the split-matching-width $\text{smw}(G)$ of a graph G based on the cut function sm defined for any graph G and $A \subseteq V(G)$ by:

$$\text{sm}(A) = \begin{cases} 1 \text{ if } (A, \overline{A}) \text{ is a split of } G \\ \text{mm}(A) = \max\{|M| : M \text{ is a matching of } G[A, \overline{A}]\} \text{ otherwise} \end{cases}$$

A cut function $f : 2^{V(G)} \to \mathbb{N}$ is said to be submodular if for any $A, B \subseteq V(G)$ we have $f(A) + f(B) \geq f(A \cup B) + f(A \cap B)$. The following very general result of Oum and Seymour is central to the field of branch decompositions.

Theorem 2 [12]. *For symmetric submodular cut-function f and graph G of optimal f-width k, a branch decomposition of f-width at most $3k + 1$ can be found in $\mathcal{O}^*(2^{3k+1})$ time.*

There is no abundance of submodular cut functions, but this result, which is proved in the full version of the paper, will be useful to us.

Theorem 3. *The cut function mm is submodular.*

3 Computing an Approximate sm-Width Decomposition

In this section we design an algorithm that given a graph G finds a branch decomposition of G having sm-width $O(\text{smw}(G)^2)$, in time FPT parameterized by $\text{smw}(G)$. The algorithm has four main steps:

1. Find a split decomposition of G into prime graphs $G_1, G_2, ..., G_q$.
2. For each G_i find a branch decomposition (T_i, δ_i) of sm-width $O(\text{smw}(G_i))$.
3. For each G_i restructure (T_i, δ_i) into (T_i', δ_i') having the property that any cut of G_i, induced by an edge of (T_i', δ_i') and having split-matching value k, is lifted, by the split decomposition of G, to a cut of G having split-matching value $O(k^2)$.
4. Combine all the decompositions (T_i', δ_i') into a branch decomposition of G of sm-width $O(\text{smw}(G)^2)$.

For step 1 there exists a well-known polynomial-time algorithm by Cunningham [6] and even linear-time ones, see e.g. [2] and see also [13] for the use of split decompositions in general. For step 2 we are dealing with a prime graph G_i, which by definition has no non-trivial splits and hence $\text{sm}(V_i) = \text{mm}(V_i)$ for all $V_i \subseteq V(G_i)$ meaning that $\text{mmw}(G_i) = \text{smw}(G_i)$. Furthermore, by Theorem 3 the cut function defining mmw is submodular so we can apply the algorithm of Oum and Seymour from Theorem 2 to accomplish the task of step 2. Step 3 will require more work. Let us first give a sketch of step 4. Suppose for each prime graph G_i of a split decomposition of G we have calculated a branch decomposition (T_i', δ_i') for G_i. If for every cut $(X, V(G_i) \setminus X)$ of G_i induced by an edge of (T_i', δ_i') we have $\text{sm}(\text{tot}(X : G_i)) \leq t$ for some value t, then we can generate a branch decomposition of G of sm-width at most t by for each pair of prime graphs G_i, G_j sharing a marker, identifying the two leaves of respectively (T_i', δ_i') and (T_j', δ_j') mapped to this marker.

What remains is step 3, covered by Theorem 8. We need to relate $\text{sm}(A)$ of a cut $(A, V(G_i) \setminus A)$ in prime graph G_i, induced by an edge of (T_i', δ_i'), to $\text{sm}(\text{tot}(A : G_i))$ of the associated cut $\left(\text{tot}(A : G_i), \overline{\text{tot}(A : G_i)}\right)$ in G. This we do by Lemma 4 and in particular Lemma 6, see below, which use the notion of a *heavy* pair of vertices: In a graph G with $smw(G) < k$ and split decomposition into prime graphs $G_1, G_2, ..., G_q$ we say that adjacent vertices $a, b \in V(G_i)$ are *heavy* if $|\text{act}(a : G_i)| \geq 3k$ and $|\text{act}(b : G_i)| \geq 3k$. The edge connecting a heavy pair is called a *heavy* edge.

Lemma 4. *Let $smw(G) < k$ and let P be a non-trivial prime graph in a split decomposition of G. If for $a \in V(P)$ we have $|\text{act}(a : P)| \geq 3k$ and $|\text{act}(N(a) : P)| \geq 9k$, then $\exists ab \in E(P)$ so that ab is heavy. Moreover, for any heavy edge $ab \in E(P)$ we have $|\text{act}(N(\{a, b\}) : P)| < k$.*

The proof follows from a series of lemmas found in the full version of this paper [15]. The second part of Lemma 4 implies that in a prime graph P, whenever $ab \in E(P)$ is heavy then since $|\text{act}(N(\{a, b\}) : P)| < k$ neither a nor b can be incident to any other heavy edge, leading to the following corollary.

Corollary 5. *For a non-trivial prime graph, its heavy edges form a matching.*

Lemma 6. *Let $smw(G) < k$ and let P be a prime graph in a split decomposition of G and let $A \subseteq V(P)$ with $2 \leq |A| \leq |V(P)| - 2$. If no heavy edges cross the cut $(A, V(P) \setminus A)$ in P and $sm(A) < t$ with respect to P, then $sm(tot(A : P))$ with respect to G is less than $9tk$.*

The idea of how to prove Lemma 6 is that for any vertex cover C in $P[A, \overline{A}]$ and vertex $v \in C$, there is a set of less than $9k$ vertices in $V(G)$ covering all edges incident with v crossing the cut of P. Since all edges crossing the cut of P are incident with a vertex of C, this means there must be a set of less than $9k|C| = 9tk$ vertices of $V(G)$ covering all edges crossing the cut in G. The entire proof can be found in the full version of this paper.

When deleting a vertex any cut that was a split remains a split or results in a cut with a single vertex on one side, and no new matchings are introduced.

Observation 7. *The sm-width of a graph G is at least as big as the sm-width of any induced subgraph of G.*

Theorem 8. *Let $smw(G) < k$ and let P be a prime graph in a split decomposition of G We can in $\mathcal{O}^*(8^k)$-time construct a branch decomposition (T'_P, δ'_P) of P so that for each cut (X, Y) of P induced by an edge of (T'_P, δ'_P), the cut $(tot(X : P), tot(Y : P))$ of G has sm-value less than $54k^2$.*

Proof. If P is a trivial prime graph, i.e. $|V(P) \leq 3|$, every cut (X, Y) of P is a split. This implies by the definition of a split decomposition that $(tot(X : P), tot(Y : P))$ in G also is a split of G. Hence, $sm(tot(X : P))$ of G equals one.

We now consider the case when P is non-trivial. Since P is isomorphic to an induced subgraph of G (this follows directly from definition of split decompositions) and $smw(G) < k$, by Observation 7, the sm-width of P is less than k. Also, since P by definition has no splits, we have $mmw(P) = smw(P) < k$. By Theorem 3 and Lemma 2, we can compute a branch decomposition (T_P, δ_P) of P with MM-width less than $3k$ in $\mathcal{O}^*(8^k)$-time. By a non-leaf edge of T_P we mean an edge with both endpoints an inner node of T_P. The cut in P induced by a non-leaf edge of (T_P, δ_P) will have at least two vertices on each side. We call such cuts non-leaf cuts of P induced by (T_P, δ_P). Note that cuts having one side containing a singleton $X = \{v\}$ are easy to deal with, either the singleton is a vertex of $V(G)$ and then $tot(X : P) = \{v\}$, or v is a marker and the cut $(tot(X : P), tot(Y : P))$ of G is a split, and thus in both cases $sm(tot(X : P)) = 1$. For the remainder we consider only non-leaf cuts.

Denote by $h(A)$ the number of heavy edges crossing the non-leaf cut $(A, V(P) \setminus A)$. If none of the non-leaf cuts of P induced by (T_P, δ_P) have heavy edges crossing them, i.e. $h(A) = 0$ for all non-leaf cuts, we apply Lemma 6 with $t = 3k$ and are done, getting for any cut (X, Y) of P induced by an edge of (T_P, δ_P) a bound of $sm(tot(X : P)) \leq 3k9k = 27k^2$. On the other hand, if some non-leaf cuts of P induced by (T_P, δ_P) do have heavy edges crossing them, we restructure the decomposition (T_P, δ_P) to a decomposition (T'_P, δ'_P) as follows: for each heavy pair a, b in $V(P)$ crossing such a non-leaf cut we remove the leaf in T_P mapping to b and make a new leaf mapping to b as sibling of the leaf mapping to a. By Corollary 5 the heavy edges in P form a matching, so this is easily done for all heavy edges of P crossing non-leaf cuts, without conflicts. Since all such heavy pairs are now mapped to leaves that are siblings of T'_P none of the non-leaf cuts of P induced by (T'_P, δ'_P) will have a heavy edge crossing them.

Let us look at how the restructuring altered the sm-value of non-leaf cuts. Note that for each non-leaf cut $(A', V(P) \setminus A')$ in (T'_P, δ'_P) there is an associated non-leaf cut $(A, V(P) \setminus A)$ in (T_P, δ_P) with $h(A)$ heavy edges crossing this cut, such that we move between the two cuts by moving $h(A)$ vertices across the cut. We have $\mathrm{mm}(A') \leq \mathrm{mm}(A) + h(A)$, as the maximum matching of a cut can increase by at most one for each vertex moved over the cut. Moreover, by Corollary 5 the heavy edges in P form a matching, which means that $h(A) \leq \mathrm{mm}(A)$, implying $\mathrm{mm}(A') \leq 2\,\mathrm{mm}(A) \leq 2 \times 3k$. We can therefore apply Lemma 6 with $t = 6k$ and this means we have $\mathrm{sm}(\mathrm{tot}(A : P)) \leq 6k9k = 54k^2$. □

Theorem 9. *Given a graph G with $smw(G) < k$, we can compute a branch decomposition (T, δ) of G of sm-width less than $54k^2$ in $\mathcal{O}^*(8^k)$-time.*

We have already given a sketch of this proof. The full proof is found in [15].

4   Dynamic Programming Parameterized by sm-width

In this section we solve MAXCUT, GRAPH COLORING, HAMILTONIAN CYCLE and EDGE DOMINATING SET on a graph G by a bottom-up traversal of a rooted branch decomposition (T, δ) of G, in time FPT parameterized by the sm-width of (T, δ). In the bottom-up traversal we encounter two disjoint subsets of vertices $A, B \subseteq V(G)$, as leaves of two already processed subtrees, and need to process the subtree on leaves $A \cup B$. There are three cuts of G involved: $(A, \overline{A})$, $(B, \overline{B})$, $(A \cup B, \overline{A \cup B})$, and each of them can be of type split, or of type non-split (also called type mm for maximum-matching). This gives six cases that need to be considered, at least if we use the standard framework of table-based dynamic programming. We instead use an algorithmic framework for decision problems where we JOIN sets of certificates while ensuring that the result preserves witnesses for a 'yes' instance. Under this framework, the algorithm for MAXCUT becomes particularly simple, and only two cases need to be handled in the JOIN, depending on whether the 'parent cut' $(A \cup B, \overline{A \cup B})$ is a split or not. For the other three problems we must distinguish between the two types of 'children cuts' in order to achieve FPT runtime, and the algorithms are more complicated.

Let us describe the algorithmic framework. As usual, e.g. for problems in NP, a verifier is an algorithm that given a problem instance G and a certificate c, will verify if the instance is a 'yes'-instance, and if so we call c a witness. For our algorithms we will use a commutative and associative function $\oplus(x, y)$, that takes two certificates x, y and creates a set of certificates. This is extended to sets of certificates X_A, X_B by $\oplus(X_A, X_B)$ which creates the set of certificates $\bigcup_{x_A \in X_A, x_B \in X_B} \oplus(x_A, x_B)$. For a graph decision problem, an input graph G, and any $X \subseteq V(G)$ we define $\text{cert}(X)$ to be a set of certificates on only a restricted part of G, which must be subject to the following constraints:

- If G is a 'yes'-instance, then $\text{cert}(V(G))$ contains a witness.
- For disjoint $X, Y \subseteq V(G)$ we have $\oplus(\text{cert}(X), \text{cert}(Y)) = \text{cert}(X \cup Y)$.

For FPT runtime we need to restrict the size of a set of certificates, and the following will be useful. For $X \subseteq V(G)$ and certificates $x, y \in \text{cert}(X)$, we say that x *preserves* y if for all $z \in \text{cert}(\overline{X})$ so that $\oplus(y, z)$ contains a witness, the set $\oplus(x, z)$ also contains a witness. We denote this as $x \preceq_X y$. A set S preserves $S' \subseteq \text{cert}(X)$, denoted $S \preceq_X S'$, if for every $x' \in S'$ there exists a $x \in S$ so that $x \preceq_X x'$. A certificate $x \in \text{cert}(X)$ so that there exists a $y \in \text{cert}(\overline{X})$ where $\oplus(x, y)$ contains a witness, is called an *important* certificate.

For a rooted branch decomposition (T, δ) of a graph G and vertex $v \in V(T)$, we denote by V_v the set of vertices of $V(G)$ mapped by δ from the leaves of the subtree in T rooted at v. With these definitions we give a generic recursive (or bottom-up) algorithm called RECURSIVE that takes (T, δ) and a vertex w of T as input and returns a set $S \preceq_{V_w} \text{cert}(V_w)$, as follows:

- at a leaf w of T INITIALIZE and return the set $\text{cert}(\{\delta(w)\})$
- at an inner node w first call RECURSIVE on each of the children nodes a and b and then run procedure JOIN on the returned input sets S_1, S_2 of certificates, with $S_1 \preceq_{V_A} \text{cert}(V_a)$ and $S_2 \preceq_{V_b} \text{cert}(V_b)$, and return a set $S \preceq_{V_a \cup V_b} \oplus(S_1, S_2)$
- at the root we will have a set of certificates $S \preceq_{V(G)} \text{cert}(V(G))$

Calling RECURSIVE on the root r of T and running a verifier on the output solves any graph decision problem in NP. Correctness of this procedure follows from the definitions. The extra time spent by the verifier is going to be $\mathcal{O}^*(|S|)$, and for an FPT algorithm we will require that all $|S|$ be $\mathcal{O}^*(f(k))$, i.e. FPT in the sm-width k of (T, δ).

In the rest of this section, we show how to solve the four problems in FPT time. We give the actual algorithm for MAXCUT, and sketch how to do the same for the other three problems. The full algorithms and proofs can be found in [15].

The problem t-MAXCUT asks, for a graph G, whether there exists a set $W \subseteq V(G)$ so that the number of edges in $(W, \overline{W})$ is at least t. For a set X, we denote by $\delta_G(X)$ the number of edges in the cut $(V(G) \cap X, V(G) \setminus X)$ (note that X does not need to be a subset of $V(G)$). For t-MAXCUT, we define $\text{cert}(X)$ for $X \subseteq V(G)$ to be all the subsets of X, and we define $\oplus(x, y)$ to be the union function; $\oplus(x, y) = \{x \cup y\}$. We solve t-MAXCUT by use of RECURSIVE and the below procedure JOIN$_{maxcut}$ with input specification as described above.

Procedure JOIN$_{maxcut}$
    Input: $S_1 \preceq_{V_a} cert(V_a)$ and $S_2 \preceq_{V_b} cert(V_b)$ with $A = V_a \cup V_b$
    Output: $S \preceq_A \oplus(\text{cert}(V_a), \text{cert}(V_b)) = \text{cert}(A)$

$S' \leftarrow \{s_1 \cup s_2 : s_1 \in S_1, s_2 \in S_2\}$ /* note $S' = \oplus(S_1, S_2)$ */
$S \leftarrow \emptyset$
$C \leftarrow$ a minimum vertex cover of $G[A, \overline{A}]$
if $(A, \overline{A})$ is a split **then for** $z = 0, \ldots, n$ **do**
$\quad c' \leftarrow \operatorname{argmax}_{c \in S'} \{\delta_{G[A]}(c) : |N(\overline{A}) \cap c| = z\}$
$\quad S \leftarrow S \cup \{c'\}$
else for all subsets $S_C \subseteq C$ **do**
$\quad c' \leftarrow \operatorname{argmax}_{c \in S'} \{\delta_{G[A]}(c) : S_C \cap A = c \cap A\}$
$\quad S \leftarrow S \cup \{c'\}$
return S

Proof of correctness of MaxCut is at the end of this section.

For HAMILTONIAN CYCLE, certificates are disjoint paths or cycles, and a witness is a Hamiltonian cycle. The important information is what neighbourhood the endpoints of each path has over the cut. For each certificate we keep track of the number of *path classes*, which are sets of paths with the same neighborhood over the cut, and the size of each such path class. The total number of path classes over all certificates is also important. For a split cut, the size of a class might be anything from 1 to n, but there will be only one class in total. For a non-split cut of sm-value k, the total number of path classes is bounded by 2^{2k} and since each path is vertex disjoint the number of paths in any important certificate is bounded by k. Based on this the JOIN operation will be able to find a FPT-sized set of certificates preserving a full set.

For t-COLORING, we note that a graph of sm-width k, unlike graphs of treewidth k, may need more than $k + 1$ colors. We let all partitions into t parts where the parts induce independent sets be our certificates. What matters for a certificate is what kind of certificates it can be combined with to yield a new certificate, i.e. inducing an independent set also across the cut. For non-split cuts, this means the number of important certificates is bounded by the number of ways to t-partition the vertices in the k-vertex cover of the cut, which is a function of k. For a split cut, what is important is the number of parts of a partition/certificate that have neighbors across the cut. The certificate minimizing this number will preserve all other certificates. Based on this the JOIN operation will be able to find a preserving set of certificates of FPT-size.

For t-EDGE DOMINATING SET the certificates are subgraphs of G and a witness is a graph $G' = (V', E')$ so that each vertex in V' is incident with an edge in E', and E' is an edge dominating set of G of size at most t. The idea of how to make an FPT JOIN-procedure is that for a vertex cover C of a cut, the number of ways a certificate can project to C is limited by a function of the size of C. Based on this we find a preserving set of FPT cardinality when $|C|$ is at most k. When $|C|$ is not bounded by k, we have a split. For splits we limit the max number of certificates needed for a preserving set by a polynomial of n. This is because almost all edges on one side of the cut affect the rest of the edges uniformly, and the other way around.

Theorem 10. *Given a graph G and branch decomposition (T, δ) of sm-width k, we can solve MAXCUT in time $\mathcal{O}^*(8^k)$, HAMILTONIAN CYCLE in time $\mathcal{O}^*(2^{24k^2})$, t-COLORING in time $\mathcal{O}^*(k^{5k})$, and t-EDGE DOMINATING SET in time $\mathcal{O}^*(3^{5k})$.*

Proof. We consider MAXCUT. In the full version of this paper we show JOIN_{maxcut} is correct and produce a preserving set S of size at most $\mathcal{O}^*(2^k)$ in time $\mathcal{O}^*(|S_1||S_2|2^k)$. So, using RECURSIVE with JOIN_{maxcut}, we know the size of both of the inputs of JOIN_{maxcut} is at most the size of its output, i.e., $|S_1|, |S_2| \leq \mathcal{O}^*(2^k)$. So, each call to RECURSIVE has runtime at most $\mathcal{O}^*(8^k)$. As there are linearly many calls to RECURSIVE and there is a polynomial time verifier for the certificates RECURSIVE produces, by the definition of $\preceq$, the total runtime is also bounded by $\mathcal{O}^*(8^k)$. By similar arguments as for MAXCUT, we get the stated runtime bounds for HAMILTONIAN CYCLE, t-COLORING, and t-EDGE DOMINATING SET from the runtime and correctness of their respective JOIN-algorithms presented in the full version of this paper. □

5   Graphs of Bounded sm-Width

We have shown that four basic problems, that cannot be FPT parameterized by clique-width unless ETH fails, are FPT when parameterized by sm-width. The sm-width parameter nevertheless shares important properties with clique-width, for example the following, which we prove in the full version of this paper ([15]):

Proposition 11. *If treewidth is bounded then sm-width is bounded $(\text{smw}(G) \leq tw(G) + 1)$ which in turn means that clique-width is bounded. If twin-cover is bounded then clique-width and sm-width is bounded $(\text{smw}(G) \leq tc(G))$. Cographs, the graphs of clique-width at most two, have sm-width one, while distance-hereditary graphs have clique-width at most three and sm-width one.*

From our procedure for computing branch decompositions of bounded sm-width, it might seem that we could instead simply have defined our parameter to be the maximum treewidth of the prime graphs of split decomposition. However, a bound on the treewidth of each prime graph does not imply a bound on the sm-width. The complexity of the four given problems for this alternative parameter is not known as far as we know.

There are several classes of graphs of bounded sm-width where no previous results implied FPT algorithms for the considered problems. We now show a class of such graphs, constructed by combining a graph of clique-width at most 3, with a graph of treewidth k and thus clique-width at most $2^{k/2}$, as follows. Let G_1 be a distance-hereditary graph and let G_2 be a graph of treewidth k. Let $X \subseteq V(G_1)$ with $|X| \leq k+1$ and $(X, \overline{X})$ a split of G_1, and let $Y \subseteq V(G_2)$ be a bag of a tree decomposition of G_2 of treewidth k. Add an arbitrary set of edges on the vertex set $X \cup Y$. The resulting graph will have sm-width at most $k+1$, a result that basically follows by taking branch decompositions of G_1 and G_2 where X and Y each are mapped as the set of leaves of a subtree, subdividing each of the two edges above these subtrees and adding an edge on the subdivided vertices to make a single branch decomposition of the combined graph.

As an intuition of how graphs of low sm-width look, we make the following observation: For any branch decomposition of sm-width less than k, we can divide the edges of the graph into two parts; one part X consisting of the edges crossing some cut of the branch decomposition of mm-width at least k (and thus a split), and one part for the remaining edges. The edges in X will disconnect the graph into components forming induced subgraphs of mm-width less than k. As the edges of X originates from splits in the decomposition, these edges connect the induced subgraphs in a way closely related to a distance hereditary graph.

References

1. Bui-Xuan, B.-M., Telle, J.A., Vatshelle, M.: Boolean-width of graphs. Theor. Comput. Sci. **412**(39), 5187–5204 (2011)
2. Charbit, P., de Montgolfier, F., Raffinot, M.: Linear time split decomposition revisited. SIAM J. Discrete Math. **26**(2), 499–514 (2012)
3. Corneil, D.G., Rotics, U.: On the relationship between clique-width and treewidth. SIAM J. Comput. **34**(4), 825–847 (2005)
4. Courcelle, B., Makowsky, J., Rotics, U.: Linear time solvable optimization problems on graphs of bounded clique-width. Theory Comput. Syst. **33**(2), 125–150 (2000)
5. Courcelle, B.: The monadic second-order logic of graphs. I. recognizable sets of finite graphs. Inf. Comput. **85**(1), 12–75 (1990)
6. Cunningham, W.H.: Decomposition of directed graphs. SIAM J. Alg. Discrete Methods **3**(2), 214–228 (1982)
7. Fomin, F., Golovach, P., Lokshtanov, D., Saurabh, S.: Algorithmic lower bounds for problems parameterized by clique-width. In: Proceedings SODA, pp. 493–502 (2010)
8. Gajarský, J., Lampis, M., Ordyniak, S.: Parameterized algorithms for modular-width. In: Gutin, G., Szeider, S. (eds.) IPEC 2013. LNCS, vol. 8246, pp. 163–176. Springer, Heidelberg (2013)
9. Ganian, R.: Twin-cover: beyond vertex cover in parameterized algorithmics. In: Marx, D., Rossmanith, P. (eds.) IPEC 2011. LNCS, vol. 7112, pp. 259–271. Springer, Heidelberg (2012)
10. Ganian, R., Hliněný, P., Nešetřil, J., Obdržálek, J., Ossona de Mendez, P., Ramadurai, R.: When trees grow low: shrubs and fast MSO_1. In: Rovan, B., Sassone, V., Widmayer, P. (eds.) MFCS 2012. LNCS, vol. 7464, pp. 419–430. Springer, Heidelberg (2012)
11. Lampis, M.: Algorithmic meta-theorems for restrictions of treewidth. Algorithmica **64**(1), 19–37 (2012)
12. Oum, S., Seymour, P.: Approximating clique-width and branch-width. J. Comb. Theory Ser. B **96**(4), 514–528 (2006)
13. Rao, M.: Solving some NP-complete problems using split decomposition. Discrete Appl. Math. **156**(14), 2768–2780 (2008)
14. Robertson, N., Seymour, P.D.: Graph minors. X. Obstructions to tree-decomposition. J. Comb. Theory Ser. B **52**(2), 153–190 (1991)
15. Sæther, S.H., Telle, J.A.: Between treewidth and clique-width. CoRR, abs/1404.7758 (2014)
16. Vatshelle, M.: New width parameters of graphs. Ph.D. Thesis, The University of Bergen (2012)

A Polynomial Turing-Kernel for Weighted Independent Set in Bull-Free Graphs

Stéphan Thomassé[1], Nicolas Trotignon[2]($\boxtimes$), and Kristina Vušković[3]

[1] CNRS, LIP, ENS de Lyon, INRIA, Université de Lyon, Lyon, France
[2] Faculty of Computer Science, School of Computing, University of Leeds, Leeds, UK
nicolas.trotignon@ens-lyon.fr
[3] (RAF), Union University, Belgrade, Serbia

Abstract. The maximum stable set problem is NP-hard, even when restricted to triangle-free graphs. In particular, one cannot expect a polynomial time algorithm deciding if a bull-free graph has a stable set of size k, when k is part of the instance. Our main result in this paper is to show the existence of an FPT algorithm when we parameterize the problem by the solution size k. A polynomial kernel is unlikely to exist for this problem. We show however that our problem has a polynomial size Turing-kernel. More precisely, the hard cases are instances of size $O(k^5)$. All our results rely on a decomposition theorem of bull-free graphs due to Chudnovsky which is modified here, allowing us to provide extreme decompositions, adapted to our computational purpose.

1 Introduction

In this paper all graphs are simple and finite. We say that a graph G *contains* a graph F, if F is isomorphic to an induced subgraph of G. We say that G is *F-free* if G does not contain F. For a class of graphs $\mathcal{F}$, the graph G is $\mathcal{F}$-free if G is F-free for every $F \in \mathcal{F}$. The *bull* is a graph with vertex set $\{x_1, x_2, x_3, y, z\}$ and edge set $\{x_1x_2, x_1x_3, x_2x_3, x_1y, x_2z\}$.

Chudnovsky in a series of papers [4–7] gives a complete structural characterisation of bull-free graphs (more precisely, bull-free trigraphs, where a trigraph is a graph with some adjacencies left undecided). Roughly speaking, this theorem asserts that every bull-free trigraph is either in a well-understood *basic* class, or admits a *decomposition* allowing to break the trigraph into smaller *blocks*. In Sect. 2, we extract what we need for the present work, from the very complex theorem of Chudnovsky. In Sect. 3, we prove that bull-free trigraphs admit *extreme* decompositions, that are decompositions such that one of the blocks is basic. In Sect. 4, we give polynomial time algorithms to actually compute the extreme decompositions whose existence is proved in the previous

---

Stéphan Thomassé and Nicolas Trotignon: Supported by *A.N.R.* under reference ANR-13-BS02-0007.

Kristina Vušković: Partially supported by EPSRC grant EP/K016423/1 and Serbian Ministry of Education and Science projects 174033 and III44006.

© Springer International Publishing Switzerland 2014
D. Kratsch and I. Todinca (Eds.): WG 2014, LNCS 8747, pp. 408–419, 2014.
DOI: 10.1007/978-3-319-12340-0_34

section. In Sect. 5, we introduce the notion of weigthed trigraphs. In Sect. 6, we give an FPT-algorithm for the maximum stable set problem restricted to bull-free graphs. Let us explain this. The notion of fixed-parameter tractability (FPT) is a relaxation of classical polynomial time solvability. A parameterized problem is said to be *fixed-parameter tractable* if it can be solved in time $f(k)P(n)$ on instances of input size n, where f is a computable function (so $f(k)$ depends only on the value of parameter k), and P is a polynomial function independent of k. We give an FPT-algorithm for the maximum stable set problem restricted to bull-free graphs. This generalizes the result of Dabrowski, Lozin, Müller and Rautenbach [8] who give an FPT-algorithm for the same parameterized problem for {bull, $\overline{P_5}$}-free graphs, where P_5 is a path on 5 vertices and $\overline{P_5}$ is its complement. Recently, Lokshtanov, Vatshelle and Villanger [17] proved that maximum independent set in P_5-free graphs can be computed in polynomial time. Also, forbidding a bull and odd holes leads to polynomial algorithm for Maximum Weight Independent Sets, see Brandstäd and Mosca? [3]. In a weighted graph the *weight* of a set is the sum of the weights of its elements, and with $\alpha_w(G)$ we denote the weight of a maximum weighted independent set of a graph G with weight function w. We state below the problem that we solve more formally.

PARAMETERIZED WEIGHTED INDEPENDENT SET

Instance: A weighted graph G with weight function $w : V(G) \longrightarrow \mathbb{N}$ and a positive integer k.

Parameter: k
Problem: Decide whether G has an independent set of weight at least k. If no such set exists, find an independent set of weight $\alpha_w(G)$.

Observe that the problem above is $W[1]$-hard for general graphs [9] and the non parameterized version is NP-complete even for triangle-free graphs [19] (and therefore for bull-free graphs).

In Sect. 7, we show that while a polynomial kernel is unlikely to exist since the problem is OR-compositional, we can prove nonetheless that the hardness of the problem can be reduced to polynomial size instances. Precisely we show that for bull-free graphs of size n one can decide if a stable set of size k exists in time $P(n)$ for some polynomial P provided that we have unlimited access to an oracle which can decide if a stable set of size k exists for bull-free graphs of size $O(k^5)$. The fact that hard cases can be reduced to size polynomial in k is not captured by the existence of a polynomial kernel, but by what is called a Turing-kernel (see Sect. 7 or Lokshtanov [16] for a definition of Turing-kernels). Even the existence of a Poly(n) set of kernels of size Poly(k) seems unclear for this problem. To our knowledge, stability in bull-free graphs is the first example of a problem admitting a polynomial Turing-kernel which is not known to have an independent set of polynomial kernels. An interesting question is to investigate which classical problems without polynomial kernels do have a polynomial Turing-kernel. This question is investigated by Hermelin et al. [15].

2   Decomposition of Bull-Free Graphs

In the series of papers [4–7] Chudnovsky gives a complete structural characterisation of bull-free graphs which we first describe informally. Her construction of all bull-free graphs starts from three explicitly constructed classes of basic bull-free graphs: T_0, T_1 and T_2. Class T_0 consists of graphs whose size is bounded by some constant, the graphs in T_1 are built from a triangle-free graph F and a collection of disjoint cliques with prescribed attachments in F (so triangle-free graphs are in this class), and T_2 generalizes graphs G that have a pair uv of vertices, so that uv is dominating both in G and $\bar{G}$. Furthermore, each graph G in $T_1 \cup T_2$ comes with a list $\mathcal{L}_G$ of "expandable edges". Chudnovsky shows that every bull-free graph that is not obtained by substitution from smaller ones, can be constructed from a basic bull-free graph by expanding the edges in $\mathcal{L}_G$ (where edge expansion is an operation corresponding to reversing the homogeneous pair decomposition). To prove and use this result, it is convenient to work on trigraphs (a generalization of graphs where some edges are left "undecided"), and the first step is to obtain a decomposition theorem for bull-free trigraphs using homogeneous sets and homogeneous pairs. In this paper we need a simplified statement of this decomposition theorem, which we now describe formally.

Trigraphs

For a set X, we denote by $\binom{X}{2}$ the set of all subsets of X of size 2. For brevity of notation an element $\{u, v\}$ of $\binom{X}{2}$ is also denoted by uv or vu. A *trigraph* T consists of a finite set $V(T)$, called the *vertex set* of T, and a map $\theta : \binom{V(T)}{2} \longrightarrow \{-1, 0, 1\}$, called the *adjacency function*.

Two distinct vertices of T are said to be *strongly adjacent* if $\theta(uv) = 1$, *strongly antiadjacent* if $\theta(uv) = -1$, and *semiadjacent* if $\theta(uv) = 0$. We say that u and v are *adjacent* if they are either strongly adjacent, or semiadjacent; and *antiadjacent* if they are either strongly antiadjacent, or semiadjacent. An *edge* (*antiedge*) is a pair of adjacent (antiadjacent) vertices. If u and v are adjacent (antiadjacent), we also say that u is *adjacent (antiadjacent) to* v, or that u is a *neighbor (antineighbor)* of v. Similarly, if u and v are strongly adjacent (strongly antiadjacent), then u is a *strong neighbor (strong antineighbor)* of v.

Let $\eta(T)$ be the set of all strongly adjacent pairs of T, $\nu(T)$ the set of all strongly antiadjacent pairs of T, and $\sigma(T)$ the set of all semiadjacent pairs of T. Thus, a trigraph T is a graph if $\sigma(T)$ is empty. A pair $\{u, v\} \subseteq V(T)$ of distinct vertices is a *switchable pair* if $\theta(uv) = 0$, a *strong edge* if $\theta(uv) = 1$ and a *strong antiedge* if $\theta(uv) = -1$. An edge uv (antiedge, strong edge, strong antiedge, switchable pair) is *between* two sets $A \subseteq V(T)$ and $B \subseteq V(T)$ if $u \in A$ and $v \in B$, or if $u \in B$ and $v \in A$.

The *complement* $\bar{T}$ of T is a trigraph with the same vertex set as T, and adjacency function $\bar{\theta} = -\theta$.

For $v \in V(T)$, $N(v)$ denotes the set of all vertices in $V(T) \setminus \{v\}$ that are adjacent to v; $\eta(v)$ denotes the set of all vertices in $V(T) \setminus \{v\}$ that are strongly adjacent to v; $\nu(v)$ denotes the set of all vertices in $V(T)\backslash\{v\}$ that are strongly antiadjacent to v; and $\sigma(v)$ denotes the set of all vertices in $V(T)\backslash\{v\}$ that are semiadjacent to v.

Let $A \subset V(T)$ and $b \in V(T)\backslash A$. We say that b is *strongly complete* to A if b is strongly adjacent to every vertex of A; b is *strongly anticomplete* to A if b is strongly antiadjacent to every vertex of A; b is *complete* to A if b is adjacent to every vertex of A; and b is *anticomplete* to A if b is antiadjacent to every vertex of A. For two disjoint subsets A, B of $V(T)$, B is *strongly complete (strongly anticomplete, complete, anticomplete)* to A if every vertex of B is strongly complete (strongly anticomplete, complete, anticomplete) to A. A set of vertices $X \subseteq V(T)$ *dominates (strongly dominates)* T if for all $v \in V(T)\backslash X$, there exists $u \in X$ such that v is adjacent (strongly adjacent) to u.

A *clique* in T is a set of vertices all pairwise adjacent, and a *strong clique* is a set of vertices all pairwise strongly adjacent. A *stable set* is a set of vertices all pairwise antiadjacent, and a *strongly stable set* is a set of vertices all pairwise strongly antiadjacent. For $X \subset V(T)$ the trigraph *induced by T on X* (denoted by $T[X]$) has vertex set X, and adjacency function that is the restriction of θ to $\binom{X}{2}$. Isomorphism between trigraphs is defined in the natural way, and for two trigraphs T and H we say that H is an *induced subtrigraph* of T (or T *contains H as an induced subtrigraph*) if H is isomorphic to $T[X]$ for some $X \subseteq V(T)$. Since in this paper we are only concerned with the induced subtrigraph containment relation, we say that T *contains* H if T contains H as an induced subtrigraph. We denote by $T\backslash X$ the trigraph $T[V(T)\backslash X]$.

Let T be a trigraph. A *path* P of T is a sequence of distinct vertices $p_1, \ldots, p_k$ such that $k \geq 1$ and for $i, j \in \{1, \ldots, k\}$, p_i is adjacent to p_j if $|i - j| = 1$ and p_i is antiadjacent to p_j if $|i - j| > 1$. Under these circumstances, $V(P) = \{p_1, \ldots, p_k\}$ and we say that P is a path *from p_1 to p_k*, its *interior* is the set $P^* = V(P)\backslash\{p_1, p_k\}$, and the *length* of P is $k-1$. We also say that P is a $(k-1)$-*edge-path*. Sometimes, we denote P by $p_1\text{-}\cdots\text{-}p_k$. Observe that, since a graph is also a trigraph, it follows that a path in a graph, the way we have defined it, is what is sometimes in literature called a chordless path.

A *semirealization* of a trigraph T is any trigraph T' with vertex set $V(T)$ that satisfies the following: for all $uv \in \binom{V(T)}{2}$, if $uv \in \eta(T)$ then $uv \in \eta(T')$, and if $uv \in \nu(T)$ then $uv \in \nu(T')$. Sometimes we will describe a semirealization of T as an *assignment of values* to switchable pairs of T, with three possible values: "strong edge", "strong antiedge" and "switchable pair". A *realization* of T is any graph that is semirealization of T (so, any semirealization where all switchable pairs are assigned the value "strong edge" or "strong antiedge"). For $S \subseteq \sigma(T)$, we denote by G_S^T the realization of T with edge set $\eta(T) \cup S$, so in G_S^T the switchable pairs in S are assigned the value "edge", and those in $\sigma(T)\backslash S$ the value "antiedge". The realization $G_{\sigma(T)}^T$ is called the *full realization* of T.

A *bull* is a trigraph with vertex set $\{x_1, x_2, x_3, y, z\}$ such that x_1, x_2, x_3 are pairwise adjacent, y is adjacent to x_1 and antiadjacent to x_2, x_3, z, and z is adjacent to x_2 and antiadjacent to x_1, x_3. For a trigraph T, a subset X of $V(T)$ is said to be a *bull* if $T[X]$ is a bull. A trigraph is *bull-free* if no induced subtrigraph of it is a bull, or equivalently, no subset of its vertex set is a bull.

Fig. 1. A homogeneous set.

Observe that we have two notions of bulls: bulls as graphs (defined in the introduction), and bulls as trigraphs. A bull as a graph can be seen as a bull as a trigraph. Also, a trigraph is a bull if and only if at least one of its realization is a bull (as a graph). Hence, a trigraph is bull-free if and only if all its realizations are bull-free graphs. The complement of a bull is a bull (with both notions), and therefore, if T is bull-free trigraph (or graph), then so is $\overline{T}$.

2.1   Decomposition Theorem

A trigraph is called *monogamous* if every vertex of it belongs to at most one switchable pair (so the switchable pairs form a matching). We now state the decomposition theorem for bull-free monogamous trigraphs. We begin with the description of the cutsets.

Let T be a trigraph. A set $X \subseteq V(T)$ is a *homogeneous set* in T if $1 < |X| < |V(T)|$, and every vertex of $V(T)\backslash X$ is either strongly complete or strongly anticomplete to X. See Fig. 1 (a line means all possible strong edges between two sets, nothing means all possible strong antiedges, and a dashed line means no restriction).

A *homogeneous pair* (see Fig. 2) is a pair of disjoint nonempty subsets (A, B) of $V(T)$, such that there are disjoint (possibly empty) subsets C, D, E, F of $V(T)$ whose union is $V(T)\backslash(A \cup B)$, and the following hold:

- A is strongly complete to $C \cup E$ and strongly anticomplete to $D \cup F$;
- B is strongly complete to $D \cup E$ and strongly anticomplete to $C \cup F$;
- A is not strongly complete and not strongly anticomplete to B;
- $|A \cup B| \geq 3$; and
- $|C \cup D \cup E \cup F| \geq 3$.

In these circumstances, we say that (A, B, C, D, E, F) is a *split* for the homogeneous pair (A, B). A homogeneous pair (A, B) is *small* if $|A \cup B| \leq 6$. A homogeneous pair (A, B) with split (A, B, C, D, E, F) is *proper* if $C \neq \emptyset$ and $D \neq \emptyset$. Note that "A is not strongly complete and not strongly anticomplete to B" does not imply that $|A \cup B| \geq 3$, because it could be that the unique vertex in A is linked to the unique vertex in B by a switchable pair.

We now describe the basic classes. A trigraph is a *triangle* if it has exactly three vertices, and these vertices are pairwise adjacent. Let $\mathcal{T}_0$ be the class of all monogamous trigraphs on at most 8 vertices. Let $\mathcal{T}_1$ be the class of monogamous trigraphs T whose vertex set can be partitioned into (possibly empty) sets

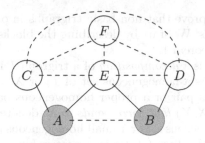

Fig. 2. A homogeneous pair.

$X, K_1, \ldots, K_t$ so that $T[X]$ is triangle-free, and $K_1, \ldots, K_t$ are strong cliques that are pairwise strongly anticomplete. Furthermore, for every $v \in \cup_{i=1}^t K_i$, the set of neighbors of v in X partitions into strong stable sets A and B such that A is strongly complete to B. Let $\overline{T_1} = \{\overline{T} : T \in T_1\}$. A trigraph is *basic* if it belongs to $T_l \cup T_1 \cup \overline{T_1}$. The following result is a direct consequence of the main result of Chudnovsky. Note that this is a simplification of a much more detailed characterization.

Theorem 1. (Chudnovsky [4–7]). *If T is a bull-free monogamous trigraph, then one of the following holds:*

- *T is basic;*
- *T has a homogeneous set;*
- *T has a small homogeneous pair; or*
- *T has a proper homogeneous pair.*

We do not know whether the theorem above is algorithmic. Deciding whether a graph is bull-free can clearly be done in polynomial time. Also, detecting the decompositions is easy (see Sect. 4). The problem is with the basic classes. It follows directly from a theorem of Farrugia [10] that deciding whether a graph can be partitioned into a triangle-free part and a part that is disjoint union of cliques is NP-complete. This does not mean that recognizing T_1 is NP-complete, because one could take advantage of several features, such as being bull-free or of the full definition of T_1 in [5]. We leave the recognition of T_1 as an open question.

3 Extreme Decompositions

The way we use decompositions for computing stable sets requires building blocks of decomposition and asking at least two questions for at least one block. When this process is recursively applied it potentially leads to an exponential blow-up even when the decomposition tree is linear in the size of the input trigraph. This problem is bypassed here by using what we call extreme decompositions, that are decompositions whose one block of decomposition is basic and therefore handled directly, without any recursive calls to the algorithm.

In this section, we prove that non-basic trigraphs in our class actually have extreme decompositions. We start by describing the blocks of decomposition for the cutsets used in Theorem 1.

We say that (X, Y) is a *decomposition* of a trigraph T if (X, Y) is a partition of $V(T)$ and either X is a homogeneous set of T, or $X = A \cup B$ where (A, B) is a small homogeneous pair or a proper homogeneous pair of T. The *block of decomposition w.r.t.* (X, Y) *that corresponds to* X, denoted by T_X, is defined as follows. If X is a homogeneous set or a small homogeneous pair, then $T_X = T[X]$. Otherwise, $X = A \cup B$ where (A, B) is a proper homogeneous pair, and T_X consists of $T[X]$ together with *marker vertices* c and d such that c is strongly complete to A, d is strongly complete to B, cd is a switchable pair, and there are no other edges between $\{c, d\}$ and $A \cup B$. The *block of decomposition w.r.t.* (X, Y) *that corresponds to* Y, denoted by T_Y, is defined as follows. If X is a homogeneous set, then let x be any vertex of X and let $T_Y = T[Y \cup \{x\}]$. In this case x is called the *marker vertex* of T_Y. Otherwise, $X = A \cup B$ where (A, B) is a homogeneous pair with split (A, B, C, D, E, F). In this case T_Y consists of $T[Y]$ together with two new *marker vertices* a and b such that a is strongly complete to $C \cup E$, b is strongly complete to $D \cup E$, ab is a switchable pair, and there are no other edges between $\{a, b\}$ and $C \cup D \cup E \cup F$.

Lemma 1. *If (X, Y) is a decomposition of a bull-free monogamous trigraph T, then the corresponding blocks T_X and T_Y are bull-free monogamous trigraphs.*

Let (X, Y) be a decomposition of a trigraph T. We say that (X, Y) is a *homogeneous cut* if X is a homogeneous set or $X = A \cup B$ where (A, B) is a proper homogeneous pair. A homogeneous cut (X, Y) is *minimally-sided* if there is no homogeneous cut (X', Y') with $X' \subsetneq X$.

Lemma 2. *If (X, Y) is a minimally-sided homogeneous cut of a trigraph T, then the block of decomposition T_X, has no homogeneous cut.*

Theorem 2. *Let T be a bull-free monogamous trigraph that has a decomposition. If T has a small homogeneous pair (A, B), then let $X = A \cup B$ and $Y = V(T) \backslash X$. Otherwise let (X, Y) be minimally-sided homogeneous cut of T. Then the block of decomposition T_X is basic.*

4 Algorithms for Finding Decompositions

The fastest known algorithm for finding a homogeneous set in a graph is linear time (see Habib and Paul [14]) and the fastest one for the homogeneous pair runs in time $O(n^2 m)$ (see Habib, Mamcarz, and de Montgolfier [13]). But we cannot use these algorithms safely here because we need minimally-sided decompositions with several technical requirements ("small", "proper") and we need our algorithms to work for trigraphs. However, it turns out that all classical ideas work well in our context.

A 4-tuple of vertices (a, b, c, d) of a trigraph is *proper* if ac and bd are strong edges and bc and ad are strong antiedges. A proper 4-tuple (a, b, c, d) is *compatible* with a homogeneous pair (A, B) if $a \in A$, $b \in B$ and $c, d \notin A \cup B$ (note that c, d must be respectively in the sets C, D from the definition of a split of a homogeneous pair).

Lemma 3. *Let T be a trigraph and $Z = (a, b, c, d)$ a proper 4-tuple of T. There is an $O(n^2)$ time algorithm that given a set $R_0 \subseteq V(T)$ of size at least 3 such that $Z \cap R_0 = \{a, b\}$, either outputs two sets A and B such that (A, B) is a proper homogeneous pair of T compatible with Z and such that $R_0 \subseteq A \cup B$, or outputs the true statement "There exists no proper homogeneous pair (A, B) in T compatible with Z and such that $R_0 \subseteq A \cup B$".*

Moreover, when (A, B) is output, $A \cup B$ is minimal with respect to these properties, meaning that $A \cup B \subseteq A' \cup B'$ for every homogeneous pair (A', B') satisfying the properties.

Lemma 4. *Let T be a trigraph and (a, b) a pair of vertices from T. There is an $O(n^2)$ time algorithm that given a set $R_0 \subseteq V(T)$ such that $a, b \in R_0$, either outputs a homogeneous set X such that $R_0 \subseteq X$, or outputs the true statement "There exists no homogeneous set X in T such that $R_0 \subseteq X$".*

Moreover, when X is output, X is minimal with respect to these properties, meaning that $X \subseteq X'$ for every homogeneous set X' satisfying the properties.

Theorem 3. *There exists an $O(n^8)$ time algorithm whose input is a trigraph T. The output is a small homogeneous pair of T if some exists. Otherwise, if G has a homogeneous cut, then the output is a minimally-sided homogeneous cut. Otherwise, the output is: "T has no small homogeneous pair, no proper homogeneous pair and no homogeneous set".*

5   Weighted Trigraphs

For the sake of induction, we need to work with weighted trigraphs. Here, a *weight* is a non-negative integer. By a *weighted trigraph with weight function w*, we mean a trigraph T such that:

- every vertex a has a weight $w(a)$;
- every switchable pair ab of T has a weight $w(ab)$;
- for every switchable pair ab, $\max\{w(a), w(b)\} \leq w(ab) \leq w(a) + w(b)$.

Let S be a stable set of T. Recall that $\nu(T)$ denotes the set of all strongly antiadjacent pairs of T, and $\sigma(T)$ the set of all semiadjacent pairs of T. We set $c(S) = \{v \in S : \forall u \in S \backslash \{v\}, uv \in \nu(T)\}$. We set $\sigma(S) = \{uv \in \sigma(T) : u, v \in S\}$. Observe that if T is monogamous, then for every vertex v of S, one and only one of the following outcomes is true: $v \in c(S)$ or for some unique $w \in S$, $vw \in \sigma(S)$. The *weight of a stable set S* is the sum of the weights of the vertices in $c(S)$ and of the weights of the (switchable) pairs in $\sigma(S)$.

From here on, T is a weighted monogamous trigraph and $\alpha(T)$ denotes the maximum weight of a stable set of T. Our main concern now is to show that deciding if $\alpha(T)$ is at least k is FPT with parameter k.

When (X, Y) is a decomposition of T, we already defined the block T_Y. We now explain how to give weights to the marker vertices and switchable pairs in T_Y. Every vertex and switchable pair in $T[Y]$ keeps its weight. If X is a homogeneous set, then the marker vertex x receives weight $\alpha(T[X])$. If $X = A \cup B$ where (A, B) is a homogeneous pair, then we give weight $\alpha_A = \alpha(T[A])$ to marker vertex a, $\alpha_B = \alpha(T[B])$ to marker vertex b and $\alpha_{AB} = \alpha(T[A \cup B])$ to the switchable pair ab. It is easy to check that the inequalities in the definition of a weighted trigraph are satisfied.

Lemma 5. $\alpha(T) = \alpha(T_Y)$.

6   Computing α in Bull-Free Graphs

In this section, we use positive weights (no vertex nor switchable pair in a trigraph has weight 0). Also, switchable pairs have weight at least 2.

Lemma 6. If T is a trigraph from $\overline{\mathcal{T}_1}$, then T contains at most $|V(T)|^3$ maximal stable sets.

We need the next classical algorithm that we use as a subroutine. For faster implementations (that we do not need here), see Makino and Uno [18].

Theorem 4. (Tsukiyama, Ide, Ariyoshi, and Shirakawa [22]). *There exists an algorithm for generating all maximal stable sets in a given graph G that runs with $O(nm)$ time delay (i.e. the computation time between any consecutive output is bounded by $O(nm)$; and the first (resp. last) output occurs also in $O(nm)$ time after start (resp. before halt) of the algorithm).*

Lemma 7. *There exists an $O(n^4 m)$ time algorithm whose input is any trigraph T and whose output is a maximum weighted stable set of T, or a certificate that T is not in $\overline{\mathcal{T}_1}$.*

Let $R(x, y)$ be the smallest integer n such that every graph on at least n vertices contains a clique of size x or a stable set of size y. By a classical theorem of Ramsey, $R(3, x) \leq \binom{x+1}{2}$. We now define two functions g and f by $g(x) = \binom{x+1}{2} - 1$ and $f(x) = g(x) + (x - 1)(\binom{g(x)}{2} + 2g(x) + 1)$. Note that $f(x) = O(x^5)$. The next lemma handles basic trigraphs.

Lemma 8. *There exists an $O(n^4 m)$-time algorithm with the following specifications.*

Input: *A weighted monogamous basic trigraph T on n vertices, in which all vertices have weight at least 1 and all switchable pairs have weight at least 2, with no homogeneous set, and a positive integer W.*
Output: *One of the following true statements.*
    1. $n \leq f(W)$;
    2. *the number of maximal stable sets in T is at most n^3;*
    3. $\alpha(T) \geq W$.

Theorem 5. *There is an algorithm with the following specification.*

Input: *A weighted monogamous bull-free trigraph T and a positive integer W.*
Output: *"YES" if $\alpha(T) \geq W$ and otherwise an independent set of maximum weight.*
Running time: $2^{O(W^5)}n^9$

7   A Polynomial Turing-Kernel

Once an FPT-algorithm is found, the natural question is to ask for a polynomial kernel for the problem. Precisely, is there a polynomial-time algorithm which takes as input a bull-free graph G and a parameter k and outputs a bull-free graph H with at most $O(k^c)$ vertices and some integer k' such that G has a stable set of size k if and only if H has a stable set of size k'. Unfortunately, the problem is OR-compositional and thus we have the following:

Theorem 6. *Unless $NP \subseteq coNP/poly$, there is no polynomial kernel for the problem $\alpha(G) \geq k$, where G is a bull-free graph and k is the parameter.*

Somewhat surprisingly, the non existence of a polynomial kernel is not related to the hard core of the algorithm (computing the leaves) but is related to the decomposition tree itself (since even complete sums cannot be handled). Indeed, our algorithm is a kind of kernelisation: the answer is obtained in polynomial time provided that we compute a stable set in a linear number of basic trigraphs of size at most k^5 (the leaves of our implicit decomposition tree). A similar behaviour was discovered by Fernau et al. [12] in the case of finding a directed tree with at least k leaves in a digraph (*Maximum Leaf Outbranching problem*): a polynomial kernel does not exist, but n polynomial kernels can be found. In our case, the leaves of the decomposition tree are pairwise dependent, hence our method does not provide $O(n^c)$ independent kernels of size $O(k^5)$. It seems that the notion of kernel is not robust enough to capture this kind of behaviour in which the computationally hard cases of the problem admit polynomial kernels, but the (computationally easy) decomposition structure does not.

Let f be a computable function. A parameterized problem has an *f-Turing-kernel* (see Lokshtanov [16]) if there exists a constant c such that computing the solution of any instance (X, k) can be done in $O(n^c)$ provided that we have unlimited access to an oracle which can decide any instance (X', k') where (X', k') has size at most $f(k)$. Thus our algorithm can be restated as:

Theorem 7. *Stability in bull-free weighted monogamous trigraphs has an $O(k^5)$-Turing-kernel.*

To turn this result into the world of graphs, we need some operations casting trigraphs into graphs. Let T be a weighted monogamous trigraph with weight function w and a switchable pair ab. We now define four ways to get rid of the switchable pair ab while keeping α the same. There are four ways because a (resp. b) can be transformed into a strong edge or a strong antiedge.

The weighted monogamous trigraph $T_{a \to S}$ (resp. $T_{b \to S}$) is constructed as follows: replace switchable pair ab with a strong edge ab; add a new vertex a' (resp. b') and make it strongly complete to $N_T(a) \backslash \{b\}$ (resp. $N_T(b) \backslash \{a\}$) and strongly anticomplete to the remaining vertices; keep the weights of vertices and switchable pairs of $T \backslash \{a\}$ (resp. $T \backslash \{b\}$) the same; assign the weight $w(a) + w(b) - w(ab)$ to a (resp. $w(a) + w(b) - w(ab)$ to b) and the weight $w(ab) - w(b)$ to a' (resp. $w(ab) - w(a)$ to b').

The weighted monogamous trigraph $T_{a \to K}$ (resp. $T_{b \to K}$) is constructed as follows: replace switchable pair ab with a strong edge ab; add a new vertex a' (resp. b') and make it strongly complete to $\{a\} \cup N_T(a) \backslash \{b\}$ (resp. $\{b\} \cup N_T(b) \backslash \{a\}$) and strongly anticomplete to the remaining vertices; keep the weights of vertices and switchable pairs of $T \backslash \{a\}$ (resp. $T \backslash \{b\}$) the same; assign the weight $w(a)$ to a (resp. $w(b)$ to b) and the weight $w(ab) - w(b)$ to a' (resp. $w(ab) - w(a)$ to b').

Note that by the inequalities in the definition of a weighted trigraph, all weights of vertices in $T_{a \to S}$, $T_{b \to S}$, $T_{a \to K}$ and $T_{b \to K}$ are nonnegative.

Lemma 9. *If T is a weighted monogamous trigraph and ab is a switchable pair of T, then $\alpha(T_{a \to S}) = \alpha(T_{b \to S}) = \alpha(T_{a \to K}) = \alpha(T_{b \to K}) = \alpha(T)$.*

It is not the case that every (integer) weighted bull-free trigraph can be interpreted as an unweighted bull-free graph with the same α. However, if we start with a bull-free *graph* and compute leaves of the decomposition tree, every switchable pair in them is obtained at some point by shrinking a homogeneous pair (A, B) of a trigraph T into a switchable pair ab of a trigraph T'. Because of the requirement that A is not strongly complete and not strongly anticomplete to B, we see that at least one of $T'_{a \to S}$, $T'_{b \to S}$, $T'_{a \to K}$ or $T'_{b \to K}$ is in fact an induced subtrigraph of some semirealization of T (and recall that a trigraph is bull-free if and only if all its semirealizations are bull-free). By Lemma 9, this allows us to represent the weighted bull-free trigraphs generated by our Turing-kernel as bull-free graphs with the same α. Finally, unweighting vertices being simply done by substituting stable sets, we have:

Theorem 8. *Stability in bull-free graphs has an $O(k^5)$-Turing-kernel.*

Acknowledgement. Thanks to Maria Chudnovsky for several suggestions. Thanks to Haiko Müller for pointing out to us [10]. Thanks to Sébastien Tavenas and the participants to GROW 2013 for useful discussions on Turing-kernels. Finally, the authors wish to thank the anonymous referees for their very helpful comments.

References

1. Alon, N., Spencer, J.H.: The Probabilistic Method. Wiley, New York (2008)
2. Bodlaender, H., Downey, R., Fellows, M., Hermelin, D.: On problems without polynomial kernels. J. Comput. Syst. Sci. **75**(8), 423–434 (2009)
3. Brandstädt, A., Mosca, R.: Maximum weight independent sets in odd-hole-free graphs without dart or without bull (preprint)

4. Chudnovsky, M.: The structure of bull-free graphs I: Three-edge-paths with center and anticenters. J. Comb. Theory B **102**(1), 233–251 (2012)
5. Chudnovsky, M.: The structure of bull-free graphs II and III: A summary. J. Comb. Theory B **102**(1), 252–282 (2012)
6. Chudnovsky, M.: The structure of bull-free graphs II: Elementary trigraphs (manuscript)
7. Chudnovsky, M.: The structure of bull-free graphs III: Global structure (manuscript)
8. Dabrowski, K., Lozin, V., Müller, H., Rautenbach, D.: Parameterized complexity of the weighted independent set problem beyond graphs of bounded clique number. J. Discrete Algorithms **14**, 207–213 (2012)
9. Downey, R.G., Fellows, M.R.: Parameterized Complexity: Monographs in Computer Science. Springer, New York (1999)
10. Farrugia, A.: Vertex-partitioning into fixed additive induced-hereditary properties is NP-hard. Electr. J. Comb. **11**(1), 9 (2004)
11. Fernández-Baca, D. (ed.): LATIN 2012. LNCS, vol. 7256. Springer, Heidelberg (2012)
12. Fernau, H., Fomin, F., Lokshtanov, D., Raible, D., Saurabh, S., Villanger, Y.: Kernels for problems with no kernel: on out-trees with many leaves. In: STACS 2009
13. Habib, M., Mamcarz, A., de Montgolfier, F.: Algorithms for some H-join decompositions. In: Fernández-Baca [11], pp. 446–457
14. Habib, M., Paul, C.: A survey of the algorithmic aspects of modular decomposition. Comput. Sci. Rev. **4**(1), 41–59 (2010)
15. Hermelin, D., Kratsch, S., Sołtys, K., Wahlström, M., Wu, X.: A completeness theory for polynomial (turing) kernelization. In: Gutin, G., Szeider, S. (eds.) IPEC 2013. LNCS, vol. 8246, pp. 202–215. Springer, Heidelberg (2013)
16. Lokshtanov, D.: New methods in parameterized algorithms and complexity. Ph.D. Thesis, University of Bergen (2009)
17. Lokshtanov, D., Vatshelle, M., Villanger, Y.: Independent set in p_5-free graphs in polynomial time, 2014. In: SODA 2014
18. Makino, K., Uno, T.: New algorithms for enumerating all maximal cliques. In: Hagerup, T., Katajainen, J. (eds.) SWAT 2004. LNCS, vol. 3111, pp. 260–272. Springer, Heidelberg (2004)
19. Poljak, S.: A note on the stable sets and coloring of graphs. Commentationes Math. Univ. Carol. **15**, 307–309 (1974)
20. Sauer, N.: On the density of families of sets. J. Comb. Theory Ser. A **13**(1), 145–147 (1972)
21. Schrijver, A.: Combinatorial Optimization, Polyhedra and Efficiency, vol. A, B and C. Springer, New York (2003)
22. Tsukiyama, S., Ide, M., Ariyoshi, H., Shirakawa, I.: A new algorithm for generating all the maximal independent sets. SIAM J. Comput. **6**(3), 505–517 (1977)

Author Index